Convergence of Knowledge, Technology and Society

Science Policy Reports

The series Science Policy Reports presents the endorsed results of important studies in basic and applied areas of science and technology. They include, to give just a few examples: panel reports exploring the practical and economic feasibility of a new technology; R&D studies of development opportunities for particular materials, devices or other inventions; reports by responsible bodies on technology standardization in developing branches of industry.

Sponsored typically by large organizations – government agencies, watchdogs, funding bodies, standards institutes, international consortia – the studies selected for Science Policy Reports will disseminate carefully compiled information, detailed data and in-depth analysis to a wide audience. They will bring out implications of scientific discoveries and technologies in societal, cultural, environmental, political and/or commercial contexts and will enable interested parties to take advantage of new opportunities and exploit on-going development processes to the full.

For further volumes:
http://www.springer.com/series/8882

Mihail C. Roco • William S. Bainbridge
Bruce Tonn • George Whitesides
Editors

Convergence of Knowledge, Technology and Society

Beyond Convergence of Nano-Bio-Info-Cognitive Technologies

Editors
Mihail C. Roco
National Science Foundation
Arlington, VA, USA

William S. Bainbridge
National Science Foundation
Arlington, VA, USA

Bruce Tonn
Department of Political Science
University of Tennessee
Knoxville, TN, USA

George Whitesides
Department of Chemistry
and Chemical Biology
Harvard University
Cambridge, MA, USA

Citation: M.C. Roco, W.S. Bainbridge, B. Tonn, and G. Whitesides, eds. 2013. *Converging knowledge, technology, and society: Beyond convergence of nano-bio-info-cognitive technologies.* Dordrecht, Heidelberg, New York, London: Springer. The convergence interviews are available on the website http://www.wilsoncenter.org/convergence.

Cover page: Using the image of a conductor, the CTKS logo on the cover page suggests coordination through societal governance of the essential knowledge and technology platforms (foundational tools, human-scale, and Earth-scale platforms) to realize a vision of their convergence to benefit society.

The U.S. Federal agencies sponsoring the study are: the National Science Foundation (NSF), National Institutes of Health (NIH), National Aeronautics and Space Administration (NASA), Environmental Protection Agency (EPA), Office of Naval Research (ONR), and United States Department of Agriculture (USDA).

Copyright 2013 by WTEC. The U.S. Government retains a nonexclusive and nontransferable license to exercise all exclusive rights provided by copyright. This document is sponsored by the National Science Foundation (NSF) under a cooperative agreement from NSF (ENG-0844639) to the World Technology Evaluation Center, Inc. The first and second co-editors were supported by the NSF Directorate for Engineering and Directorate for Computer and Information Science and Engineering, respectively. The Government has certain rights in this material. Any writings, opinions, findings, and conclusions expressed in this material are those of the authors and do not necessarily reflect the views of the United States Government, the authors' parent institutions, or WTEC.

Additional material to this book can be downloaded from http://extras.springer.com

Video 1 – produced by Bruce Tonn from NSF and the Wilson Center.
Video 2 – produced by Clement Bezold from NSF and the Wilson Center.
Video 3 – produced by Mark Lundstrom from NSF and the Wilson Center.
Video 4 – produced by Robert Urban from NSF and the Wilson Center.
Video 5 – produced by George Whitesides from NSF and the Wilson Center.

ISSN 2213-1965   ISSN 2213-1973 (electronic)
ISBN 978-3-319-02203-1   ISBN 978-3-319-02204-8 (eBook)
DOI 10.1007/978-3-319-02204-8
Springer Cham Heidelberg New York Dordrecht London

Library of Congress Control Number: 2013956823

© Springer International Publishing Switzerland 2013
This work is subject to copyright. All rights are reserved by the Publisher, whether the whole or part of the material is concerned, specifically the rights of translation, reprinting, reuse of illustrations, recitation, broadcasting, reproduction on microfilms or in any other physical way, and transmission or information storage and retrieval, electronic adaptation, computer software, or by similar or dissimilar methodology now known or hereafter developed. Exempted from this legal reservation are brief excerpts in connection with reviews or scholarly analysis or material supplied specifically for the purpose of being entered and executed on a computer system, for exclusive use by the purchaser of the work. Duplication of this publication or parts thereof is permitted only under the provisions of the Copyright Law of the Publisher's location, in its current version, and permission for use must always be obtained from Springer. Permissions for use may be obtained through RightsLink at the Copyright Clearance Center. Violations are liable to prosecution under the respective Copyright Law.
The use of general descriptive names, registered names, trademarks, service marks, etc. in this publication does not imply, even in the absence of a specific statement, that such names are exempt from the relevant protective laws and regulations and therefore free for general use.
While the advice and information in this book are believed to be true and accurate at the date of publication, neither the authors nor the editors nor the publisher can accept any legal responsibility for any errors or omissions that may be made. The publisher makes no warranty, express or implied, with respect to the material contained herein.

Printed on acid-free paper

Springer is part of Springer Science+Business Media (www.springer.com)

Acknowledgements

We at WTEC wish to thank all of the participants for their valuable insights and their dedicated work in conducting this international study of the transforming tools of emerging and converging technologies for societal benefit. Appendix B has a complete list of the many experts around the world who shared their valuable time with us. Here I have only the space to recognize a few people. The contributors who took the lead in writing these chapters deserve special mention. They include William Bainbridge, Jian Cao, Mamadou Diallo, Mark Lundstrom, James Olds, Mike Roco, Bruce Tonn, Robert G. Urban, George Whitesides, and H.-S. Philip Wong.

For making this study possible, our sincere thanks go to those who helped arrange research funding, including many at NSF in the Directorates of Engineering (Kesh Narayanan), Computer and Information Science and Engineering (Sankar Basu), Biological Sciences (Chuck Liarakos), Mathematical & Physical Sciences (Celeste Rohlfing) and Social, Behavioral & Economic Sciences (Frederick Kronz). Nora Savage at EPA, Michael Meador at NASA, Lawrence Kabacoff at the Office of Naval Research, Piotr Grodzinski and Larry Nagahara at the NIH National Cancer Institute, and Hongda Chen of the U.S. Department of Agriculture also contributed. David Rejeski at the Woodrow Wilson International Center for Scholars created video interviews with leading experts on convergence. Our international hosts supported the costs of the workshops abroad and provided unparalleled hospitality. Among others, they included Esper Abrão Cavalheiro of the University of Sao Paulo (Brazil), Christos Tokamanis of the European Commission (EU), Gilbert Declerck of IMEC (Belgium), Chang Woo Kim of the National Nanotechnology Policy Center and Jo-Won Lee of Hanyang University (Korea), Kazunobu Tanaka of Japan Science and Technology Agency (Japan), Chunli Bai of the Chinese Academy of Sciences and Chen Wang of the National Center for Nanoscience and Technology (China), and Chennupati Jagadish of Australia National University (Australia).

And, of course, Mike Roco provided the guiding light for the whole effort, very much engaged in coordination, writing, and editing, and in pushing everyone along to get the best possible results.

President, WTEC   R. D. Shelton
Baltimore, MD, USA

WTEC Mission

WTEC provides assessments of international research and development in selected technologies under awards from the National Science Foundation (NSF), the Office of Naval Research (ONR), and other agencies. Formerly part of Loyola College, WTEC is now a separate nonprofit research institute. Kesh Narayanan, Deputy Assistant Director for Engineering, is NSF Program Director for WTEC. WTEC's mission is to inform U.S. scientists, engineers, and policymakers of global trends in science and technology. WTEC assessments cover basic research, advanced development, and applications. Panelists are leading authorities in their field, technically active, and knowledgeable about U.S. and foreign research programs. As part of the assessment process, these peer review panels visit and carry out discussions with foreign scientists and engineers abroad. The WTEC staff helps select topics, recruits expert panelists, arranges study visits to foreign laboratories, organizes workshop presentations, and finally, edits, and publishes the final reports. See http://www.wtec.org for more information, or contact R. D. Shelton, shelton@wtec.org.

Study Team and Report Authors

Study Coordinators

Mihail C. Roco, Ph.D.	**William S. Bainbridge, Ph.D.**
Senior Advisor for Nanotechnology	Division of Information and Intelligent Systems
National Science Foundation	National Science Foundation
4201 Wilson Boulevard	4201 Wilson Boulevard
Arlington, VA 22230, USA	Arlington, VA 22230, USA

WTEC Panel and Other Contributors

George Whitesides, Ph.D. (Chair)	**Bruce Tonn, Ph.D. (Co-chair)**	**Jian Cao, Ph.D.**
Harvard University	University of Tennessee, Knoxville	Northwestern University
12 Oxford St.	McClung Tower, Room 1018	Dept. of Mechanical Engineering
Cambridge, MA 02138, USA	Knoxville, TN 37996, USA	2145 Sheridan Rd.
		Evanston, IL 60208, USA
Mamadou Diallo, Ph.D.	**Mark Lundstrom, Ph.D.**	**James Murday, Ph.D.**
California Institute of Technology (USA) and KAIST (Korea)	Purdue University School of Electrical and Computer Engineering	USC Office of Research Advancement
1200 East California Blvd.	465 Northwestern Ave.	701 Pennsylvania Ave. NW
Mail Stop 139-74,	Suite 207 S	Suite 540
Pasadena, CA 91125, USA	West Lafayette, IN 47907, USA	Washington DC 20004, USA
James Olds, Ph.D.	**Robert G. Urban, Ph.D.**	**H.-S. Philip Wong, Ph.D.**
George Mason University	Johnson & Johnson Boston Innovation Center	Stanford University
Krasnow Institute for Advanced Studies	One Broadway	Paul G. Allen Building 312X
4400 University Dr.	14th Floor	420 Via Palou
Mail Stop 2A1	Cambridge, MA 02142, USA	Stanford, CA 94305, USA
Fairfax, VA 22030, USA		

(continued)

(continued)

Piotr Grodzinski, Ph.D.
National Cancer Institute
Building 31, Room 10A52
31 Center Drive, MSC 2580
Bethesda, MD 20892, USA

Barbara Harthorn, Ph.D.
Center for Nanotechnology
  in Society
University of California Santa
  Barbara
Santa Barbara, CA 93106, USA

Todd Kuiken, Ph.D.
Woodrow Wilson International
  Center for Scholars
One Woodrow Wilson Plaza
1300 Pennsylvania Ave., NW
Washington, DC 20004, USA

Robert Langer, Ph.D.
MIT
Langer Lab
77 Massachusetts Ave.
Room 76-661
Cambridge, MA 02139, USA

David Rejeski, Ph.D.
Woodrow Wilson International
  Center for Scholars
One Woodrow Wilson Plaza
1300 Pennsylvania Ave., NW
Washington, DC 20004, USA

R. Stanley Williams, Ph.D.
Hewlett-Packard Laboratories
Memristor Research Group
1501 Page Mill Rd.
Palo Alto, CA 94304, USA

External Reviewers for the Written Report

Toshio Baba, Ph.D., Science and Technology Agency, Japan

Gerd Bachmann, Ph.D., VDI Technology Center, Germany

Robbie Barbero, Ph.D., Office of Science and Technology Policy, Executive Office of the President (EOP), USA

Dawn Bonnell, Ph.D., University of Pennsylvania, USA

David Guston, Ph.D., Arizona State University, USA

Tony Hey, Ph.D., Microsoft Corporation, USA

Bruce Kramer, Ph.D., National Science Foundation, USA

Neal Lane, Ph.D., James A. Baker III Institute for Public Policy, Rice University, USA

Robert Langer, Massachusetts Institute of Technology, USA

Jo-Won Lee, Ph.D., Hanyang University, Korea

Stuart Lindsay, Ph.D., Biodesign Institute, Arizona State University, USA

Mira Marcus-Kalish, Ph.D., Tel Aviv University, Israel

J. Tinsley Oden, Ph.D., The University of Texas at Austin, USA

Robert Pohanka, Ph.D., National Nanotechnology Coordination Office, EOP, USA

Françoise Roure, Ph.D., OECD and High Council for Industry, Energy, and Technologies, France

George Strawn, Ph.D., National Coordinating Office for Networking and Information Technology Research and Development, EOP, USA

Thomas Theis, Ph.D., IBM Corporation and Semiconductor Research Corporation, USA

James J. Valdes, Ph.D., U.S. Army

World Technology Evaluation Center, Inc. (WTEC)

R. D. Shelton, President

Patricia Foland, Vice President for International Operations, Project Manager

Geoffrey M. Holdridge, Vice President for Government Services

Patricia M.H. Johnson, Copy Editor

Matthew Henderson, Assistant Project Manager

Convergence of Knowledge, Technology and Society: Beyond Convergence of Nano-Bio-Info-Cognitive Technologies

Executive Summary[1]

A general process to improve creativity, innovation, and outcomes

Convergence as a Fundamental Principle of Progress

Convergence of knowledge and technology for the benefit of society (CKTS) is the core opportunity for progress in the twenty-first century. It is defined in this study as the escalating and transformative interaction among seemingly distinct scientific disciplines, technologies, communities, and domains of human activity to achieve mutual compatibility, synergism, and integration, and through this process to create added value and branch out into emerging areas to meet shared goals. Convergence is as essential to our future knowledge society as engines were to the industrial revolution. CKTS allows society to answer questions and resolve problems that isolated capabilities cannot, as well as to create *new* competencies, technologies, and knowledge on this basis. This concept is centered on the principles presented in this report: of interdependence in nature and society with application to the essential platforms of the human activity system; the enhancement of creativity and innovation within knowledge and technology through convergence–divergence (spin-off) evolutionary processes; a holistic system deduction approach being applied in decision analysis; the value of higher-level cross-domain languages to generate new solutions and support transfer of new knowledge; and the value of vision-inspired basic research endeavors.

Based on these principles, this report suggests solutions for key societal challenges in the next decade, including: (a) accelerating progress in foundational emerging technologies and creating new industries and jobs at their frontiers and interfaces in the economic, human-scale, Earth-scale, and societal scale

---

[1] *Note*: The Executive Summary represents the findings of the study group as a whole, whereas case studies in the report represent individual perspectives.

platforms; (b) increasing creativity, innovation, and economic productivity through convergence of knowledge and technology, including developing a universal domain of information exchange and interaction; and (c) improving lifelong wellness and human potential, including advancing a cognitive society, achieving individualized and integrated healthcare and education, and securing a sustainable quality of life for all. The report also suggests developing a knowledge-based platform for decision-making to implement the most effective convergence methods to valuate and assemble individual theories and technologies and find integrated solutions for societal challenges.

This Study of Convergence

CKTS builds on previous stages of convergence, beginning with the integration of disciplines at the nanoscale, followed by convergence of nanotechnology, biotechnology, information, and cognitive (NBIC) technologies. This study addressed the third level of convergence, that between emerging NBIC technologies and the essential platforms of human activity (technology, human-scale, Earth-scale, and societal-scale platforms), and sought to identify the general convergence processes that characterize all three stages. Longer-term views of convergence, looking up to 40 years ahead, that were outlined in previous NBIC studies provide reference points for this study. The main goals of this study were (1) to understand how convergence works and how it can be improved and implemented, (2) to chart trends for the next decade, and (3) to identify opportunities for key transformative actions to improve societal outcomes through convergence.

This study sought the input of leading experts in the United States and other nations, from academia, industry, government, and NGOs, in five brainstorming meetings held in the United States, Latin America, Europe, and Asia between October 2011 and November 2012, with a final U.S. presentation in December 2012. The study was supported by the National Science Foundation, the National Institutes of Health, the National Aeronautics and Space Administration, the Environmental Protection Agency, the Department of Defense, the U.S. Department of Agriculture, and international partners.[2]

Principal Study Findings

Convergence has been increasing by stages over the past several decades. In the first stage, concerted efforts to research and develop nanotechnology called attention to the convergence of many formerly separate scientific and engineering disciplines (biology, chemistry, condensed matter physics, materials science, electrical engineering, medicine, and others) when applied to the material world, based on

---

[2] The full list of sponsors is given in Appendix A.

Executive Summary xv

growing understanding of atomic and nanoscale structures. "NBIC" convergence was the second stage, connecting emerging technologies based on their shared elemental components such as atoms, DNA, bits, and synapses (all with shared abstractions from information science and system theory), hierarchically integrated across technology domains and scales. CKTS is the next stage in convergence; it expands emerging technologies at their interfaces and frontiers and intimately introduces them into the human-scale, Earth-and societal-scale platforms. It brings together the relevant areas of human, machine, societal, and natural resource capabilities to attempt to answer questions and resolve problems that isolated capabilities cannot, as well as to create and disseminate new competencies, technologies, industries, products, and solutions for human well-being. This report provides a status survey of CKTS, the hierarchical structure of the evolutionary human activity system, methods to improve and expedite knowledge and technology convergence within this structure, a vision for the next decade, and implementation opportunities.

The experts around the world who were consulted in this study recognize CKTS as a timely engine of change that has the potential to provide far-reaching solutions to achieve improved economic productivity, new industries (new jobs) and products (such as smart phones), increase human physical and cognitive potential, and secure a sustainable quality of life. The study identified barriers to progress; this report proposes a framework, methods, and possible actions to overcome them. The proposed convergence methods have wide applications, for example, to improve the "innovation chain", evaluate commercial network performance, support planning for emerging technologies, and study cognitive processes. To effectively take advantage of this potential requires immediate action, particularly because of the current environment of limited resources. Principal opportunities for immediate action are (1) creation of a global convergence network to build and connect the convergence efforts of various regions, (2) creation of a U.S. CKTS initiative coordinated by the Federal Government, (3) development of the EU multi-annual 2014–2020 framework program of research and innovation activities called "Horizon 2020," and (4) enactment of a South Korean program for a "Convergence Research Policy Development Center."

Organization of the Report

The four essential and interdependent platforms for convergence are defined in Chaps. 1, 2, 3, and 4: (1) *"NBIC" foundational tools*, (2) *human-scale activities*, (3) *Earth-scale environmental systems*, and (4) *societal-scale platform*. Chapters 5, 6, 7, 8, 9, and 10 define and illustrate the main implications of and proposed responses to converging sciences and technologies in terms of (5) *human health and physical potential*, (6) *cognition and communication*, (7) *productivity and societal outcomes*, (8) *education and physical infrastructure*, (9) *societal sustainability*, and (10) *innovative and responsible governance*. The Overview and Recommendations section, which follows the Executive Summary, provides details both on the intellectual framework for convergence activities and on the kinds of Federally led actions that the panel believes will take best advantage of the inevitable progression of convergence in science and technology. As a whole, the report aims to show how concerted efforts that acknowledge and assist the CKTS processes as a fundamental principle of progress can achieve major goals that support broad societal benefits.

How Convergence Works

Convergence is actually part of a dynamic and cyclical *convergence–divergence process* that originates organically from brain functions and other domains of the global human activity system. This process can provide a structure and specific improvement methods for the creative-innovation-production chain. The convergence phase consists of analysis, making creative connections among disparate ideas, and integration. The divergence phase consists of taking these new convergences and applying them to conceptual formation of new systems; application of innovation to new areas; new discoveries based on these processes; and multidimensional new outcomes in competencies, technologies, and products. This convergence–divergence process is reflected in the coherent chain of ideas from the ancient to modern eras, in the evolution over time of knowledge and technology, and in the development of human organizations and industries. The model proposed in this report suggests that creativity and innovation rates are increasing with respect to the convergence domains of various activities and the speed of movement between the convergence and divergence cycles. To conceptualize the influence of convergence on creativity and innovation, this study (see Chap. 4) has defined an index of innovation rate (I) that is a function of the size (S) of the convergence domain $(I \sim S^2)$ and time scale (T) of the convergence–divergence cycle $(I \sim 1/T^3)$.

There are five general approaches that are at the core of the CKTS concept Chap. (4):

(a) *Added-value decision-making and knowledge transformation* based on the convergence–divergence (spin-off) evolutionary processes in science, technology, and applications.
(b) *A holistic systematic deduction approach*, beginning from the global evolutionary system of human activity and considering hierarchical interconnections among knowledge, technologies, and societal systems.
(c) *Establishment of higher-level languages* (multidomain, convergent) using knowledge, technology, and cultural integrators—such as unifying theories, benchmarking, multidisciplinary nanotechnology, informatics, knowledge mapping, similar fractal patterns, and music—that can allow construction of shared terminology and concepts that are common and essential to multiple domains.

(d) *Focusing efforts on vision-inspired basic research and grand challenges* enabled by convergence. To efficiently and responsibly achieve the benefits of research in emerging areas, convergence processes will be used to identify the vision and then its corresponding basic research strategic areas, changing priorities periodically as interdependencies change. The proposed "Vision-Inspired Basic Research" quadrangle will extend the Stokes diagram beyond the Pasteur quadrangle.[3]

(e) *Proactively encouraging coherent public and private efforts* that currently contribute to the unguided convergence of knowledge and technology to use a systematic approach to convergence that amplifies the most beneficial endeavors for society to consider undertaking.

Examples of coincidental, rapidly evolving, and valuable convergences in knowledge, technology, and society can be seen in NBIC emerging technologies; universal databases; cognition and communication developments; cloud computing; human–robotics systems; mind-cyber-physical systems; platforms for unmanned vehicles; the space program; the research program on fundamental particles (Higgs et al.); the birth of entirely new disciplines such as synthetic biology, quantum communication, nanophotonics, and nanofluidics; and the integration of biomedicine with physics and engineering that is already effecting transformations in human healthcare systems. One specific illustration of the convergence–divergence process is the cell phone platform, which began with the creative assembling of a wide range of technologies and cognitive and human–computer interface sciences, all of which converged to create the "smart phone" about a decade ago. This is now diverging into thousands of applications scarcely imagined 10 years ago that have profound "cascade" implications on areas as diverse as national security, education, and cognitive science.

The main barriers to overcome for such bold technologies are insufficient synergistic methodologies and interconnections between NBIC technologies and the human-scale, Earth-scale, and societal-scale activity platforms. The methods and domains of convergence discussed in the report provide a framework for solutions to overcome such barriers.

Emerging Paradigms of Convergence

This report identifies a number of convergence trends with goals that are poised for radical paradigm transformations in human endeavors:

- *Support the emerging and converging technologies* by bottom–up discovery-driven and top–down vision-driven programs. A grand challenge is to accelerate progress in the foundational NBIC technologies and *create new industries and jobs at their frontiers and interfaces* (Chaps. 1, 7, and 10).

---

[3] Stokes, D.E. 1997. *Pasteur's quadrant: Basic science and technological innovation.* Washington, DC: Brookings Institution Press.

- *Expand human physical and cognitive potential* through convergence of NBIC technologies with the human-scale platform (see Chaps. 1, 2, 4, 5, 6, and 8). A grand challenge is to achieve coordinated improvements *in lifelong wellness, cognitive technologies, and human development.*
- *Achieve higher societal productivity and economic efficiency* through convergence of human activity platforms focused on societal governance (see Chaps. 1, 4, 7, 8, 9, and 10). Improving economic productivity is envisioned through facilitating the circuit of creativity and innovation and a vision-driven expansion into new knowledge and technology fields. Challenges are in developing resource-efficient production systems; establishing information technology with capability far beyond silicon integrated circuits; creating a universal domain of information exchange, including a database for all disciplines and industry sectors; and developing a flexible and efficient transportation system, complementary human–robotics systems, brain–computer communication (e.g., for neuromorphic engineering and prosthetics), cognitive computing, and mapping of brain activity and brain–behavior interdependencies. A grand challenge is to use converging technologies in manufacturing methodologies.
- *Secure a sustainable quality of life for all* through convergence of the Earth-scale and human-scale platforms (see Chaps. 2, 3, 4, 5, 6, 9 and 10). This includes providing equitable access to knowledge, natural resources, food, healthcare, and safety in the face of increasing population, bounded Earth resources, higher carbon and nitrogen levels in the atmosphere, and climate change (Chap. 3). A grand challenge is to achieve societal sustainability, including efficient solutions to interconnected water, energy, and materials needs, and sustainable urban communities (Chap. 9).
- *Empower individuals and groups* through integrated education, use of the spiral circuit of creativity and innovation within larger domains—including multidomain design, expansion of human knowledge and cognitive capabilities—and added-value decision analysis, made possible by convergence (Chaps. 4, 6, 7, and 8). A grand challenge is to build a CKTS-based individualized lifelong education system.
- Advance societal progress through integration of convergence methodologies, ethical aspects, citizen participation, and management for responsible development into *a new governance model* (Chap. 10). A grand challenge is to change the framework and improve the efficiency of the methods of societal governance by applying CKTS principles.

Action Opportunities

The study panel sees an international opportunity to develop and apply CKTS in ways that will produce synergies leading to technological, economic, environmental, and societal benefits. An international network to collaboratively advance the methods and applications of convergence is proposed. The panel also proposes a proactive, integrative program in the United States to focus disparate R&D

energies into a coherent activity that distills the best of our knowledge and abilities to the greater good of the national and global human communities. Concerted efforts would be required to develop a CKTS initiative that could take advantage of convergence as a fundamental opportunity for progress. This initiative could cooperate with and maintain U.S. competitiveness with parallel efforts already underway in other economies, including in the European Union, Korea, Japan, Brazil, and China.

The proposed U.S. CKTS initiative would incorporate convergence innovations in five modes of support:

- **Centers** for establishing new, creative, and innovative socioeconomic models based on convergence, including methods, education, research, standards, informatics, and biomedicine
- **Technology platforms** for addressing societal grand challenges, including distributed and connected NBIC manufacturing and global virtual factories, use of converging cognitive technologies in society, brain-mapping activities, and cognitive computing
- **Programs** for creating a shared universal convergence database and methods to evaluate convergence, risk governance, and integration of science into society
- **Organizations** to monitor and accelerate increases in human potential, societal sustainability, improved decision analysis using the convergence–divergence cycle, and conflict resolution
- **Government coordination** (a "Federal Convergence Office") for supporting convergence in science, technology, investment planning and policies, decision-making, wellness and long-term human development activities, supporting our aging society, and sustaining Earth systems, as well as advancing ethical, legal, and public participation aspects of convergence

The primary initiative opportunities for government involvement are in the application of the five CKTS approaches (convergence–divergence, system deduction applied to decision-making, higher-level languages, vision-inspired basic research and grand challenges, and proactive channeling of public–private R&D efforts toward CKTS) in the emerging paradigms of convergence identified above. Promising high-impact areas for the next decade are in improving productivity using converging technologies in personalized, distributed, and connected manufacturing and services; promoting a cognitive society to increase human potential; improving quality of life through wellness initiatives, support for a sustainable society, and biology- and healthcare-centered convergence; developing integrated education devoted to promoting creativity and innovation; and advancing coordinated and responsible management of convergence through Federal convergence offices dedicated to critical areas of science, technology, and investment policy and planning. Detailed analysis in the areas of interest is needed for constructive and beneficial application of CKTS.

Several areas with immediate societal benefits are opportunities for pilot projects in the CKTS framework: "converging revolutionary technologies for individualized services (CORTIS)" (where individualized services include providing and receiving personalized education, medicine, cultural, productive, and general services); a distributed cyber nanobiomanufacturing network; R&D for emerging logic devices and new information carriers for nanoelectronics; creating a universal domain of information exchange between converging areas; cognitive computing; physical and mathematical modeling of the brain from the nanoscale (synapse-level); biomedicine-physical-engineering convergence; convergence–divergence decision-making in R&D planning; a CKTS summary document for policymakers on innovation policy; and a "Federal Convergence of Knowledge and Technology (CONEKT) Office" to identify the best approaches and areas of implementation.

CKTS is a foundational transformational approach to connect, synergize, and valuate existing and emerging technologies in much the same way as Higgs-Boson has become an essential standard model particle that provides connections and interactions (mass, forces) with previously discovered fundamental physics particles. CKTS promises to become a key science and technology field of the same level of importance as the key technologies it promises to connect and synergize.

With the increasingly complex interactions in our knowledge society, and limited resources, a systematic CKTS approach would most efficiently bring about the kinds of beneficial societal transformations explored in this report in the same way as selective "small changes" have been shown to yield major changes in large complex systems.

Vision for Societal Convergence

The concept of "societal convergence" in this report encompasses the involvement of society at all stages of supporting the progressive convergence of scientific knowledge, its technological applications, and democratic principles of equal opportunities for progress. This concept provides a rich resource that can lead to revolutionary advances in sustainable global development, economic productivity, human potential, and national security. Over the next decade, proactive and systematic convergence can fundamentally improve the quality of our daily lives, transforming the ways we and our descendants learn, work, thrive, and age, and protecting the many integrated natural and social systems that support our human activities.

Convergence of Knowledge, Technology and Society: Beyond Convergence of Nano-Bio-Info-Cognitive Technologies

Overview and Recommendations

Science, engineering, and technology are recognized to permeate nearly every facet of modern life and to hold the key to solving many of humanity's most pressing current and future challenges (NRC 2012). In their recent book *That Used to Be Us*, Friedman and Mandelbaum (2011) identify one of the main challenges for the United States as bringing activities together driven by a higher national purpose—to "act collectively for the common good." Our report aims to address that challenge by identifying the basic mechanisms of human activity—including knowledge creation and technological innovation—and proposes an approach to better understand and use these mechanisms for a holistic purpose. The report advances convergence in science and technology to benefit society; evaluates methods to improve its transforming tools and governance; and identifies long-term trends in the application of converging technologies. The project has gathered information on interdisciplinary research, development, application projects, and trends from the Americas, the European Union, Asia, and Australia.

Some important themes pervade science, mathematics, and technology and appear over and over again, whether we look at an ancient civilization, the human body, or a comet. They are ideas that transcend disciplinary boundaries and prove fruitful in explanation, in theory, in observation, and in design (AAAS 1989). This report is centered on five such ideas focused on convergence of knowledge, technology and society (CKTS): (1) *the principle of holistic system interdependence in nature, knowledge, and society with application to the economic, human-scale, Earth-scale, and societal-scale platforms*, (2) *the convergence–divergence (spin-off) evolutionary process in science and technology,* (3) *the cultivation of higher-level (multidomain) languages to support fruitful transfer and application of new knowledge,* (4) *the need for vision-inspired basic research and grand challenges to efficiently achieve desired outcomes*, and (5) *the merit of systematic channeling of public and private efforts to promote convergence activities.*

CKTS is the natural extension of precursor unifying principles in science and technology (S&T), notably the wave of integration across "NBIC" fields that began with nanotechnology, biotechnology, and information technology and quickly expanded to include technologies based on and enabling cognitive science (Roco and Bainbridge 2003). It is reflected in bridging the divide between research, education, innovation, and production needs in national and regional activities. The dynamic integration of knowledge, technologies, and society is a fundamental opportunity for human progress as expansion in harnessing Earth's resources and availability of larger investments have reached limitations.

Intellectual unification of the diverse branches of science and engineering should progress hand-in-hand with expanding the cultural unification and economic effectiveness of the world as an interdependent system. However, convergence should increase rather than diminish the importance of the individual human being. Since the invention of the wheel and the lever, effective technologies have increased human potential, but now capability augmentation of entirely new kinds is becoming possible. The challenge will be to increase the capabilities of both individuals and societies in a manner conducive to collective and individual human well-being and freedom, and based on respect for commonly held ethical principles.

This report provides an overview of CKTS, methods to improve and expedite knowledge and technology convergence in this system, possible solutions for implementation of key opportunities, and a vision for the next decade. A longer-term vision for converging technology and human progress was provided in previous studies (Roco and Bainbridge 2003, 2006); these and other earlier works on a holistic approach (Wilson 1999) and development of a technology-driven society (Kurzweil 1999) have been used as references.

1 Convergence: A Fundamental Principle and Timely Opportunity for Progress

1.1 Principle for Progress

Interdependencies between the human mind and the surrounding natural system determine a *coherent convergence–divergence evolutionary process in the interconnected knowledge, technology, and societal development* that leads to creation of added value and progress. Convergence includes bringing together all relevant areas of human, machine, and natural resource capability that enable society to answer questions and resolve problems that isolated capabilities cannot; divergence creates and disseminates new competencies, technologies, and products. The convergence–divergence process aims at what is essential in a system: synergism, new pathways, innovation, efficiency, and simplicity. CKTS provides a systematic approach to connect and enable other emerging technologies. Convergence processes already have begun between several specific domains of science and technology (Sharp and Langer 2011; Roco 2012).

Nature is a single coherent system, and diverse methods of scientific and engineering investigations should reflect this interlinked and dynamic unity. Accordingly, general concepts and ideas should be developed systematically in interdependence, with cause-and-effect pathways, for improved outcomes in knowledge, technology, and applications. At the same time, industrial and social applications rely on integration of disciplines and unification of knowledge. This report identifies essential knowledge and technology platforms of human activity in society and methods for reaching convergence within and among them to achieve societal benefits.

1.2 Timely Opportunity

"It was the best of times, it was the worst of times." So Dickens described the revolutionary changes of a past century. But the words aptly describe the collection of scientific and social revolutions currently raging today, some admirable—such as increased human connectivity enabled by new digital technologies, and the eradication of starvation and epidemic diseases in nation after nation facilitated by medical and social innovations—and some lamentable, such as global economic crises and bloodshed fueled by ethnic and ideological intolerance. The most powerful creations of the human mind—science, technology, and ethical society—must become the engines of progress to transport the world away from suffering and conflict to prosperity and harmony. Today, because science and society are already changing so rapidly and irreversibly, the *fundamental principle for progress must be convergence*, the creative union of sciences, technologies, and peoples, focused on mutual benefit.

The discrepancy in most of the world's economies between the accelerated quasi-exponential growth of knowledge (discoveries and innovations) since 2000 and the relatively slow quasi-linear economic growth underlines the unmet potential to improve technology deployment and governance. This report shows that convergence offers added-value solutions in many application areas *to address the knowledge–outcomes gap* and eventually accelerate economic and societal development. This is timely, as recent statistics show a deceleration of GDP per capita growth rate in developing countries, including in the United States.

Furthermore, in the last decades, there has been a gap between the rate of growth of R&D investments—which indicate future trends in knowledge growth—in several countries and the rate of GDP growth. For example, in the United States, the Federal spending for research and development grew at just 1.3 % annually from 1989 to 2009, while gross domestic product rose annually 2.4 % (NSF statistics; Reif and Barrett 2013). Because of this investment slow-down, finding cost-effective means to address the knowledge–outcomes gap discussed in the previous paragraph is an even more timely opportunity.

The United Nations studies on human development (2011, 2012c) provide a well-considered blueprint of the global status of human society and desirable, immediate goals: inclusive economic development, inclusive social development, environmental sustainability, and peace and security. The concept of the convergence of knowledge, technology and society explored in this report is an approach to addressing human challenges through the synergism it generates in creativity, innovation, and decision-making.

The study received input from leading academic, industry, and government experts from four continents and a number of major economies, including the United States, Australia, China, the European Union, Japan, Korea, and Latin America, in the context of the five workshops held during the year of the study. Convergence of knowledge and technology has been identified as an emerging field around the world, with specific actions recommended to accelerate progress and to benefit from it as promptly as possible. For example, the European Union draft program HORIZON2020 using convergence principles (Sect. 10.8.8) was the subject of EU–U.S. workshop deliberations, and the South Korea draft program "Convergence Research Policy Development Centre" was the subject of U.S.–South Korea–Japan workshop deliberations (Sect. 10.8.9). From the interactions among study participants, it is evident that building an international network to collaboratively advance the methods and applications of convergence would be valuable to all nations. The primary conclusion of the U.S. panel's deliberations is that a Federal CKTS initiative could best take advantage of this fundamental opportunity for progress in the United States (see Sect. 8 below).

2 Goals of the Study

This study of converging knowledge, technologies, and society has aimed to document the most important accomplishments worldwide in the last 10 years in research, development, and application projects, as well as trends in the United States and abroad, regarding convergence of knowledge and technology, and ideas from leading researchers regarding the greatest opportunities in the next decade, focused on ways these trends and opportunities can be harnessed to address human problems and needs. The study took place over the course of five international workshops held from November 2011 to October 2012 in Latin America (Brazil), the United States, Europe (Belgium), and Asia (Korea and China), with a final presentation of findings in December 2012. It was sponsored by the National Science Foundation, the National Institutes of Health, the National Aeronautics and Space Administration, the Environmental Protection Agency, the Department of Defense, the U.S. Department of Agriculture, and international partners. The principal expert panelists and report reviewers are listed in the front of the report; locations, sponsors, and other details of the workshops are listed in Appendix A; and all workshop participants and report contributors are listed in Appendix B.

The main goals of the study were to advance the convergence of science and technology for the benefit of society by:

- Better understanding the process of convergence and the methods for improving convergence through transforming approaches and governance
- Identifying a vision and trends for the next decade and beyond
- Suggesting opportunities for key transformative actions to improve societal outcomes through added-value convergence–divergence processes in various sectors

The study targeted three overarching societal goals:

- Improved economic productivity
- Increased human potential
- Securing a sustainable quality of life for all (i.e., access to natural resources, food, healthcare, knowledge, and safety)

The challenges faced by humanity require a significant and sustained effort in the above three areas that begins now and is clearly charted into the future, at a minimum into the next decade.

3 Organization of the Report

Convergence brings people with compatible capabilities together so that they can move forward coherently. It is an iterative and progressive process in which people use science and technology to address societal needs by means of transformative tools that themselves are transformed by the advances they facilitate. While convergence is a beneficial unification process, it cannot be fully comprehended without some understanding of its components and the processes that bring them together. Thus, this report is organized in terms of general platforms for convergence—*foundational tools, human-scale, Earth-scale, and societal-scale platforms* (Chaps. 1, 2, 3, and 4)—and the main implications of CKTS convergence for:

- *Human health and physical potential* (Chap. 5)
- *Cognition and communication* (Chap. 6)
- *Productivity and societal outcomes* (Chap. 7)
- *Education and physical infrastructure* (Chap. 8)
- *Society and sustainability* (Chap. 9)
- *Innovative and responsible societal governance* (Chap. 10)

Each chapter presents the panel's vision in the respective area, goals for the next decade, priorities, and case studies of paradigm changes. Together, the chapters explain how concerted efforts can achieve major goals for the well-being of humanity.

A set of video interviews for the public related to this study have been produced in collaboration with the Woodrow Wilson International Center for Scholars. The videos are available on website http://www.wilsoncenter.org/convergence.

4 An Evolving Definition of Convergence

Three successive levels of convergence have been described by U.S. Government-sponsored studies described in Fig. 1 and below, including the more holistic CKTS approach that is considered in this study for transformative knowledge and technology programs.

- First, in the late 1990s moving into the 2000s, *nanotechnology* provided integration of disciplines and technology sectors of the material world building on new knowledge of the nanoscale (see *Nanotechnology Research Directions*, Roco, Williams, and Alivisatos 1999).
- Second, in the 2000s, converging *nanotechnology, biotechnology, information technology, and cognitive* ("*NBIC*") *technologies*—starting from basic elements, atoms, DNA, bits, and synapses, as well as a system approach—led to foundational tools that integrated (both horizontally and vertically) various emerging technologies into multifunctional systems (see Roco and Bainbridge 2003).
- Third, moving into the 2010s and beyond, *CKTS* (also referred to in this study as "beyond-NBIC" or "NBIC2") is integrating essential human activities in knowledge, technology, human behavior, and society, distinguished by a purposeful focus on supporting societal values and needs.

In this report, each level of convergence is defined by an application domain and a specific process of integration. The respective CKTS domains and processes are discussed in the following two sections.

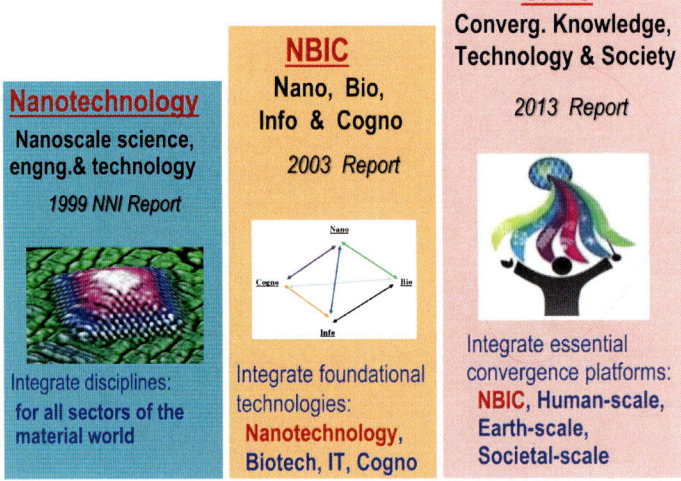

Fig. 1 Successive levels of convergence for added value, as described in three reports: (from *left* to *right*) *nanotechnology* integrating disciplines for all sectors of the material world (Roco et al. 1999); *NBIC* recognizing the novel synergies among different sciences and technologies (Roco and Bainbridge 2003); and *CKTS* looking holistically at the interconnections between science, technology, and society (this volume)

5 The Evolutionary System of Human Activity

There are hierarchical levels of the convergence domains, the top level being the global and evolutionary human activity system driven by societal values and needs for progress in human development (Fig. 2). Under this, the four essential convergence platforms are each defined by a core system with key players (individuals,

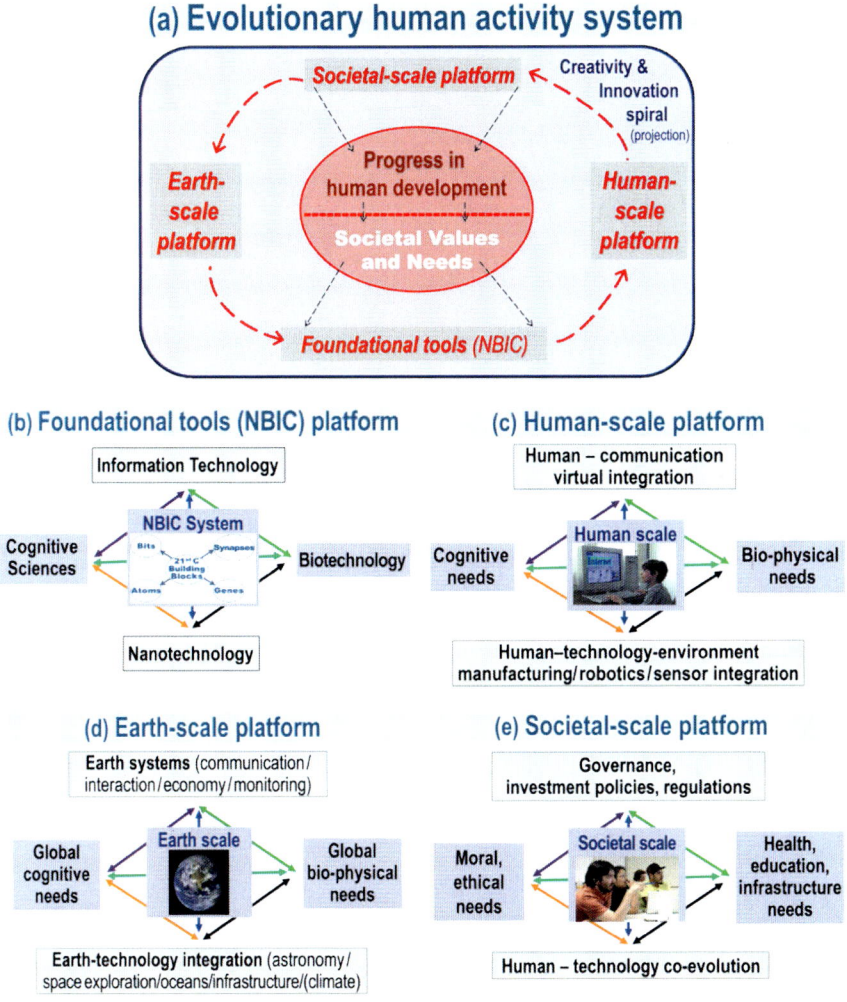

Fig. 2 Four general convergence platforms of the evolutionary human activity system. (**a**) Global interactions are driven by societal values and needs leading to progress in human development. (**b**) Foundational tools (NBIC) platform (see Chap. 1 in this volume). (**c**) Human-scale platform (Chap. 2). (**d**) Earth-scale platform (Chap. 3). (**e**) Societal-scale platform (Chaps. 4, 5, 6, 7, 8, 9, and 10)

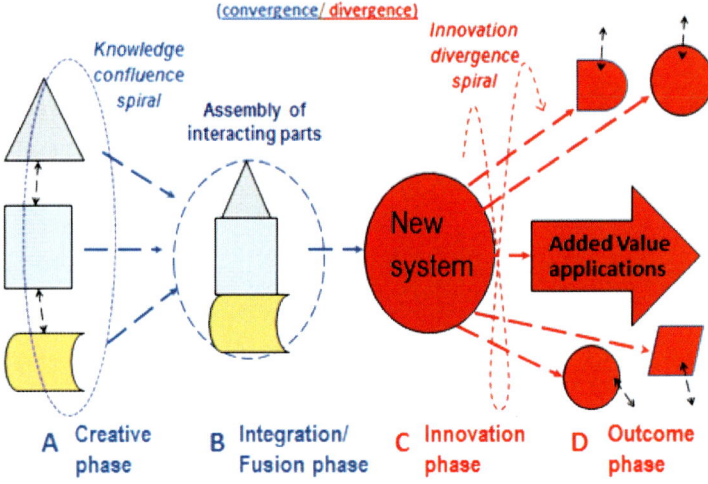

Fig. 3 The convergence (*A*, *B*)–divergence (*C*, *D*) cycle, initially formulated for megatrends in science and engineering (Roco 2002), is extended here, as a typical process, to knowledge, technology, and societal domains. It includes (*A*) a creative phase dominated by synergism between multidisciplinary components leading to discovery, (*B*) integration and/or fusion into a new system for known applications, (*C*) an innovation phase leading to new competencies and products, and (*D*) an outcome (spin-off) phase consisting of new applications and new inputs into the creative phases (*A*) of successive cycles

organizations, and technologies), specific interactions and characteristics, tools, and outcomes, as outlined below:

- *Foundational tools* (NBIC): These each start from a basic element (atom, bit, DNA, or synapse) and are integrated into systems.
- *Human-scale platform:* This is characterized by the interactions between individuals, between humans and machines, and between humans and the environment.
- *Earth-scale platform*: This is the environment for human activities, including global natural systems, communication systems, and the global economy; there are limitations for human intervention.
- *Societal-scale platform*: This is characterized by the activities and systems that link individuals and groups on several larger scales; it consists of collective activities, organizations, and procedures, including governance.

In the evolutionary human activity system (Fig. 2a), there is a feedback interaction loop where there is a central core (progress in human development) that is in a perpetual state of advancement, and a circumferential information exchange loop that crosses multiple domains and represents the projection of a three-dimensional spiral of creativity and innovation (shown in Fig. 3).

5.1 Key Attributes

The first phase of NBIC convergence integrated foundational, emerging technologies that are defined from elementary components and interact at all scales and levels of complexity. CKTS builds on that foundation to achieve integrated knowledge and capability across all fields of science and engineering, based on common convergence platforms and concepts, to achieve societal benefits. The core ideas are that CKTS:

- Advances an integrative approach across human dimensions, encompassing value systems, operating at societal and global scales, while remaining valuable for each individual person
- Is based on the material unity at the nanoscale, integrated systems, and information universes, connected via human behavior and other integrators
- Is best facilitated by a holistic approach with shared methodologies, theories, and goals, which is quite different from traditional forms of collaboration in which a division of labor separates disciplines from each other
- Renews the focus on people's capabilities and human outcomes, rather than allowing decisions to be technology-driven, and seeks to transcend existing human conflicts to achieve vastly improved conditions for work, learning, aging, physical and cognitive wellness, and to achieve shared human goals

6 The Convergence–Divergence Evolution Process

In this report, the convergence process is defined as the escalating and transformative interactions among seemingly different disciplines, technologies, and communities to achieve mutual compatibility, synergism, and integration, and thus to create added value to meet shared goals. This societal convergence definition expands on convergence–divergence concepts in science and engineering megatrends (Roco 2002) and applies them to unprecedented interconnected advancements in knowledge, technology, and social systems. The result of societal convergence thus broadly conceived is expected to be numerous new applications of science and technology with significant added value to society. Convergence is not a simple, unidirectional process. The convergence–divergence processes follow in cycles of various intervals (the interval between A and D in Fig. 3) and are applicable to the various human activity platforms noted earlier.

Convergence as a process unites knowledge, technology, and applications, both across traditionally separate disciplines, and across multiple levels of abstraction and organization. It has a creative phase (A in Fig. 3) from where the knowledge confluence spiral leads to an integration/fusion phase (B). After integration, a new system is created, which leads to new outcomes in various areas (innovation phase, C).

Divergence as a process starts after the formation of the new system and leads to new competencies, products, and application areas for the knowledge achieved in the convergence process (outcome phase, D), on that basis following an innovation spiral. The outcomes (D) become seeds for new convergence processes, as the figure suggests. The exponential random graph (Wasserman and Robin 2005) and successive cascade divergence are examples of models for stage D.

The rapid pace of change in science and the emergence of new technologies require new approaches that manage complexity, achieve sophisticated functionality, and yet are intelligible for ordinary human users of the applications. Convergence thus extends beyond science and technology, even beyond new applications, to include harmonious unification of activities across the entire societal spectrum.

One way to conceptualize this is in terms of the complementary functions of the human brain, recognizing that in fact brain lateralization is only modest in degree, and individuals differ in the specialization of their brain hemispheres. The convergence phase—characterized by analysis, creative connections, and integration—is perhaps best done by the left half of the brain. The divergence phase—characterized by synthesis or formation of the new system, innovation in applying to new areas, finding new things (outcomes in knowledge, technology, and society), and multidimensional outcomes—may best be done by the right half of the brain. Whatever model or metaphor one prefers, the convergence–divergence process reflects *two complementary roles of brain functions*. Decision-making follows a convergence–divergence process *driven by the need for improvement and added value* that is at the core of human thought and behavior, which also is reflected in group and organization actions.

Each convergence–divergence process is cause-and-effect connected to upstream and downstream related processes and is in coherence with other simultaneous CKTS processes at various domain and temporal scales. One way to consider and better use such longer-range coherence is by engaging the brain functions of contemplation and reflection ("mindfulness") (Fishman 2004; Langer 1997) to focus on making the most beneficial decisions by considering the relevant events and by anticipating and avoiding unintended harmful aspects of the new powerful and connected technologies. Mindfulness is distinguished by expanded perspective, context-relevant interpretations, and receptiveness that can lead to more long-range, discriminating, and holistic innovation and solutions. Contemplation and reflection in research and education will also help provide a foundation for enhancing human capacity and unity of purpose in terms of addressing longer-term aspects of wellness, creativity, and innovation.

While engaged in study and dialog about convergence of science, technology, and society, the participants in the several international meetings of this study panel were inspired by the simple diagram in Fig. 4, which shows the conductor of an imaginary orchestra, perhaps playing Kepler's *Music of the Spheres* or *The Planets* by Holst, in which the conductor represents the coordination required to orchestrate the complex vision, with each converging field of science and technology functioning like one of the instruments in the orchestra, playing together in different combinations for various passages in the musical score. Like the conductor, societal governance

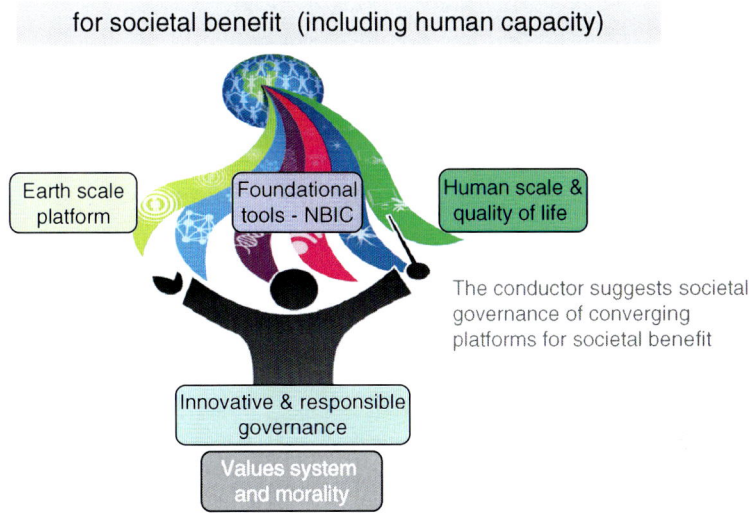

Fig. 4 Schematic of the orchestration of the essential convergence platforms

must actively and harmoniously engage in coordination of the total composition of complex scientific, technical, and societal convergences that are taking place and gaining momentum.

6.1 Examples of the Convergence–Divergence Process

One specific example of convergence and of the convergence–divergence process described above is the development of the cell phone platform, where a wide range of technologies including high-frequency communications and packet switching protocols (for connections to global networks); materials science and nanotechnology (for CPUs, data storage, touch screens, antennas, etc.); and cognitive science and human–computer interface technologies (for the user interface) *converged* to create the "smart phone" about a decade ago. This is now *diverging* into thousands of applications scarcely imagined 10 years ago, from social networks to controlling swarms of very inexpensive miniaturized satellites, and many other examples, too many to list, affecting virtually every aspect of our society. These impacts in turn have profound implications for and secondary impacts on areas as diverse as national security, education, and cognitive science.

Another more general example of the convergence–divergence process is in creating S&T programs and organizations that seek to deliver solutions by linking the creative phase of scientific excellence available in society to generate converging knowledge and technologies that can then be used to address technology needs and societal grand challenges.

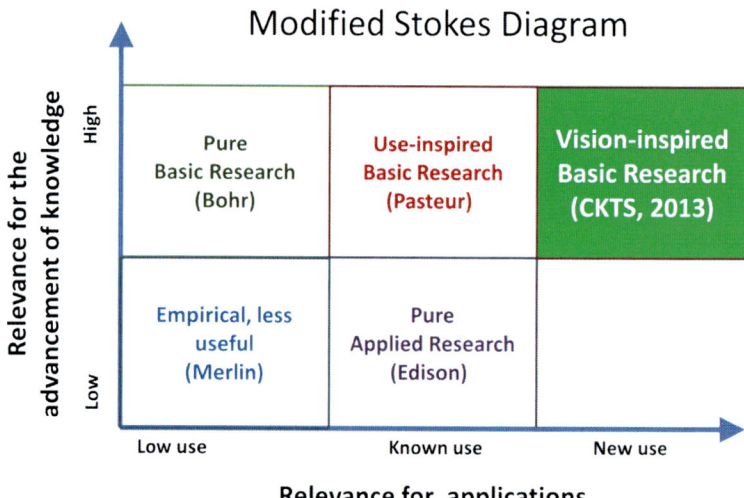

Fig. 5 Schematic for the proposed "vision-inspired basic research" domain in the modified Stokes diagram (Courtesy M. C. Roco)

6.2 The Innovation Component of CKTS

To conceptualize the influence of convergence on creativity and innovation, this report defines an index of innovation rate (I) (Chap. 4). This index is estimated to be in direct proportion to the size S of the convergence domain from where information is collected (the domain circumscribed by the innovation spiral, or the number of disciplines or application areas intersected by the circumferential spiral, in Fig. 2a); the speed of information exchange supporting innovation in that platform (S / t); the speed of the convergence–divergence cycle (1/T) where T is the duration of the cycle (from phase A to phase D, in Fig. 3); and the divergence angle in realizing the outcomes (~ O/T):

$$I \sim k(S)(S/t)(1/T)(O/T) \sim kS^2O/(T\ T\ t) \sim k'S^2O/T^3 \qquad (1)$$

where $t \sim T$ (the time scale t for the circumferential projection is proportional with the time scale T for the convergence–divergence cycle, both being projections of the same three-dimensional spiral), O is the outcome size, and k and k' are coefficients of proportionality. This qualitative correlation underlines the importance of the size of the convergence domain (S^2) and time scale of the convergence–divergence cycle ($1/T^3$) on the innovation rate. The (S^2) term in the equation agrees nicely with the well-known "Metcalf's Law" (which states that the value of a network scales as the square of the number of nodes in the network; Shapiro and Varian 1999). The (O/TT) term agrees with the exponential growth of technological developments illustrated by the well-known

Moore's law in the semiconductor industry. The remaining (1/T) term is proportional with the rate of technology diffusion.

7 Vision-Inspired Basic Research and Grand Challenges

The first step in "thinking outside the box" is to recognize that the box exists. Indeed, science and engineering are collections of boxes, and multidisciplinary work is one way to escape some of them. But there also exist more general conceptual boxes that may be more difficult to perceive and thus to escape.

One of the most influential conceptual frameworks for thinking about the goals of research was proposed by Stokes (1997), who advocated research in what he called Pasteur's quadrant. This is defined as research motivated in such a way that both of the following questions are answered in the affirmative: "Are considerations of the practical utility of the results crucial?" and "Is the research a quest for fundamental understanding?" Visionary research may also fit both of these criteria, but adds two others: "Is the research leading to emerging uses beyond known applications?" and "Is the work transformative in the sense that entirely new ideas are being explored and invented?"

The CKTS methods may be used to set up the connection between long-term science and technology visions and basic research activities (Chap. 4). As shown in Fig. 5, this proposed "Vision-Inspired Basic Research" domain will expand the existing four domains of the Stokes diagram to a new quadrangle dedicated to basic research for new emerging applications inspired by a vision (new use) beyond the Pasteur quadrangle.

Different basic research approaches are needed as a function of the phase in the knowledge, and technology convergence–divergence process (refer back to Fig. 3): the Bohr approach for the creative phase (A), the Pasteur approach for the integration/fusion phase (B) and vision-inspired basic research for the divergent phase (C and D).

8 Timeline

Because convergence is a process operating on a vast scale and along many dimensions, as it is achieved, its own focus and characteristics will continue to evolve. We can identify three successive and overlapping phases in the convergence of science, technology, and society, as described in Table 1.

Building on previous NBIC studies, this report seeks to provide a vision for the future of societal convergence and to define transformational actions for key stakeholders. This includes understanding convergence mechanisms and methods to improve the outcomes in various human activity areas on this basis, defining the roles of national science and technology strategies, and open governance approaches. The study's key findings on future trends (in Sect. 6 of Chaps. 1, 2, 3, 4, 5, 6, 7, 8, 9, and 10), methods for convergence (Chap. 4), and overarching opportunities for transformative actions are presented in the following sections.

Table 1 Three phases of CKTS convergence

Time frame	Phase	Characteristics
2001–2010	Reactive convergence	Coincidental, based on ad hoc collaborations of partners or individual fields for a predetermined goal
2011–2020	Proactive convergence	More principled and inclusive, approaching convergence through more explicit decision analysis; the immediate future of CKTS
After 2020	Systemic convergence	Holistic, with higher-level (multidomain) purpose, with input from convergence / governance organizations

9 A Vision for the Future: Major Emerging Paradigms of Convergence

Convergence of knowledge, science, and technology can benefit human society in many significant ways. In a sense, convergence can be viewed as the ultimate grand challenge, which if successfully achieved, can pave the way to achieve the numerous other grand challenges facing society. This is because convergence provides added value and synergistic benefits to human endeavors to:

- Improve wellness and human development
- Increase productivity and promote economic development
- Achieve societal sustainability
- Empower individuals and communities
- Expand human knowledge and education
- Achieve an innovative and equitable society

9.1 Wellness and Human Development

Improving human potential is a main societal goal. All forms of scientific and technological progress will depend upon the ability of human beings to have the intelligence, wisdom, and health to invest energy into new projects and make carefully considered decisions about convergence and developments that fundamentally impact their lives. It is realistic to believe that physical and cognitive potential can indeed be increased at the individual and collective levels to a significant extent, and this observation should not be mistaken for a fanciful dream but as a logical extension of advances in education, communication and information technologies, and cognitive science.

A holistic approach for healthcare is a main wellness component. Hood and Flores (2012) called for a predictive, participative, preventive, and personal "P4" medicine that is founded on systems biology. People cannot achieve their full potential if they suffer physical or mental health problems or must cope with chronic disabilities. Converging technologies have the potential to improve cancer detection and treatment, develop regenerative medicine, and harness the human immune system in new ways to maintain vigor and increase longevity (Chap. 5). They also can improve human life through all of its stages, toward achievement of total wellness, based on reliable information about an individual's particular needs in nutrition, exercise, medical treatment, and emotionally positive life experiences. One essential tool for health and well-being is comprehensive data, including genomic information about the individual, provided both to the individual in a manner intelligible without technical training, and in a reliable and comprehensive manner to health professionals who treat the individual. All of these advances can benefit from improved understanding of what recently has been called the *cognome*, the cognitive equivalent of the genome, defining the principles controlling mental functions in the individual and in the community to which the individual belongs (Chap. 6).

Improving human wellness is particularly important in child development. It is well understood that prevention of disease is at least as important as cure, yet achieving healthy lifestyles is often difficult for individuals or communities to achieve, especially as medical specialties and leisure activities continue to go separate ways. It is plausible that holistic approaches to wellness and human development can contribute greatly to the solution of this problem, but only if they are based in converging technologies and an increasingly holistic understanding of physical and cognitive health needs.

High-quality and affordable healthcare in the future will require a partnership between medical professionals, their patients, and scientists and engineers. Medicine must evolve from a passive coincidence model to an active convergence model, from reactive to proactive practice, from waiting for a crisis to taking preventive approaches. Its aim should include reducing economic costs for medical treatments and lost productivity. Biomedicine-based convergence is of immediate importance to society. Five health wellness goals that are largely achievable in the coming decade illustrate the range of directions convergence can take:

- Improved methods will detect cancer and chronic diseases earlier and identify the appropriate treatment for the particular condition and patient, with reduced side effects.
- Improved collection, analysis, and delivery of data will monitor general health, as well as a myriad of specific conditions, toward wellness and prevention of disease.
- Given more effective ways to halt the progress of a disease, the emphasis will shift to repairing damage and restoring healthy functioning, including tissue regeneration and advanced prosthetics.
- Radically improved understanding of the complex human immune system, including swift and efficient analysis of biomarkers in individual patients, will harness the immune system as a constant monitor of health and disease, and enhance its curative powers.
- Reduction in the time needed to detect and treat the emergence of highly infectious diseases at the global level will occur, along with design and development of cost-effective medical responses to the threats, repurposing of drugs, the tailoring of drug and vaccine use to the individual, and seeking innovative cures for currently untreatable conditions.

9.2 Productivity and Economic Development

Convergence is expected to help improve human capabilities, decision processes, and infrastructure, all supporting higher productivity and creation of new types of useful, meaningful, and well-paying jobs.

NBIC2 workshops held in Europe, Asia, and South America indicate that convergence is evolving to play a strong, strategic role in economic development programs worldwide. Many of these efforts are building upon years of investments in the individual NBIC areas of nano-, bio-, information, and cognitive technologies. South Korea is one of the countries that have launched convergence programs, in its case creating educational institutions and government organizations that have convergent themes and linking these institutions to industry leaders. The European Union is investing substantial resources in NBIC, focusing these resources in certain geographic areas to create agglomeration benefits, and linking these R&D institutions to established private sector firms. Under the draft framework of the EU "Horizon 2020" plan, the convergence of (1) societal challenges (e.g. health, well-being, ageing), (2) industrial competitiveness, and (3) scientific excellence is considered pivotal: developing new knowledge, technologies, products, and applications that bridge the divide between research and innovation needs is a crucial EU 2020 objective (*Horizon 2020 Strategic Programme 2014–16, Draft*). In the joint NBIC2 workshop in Leuven the EU participants had slightly different terminology to express a similar approach identified in this CKTS report leading towards knowledge-, capital-, and skill-intensive innovation cycles.

One could argue that the convergence of science and technologies is ideally suited to American-style economic development, through the entrepreneurism of start-up companies. NBIC technologies can also be worked into strategies to promote economic development in traditionally economically distressed communities.

Convergence generally implies an increase in diversity, not in uniformity, within a set of rational overarching principles. Thus, development of a global economy can be compatible with local autonomy; for example, the growth of largely self-sufficient urban communities can rely heavily upon renewable resources such as solar energy and conducting much manufacturing locally, even as information is shared worldwide. Convergence approaches could also strengthen the organizational structures of corporations and other agencies, which would have positive effects on group productivity. Strength is not the same thing as rigidity, so organizations would need to understand the value of a diversity of viewpoints and appropriate flexibility, as encouraged by the divergence part of the convergence–divergence cycle (Janis 1982). The benefits of the convergence–divergence process are supported by other studies such as the theory of increasing returns (Arthur 2009).

Integrated and Distributed Production Systems

Advancing manufacturing through converging NBIC technologies will be essential to the progress of the economy and quality-of-life initiatives (Chaps. 1 and 7), because manufacturing is a means for wealth creation. It transforms the geometry, composition, and functionality of incoming materials through utilization of natural resources and converging technologies. Consequently, it improves quality of life at the human-scale platform due to the development of new machines and devices; directly impacts the environment at the Earth-scale platform; and poses challenges at the societal-scale platform as mega cities develop as a result of concentrated manufacturing. As we gain understanding of the converging technologies, three areas show promise (Chap. 7):

- *Distributed and connected manufacturing*—enabled by process flexibility, modularity, in-process metrology, predictive sciences and technologies, and human–machine interaction—could greatly improve the ease and efficiency of making products available, customizing designs (from performance-specific wearable sensors to replacement body parts for surgery or individual art works), and scalability from small- to large-batch production. Connecting cyber and physical systems will be essential.
- *"Manufacturing process DNA"* is a metaphor suggesting that the design of manufacturing processes could be structured in a manner like the genetic code, allowing the rapid switching in and out of production steps and system components, and allowing precise control over product parameters with very little effort.
- *Integration of the social and physical sciences* could optimize design and production methods, not merely relying upon the functioning of market mechanisms to determine what products are produced, but also incorporating sustainability and general human welfare into decision-making at all stages of manufacturing towards knowledge-, capital-, and skill-intensive innovation cycles.
- An example where the above areas will be needed is in developing capabilities for integrated electronic nanosystems that extend to powerful and logical memory platforms capable of coordinating an increasing variety of nanoscale devices: sensors, actuators, energy-harvesting devices, programmable resistors, nonvolatile memories, user-driven systems, etc. (Chap. 1).

Integrated Organizations

Organizations will need to adapt to and facilitate the new knowledge creation and manufacturing models. Transformation of organizations and businesses would guide convergence by higher-purpose criteria such as improvement of economic productivity, human potential, and life security, including sustainable development. Both international coordination and international competitiveness are necessary components. Such transformation is required for traditional institutions to adapt to diminished roles as they are bypassed by social-media-enabled movements, and to address the opportunities and threats arising from changes in technology and governance roles.

9.3 Societal Sustainability

Most concepts of sustainability focus on maintaining and improving the quality of life enjoyed by humans into the distant future. It is recognized that current energy consumption, water usage patterns, agricultural practices, manufacturing practices, climate stability, clean environment, communities, and other components of modern life and economies are not currently sustainable (Chap. 9) within the Earth's boundaries (Chap. 3). Emissions from the modern economy threaten human and ecological health and are a major cause of climate change.

In the absence of sustainability in any of these areas, humanity faces not only a degraded natural environment but also severe threats to world peace, as economic crises bring undemocratic forces to power in nations suffering from one or more unsustainable conditions. Conflict would further degrade the world's ability to deal with these problems; thus, it is essential to bind the people of the world together, and scientific convergence can actually play a key role in that global unification, as explained next. There is enormous potential for converging knowledge, science, and technologies to help achieve sustainability goals. Scientific understanding of society's needs (Chap. 10) in this field is crucial. For example, convergence is resulting in the development and production of sustainable materials. Convergence is a key to the development of new sustainable technologies, ranging from small-scale water purification systems to large-scale desalinization systems. Convergence can produce technologies to separate rare materials from wastewater and seawater (Chap. 9). Convergence is also leading to the development and deployment of arrays of sensors designed to measure in real time a wide range of sustainability indicators, from air and water quality to the migrations and health of key species.

One can imagine a plethora of NBIC2 technologies converging on platforms such as megacities to foster sustainability. The convergence of intelligent software, ubiquitous sensors, and mobile devices can substantially increase the efficiency of megacities with respect to transportation, energy production and consumption, and manufacturing. NBIC2 solutions can also be used to improve the resiliency of these types of megacity systems to natural disasters, acts of terrorism, and cascading systems failures.

Beyond the scale and scope of megacities, one can argue that planet Earth is a single ecosystem, upon which a unified economic system of production, distribution, and resource use is consolidating. Thus it must now be considered as one complex system in which each change affects all the other variables, and wise management of the entire system is essential. Among the goals must be to improve overall efficiency in using natural resources, increasing productivity, and protecting environmental and human health at this Earth scale.

Climate change is one such Earth-scale issue that requires particular attention. A fresh approach is needed, one that is actually even more ambitious than the control and adaptation strategies common in the past, where the goal for the future is integrating environmental preservation into a coherent effort to create a new world system. One can envision the convergence of NBIC2 technologies as laying the groundwork for fresh approaches to reducing fossil fuel use and sequestering carbon from the atmosphere (Chap. 3).

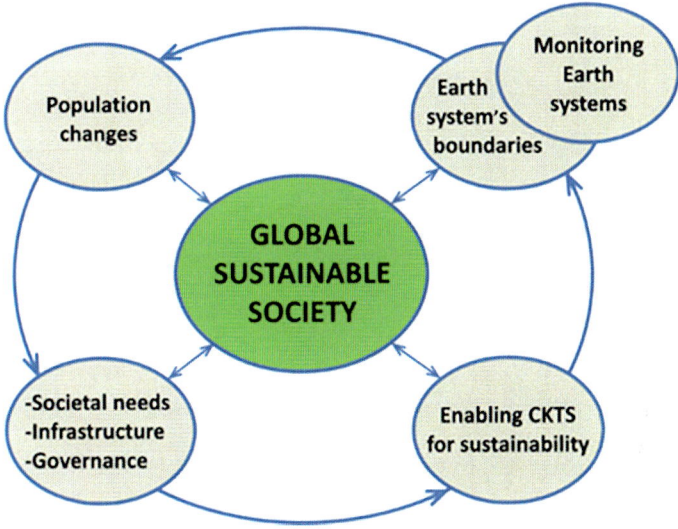

Fig. 6 Sustainability and interacting Earth systems (a part of the Earth-scale platform)

Monitoring of Global Data and Information Infrastructures

Given that the world is an integrated natural, human, and technological system, it will be impossible to achieve sustainability without *a global data and information infrastructure*. The panel advocates creating an *"Earth monitoring center"*—an immersive data visualization environment for collaborative interdisciplinary research, modeling, and education focused on understanding the whole Earth system environment, the dynamic human impact on it, and decision options for effecting long-term planetary sustainability (Chaps. 3 and 9). This will enhance the coordination done under the Global Earth Observation Systems of Systems (GEOSS, http://www.earthobservations.org/geoss.shtml). Specialized data systems and researcher slots would be available for experts from the wide variety of scientific and research disciplines (climate scientists, oceanographers, modelers, energy experts, nanotechnologists, systems biologists, information scientists, data visualization experts, economists, cognitive scientists, behavioral sociologists, political scientists, policy analysts, and many more) required for understanding options for long-term planetary sustainability and economic development. The project is suited for public–private and international partnerships. Dissenting opinions can be included in the convergence phase of the convergence–divergence cycle. It also will include components for global catastrophic and existential risk assessment, global megacities, and global material resources.

Future activities need to be focused on the sustainability of global society by considering demographics, societal needs, and governance, and enabling CKTS solutions for mitigation and life security within the Earth system's boundaries (Fig. 6).

9.4 Empowered Individuals and Communities

Convergence of knowledge and science has the potential to empower individuals and communities to explore new pathways to improving the quality of their lives. NBIC2 technologies can support the development of tools for life-long learning, participation in public policy decision-making, and peer production of information resources, software, and other goods and services. It is important, though, that these tools be designed with respect for human cognitive capabilities and limitations.

Envisioning the Cognitive Society

One can envision a major convergence field, cognitive science, as being essential for advancements to improve the empowerment of individuals and communities (Chap. 6). As the existence of language demonstrates, cognition is largely social, because information and concepts are exchanged between people and assembled over time into cultures. *Basic components* of this new cognitive model of the social mind include the global cognitive society, a cognome project to understand our evolving mentality, cognitive computing, cognition and communication in team science, and a cognitive science of science.

The traditional concept often used to describe the way a community thinks is *culture,* although some writers today use terms like *social intelligence* and *social cognition.* Thus, by analogy with how cognitive science came into being, it may be necessary to create a new field, *cultural science*, which includes the most rigorous parts of cultural anthropology, sociology, and social psychology—with additions from other relevant fields that were left out of the formation of today's cognitive science. More generally, cognitive science can help identify underdeveloped areas of science, as well as opportunities for conceptual unification of separate highly developed fields, thus facilitating the wider progress of convergence.

The cognitive concept in governance would bring the power and catalysis of convergent technologies to bear on enhancing societal cognition, for both individuals and groups. The rationale for this rests upon the key role for human cognition in charting a successful course for humans and their societies as they face unprecedented challenges in areas ranging from public health to global climate change. Human cognition is central to decision-making at all levels, and decision-making increasingly is enormously consequential, not only for humans, but for the planet. One can imagine a cognitive focus on empowerment bringing together agencies and institutions—public and private—to invest in a positive future for humans, one the panel terms "The Cognitive Society". Such a society will be characterized by the ubiquity of convergent cognitive technologies that are leveraged to enhance human decision-making, human well-being, and public health. It will be marked by cognitive contributions to life-long human learning, natural language-based information flows between machines and people, and wise decisions at all levels of society. Convergence may catalyze the transformation of our current challenges from intractable problems into planet-scale solutions.

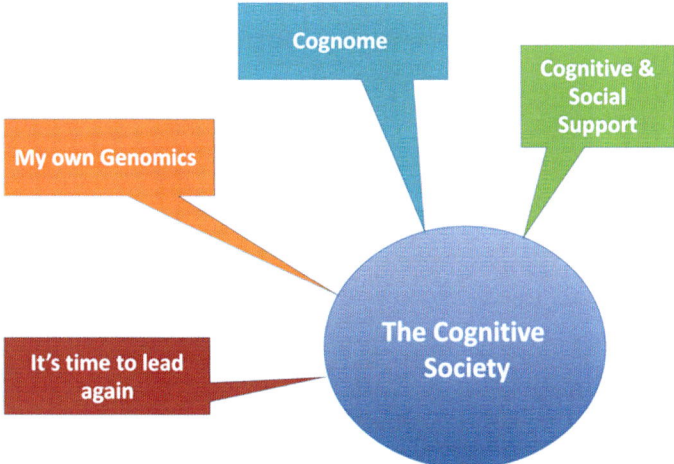

Fig. 7 Playing a role in creating a Cognitive Society are four goals: Creating "Cognitive and Social Support," the "Cognome," "My Own [individualized] Genomics", and "It's time to lead again" (a reference to reestablishing a global leadership role for the United States through proactively encouraging convergence and the Cognitive Society) (see Chap. 6)

All of these advances can benefit from improved understanding of what recently has been called the *cognome*, the cognitive equivalent of the genome, defining the principles controlling mental functions in the individual and the community to which the individual belongs.

Cognitive science came into being as the convergence of a half-dozen subfields of different sciences, and now that it is well established, it has the potential to contribute to all fields of science and engineering, because both *discovery and invention are cognitive processes* (Fig. 7).

Cognitive scientists should do research on the forms of thinking done by scientists and engineers in various fields, resulting in (1) better designs for the tools used by the scientists, (2) improved education of future scientists, (3) insights that can help the scientists understand their own mental processes, and (4) new principles for convergence between sciences based on much better understanding of how the various different kinds of scientists think (e.g., inductive versus deductive, qualitative versus quantitative, categorizing versus model-building). Convergence methods need to be developed and evaluated in conjunction with the development of a science of convergence.

Systematic Convergence in Education and Infrastructure

Integrated education systems are proposed to support convergence (Chap. 8). The changing nature of technology and the global economy require serious redesign of education, in grade school, college, the workplace, and across the full span of each human life. This is a need for *developing means for flexible adaptation to change through convergence, and even getting ahead of trends, to transform education into an engine of creativity and innovation.* Education can be integrated across disciplines, based on new insights into cognition, and utilizing new nanotechnology-enabled digital education assets toward more effective, interactive approaches to personalized education. Information technology can contribute new or improved interaction capabilities—such as 3D video/text display, oral (including interactive dialog and other speech-related media), motion capture whether haptic or large-motor, and virtual or augmented reality—all tailored to the personal learning modes and motivations of each student.

Successful development of *person-centered education* will require clear awareness of each person's present state of knowledge and comprehension, perhaps gained through privacy-protected automatic monitoring of the environment and the student, and through constant assessment of comprehension, to ensure an efficient rate of progress.

Consideration should be given to changing the foundation of learning in universities and trade schools, beginning with broader concepts of nature, foundational NBIC tools, and converging platforms in the freshman year, instead of beginning with introductions to narrow disciplines.

While one new focus will be on the individual student, it will also be essential to develop a *global educational environment.* For example, all educational materials could be downloaded from a central storage site into a local memory for real-time accessibility; it would be necessary to create a distributed international *"converging knowledge and technology network"* with a multidomain database, education modules, and facilities. The concepts of geographically distributed user facilities, such as the National Nanotechnology Infrastructure Network and the Network for Computational Nanotechnology, could be expanded to form converging technologies clusters.

Progress in educational convergence requires *supportive infrastructure* of at least three kinds: *physical infrastructure* (including laboratories and prototype production facilities corresponding to various converging platforms), *information infrastructure* (such as "universal" massive databases and the tools for using them), and *institutional infrastructure (*such as professional associations, industry partnerships, and government agencies).

A distributed national converging knowledge and technology network might be constructed with a multidomain database, education modules, and facilities. There is an intense need for manufacturing user facilities and test-beds plus remote access to instrumentation to support education in multiple disciplines and fields of relevance.

Citizen Science

The convergence of technologies and tools has the potential to greatly empower the contributions individuals can make to science. The term *citizen science* refers to the serious involvement of non-specialists in scientific research and engineering development. It has always been the case that some ordinary citizens have contributed their time and intelligence to science, for example, in collecting fossils and bringing them to a paleontologist for study, in volunteering to be research subjects in medical experiments, and by answering social science questionnaires. But very recently, such opportunities have expanded greatly using the new information technologies, for example having amateurs classify galaxies in Hubble images, compiling tens of thousands of observations by bird watchers, and running online experiments in systems like Mechanical Turk (https://www.mturk.com/mturk/). A prime example in *citizen engineering* is open source software development, through which volunteers have created tens of thousands of useful computer programs, including even the widely used Linux operating system. Early citizen science exploited the free labor of volunteers, but their involvement is becoming increasingly sophisticated, to the point that science can converge with society, and different sciences can build bridges between themselves via the wider personnel involvement encouraged by citizen science.

Many challenges must be met in order to achieve the full potential of citizen science and citizen engineering. Effective tools for collaboration must be developed—reliable, easy to use, and capable of being modified and expanded to be useful far beyond their original application areas. Sharing information and scientific education at many levels must be integrated with the research goals, enhancing the possibilities for scientific discovery rather than burdening the project with excessive tutoring demands. In most cases, citizen scientists are not paid for their work, so other motivations must be found to make involvement valuable for all participants without in any way distorting the results. Keeping the scientific goals in focus has shown to be essential. Scientists themselves are citizens, and citizen science can thus serve to link sciences with each other. This remarkable trend has great implications for public education and for creating a well-informed citizenry that can participate positively in the governance of science and engineering development.

9.5 *Human Knowledge and Education*

The prospects for convergent science and technology development to foster the growth, evolution, and potential revolution in human knowledge are virtually unlimited. We are already witnessing converging disciplines, design principles, and transformative tools. Transdisciplinary approaches are linking together disparate knowledge bases to produce new knowledge. One can imagine convergence as the engine to achieve the consilience of knowledge linking together the humanities and natural and social sciences.

Innovation occurs best in the modern world through the integration into a single system of science, engineering, and society. In the convergence–divergence cycle (Figs. 2 and 3) knowledge input may be from any CKTS area, and the outcomes can be in all CKTS areas. In particular, more emphasis should be given in R&D to "vision-inspired basic research" (Fig. 5) in strategic areas, while discovery-based research continues to be practiced and protected.

The convergence process has two fundamental phases: (1) the novel, creative assembling of previous knowledge and technology components into a new system, and (2) diffusion of innovations from the original application to others that may be far removed in materials, goals, and design traditions, and which leads to precursors of new assemblies. Research focused on the translations of innovations across fields can not only strengthen convergence of design across disciplines but also allow disciplines to explore entirely novel approaches on the basis of solid knowledge about the range of field-independent approaches that have already been developed. The goal is finding fundamental principles that can unite previous separate fields in a rigorous manner. New communication and design platforms—hardware, software, cognition, and culture—can accelerate the spirals of creativity and innovation by crossing faster, more, and broader platforms.

Convergence implies an increasing level of multidisciplinarity to solve scientific and engineering problems, design methods, and develop productive tools. In addition to the integrated educational system discussed as part of infrastructure, three opportunities are computational tools for convergence, education for convergence, and knowledge resources for convergence.

Computational tools for convergence. Powerful scientific and engineering computer simulation and material processing tools have been developed over the past few decades in each of the four NBIC disciplines and in other areas. One now needs a new class of tools to address problems at the interface of disciplines. For example, tools to design brain–machine interfaces must incorporate understanding of both electronics design and cognitive science. Efforts to develop these new simulation capabilities should be supported.

Education for convergence. Exciting new initiatives such as Coursera and EdX are bringing low- or zero-cost, high-quality educational resources to a global audience via so-called MOOCs (massively open online courses). A similarly ambitious initiative is needed to develop new approaches to education that are specifically designed for the era of convergence, which will require new types of computer simulation tools and new approaches to education. These should be widely disseminated at low cost as resources for practicing scientists and engineers and as models for universities as they evolve to address convergence.

A broader desired outcome of education for convergence is advancing collaboration and conflict avoidance in human behavior and culture. This is important because natural resources are limited, and destructive means are more powerful in an increasingly crowded planet.

Knowledge resources for convergence. For hundreds of years, scientists and engineers have been using publications and archives to share information, but often separately within each narrow field, whereas convergence requires a scientist or engineer in any one field to have access to information from any other. The arXiv.org electronic preprint service has pioneered swift sharing of publications in a set of rather mathematical fields, but many other disciplines have so far failed to emulate it successfully. Thus, convergence requires a *major transformation of current scientific publishing practice* that poses substantial economic, organizational, and technical challenges.

The problem of *sharing data and publications* may be relatively simple to solve, although free sharing runs contrary to the commercial interests of current organizations and the journal publishing industry. If all science and engineering journals were freely available online, there would still be much work to achieve convergence, because each field has its own terminology, and nations with many different languages now conduct excellent research. Thus, more effective tools for finding, translating, and annotating publications would need to be developed, in connection with the development of overarching conceptualizations achieving intellectual unification.

Increasingly, human activities are becoming embedded in complex systems that mix social with technical components that interact in complex and dynamic manners (Chaps. 3, 6, and 7). Given that some sectors of society are already organized as complex socio-technical systems, the questions arise, which others are evolving in the same direction, what innovations are required to make each different system function well, and do we need to consider means to manage all of society as single technically convergent system? *Mathematical modeling and computer simulation tools become integrators across all the fields of application* in the growing cognitive society.

Information sharing is one factor fostering the growing effectiveness of collaboratories and virtual organizations (VOs), organized around shared instrumentation and data archives, in many specific fields of science. One barrier to communication across them is that they employ different standards for hardware, data, and metadata, a problem that was seen a few years ago in archives of brain scan images because different laboratories were following different procedures and standards. Another is that few existing VOs are in immediately adjacent fields, such that building a bridge between them would be relatively cheap, although this problem may diminish as many more scientific VOs are established. It is possible that the Federal Government may need to create overarching digital libraries, comparable to the National Library of Medicine and the Library of Congress. Scalable universal databases are needed for convergence. Often, separate scientific associations set standards for data and analysis methodologies, and logically this could be done on a higher level of aggregation, although to do so might aggravate the problem of locking science into well-established but inflexible category systems.

9.6 *Innovative and Equitable Society*

Challenges to governance of convergence have included developing the multidisciplinary knowledge foundation, strengthening the innovation chain from priority-setting and discovery to societal use, addressing broader societal implications including risk management, and overall, creating the tools, people, and organizations to responsibly develop and equitably distribute the benefits of the new technologies (Chap. 10).

The governance of knowledge and technology currently involves very complex economic, legal, and management systems, and it seems unlikely that the institutions developed decades ago are perfectly designed for today's rapidly changing circumstances. There are clear signs that problems are endemic in the current system, as illustrated by the public controversies about climate change, healthcare, and national defense investments, all of which have substantial science and engineering components.

The social sciences will have a crucial role to play, but perhaps not a role that has been traditionally scripted for them. Fundamental reconceptualization may be needed. For example, it is easy to say that social scientists should conduct intensive, diverse, but coordinated studies of the second-order implications of new technologies, to identify both unexpected positive benefits that need to be exploited and unintended negative effects that need to be avoided. Yet, how exactly can they do this without delaying progress and without imposing biases on the results? Similarly, recent advances in communication technology could support far greater public involvement in decision-making, yet this would risk basing decisions on popular illusions or myths promulgated through the mass media rather than on professional expertise. Two approaches can be identified now, and perhaps others can be devised. First, modern methods based on traditional opinion polling and conducted online could measure the changing values of the general public, to identify the goals that are important to people, and then experts in the relevant fields of science and technology could determine how to achieve those public priorities. Second, in each area of technical decision-making, ordinary citizens could select professionals, such as academics or leaders in industry, who would serve as their representatives in deliberations in the given area.

The transformative power of convergence must be used for human benefit, which means more than merely having good intentions, but also possessing the fact-based wisdom to avoid negative unintended consequences. The emphasis on converging technologies, and on Pasteur Quadrant basic research in strategic areas, should not detract from the importance of pure research to gain knowledge about the universe in which we dwell. The fact that some technologies are ready for near-term deployment should not cause us to invest less in emerging technologies that have not yet reached that level of maturity. Indeed, the "vision-inspired basic research" quadrant (Fig. 5) may be especially valuable in areas that have not really begun to emerge yet, and where effective paradigms for assessing ethical implications may not yet exist. Researchers in both biotechnology and nanotechnology have traditionally accepted formal responsibility for the ethical, legal, and social implications (ELSI) of their innovations—although this has not generally been the case with information technology—and now this principle must be extended across all the convergent areas (Chap. 10).

There are multiple visions of both past and future, depending upon whether one sees converging technologies as a field of opportunities, or a quagmire of risks, or various points in between. Framing will be key to communicating with the public. It will be important to get all visions out on the table for discussion, after which some empirical validation and calibration will be in order. It is vital to consider our society's previous experience with earlier convergences and with current disturbance factors that are accompanied by major social change.

The overarching challenge is to *support rational governance of visionary ideas* that benefit from their positive possibilities, without being deceived by fads or abandoning those practices of the past that still function well (see list of visionary ideas in Appendix D).

10 Opportunity for Action

The panel identified synergistic opportunities for international collaboration in CKTS projects leading to societal benefits. It has proposed the establishment of an international network to identify and deliberate methods for convergence.

The envisioned benefits of convergence suggested above can be brought to life through a **CKTS Initiative**. This study offers new ideas how to improve the decision-taking and transformation approach based on (a) the *convergence–divergence process* in science, technology, and applications; (b) a *holistic deductive approach* based on the bottom-up, top-down and interdisciplinary horizontal interconnected knowledge, technology, and societal system (Fig. 2); (c) establishment of higher-level (multidomain, convergent) languages in planning and management; and (d) vision-inspired basic research and grand challenges enabled by convergence; furthermore, (e) it would be beneficial to channel public and private efforts that are now contributing to *de facto,* unmanaged convergence of knowledge, technology and society into a *proactive, systematic approach* that can focus disparate energies into a coherent activity that distills the best of our knowledge and abilities to the greater good of the national and global human communities. In R&D, an adaptive process based on these five ideas is needed to support vision-inspired basic research and grand challenges.

Concerted efforts would be required to develop the CKTS Initiative in the United States that could take advantage of convergence as a fundamental opportunity for progress, using five multifaceted modes of support (Centers, Technology Platforms, Programs, Organizations, and Government Coordination) as described below.

10.1 Centers

National Convergence Centers and Networks would be established in research and education institutions to address the formation and dissemination of CKTS. Key priorities are:

- **CKTS Theory and Methods**. These centers would address measurement, evaluation, and informatics approaches for convergence platforms and processes. The centers would explore development of theory, models, methods, institutions, and other conditions that tend to foster or constrain convergence (Chap. 4). Methods might include mathematical and social models for convergence, development of multidomain languages and tools, evolutionary systems approaches (Dunbar 2003), open collaboration, converging design methods across disciplines, foresight, and use of socio-technical laboratories for prototyping entirely novel forms of the system. The centers would develop a set of practices and strategies for implementing convergence science, engineering, and production, as well as identifying the most beneficial convergence applications.
- **CKTS-Education**. This distributed network would address the education "grand challenge" for CKTS, including horizontal (across disciplines) and vertical

(along years of a person's life) system integration to determine how much knowledge is necessary to build and maintain the technological infrastructure required to support a satisfactorily high quality of life (Chap. 8; Jones 2009). The unity of knowledge is essential in education (Gregorian 2004). The network would aim at creating a convergence-based creative and innovative ecosystem. It would be supported by multidomain databases, convergence education modules, and user facilities. It would propose new programs and policies to educate new generations of scientists and practitioners.

- **CKTS-Biomedicine**. These convergence centers would bring together biology, medicine, science, and engineering for biomedical applications (Chap. 5). This topic is particularly accessible for convergence because of the already vibrant scientific progress and societal relevance of this field (NRC 2010; Sharp and Langer 2011). This topic has potential for significant impact in both medium and long terms, and it has immediate benefits for wellness and healthcare systems.

10.2 Technology Platforms

Converging knowledge, technology, and society research platforms, created in areas of national interest, could improve R&D focus on priority topics. The first consideration would be given to the major emerging paradigms of convergence outlined in Sect. 7. Key priorities are:

- **CKTS-Products**. *Sustainable and distributed converging technologies (NBIC) manufacturing research*—such as cyber-enabled distributed manufacturing, nanobiotechnology, and mind-cyber-physical systems—could tie devices created by translational research together with the needs of several platforms such as families, communities, and megacities. The focus on emerging technologies will allow for high-value-added production (Chap. 7).
- **CKTS-Sensors**. More specific platforms may be created for similar tools or types of products. For example, *ubiquitous mobile sensing and environmental conditions reporting devices* could collect information from the environment for direct reporting to their human owners and also to be added to the cloud of globally integrated data. Aggregated information from the cloud (e.g., local air quality) could be reported to individuals in real time. Research on such devices could tie together translational research, universal databases, and could also support research on human platforms. Boosting industrial systems by developing new knowledge, technologies, products, and applications that bridge the divide between research and innovation is a crucial global objective for the future, with high potential for innovation, growth, and competitiveness.
- **CKTS-Cognitive Society**. The *Cognitive Society* concept (Chap. 6) in governance would bring the power and catalysis of convergent technologies to bear on enhancing societal cognition, both for individuals and groups. Such a society would be characterized by ubiquitous convergent cognitive technologies that are

leveraged to enhance human decision-making, human well-being, and public health. It would be marked by cognitive contributions to life-long human learning, natural language-based information flows between machines and people, and thoroughly considered ethics-based decisions at all levels of society. The CKTS-Cognitive Society platform would aim to catalyze the transformation of our current challenges from "intractable problems" into planet-scale solutions.
- **CKTS-Brain-Mapping Activity.** This is a technological challenge that would aim to reconstruct the human brain's neural activity across complete neural circuits, which could prove to be a step toward understanding fundamental and pathological brain processes (Alivisatos et al. 2012). The focus in understanding the brain should shift from considering a neuron to a synapse as the basic element in brain function (Kuzum et al. 2012; Chaps. 1 and 6).
- **CKTS-Cognitive Computing.** This would require a new computation paradigm in which cognitive systems and algorithms recognize context and intent without having to be programmed by experts in arcane software languages, but rather learn in an uncertain and changing environment (Chaps. 1, 6, and 7). This is likely to be realized early in the next decade. The one successful example we know that is currently capable of such analysis is the human brain.
- **CKTS-Sustainable Water, Energy, and Materials Nexus.** This would entail using wastewater, brackish water, seawater, and brines as sources of clean water, energy, nutrients, and critical materials. It would apply CKTS research to develop sustainable processes and systems to (1) reuse wastewater and (2) desalinate brackish water and seawater while recovering (a) energy (e.g., salinity gradient-based power generation and hydrogen production using saline water splitting), (b) nutrients (nitrogen and phosphorous) and (c) valuable/critical metal elements (e.g., lithium, magnesium, uranium, gold, silver, etc.) (Chap. 9).
- **CKTS-Sustainable Urban Communities.** *Building the sustainable urban communities of the twenty-first century* would utilize CKTS research to reconfigure existing megacities and configure future ones into smarter cities with (1) more affordable/resilient housing and distributed energy and water infrastructures, (2) more efficient agriculture/food production and delivery systems, (3) more energy-efficient transportation systems, and (4) distributed healthcare infrastructures, more livable environments, and employment opportunities for all (Chap. 9).

10.3 Programs

Societal convergence data, systemics, and information research could achieve integration of data and information across topics, scales, and time. R&D programs are needed for foundational aspects. Key priorities are:

- Evaluations of methodologies for the performance of a convergence ecosystem.
- Integration of databases from the area of a human brain, at one end, with Earth-scale systems data at the other end.

- Revision of rules and regulations to advance individual and group creativity and innovation in convergent processes in the economy as a critical condition for competitiveness.
- Implementation of effective risk governance of emerging and converging technologies.
- Development of new paradigms for understanding science and communicating about it.

10.4 Organizations

Transformation of organizations and businesses are an essential precondition for convergence. Key priorities are organizations that can:

- *Monitor increasing human potential*, physical and mental, at the individual, group, and society levels by using convergence tools.
- *Improve decision analysis* using the convergence–divergence cycle as a guide, which requires data collection and analysis tools that operate at multiple levels of differentiation, and development of theories to understand systems that can achieve homeostatic balance along some dimensions, while progressing along others.
- *Support and expand citizen science* by providing support to facilitate interactions and convergence methods such as general-purpose databases. This would mean informing, helping to educate, and treating volunteers more as colleagues of scientists, thereby achieving better scientific work, improving education, and facilitating the convergence of science with society.

10.5 Government Coordination

The Federal Government would have the role of supporting and evaluating the opportunities for convergence in the Federal Government, and between Federal and local governments. These opportunities include areas such as wellness and aging where the role of government is essential. Key priorities are in the following areas:

- **CKTS-Federal Convergence Office**. This office would have the responsibility to identify, facilitate, and coordinate opportunities of convergence in the Federal Government, and between Federal and local governments, the private sector, and civic organizations. It would include setting the vision, long-term planning, investment policies, organizational outlines, and evaluation. It would revise rules and regulations to advance individual and group creativity and innovation in convergent processes in the economy as a critical condition for competitiveness. It would promote integration of the five CKTS components (convergence centers, technology platforms, programs, and both private organization and government involvement) with existing coordinating offices and programs, including for the

National Nanotechnology Initiative (NNI), National Information Technology Research and Development (NITRD), the Global Change Research Program (GCRP), and similar existing efforts. A section of this office would be dedicated to "convergence of national research and development programs" to support the policy and decision-making organizations.[1]

- **CKTS-Wellness.** Global and longitudinal approaches to achieving individual physical and mental wellness, human development, and societal wellness are envisioned. Models of medical care would need to shift from being reactive—treating the disease after the patient has already succumbed to it—to proactive and preventative. Wellness issues are growing in importance with the expansion of societal interactions, aging populations, and the recognition of need for environmental sustainability. This includes mental satisfaction in family, interpersonal relationships, and society at large. *Technology for aging societies* could tie integrated data resources and platform research to humans, families, and communities to help societies across the globe deal with their aging populations. Excellent opportunities exist for collaborative research involving the United States, the European Union, and Asian countries, including Korea, Japan, and China.
- **CKTS-Earth.** Sustainable Earth systems R&D would be dedicated to evolutionary interactions between natural Earth systems, communities, converging technologies, and population size. Monitoring and mitigating sustainable Earth systems would be enabled by creating a single coordinating system for all Earth-scale activities. A focus would be on effective methods to maintain biodiversity and optimal levels of nitrogen and carbon in the environment. This would require extensive cooperation at the international level, as well as across fields and sections of the economy, thus achieving full societal convergence.
- **CKTS-Convergence in Society.** This program area would address ethical, legal, and public participation aspects of societal convergence. Progress must be achieved in a balanced way (for example, enhancing equality even as it generates profits (Mouw and Kalleberg 2010), which would require unbiased assessment of the broader implications of each major innovation within the complex system that is our planet, not to retard change but to channel it in the best direction for humanity. Converging technologies require advance deliberations to evaluate potential risks to consumers and the general population. *Public participation and neutral observatories are needed for decision-making.* Both in advanced democratic societies and in more authoritarian nations of the developing world, government has come under severe criticism, and there is reason to believe that traditional governance institutions are not capable of dealing with the rapidly evolving challenges of the future. The great danger is that frustration will produce even worse outcomes; therefore, rigorous, objective research must carefully determine which precious traditions need to be preserved, and what kinds of new methods can be implemented for improved public decision-making.

---

[1] South Korea currently is considering establishing such an organization.

Integration of the five main CKTS Initiative components would benefit from the results of basic NBIC research emerging from the National Nanotechnology Initiative, Networking and Information Technology Research and Development Program, Global Climate Change Initiative, GEOSS, and similar existing efforts. In return, the CKTS Initiative would support these and other existing initiatives.

This report provides a framework for advancing convergence. The primary opportunities to stimulate a new government program in convergence are those centered on the role of convergence in areas of national interest:

- "Converging revolutionary technologies for individualized services (CORTIS)", where individual services include providing and receiving personalized education, medicine, production, and general services (web-based or not), and creating personalized smart environmental and cultural surroundings
- Cognitive society and lifelong wellness
- Distributed NBIC manufacturing (production-centered convergence)
- Biomedicine-centered convergence (life sciences, physical sciences, and engineering)
- Increasing human potential (human capacity-centered convergence, including human–machine interaction, computer-supported collaboration, complementary robotics, brain-to-brain communication, cognitive computing, and brain activity mapping)
- Sustainable Earth systems (new trends in resources-technology-communities convergence, including monitoring global Earth dynamic systems and methods to influence them, urbanization, and other population displacements)
- Fostering creativity, innovation, and added-value decision analysis
- A "Federal Convergence of Knowledge and Technology (CONEKT) Office" focused on convergence approaches, and other Federal offices for priority convergence platforms (e.g., for programs in government for science, technology, and investment planning.)

11 The Future Converging Society

CKTS can manifest at all levels in knowledge, technology, and society, and typically is an outcome of bottom-up input and multidomain reasoning. Convergence is at the forefront of scientific discovery and technology development, promising to become a foundational and integrating knowledge and transforming field, as information technology and nanotechnology already have done. The CKTS Initiative described above has the potential to impact every sector of society, from improving education to enhancing wellness, from achieving environmental sustainability to promoting innovative economic development. The prospect of new knowledge, ideas, materials, and technologies that will emerge from convergent activities is profoundly exciting. Their impact on everyday lives is expected to be extraordinarily beneficial in terms of the way we and our descendants learn, work, thrive, and age.

Societal convergence has the potential to greatly and efficiently improve human capabilities, economic competitiveness, and life security. There is an urgent need to nationally and internationally take advantage of this opportunity and to take concrete steps to implement convergence in a timely way to deal most effectively with the serious problems facing humanity today.

M.C. Roco
W.S. Bainbridge
B. Tonn
G. Whitesides

References

AAAS (American Association for the Advancement of Science): Science for all Americans. Project 2061. Oxford University Press, New York. Available online: http://www.project2061.org/publications/sfaa/online/sfaatoc.htm (1989)

Alivisatos, A.P., Chun, M., Church, G.M., Greenspan, R.J., Roukes, M.L., Yuste, R..: The brain activity map project and the challenge of functional connectomics. Neuron **74**(June 21), 1–5 (2012)

Arthur, B.: The Nature of Technology: What it is and How it Evolves. Free Press, New York (2009)

Bainbridge, W.S., Roco, M.C. (eds.): Managing Nano-Bio-Info-Cogno Innovations. Springer, Dordrecht/New York (2006)

Dunbar, R.I.M.: The social brain: mind, language, and society in evolutionary perspective. Annu. Rev. Anthropol. **32**, 163–181 (2003)

Fichman, R.G.: Going beyond the dominant paradigm for information technology innovation research: emerging concepts and methods. J. Assoc. Inf. Syst. **5**(8), 314–355 (2004)

Friedman, T.L., Mandelbaum, M.: That Used to be us: How America Fell Behind in the World it Invented and How we Can Come Back. Picador/Farrar, Straus and Giroux, New York (2011)

Global Future 2045 International Congress: 2045: a new era for humanity (video). Available online: http://www.youtube.com/watch?v=01hbkh4hXEk (2012)

Gregorian, V.: Colleges must reconstruct the unity of knowledge. The Chronicle of Higher Education, 4 June 2004

Hey, T., Tansley, S., Tolle, K. (eds.): The Fourth Paradigm, Data-Intensive Scientific Discovery. Microsoft Research, Redmond (2009)

Hood, L., Flores, M.: A personal view on systems medicine and the emergence of proactive P4 medicine: predictive, preventive, personalized and participatory. N. Biotechnol. **29**(6), 613–724 (2012)

Jones, B.F.: The burden of knowledge and the "death of the Renaissance Man": is innovation getting harder? Rev. Econ. Stud. **76**(1), 283–317 (2009). http://dx.doi.org/10.1111/j.1467-937X.2008.00531.x

Kurzweil, R.: The Age of Spiritual Machines: When Computers Exceed Human Intelligence. Viking/Penguin Group, New York (1999)

Kurzweil, R.: Video interview about his book *How to create a mind: the secret of human thought*. http://www.youtube.com/watch?v=dwgkbhDJKno (2012).

Kuzum, D., Jeyasingh, R.G.D., Lee, B., Wong, H.-S.P.: Nanoelectronic programmable synapses based on phase change materials for brain-inspired computing. Nano Lett. **12**(5), 2179–2186 (2012)

Langer, E.J.: The Power of Mindful Learning. Addison-Wesley, Reading (1997)

Mouw, T., Kalleberg, A.L.: Occupations and the structure of wage inequality in the United States, 1980s to 2000s. Am. Sociol. Rev. **75**(3), 402–431 (2010)

National Intelligence Council: Global Trends 2030: Alternative Worlds. NIC 2012-001, Washington, DC. Available online: http://www.dni.gov/nic/globaltrends (2012)

NRC (National Research Council of the National Academies): Research at the intersection of the physical and life sciences. The National Academies Press, Washington, DC (2010)

NRC (National Research Council of the National Academies): Research universities and the future of America: ten breakthrough actions vital to our nation's prosperity and security. The National Academies Press, Washington, DC (2012)

PCAST (President's Council of Advisors on Science and Technology): Transformation and Opportunity: The Future of the U.S. Research Enterprise. PCAST, Washington, DC. Available online: http://www.whitehouse.gov/administration/eop/ostp/pcast/docsreports (2012)

Reif, R., Barrett, C.: Science must be spared Washington's axe. Financial Times, 25 February 2013

Roco, M.C.: Coherence and divergence of megatrends in science and engineering. J. Nanopart. Res. **4**, 9–19 (2002)

Roco, M.C.: Chapter 24: Technology convergence. In: Bainbridge, W.S. (ed.) Leadership in Science and Technology, pp. 210–219. Sage Publications, Thousand Oaks (2012)

Roco, M.C., Bainbridge, W.S. (eds.): Societal Implications of Nanoscience and Nanotechnology. Kluwer Academic, Dordrecht. Available online: http://www.wtec.org/loyola/nano/NSET.Societal.Implications/nanosi.pdf (2001)

Roco, M.C., Bainbridge, W.S. (eds.): Converging Technologies for Improving Human Performance: Nanotechnology, Biotechnology, Information Technology and Cognitive Sciences. Springer, Dordrecht/New York (previously Kluwer). Available online: http://www.wtec.org/ConvergingTechnologies/Report/ (2003)

Roco, M.C., Bainbridge, W.S. (eds.): Nanotechnology: Societal Implications. Springer, New York (2007)

Roco, M.C., Mirkin, C.A., Hersam, M.C.: Nanotechnology research directions for societal needs in 2020: retrospective and outlook Available online: http://nano.gov/sites/default/files/pub_resource/wtec_nano2_report.pdf. Also, Springer, Dordrecht/New York (2010) (Springer version is 2011)

Roco, M.C., Montemagno, C.: The co-evolution of human potential and converging technologies. Annal. NY Acad. Sci. **1013** (2004)

Roco, M.C., Williams, R.S., Alivisatos, P. (eds.): Nanotechnology Research Directions: Vision for the Next Decade. IWGN Workshop Report 1999. National Science and Technology Council, Washington, DC. Also published in 2000 by Springer. Available online: http://www.wtec.org/loyola/nano/IWGN.Research.Directions/ (1999)

Shapiro, C., Varian, H.R.: Information Rules. Harvard Business School Press, Boston (1999)

Sharp, P.A., Langer, R.: Promoting convergence in biomedical science. Science **222**(6042), 527 (2011)

Siegel, R., Hu, E., Roco, M.C. (eds.): Nanostructure Science and Technology: A Worldwide Study. National Science and Technology Council Interagency Working Group on NanoScience, Engineering and Technology (IWGN), Washington, DC (1999)

Stokes, D.E.: Pasteur's Quadrant: Basic Science and Technological Innovation. Brookings Institution Press, Washington, DC (1997)

UN (United Nations): Sustainability and Equity: A Better Future for all. United Nations, New York (2011)

UN (United Nations): Millennium Development Goals Report 2012. United Nations, New York (2012a)

UN (United Nations): Realizing the Future we Want for all. United Nations, New York (2012b)

UN (United Nations): Report of the United Nations Conference on Sustainable Development. United Nations, New York (2012c)

UN (United Nations): Building a Sustainable and Desirable Economy-in-Society-in-Nature. United Nations, New York (2012d)

Wasserman, S., Robins, G.: An introduction to random graphs, dependence graphs, and p*. Model. Methods. Soc. Network. Anal. **27**, 148–161 (2005)

Wilson, E.O.: Consilience: The Unity of Knowledge. Random House, New York (1999)

Contents

1. **Convergence Platforms: Foundational Science and Technology Tools**........................... 1
   Mark Lundstrom and H.-S. Philip Wong

2. **Convergence Platforms: Human-Scale Convergence and the Quality of Life**........................... 53
   Donald MacGregor, Marietta Baba, Aude Oliva, Anne Collins McLaughlin, Walt Scacchi, Brian Scassellati, Philip Rubin, Robert M. Mason, and James R. Spohrer

3. **Convergence Platforms: Earth-Scale Systems**........................... 95
   Bruce Tonn, Mamadou Diallo, Nora Savage, Norman Scott, Pedro Alvarez, Alexander MacDonald, David Feldman, Chuck Liarakos, and Michael Hochella

4. **Methods to Improve and Expedite Convergence**........................... 139
   Mihail C. Roco, George Whitesides, Jim Murday, Placid M. Ferreira, Giorgio Ascoli, Chin Hua Kong, Clayton Teague, Roop Mahajan, David Rejeski, Eli Yablonovitch, Jian Cao, and Mark Suchman

5. **Implications: Human Health and Physical Potential**........................... 185
   Robert G. Urban, Piotr Grodzinski, and Amanda Arnold

6. **Implications: Human Cognition and Communication and the Emergence of the Cognitive Society**........................... 223
   James L. Olds, Philip Rubin, Donald MacGregor, Marc Madou, Anne McLaughlin, Aude Oliva, Brian Scassellati, and H.-S. Philip Wong

| 7 | Implications: Societal Collective Outcomes, Including Manufacturing | 255 |

Jian Cao, Michael A. Meador, Marietta L. Baba, Placid Mathew Ferreira, Marc Madou, Walt Scacchi, James C. Spohrer, Clayton Teague, Philip Westmoreland, and Xiang Zhang

| 8 | Implications: People and Physical Infrastructure | 287 |

James Murday, Larry Bell, James Heath, Chin Hua Kong, Robert Chang, Stephen Fonash, and Marietta Baba

| 9 | Implications: Convergence of Knowledge and Technology for a Sustainable Society | 371 |

Mamadou Diallo, Bruce Tonn, Pedro Alvarez, Philippe Bardet, Ken Chong, David Feldman, Roop Mahajan, Norman Scott, Robert G. Urban, and Eli Yablonovitch

| 10 | Innovative and Responsible Governance of Converging Technologies | 433 |

Mihail C. Roco, David Rejeski, George Whitesides, Jake Dunagan, Alexander MacDonald, Erik Fisher, George Thompson, Robert Mason, Rosalyn Berne, Richard Appelbaum, David Feldman, and Mark Suchman

Appendices

A	List of U.S. and International Workshops	493
B	List of Participants and Contributors	497
C	Abstract of the Converging Technologies Workshop in São Paulo, Brazil, November 2011	513
D	Review of NBIC Visionary Goals	515
E	Selected Reports and Books About Convergence	529
F	The Wilson Center Video Interviews: "Leading Scientists Discuss Converging Technologies"	541
G	Glossary of Selected Terms and Concepts	543
H	List of Acronyms	553

Chapter 1
Convergence Platforms: Foundational Science and Technology Tools

Mark Lundstrom and H.-S. Philip Wong

As the modern scientific era began, breadth gave way to specialization, which produced astounding progress in science and technology in that phase of development, resulting in an enormous improvement in the human condition. The initial work in nanotechnology, biotechnology, information technology, and cognitive science is testament to the success of this paradigm. Today, however, we face an array of new challenges in scientific complexity, as well as in jobs, health, environment, and security (NAE 2008) that can only be addressed with combinations of the great technologies developed in the twentieth century. Multidisciplinary research and development addresses challenges by pulling together concepts and techniques from different fields. Converging knowledge and technologies for society (CKTS) is different; it includes all relevant areas of human and machine capability that enable each other to answer questions and to resolve problems that isolated capabilities cannot, as well as to create *new* competencies, technologies, and products to benefit society on that basis (see Chap. 4). Convergence is multidisciplinary research with transformative interactions that create new outcomes. Nanotechnology, biotechnology, information technology, and cognitive science (NBIC) are foundational emerging and converging tools, which together form one of the four general CKTS platforms besides those at the human scale,

---

With contributions from Barry D. Bruce, Frederica Darema, Eric Fisher, Sangtae Kim, Marc Madou, Michael Meador, Abani Patra, Robin Rogers, James Olds, and Xiang Zhang.
Corresponding editors M.C. Roco (mroco@nsf.gov) and W.S. Bainbridge (wbainbri@nsf.gov).

M. Lundstrom
Purdue University, West Lafayette, IN, USA

H.-S.P. Wong
Stanford University, Stanford, CA, USA

Earth scale and societal scale. As a consequence of convergence, advances in these foundational NBIC sciences and technologies will increasingly be driven by requirements of and contributions from other fields (Roco and Bainbridge 2003), and new fields at the intersection of these disciplines will emerge.

Beginning with the telescopes and microscopes of the Renaissance, tools have played a critical role in advancing science and technology. Scientific tools provide us with abilities to observe and understand nature. Design tools translate scientific concepts into practical technologies. Increasingly sophisticated and integrated manufacturing tools drive production. Foundational scientific and engineering theories are conceptual tools that help explain a class of phenomena or guide a field of technology. These kinds of scientific and engineering tools have played a critical role in the development and application of the specific NBIC technologies, which provide in their turn foundational emerging tools in the application of all areas of CKTS. NBIC tools will evolve in response to the needs of specific disciplines as well as to broader inputs from other disciplines, but in the era of converging knowledge and technologies, new types of tools will also be needed. These transdisciplinary tools will play a critical role in advancing science and technology at the intersection of the NBIC disciplines. Finally, the importance of education should be stressed. Education is an essential tool for training a new generation of scientists and engineers to realize the promises of NBIC and CKTS, respectively, and it can also be a driver for convergence. Multidisciplinary research uses methods from different disciplines to provide ad hoc or new solutions to problems. Discovering the underlying general principles and conceptual frameworks and articulating them in a way that can be taught to students and practicing scientists and engineers can be a driver for the transformative interactions that convergence is all about.

1.1 Vision

1.1.1 Changes in the Vision Over the Past Decade

The rise of multidisciplinary research and converging technologies has been underway for more than a decade. Solutions to problems increasingly integrate a suite of technologies, as do the products we use every day. The cell phone has become a smart phone and includes imagers, gyroscopes, microelectromechanical (MEMS) devices, speakers, microphones, etc., in addition to electronics. Tremendous progress has occurred in NBIC disciplinary tools (Roco 2012), but converging technologies are also producing dramatic advances in the capabilities of tools. Gene sequencing is an example, where a decade ago Sanger-chemistry-based techniques were on a Moore's Law trajectory (Fig. 1.1). The introduction of approaches from electronics—such as micro-arrays, massive parallelization, and even wholesale adoption of mainstream CMOS (complementary metal-oxide semiconductor) technology—has led to a dramatic decrease in the cost of sequencing (Mardis 2008, 2011; Metzker 2010; Rothberg et al. 2011).

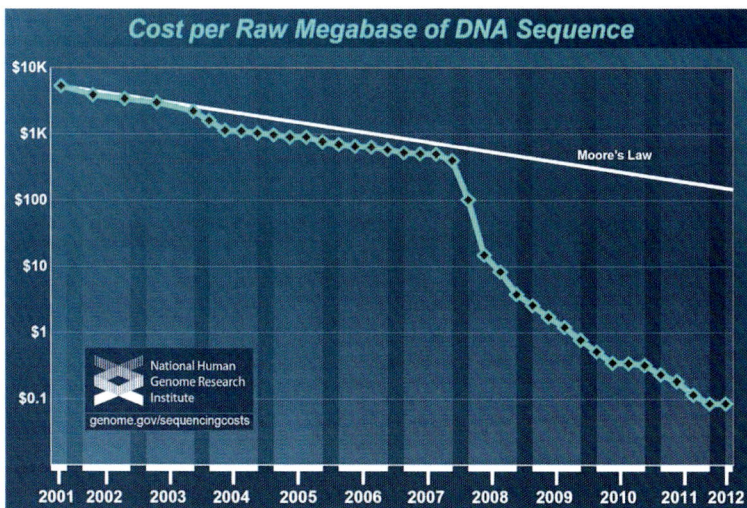

Fig. 1.1 Cost of sequencing a megabase of DNA vs. time (Source: NIH/NHGRI, K.A. Wetterstrand, http://www.genome.gov/sequencingcosts/, DNA sequencing costs: data from the NHGRI Genome Sequencing Program (GSP); accessed January 2013)

In addition to the enhanced capabilities of tools for the NBIC disciplines, some broader changes and trends can be identified that have implications for the CKTS tools of the future. One trend is that things that were once expensive are becoming inexpensive and, therefore, pervasive. Examples of this are the gene sequencing mentioned above; ability to manipulate and measure objects at the nanoscale; and ease and capacity with respect to storing, transmitting, and searching for information. How to effectively use such capabilities and the enormous amount of data that is being generated and stored has become a major research topic. Another trend is that materials properties that were otherwise unattainable in nature now can be engineered (optical metamaterials is one example). Other trends include the increasing attention paid to societal impact in early-stage research, more attention directed to sustainability and green manufacturing, the increasing globalization of scientific/engineering research, and more attention given to use-inspired research. A recent trend in education is the ready availability of high-caliber educational resources online at little or no cost to the student (Lewin 2012).

1.1.2 The Vision for the Next Decade

Subsequent chapters will articulate a vision and new directions for addressing society's challenges in human health and human potential, human cognition, manufacturing, a sustainable society, etc. Converging knowledge and technologies will play a critical role in realizing this vision. The multidisciplinary application of converging technologies will increase, but our vision is to go beyond multidisciplinarity to transdisciplinarity. Multidisciplinary research adopts tools and techniques from many disciplines to provide innovative solutions to problems. Transdisciplinary research transcends disciplines through transformative interactions that change the evolution of disciplines and create new fields at the intersections of disciplines. Instead of ad hoc solutions to problems, general frameworks are created for addressing new classes of problems. Examples from the past include materials science, computer science, and semiconductor electronics. Our vision is to address society's challenges through multidisciplinary research that transcends disciplines and sets the stage for the creation of new disciplines.

In this chapter, convergence involves both NBIC knowledge and NBIC tools. Our vision for convergence NBIC knowledge is that these disciplines will continue to evolve, becoming more and more interdependent and powerful. Increasing multidisciplinary research will lead to the increasing use of tools and methodologies originally developed for specific disciplines by experts outside those disciplines. The use of converging technologies also provides opportunities to develop new tools that dramatically advance specific disciplines. The gene sequencing example discussed earlier illustrates the possibilities. An example of how converging technologies can dramatically improve tools for imaging and interacting with the human brain is discussed in Sect. 1.3.2.

Our vision for converging tools mirrors the vision for converging knowledge. We foresee the emergence of a new class of tools not tied to a single NBIC discipline. These "converging technology tools" will integrate NBIC technologies in pursuit of common scientific and technological goals. They will support research and engineering at the interfaces of the NBIC disciplines. This will support the new fields of science and technology developed through transdisciplinary research; examples of such tools will be presented in Sect. 1.3.

New educational paradigms are an integral part of the vision. New approaches to education must be developed to give students and practicing engineers and scientists the depth and the breadth needed to capitalize on the promise of converging technologies.

1.2 Advances in the Last Decade and Current Status

Addressing societal challenges with converging technologies also requires the convergence of knowledge and the convergence of tools. Galileo's telescope gave us the opportunity to see our solar system in a new way and led to the rise of modern science in the Renaissance. About a half century ago, in his famous lecture, "There is Plenty of Room at the Bottom" (Feynman 1960), Richard Feynman asked what would happen if we had tools to enable us to see and manipulate matter at the atomic scale. The scanning tunneling microscope provided this capability and led to the rise of nanotechnology as a field (Binnig et al. 1982; Binnig and Quate 1986). Over the past decade, nanoelectronics became the first, large-scale nanomanufacturing technology, and advances in manufacturing tools (e.g., rapid prototyping, maskless lithography with digital light processors, desktop factories, and robotics) began to change manufacturing. Biotechnology has transformed agriculture and medicine, and information technology is rapidly reshaping science and society. Cognitive science is expanding our understanding of ourselves. The synergistic advances in NBIC foundational knowledge and in NBIC tools have fueled modern science and technology.

1.2.1 Advances in NBIC Foundational Knowledge

Over the past 50 years or so, the NBIC disciplines have advanced to a remarkable degree of sophistication. These advances are a testament to the success of the traditional approach to science—the divide and conquer, division of labor, specialization paradigm. As discussed below, advances in these fields are intimately connected to advances in tools, with advances in one driving progress in the other.

1.2.2 Advances in NBIC Tools

During the past decade, improved characterization tools and techniques for **nanotechnology** provided new capabilities to "see" at the nanoscale—with improved time resolution as well (Fang et al. 2005; Liu et al. 2007). Our ability improved to probe single-charge, single-spin, spin excitation, and bond vibrations at the atomic scale, and new capabilities were developed for measuring "continuum" properties (e.g., dielectric function, work function, etc.) with atomic resolution. Coupled with these new capabilities were advances in atomic- and molecular-level simulations and the emergence of the non-equilibrium Green's function method as an engineering tool for simulating electronic transport in nanoelectronic devices (Lake et al. 1997). Along with advances in imaging and simulation at the nanoscale have been advances in synthesis and fabrication such as new techniques for soft lithography, development of synthesis and separation strategies for monodisperse nanomaterials (e.g., nanocrystals and quantum dots) and macromolecules, and progress in self-assembly, directed assembly, and molecular recognition, to name a few. Advances in electronics, optics, photonics, plasmonics, and nanomaterials and metamaterials are leading to new applications in biomedicine, energy, and information technology (Srituravanich et al. 2008; Valentine et al. 2008). For a more extensive discussion of advances in tools for nanotechnology over the past decade, see Roco et al. 2011.

During the twentieth century, **biotechnology** transformed agriculture and medicine, and more recently, synthetic biology congealed as an emerging technology, transforming biology from primarily an observational science to a field of research that is beginning to resemble the physical sciences. Over the past decade, the rise of "-omics" (e.g., genomics, proteomics, metabolomics, cognitive genomics, etc.) has been a significant development. Tools have played an important role in these advances. One example is in biology and medicine where hardware tools and theoretical constructs that were traditionally used in the engineering of physical systems have begun to make important contributions to advances in biological sciences and the delivery of healthcare. Dramatic advances in the capabilities and speed of DNA/gene sequencing continue, and new diagnostic methods sensitive to picomole and attomole levels have been developed (Roco 2012). Three-dimensional tracking (at the single-molecule level) of protein motors, enzymes, liposomes, and other bio-nanostructures is now possible (Roco 2012), and optical technology shows promise for parallel, remote control of neuronal activity with high spatial and temporal resolution.

Computing hardware and software advances continue to propel **information technology**. These advances, along with the global deployment of fiber and wireless communications systems are transforming science, engineering, and society. Advances in computing speed and memory capacity enable simulations with improved physical fidelity. The increasing use of parallel programming software such as OpenMP (http://openmp.org/wp/) has been an important factor in harnessing the power of today's multicore processors. Over the past decade, data mining has emerged as standard operating practice across science and engineering because the acquisition of data becomes essentially free due to advances in nanoelectronics. Increasingly powerful search engines have had great impact. Cyber-infrastructure (with open source platforms such as HUBzero (McLennan and Kennell 2010) has also become pervasive and became, over the last decade, the information and communication tool for everyone, not just for those who are technically trained. Together with advanced computation and communication, the advent of ubiquitous instrumentation (sensing) has created opportunities to vastly increase our ability to monitor, understand, and predict the behavior of systems and systems of systems. New paradigms that couple computation and instrumentation to execute applications that are dynamically steered in a dynamic feedback loop provide new capabilities for understanding and predicting the behavior of complex systems.

Tools have played an especially important role in the advance of **cognitive science** over the past decade. Particularly significant have been tools for noninvasive brain imaging, such as positron emission tomography (PET), magnetoencephalography (MEG), electroencephalography (EEG), and functional magnetic resonance imaging (fMRI) (Bandettini 2012). Tools for intervention and interaction with brain activity have also played an important role. Examples include deep brain stimulation electrodes, transcranial magnetic stimulation (tCMS), direct current brain stimulation (DCBS), and optogenetic approaches (Fehrentz et al. 2011; Theogarajan 2012). Another important class of tools is those for cellular imaging. Examples include fluorescent probes (voltage-sensitive and calcium indicator dyes) and optical stimulation and inhibition of neural activity (Banghart et al. 2004; Boyden et al. 2005). Along with these tools to understand human cognition, human–machine interfaces have begun to appear (Wang et al. 2007). Such interfaces will broaden the use of machines in a human-centric context.

The work on cyber-physical systems (CPS) is an example of a tool for economic productivity combining cyberinfrastructure use with software development and physical manufacturing units. The investments made in this converging field of computing, networking, and physical systems have resulted in economic returns exploiting advances in micro- and nanoelectronics. Their impact in manufacturing also is significant.

1.2.3   Converging NBIC Knowledge and Tools

As the individual NBIC fields and their tools have advanced, the beginnings of knowledge and technology convergence can also be seen. The twentieth century witnessed the convergence of disciplines to produce materials science, computer science, and semiconductor electronics. Examples of converging knowledge from the recent past include the merging of mechanics and electronics in MEMS systems, now widely employed in sensors and actuators from automobiles to electronic game consoles and cell phones. Other examples are the convergence of computer science and genomics, and the evolution of microelectronics to nanoelectronics; both of these trends have been greatly aided by converging technologies. Nanofabrication has enabled the development of sensors and devices for biomedical applications (Hall et al. 2010). A very recent example of things that may come is a "genetic hard drive" that encodes digital information in DNA to produce a memory much denser than is possible with traditional technologies (Waltz 2012).

The past decade has also witnessed an increasing use of tools across disciplines by experts in other disciplines. Tools that support NBIC convergence are also beginning to appear. One example is circuit simulation-compatible electronic device models for nanobiosensors that augment electronic systems with new capabilities for medical diagnostics. The open-source AFNI package (a set of C programs for processing, analyzing, and displaying fMRI data) is an example of pervasive application of tools of information technology in biotechnology (see http://afni.nimh.nih.gov/). Emerging examples of converging technology tools include design tools that connect electronic systems to the life sciences (e.g., for applications in medicine, health, human–machine interfaces, etc.) and optical platforms for parallel, remote control of neuronal activity (Wang et al. 2007).

1.2.4   Integration of Social Sciences in Knowledge and Technology Development

Societal support of research comes with an implicit expectation that advancing knowledge will help improve the human condition. Today, the global challenges are jobs, health, and security. The United States has transitioned from a manufacturing-based economy to a knowledge-based economy. The traditional image of factory workers in steel and automobile factories has gradually been replaced by cadres of computer programmers or data analysts sitting in front of rows of computer terminals (Fig. 1.2) and driving manufacturing automation to new levels. In a knowledge-based economy,

Fig. 1.2 A "factory" in the modern knowledge economy (Photo credit: Peter DaSilva, *The New York Times*/Redux, http://www.nytimes.com/2012/08/24/technology/facebook-rewrites-its-code-for-a-small-screen-world.html?_r=1; used by permission)

depth and breadth of knowledge becomes the "steam-engine" of a century ago. Education becomes an essential tool in a knowledge-based society; and human–machine interaction is not only unavoidable in daily life but also necessary for knowledge and technology development.

The increasing interaction of humans and machines calls for a broader and deeper understanding of the dynamics of societal interactions. An example is the impact brought on by social networks, which are enabled by advances in information technology and nanotechnology. The field of social networks is as much a field of study in computer science and information theory as it is a field of study in psychology and socio-economics. Will the pervasive use of information technology bring social groups together or push them further apart? Advances in various "-omics" offer new opportunities in personalized medicine, but how can we manage the privacy and potential class-segregation issues? Will a knowledge-based economy impede or enhance social mobility? Will new education modalities (as discussed in Sect. 1.4.4) level the playing field for the haves and the have-nots? CKTS brings to the fore these new areas of studies in the social sciences.

1.3 Goals for the Next Decade

Examples of current research in NBIC include low-cost, energy-efficient nanomanufacturing, pathogen-safe biomanufacturing, and the broadening charter ("more than Moore") for nanoelectronics. The discussions on biotechnology at the U.S. "NBIC2" conference focused on advancing synthetic biology and personalized medicine. Research to fully integrate large data and large-scale computing was identified as an area with potentially high impact across NBIC. The development of flawless language translators, natural language machine–human interfaces, noninvasive biophotonic human–machine interfaces, and robots difficult to distinguish from humans was discussed. Finally, there were significant discussions on the need for new educational platforms and paradigms to support NBIC. Section 1.8 gives examples of current NBIC research.

Although the scale of these efforts is still modest, the level of activity in converging technologies is increasing. Several goals (or grand challenges for humanity) for the next decade are discussed in subsequent chapters. Our discussions identified three specific goals that relate to NBIC: (1) advancing foundational NBIC tools for economic productivity; (2) NBIC for new tools that advance human health and physical capacity; and (3) security of computers, networks, software, and data, which increasingly pervade all areas of human activity.

1.3.1 Goal: Tools and Foundational Science for Advanced Electronic Manufacturing

The problem. Economic productivity with its concomitant jobs requires innovation through technology. Chapter 7 addresses these issues from a broad manufacturing perspective, where prospects for complementing "mass production" with "customized designs" enabled by new prototyping tools such as 3D printing and desktop manufacturing are discussed. The augmentation of centralized manufacturing by customized designs emphasizes again the need for a knowledge-based workforce.

This section illustrates how electronics will play a pervasive role in the application of converging technologies to societal challenges. Integrated nanosystems will continue to make use of electronic/photonic/magnetic devices scaled to nanoscale dimensions. The ready availability of a powerful and versatile nanoelectronics platform presents yet another opportunity. The well-developed nanoelectronics manufacturing base (CMOS) technology will be complemented via "hybrid processing" with novel nanodevices (e.g., biosensors, programmable resistors, nonvolatile memory, energy harvesters, etc.) to achieve new systems that combine increasingly powerful information processing and communications technology. These new systems will have new capabilities for applications in biomedicine, energy, security, as well as for developing tools that help advance foundational knowledge in fields such as cognitive science. To realize these possibilities, new design tools are needed, along with new hybrid fabrication facilities that combine powerful CMOS capabilities with those of new materials and devices.

Current capability. A well-established ecosystem exists for designing and manufacturing highly sophisticated micro- and nanoelectronics products based on CMOS technology. These facilities have evolved to be highly efficient—and

Fig. 1.3 The current electronics manufacturing paradigm is exemplified by very large-scale, highly efficient, but very expensive semiconductor "fabs" (Courtesy of Global Foundries via James Murday, 2012, private communication)

expensive[1]—but they are not flexible for supporting research and prototyping or for custom applications. Although the cost of manufacturing facilities receives a lot of attention, it is important to recognize that *design costs* outweigh manufacturing costs. Over the past 40 years, a powerful, multiscale design infrastructure has been created to support the design of integrated circuits. At the fundamental end are first-principles materials simulations and physically detailed device simulators. Physics-based "compact models" provide computationally efficient descriptions of device characteristics in a form suitable for circuit simulation. High-level behavioral models simulate the performance of complete systems. Associated tools verify performance, lay out devices, wire-up circuits, create the mask used for manufacturing, and perform tests for functionality and reliability. The result is a complete end-to-end infrastructure for designing and manufacturing integrated circuits for computing and communications.

During the past 10–15 years, the semiconductor industry has consolidated into a few large companies that have multibillion-dollar manufacturing facilities (Fig. 1.3). Concurrent with this trend is the significant reduction and near-disappearance of corporate and smaller research laboratories from companies. These R&D labs used to be the breeding grounds for experiments in new materials, new tools, and new

---

[1] Costs are roughly $3 billion per facility (Wikipedia: http://en.wikipedia.org/wiki/Semiconductor_fabrication_plant).

manufacturing techniques. Corporate research and development are increasingly being performed right at the manufacturing facilities to shorten the development-to-manufacturing cycle. Facilities at major universities currently support a variety of fabrication modalities, and the National Nanotechnology Infrastructure Network (NNIN) funded by the National Science Foundation (NSF) has been a very good model for supporting NBIC research such as hybrid processing, both in academia and for industry, especially small businesses. The national laboratories such as Lawrence Berkeley National laboratory (LBNL) and Brookhaven National Laboratory (BNL) also have facilities that support academic and small business R&D. Funding of large projects from fundamental research to manufacturing is illustrated by the EU Future of Emerging Technologies flagship award made on graphene in 2003 (http://graphene-flagship.eu/GF/reports.php).

What's missing. Although shared user facilities are available to academia and small businesses, these facilities have already fallen behind in the level of sophistication and precision required of tomorrow's R&D. For example, very few R&D facilities can accommodate 200 mm wafer sizes, let alone 300 mm wafers (the industry standard today). The problem is not the size of the wafers per se (which tend to be large for cost-effective manufacturing); the issue is that advanced manufacturing techniques are only available for tools that use these large wafer sizes. As a result, R&D labs in universities and national laboratories currently do not have access to the most up-to-date manufacturing capabilities.

In the era of converging technologies, integrated electronic nanosystems extend the capabilities of powerful CMOS platforms with an increasing variety of nanoscale devices such as sensors, actuators, energy harvesting devices, programmable resistors, nonvolatile memories, etc. These new technologies entail the use of an increasingly diverse set of new materials that are not conventionally used in the semiconductor industry, which presents a challenge for contamination control and compatibility of the materials and tool sets. For example, the use of advanced semiconductor techniques for biomedical research is currently hampered by the lack of access to advanced devices that offer superior performance. Finally, the current design infrastructure has limited capability to accommodate a wide variety of new devices. The design infrastructure needs to be extended to encompass a growing variety of devices and system architectures used in conjunction with traditional CMOS platforms.

Why now? The era of VLSI (very large scale integrated) circuits began in the 1980s when circuits began to contain tens of thousands of transistors. Circuit designers quickly recognized that circuit research could not be performed by fabricating the circuits in university laboratories. In response to this challenge, MOSIS (http://www.mosis.com/) was formed to combine circuit designs from many researchers into one composite design for which the chips could be manufactured by a semiconductor "foundry" and distributed to the researchers. MOSIS propelled circuit research to new heights and contributed to the emergence of fabless design companies such as Apple, Nvidia, Broadcom, Qualcomm, Xilinx, and Altera. The same needs to be done for physical technologies and hybrid processing today, because sophisticated manufacturing techniques are currently out of reach for university researchers and small businesses. Research and innovation will be stifled if access to advanced fabrication techniques is available only to a select few.

Progress in nanoscience has identified a number of interesting new devices that have great potential for realizing new types of systems. To fulfill the promise, there is a need for sophisticated manufacturing technologies that are out of the reach of university researchers and small businesses. Also needed are capabilities for hybrid processing, since many of these new technologies will be used in conjunction with conventional CMOS nanoelectronics. Suitable models are needed for novel nanodevices, in a form that is compatible with the existing design infrastructure, in order to transform designs into hardware.

Desired outcomes/new capabilities. The semiconductor foundry paradigm enabled university research and innovation by small and large companies. By providing new types of fabrication and design capabilities, a new era of custom electronics enabled by advanced nanodevices can be realized. Fabrication facilities that provide the most advanced device fabrication capability should be established at the national level and made accessible to academic researchers and small businesses. In Belgium, IMEC (formerly the Interuniversity Microelectronics Centre; http://www2.imec.be/be_en/about-imec.html) is an example. It performs world-leading research in nanoelectronics with applications in information and communication technologies, healthcare, and energy. Such a facility would serve the material and device research community, but also important will be the development of a design infrastructure to help university researchers and small businesses create new products based on novel nanoelectronics—either alone or in conjunction with standard CMOS technology.

Expected benefits. Just as a move from 2D printing to 3D printing will bring new capabilities to advanced manufacturing, hybrid nanoelectronics will produce 3D integrated nanosystems with new capabilities. By providing device and processing technology researchers with baseline semiconductor technology and advanced toolsets to explore new possibilities, hybrid nanoelectronics will facilitate solutions to a variety of societal problems such as healthcare, energy, and information technology. By providing designers with a much wider "palette" of devices to design with, new possibilities will be realized to enable hybrid, CMOS+ nanosystems to address a wide range of problems (e.g., Fig. 1.4). Innovation will be fostered by making sophisticated manufacturing capabilities available to individual researchers and small businesses.

1.3.2 Goal: New Tools to Measure, Study, Model, and Interact with Brain Activity with Vastly Improved Spatial and Temporal Resolution

The problem. Developing better tools to measure and interact with human brain activity is essential to the further advances in cognitive science that are necessary to enhance human capability. The development of these new tools will require the convergence of multiple technologies that include, for example, sensors, nanotechnology,

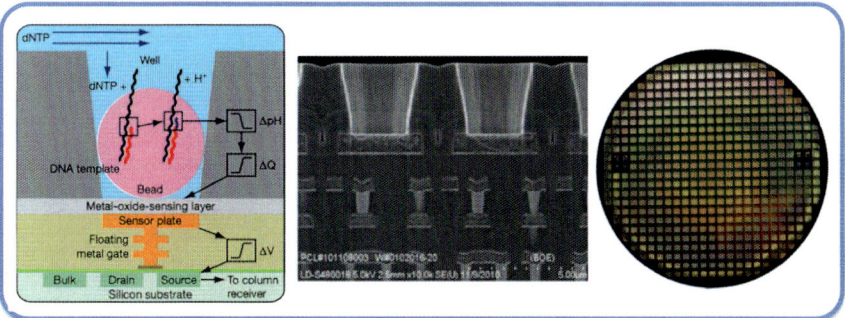

Fig. 1.4 All-electronic gene sequencing (Ion Torrent). Illustration of 3D electronics: an emerging new era in integrated nanosystems built on a powerful CMOS technology platform. Ion-sensitive field-effect transistors (FETs) are integrated on silicon metal-oxide field-effect transistors (MOSFETs) (*left*), and integrated on CMOS chips to provide electronic readout and information processing (*middle*), resulting in a technology that can leverage high-volume, 200–300 mm silicon manufacturing technology (*right*) (Adapted from Rotherberg et al. 2011, p. 349, Figure 1, ©*Nature*; reuse cleared through Rightslink)

VLSI electronics, materials fabrication, signal analysis, and genetic therapy. The overarching goals are to develop methods for reading out and manipulating the spike code with sufficient spatial sensitivity to "reach out and touch" individual neurons, and to develop methods and models to describe the collective behavior of neurons at the functional level.

Current capability. Noninvasive brain imaging studies have been crucial for revealing new insights into human decision-making (Studer et al. 2012), brain diseases (Johnson et al. 2012), and the relationship between functional brain activity and human behavior (Fox et al. 2005). The available twentieth century technologies include positron emission tomography, magnetoencephalography, and electroencephalography. More recently, the literature has been dominated by functional magnetic resonance imaging studies (Bandettini 2012). These tools have played a critical role in the development of cognitive science in the twentieth century and during the first decade of the twenty-first century.

Current methods for intervention and interaction with human brain activity include deep brain stimulation electrodes, transcranial magnetic stimulation, and direct current brain stimulation. The spatial and temporal resolution of cortical surface electrode arrays has improved significantly over the past decade, and optogenetic approaches are showing promise in animal models (Fehrentz et al. 2011; Theogarajan 2012).

At the level of cellular imaging, fluorescent probes such as molecularly engineered reporter genes and voltage-sensitive and calcium indicator dyes have been among the most useful tools for accurately recording electrical activity in neurons remotely. In addition, the established methods for optical stimulation and inhibition of neural activity have now made remote and noninvasive control possible (Banghart et al. 2004; Boyden et al. 2005). Advances in optical microscopy such as two-photon and laser-scanning confocal imaging enable all of the above techniques.

What's missing. The general consensus among neuroscientists is that neural information is spike-encoded. Each of the current brain imaging technologies has specific advantages and disadvantages, but all have the shared disadvantage of not reaching down to the spatial and temporal dimensions of the action potential, or "spike". Therefore, current functional brain studies can only reveal severely degraded information, "blurred" in both the spatial and time domains by at least three orders of magnitude (Rolls and Treves 2011; Bandettini 2012). Current noninvasive electrophysiological techniques such as EEG and MEG are temporally matched to spikes, but with spatial resolution poorer than that of fMRI. While such spatio-temporal matching to the neural code exists for neurons in cell culture or in chronic animal model preparations, the challenge of achieving such resolution noninvasively in human subjects is a major challenge.

Similarly, current methods for intervention and interaction with human brain activity also lack sufficient spatial and temporal resolution, in addition to being too invasive. Deep brain stimulation electrodes are invasive and additionally suffer functional degradation over time (Gimsa et al. 2005). Transcranial magnetic stimulation and direct current brain stimulation (DCBS) are limited both in spatial specificity and penetration depth within the brain (Foucher et al. 2007). Cortical surface electrode arrays, while having improved spatial and temporal sensitivity, are similarly depth-limited. Optogenetic approaches, while promising in animal models, require gene therapeutic approaches, which haven't yet reached maturity for human subjects (Fehrentz et al. 2011; Theogarajan 2012).

In the case of brain stimulation, central challenges are the opposing needs of proximity to neuronal targets and the desirability of avoiding surgical intervention. Development of human optogenetic approaches may be initially limited to sensory systems such as cochlear hair cells and retina, even with the maturation of gene therapies, due to the need for photonic stimulation.

Above and beyond the need to interact with brain signals is the parallel imperative to make sense of the actual neural code (i.e., the relationship between the ensemble spike train and cognitively meaningful outputs such as concepts or behavior). Progress along this neurocryptological trajectory will also require convergent technologies such as nanotechnology and information technology in addition to those necessary for interacting with neurons that are described above.

In parallel with advances in biological measurements, it is now possible to use advances in nanoelectronics to emulate the functions of the brain. It is now possible to build very-small-scale electronic models of the functions of neurons and synapses using VLSI technology (Merolla et al. 2011) and novel use of nanomaterials (Seo et al. 2011; Kuzum et al. 2012; Jo et al. 2010; Ohno et al. 2011). What is missing is a concerted effort to scale-up these small-scale demonstrations to larger systems that can exhibit emergent behavior. Because one can easily change the behavior of electronic systems, it is possible to use these electronic systems as a tool to explore "what if" questions related to neuroscience and enhance our understanding of the cognome.

At the neuronal level, a primary challenge of neuroscience is to understand how groups of cells in the massive neural networks of the brain communicate and dynamically regulate their connections. Optical recording and stimulation do not require physical contact with cells and are inherently noninvasive. Control of neuron activities with light has drawn substantial interest in various fields including optical science, genetics, and neural science (Wang et al. 2007; Abrams et al. 2010). Thus, developments on parallel stimulation and inhibition of neural activity are crucial to creating an interface potentially capable of bridging the massive information flow between the computer and the brain.

Why now? The human brain has been called the most complicated machine we know of in the Universe. Understanding how the activity of the brain produces higher cognition (including subjective experience) and behavior is important, not only at the level of the individual, but also in understanding the emergent properties of human interactions at multiple scales. The problem is magnified by its recursive nature: we are trying to understand the "machine" that is ourselves.

In the twentieth century, brain science was largely limited to animal models. It is only in the last decade that convergent technologies have made it possible for hypothesis-based science to query the conscious human brain *in situ*. These convergent technologies center on noninvasive methods for detecting and manipulating the activity of neurons. In general, no single approach has been optimized to the actual neural code; however, taken together, these approaches have revealed a great deal about the complex relationships between functional brain activity and cognition.

Going forward, there are significant opportunities for these convergent approaches to be further improved in both spatial and temporal domains, thus offering a much better picture of the conscious human brain as it perceives its environment. Examples of new projects to take advantage of these developments are the European Union's 2013 funding of a new Future of Emerging Technologies flagship project for measurement and modeling of brain research, the Human Brain Project (http://www.humanbrainproject.eu/), and the Administration's BRAIN (Brain Research through Advancing Innovative Neurotechnologies) Initiative proposed for start in FY2014.

Desired outcomes/new capabilities. Within a decade, magnetic resonance imaging will deliver far greater information about the human brain than is possible today (Fig. 1.5). Specifically, magnetic resonance spectroscopy (MRS; nuclear magnetic resonance spectra of brains on a voxel-by-voxel basis) will have advanced both in spatial and temporal resolution. Such MRS data will reveal the functional dynamics of brains at the molecular level, most importantly allowing the differentiation of excitatory and inhibitory neurotransmission (Manganas et al. 2007), in great contrast to current fMRI methods.

Advances in gene therapy should also allow the deployment of optogenetic methods in humans, particularly at the level of sensory prosthetics, where copper wire would be replaced by photonics.

Finally, data-fusion techniques may be brought to bear that optimize data gathering from multiple existing convergent methods. Current examples of such fusion range from hyper-scanning (multiple fMRI scanners for interacting human subjects) to EEG embedded in fMRI (taking advantage of EEG's superb temporal resolution). This type of fusion might reasonably be expected to expand greatly in the next 10 years.

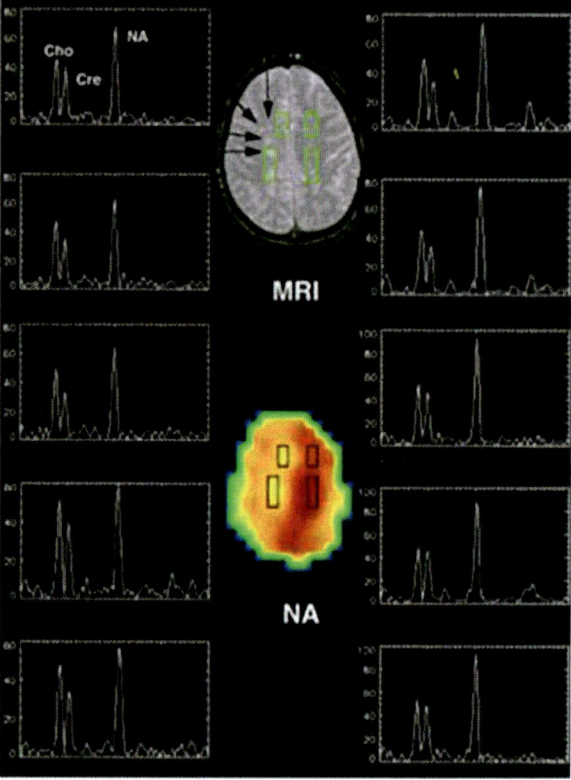

Fig. 1.5 Weighted MRI, a section of the N-acetylaspartate (NA) image, and ten representative spectroscopic voxels at a level slightly above the centrum semiovale of a Fabry disease patient. Concentration of NA, as revealed by MRS has proven to be a good biomarker for neurological disease (Tedeschi et al. 1999; ©*Neurology*; reuse cleared through Rightslink)

The development of optical platforms to advance parallel noninvasive approaches for interrogating and controlling live cell activities in real time will allow targeting specific cellular functional motifs (Wang et al. 2007; Abrams et al. 2010). With these unique capabilities and resources, we will be able to explore the dynamic state of the neuron network responses to external stimulations, such that we will bring the understanding of molecular biological science to a different dimension and facilitate the progress of development of new human–machine interfaces that can ameliorate the symptoms of brain injury and disease. This new platform can potentially significantly enhance the reporting and control signals for more effective human–machine communication.

Benefit. Cognitive science will benefit humankind because the mapping between brain activity and human cognition will be increasingly revealed. Understanding the neural representation of cognition and behavior is a *sine qua non* to developing a general theory of cognition. Human society will benefit because a greater understanding of human cognition is required before human cognitive diseases can be cured. Over time, human cognition might even be augmented.

1.3.3 Goal: Open Source Software, Cyber Security, and Risk Management Tools

The problem. Wikipedia contains the following definition for "open source": "In production and development, open source is a philosophy or pragmatic methodology that promotes free redistribution and access to an end product's design and implementation details" (http://en.wikipedia.org/wiki/Open-source). The concept has roots that predate the computer and Internet eras. In the current context of the NBIC roadmap, many converging elements of NBIC rely on the open source movement to drive rapid progress, and it is an important enabler of NBIC knowledge and technology convergence. Open access, modification, and redistribution are the central concepts of the philosophy of open source.

The inherent advantages of open source development for fostering rapid innovation are tempered by the sobering security issues arising from deliberate insertions of malicious components in the systems (in the semiconductor material substrates or in the software) or from unintended consequences of defective contributions. Assuring that open-source software is free from intentional or accidental vulnerabilities will be critical to its success for NBIC.

Current capability. Several prominent examples of open source methodologies in the NBIC landscape go well beyond code development:

- HUBzero, the cyber-infrastructure framework for the nanoHUB (http://nanohub.org/), has embraced the open source distribution model for its user community (Klimeck et al. 2008; McLennan and Kennell 2010). The HUBzero/nanoHUB model goes beyond the distribution of open source codes to provide the NBIC community with open simulation services.
- The Observational Medical Outcomes Partnership (OMOP; http://omop.fnih.org/), a public–private partnership[2] "established to inform the appropriate use of observational healthcare databases for studying the effects of medical products," will use the open source model to develop and distribute its informatics tools (FNIH 2012).
- An example of software assurance in the commercial world is the certification process of the App Store of Apple.
- Returning the open source model full circle to its roots in manufacturing, the desktop manufacturing community is embracing open source to accelerate innovation.
- Open-source tools such as AFNI (analysis of functional neuroimages) are becoming increasingly important for creating informatics in the domain of noninvasive human brain imaging. Because functional brain studies are extremely expensive, the sharing and reuse of data between research groups has become a priority.

---

[2] OMOP partners are PhRMA (a membership organization of U.S. research-based pharmaceutical and biotechnology companies), the FDA, and the Foundation for the National Institutes of Health.

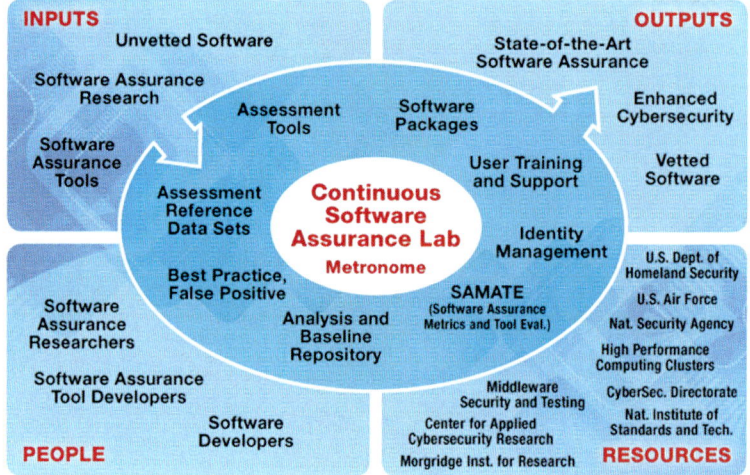

Fig. 1.6 Resources configuration of the Software Assurance Market Place (SWAMP), a national cybersecurity resource offering continuous software assurance (Courtesy of Miron Livny, University of Wisconsin–Madison)

In addressing the security of open source software, the open source communities in NBIC can benefit from the latest exploratory efforts of the cyber community in the form of "software assurance" (SwA) for open source software (Miller 2010). Wikipedia defines SwA as, "the level of confidence that software is free from vulnerabilities, either intentionally designed into the software or accidentally inserted at anytime during its lifecycle, and that the software functions in the intended manner" (http://en.wikipedia.org/wiki/Software_assurance). The Department of Homeland Security (DHS) has taken this a step further by establishing the SwA Market Place (SWAMP) initiative (Fig. 1.6; DHS 2011). The goal of the SWAMP initiative is to create a national cyber-infrastructure with "test and build" capabilities that address:

- Trustworthiness: No exploitable vulnerabilities exist, either maliciously or unintentionally inserted
- Predictable execution: Justifiable confidence that software, when executed, functions as intended
- Conformance: A planned and systematic set of multidisciplinary activities that ensure software processes and products conform to requirements, standards, and procedures

What's missing/why now? The increasing frequency of cyber-attacks is headline news. Currently, the open source approach is not used for sensitive and mission-critical tasks, because the security and trustworthiness of open-source environments currently cannot be taken for granted. For new NBIC initiatives, SwA should be a central part of the foundational planning and strategy. For existing major NBIC projects and their mission critical components, SwA should be elevated to a key deliverable with milestones that underscore the seriousness of the objectives.

Desired outcomes/new capabilities. The open source approach is expected to play a major role in driving the convergence of NBIC knowledge and technologies in both the narrow context of code development and in the broader context as an organizational principle. Because of its importance and ubiquity, open source security risks must be managed and mitigated by embracing new ideas arising from the software assurance community. The overarching goals of NBIC convergence will be advanced by building strong bridges between the NBIC and SwA communities.

Benefit. As noted above, achievement of many of the goals within NBIC will require extensive use of open source code, so secure open source software is critical to and benefits NBIC.

1.4 Infrastructure Needs

Scientific and engineering research is currently supported by a well-established infrastructure, and supporting, maintaining, and enhancing this existing infrastructure will require ongoing planning, resources, and commitment. In addition to the existing infrastructure, however, we foresee the need for new types of infrastructure to support converging knowledge and technologies. Four examples of new infrastructure needed for NBIC are listed below and described in the text that follows:

1. Major, open-use materials processing and advanced device fabrication facilities for hybrid processing (e.g., electronics + bio, etc.) complemented by an electronic design infrastructure specifically to support electronics for converging technologies
2. Instrumentation facilities for noninvasive brain imaging matched to space–time constraints of neural code and for rapid imaging, noninvasive imaging, and manipulation at the nanoscale
3. Software/cyber-infrastructure/cyber security institutes as enablers for convergence
4. Educational infrastructure for converging knowledge and technologies

1.4.1 Open-Use Processing and Design Facilities for Hybrid "CMOS +" Processing

Section 1.3.1 discussed the opportunity to develop a new era of electronics that couples the powerful semiconductor manufacturing and design capabilities that have been developed for information processing and communications with new materials and devices that enable new applications. The nanoscale dimensions of state-of-the-art CMOS technology provide a natural connection to the biological world. To support this vision, facilities are needed that have the advanced

CMOS manufacturing capabilities and tool sets and yet are flexible enough to accommodate the use of new materials and fabrication techniques. An example (referenced earlier) of such a facility is the bio-CMOS fabrication line at IMEC in Leuven, Belgium. In addition to fabrication facilities, the current design infrastructure needs to be extended to encompass a growing variety of devices and system architectures used in conjunction with traditional CMOS platforms.

This new infrastructure could leverage current investments by extending the missions of the NNIN and the user facilities at the U.S. national laboratories to accommodate the new activities. It is also possible to leverage some of the underutilized facilities of companies that have reprioritized their investments. A substantial investment of the order of hundreds of millions of U.S. dollars is required. An infrastructure to support ongoing operating costs is mandatory. The development of such an infrastructure should broadly involve the electronics, communications, energy, and healthcare industries, and Federal funding agencies such as NSF, NIH, the national laboratories, DOE, and DOD, as well as partnerships with major research universities. The success of IMEC is due in large part to the intellectual leadership provided by academics, especially those in the nearby Catholic University of Leuven.

1.4.2 Instrumentation Facilities for Noninvasive Brain Imaging

As discussed in Sect. 1.3.2, current human noninvasive brain imaging cannot capture the true complexity of neural dynamics due to limitations in both spatial and temporal sensitivity of current methods. Current noninvasive brain imaging techniques may well have reached technical limits. New approaches will be needed to reveal the basis of human cognition and behavior.

Desired outcomes/new capabilities: New imaging technology is needed that is better matched to the neural code of the brain and that has the ability to reveal the molecular substrates of human cognition. An MRS technology with resolutions of 1 ms, 10 µm, and 1 mM (chemical concentration) might be a suitable goal.

Benefit: Such a technology would allow many of the recent advances in molecular neuroscience to be leveraged into a better understanding of human cognition and behavior.

Costs: The research and development costs for the next generation of noninvasive human brain imaging technologies are likely to be high. If the approach involves detecting atomic resonance signals (as would be true for MRS advances), the investments are likely to involve better magnets, better radio frequency detection systems, and more sophisticated high-performance computation. These costs could easily be on the order of $100 million.

Assignment of responsibilities: The creation and operation of such an infrastructure should be the shared responsibility of the agencies that fund research in NBIC fields.

1.4.3 Software Institutes, Including Cyber Security Infrastructure

Infrastructure in the form of "software institutes" can play an important role in addressing the goals for open source software, cyber security, and open source risk management tools discussed in Sect. 1.3.3.

The problem. Recent studies have articulated the need for a sustained emphasis on software development. For example, a 2009 International Assessment of R&D in Simulation-Based Engineering and Science (Glotzer et al. 2009) notes that, "the practical applications of computational engineering and science are built around effective utilization of the products resulting from *software development*" (Head-Gordon 2009, 65). Moreover, given the increasing complexity of the state of the art in computational models, the traditional academic model for software development, namely code development as a side activity of dissertation research by a succession of graduate students, is being replaced by a new paradigm featuring a professional team of software engineers working closely with research experts from the scientific domains. The research community is now poised (with encouragement and support from the major funding agencies) to establish software institutes as a concerted shift to the new paradigm.

The pending creation of software institutes also addresses the key problem of cyber security and software assurance in the open source model of software development. The philosophy of open source is an important enabler of NBIC convergence. However, the inherent advantages of open source for fostering rapid innovation are tempered by the challenging security issues arising from deliberate insertions of malicious materials or the unintended consequences of defective contributions. Assuring that open-source software is free from intentional or accidental vulnerabilities will be critical to the success of NBIC.

Current capability. Software institutes as described above do not exist yet, but several agencies (most notably the NSF Office of Cyberinfrastructure and the DOE Office of Science) are on the cusp of establishing such efforts as a culmination of almost a decade of sustained input from the research community. New initiatives such as NSF's Software Infrastructure for Sustained Innovation (SI^2) and DOE's Scientific Discovery through Advanced Computing (SciDAC) institutes represent positive momentum in terms of enhancing near-term capabilities.

Cyber security aspects have not been a priority within the open source movement given the "grass roots" aspects of the open source communities, so current capabilities in this respect are lacking. However, the funding agencies, most notably DHS, have recognized this gap, and facilities and infrastructure efforts are now emerging under the rubric of software assurance (DHS 2011).

What's missing. The software institutes are just beginning to be established. CKTS will increase the need for such institutes. It is important to maintain the positive momentum after a decade-long effort by the research community to highlight the software institutes as a priority. From the perspective of NBIC, there is a need for an initiative at the appropriate scale to leverage current plans for software institutes and SwA (cyber security) so as to insure that these capabilities are linked to efforts in NBIC software development.

Why now? The issue is timely from a cyber security perspective because converging knowledge and technologies in NBIC are being developed with essentially universal adoption of open source as the organizational model. Early incorporation of SwA and cyber security principles would be far more cost-effective than remedial actions at a later stage.

Desired outcomes/new capabilities. These include an infrastructure/facility on the scale of SI^2, SCIDAC institutes, and DHS SWAMP, with strong linkages to the NBIC fields and engaging the best NBIC experts and practitioners in the country.

Benefits. Such foundational elements in the software infrastructure will greatly accelerate the development of trustworthy software for converging technologies.

Costs. Incremental operation and production costs can be fairly modest (by augmentation of emerging software institutes), perhaps on the order of a few million dollars a year.

Assignment of responsibilities. The augmentation strategy for such an infrastructure should be the shared responsibility of all the agencies that fund research in NBIC fields, so the National Nanotechnology Coordination Office (NNCO), with appropriate help from the NSF Office of Cyberinfrastructure, could take an oversight role on the supported facility to insure that the operational activities achieve the desired software development outcomes for NBIC.

1.4.4 Educational Infrastructure for Converging Technologies

The problem. The specialization that played a crucial role in the remarkable advances in science and technology during the twentieth century led to highly specialized academic programs. Individuals working on NBIC will continue to require deep, specialized knowledge complemented by a breadth that is reminiscent of Renaissance scientists and engineers.

Current capability. Highly specialized graduate programs abound, and interdisciplinary education is increasingly common. Web-based technologies such as Coursera (https://www.coursera.org), OpenCourseWare (http://ocw.mit.edu/index.htm), and edX (http://www.edxonline.org) make high-quality instructional material and even complete courses available online at no cost to users. An early example of an initiative, specifically directed education for CKTS, is nanoHUB-U (http://www.nanoHUB.org/u), which makes available 5-week courses designed for those without specialized, discipline-specific knowledge (Fig. 1.7).

1 Convergence Platforms: Foundational Science and Technology Tools

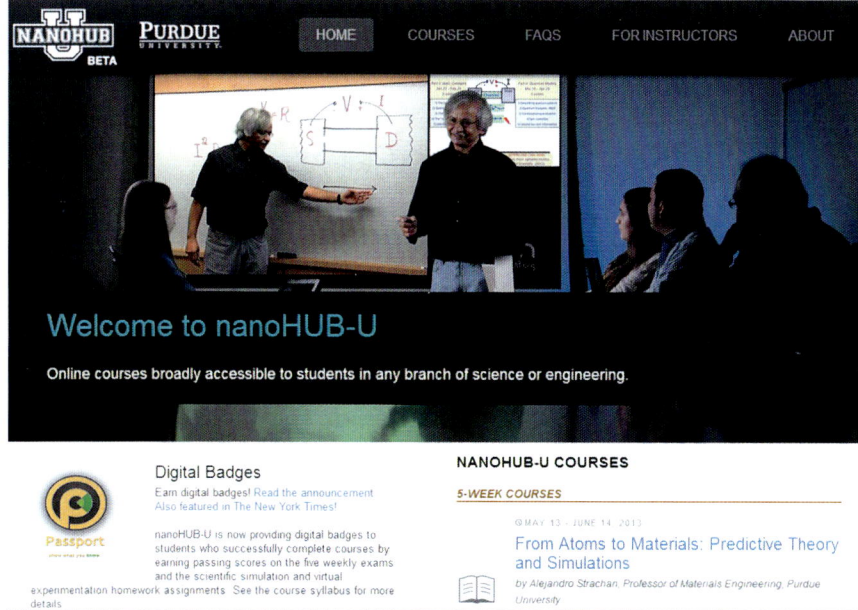

Fig. 1.7 A nanoHUB-U landing page (2013). This initiative provides graduate-level short courses on NBIC topics that are broadly accessible to anyone with an undergraduate science or engineering degree (https://nanohub.org/groups/u; ©nanoHUB-U; used by permission)

What's missing. Still lacking is an initiative at the appropriate scale to provide those having or obtaining specialized knowledge with the additional cross-disciplinary knowledge needed for working in converging NBIC fields. The required online delivery technologies exist and are rapidly evolving, but the challenges of developing these new kinds of courses and effective teaching strategies are nontrivial and must engage the best and brightest NBIC practitioners to create a robust infrastructure available to students and practicing scientists and engineers.

Why now? Converging knowledge and technologies in NBIC require new knowledge and skills. In an earlier era, interdisciplinary work involved mostly learning the language of the relevant fields to enable collaboration. The work itself is still performed by those conventionally trained in that discipline. Today, interdisciplinary work increasingly requires that the person doing the work is proficient in performing in more than one field of study and not just communicating with the collaborator. The levels of training and skill required for such interdisciplinary work is qualitatively different from those of an earlier era.

Desired outcomes/new capabilities. It would highly desirable to have an infrastructure similar to Coursera, OpenCourseWare, or edX, but one that is specifically devoted to graduate education in the NBIC fields. This initiative should engage the best faculty in the country and be available to both students and practicing scientists and engineers.

Benefit. Such an infrastructure can greatly accelerate the application to converging technologies to the grand challenges that humans face, and it is one that will capture the interest of students and inspire them to prepare for careers in NBIC and overall in CKTS.

Costs. Operation and production costs are fairly modest: a significant infrastructure facility could be operated for a cost on the order of $10 million/year. However, the costs of developing these new kinds of educational resources (mostly faculty time) should not be underestimated and could be significantly higher.

Assignment of responsibilities. The creation and operation of such an infrastructure should be the shared responsibility of the agencies that fund research in NBIC fields.

1.5 R&D Strategies

Over the course of the next decade, significant sustained investments are needed in the following:

- Programs that fill the "Bell Labs gap." Models may need to differ but still be effective at innovation; innovation tends to suffer in systems that over emphasize "outputs and accountability." Possibilities include reorienting national labs, creating major institutes, and building more effective public–private and industry–university partnerships. Funding agencies have to be willing to let creative people try and fail without penalty (e.g., BIO's Ideas Labs).
- Funding strategies specifically designed to support converging knowledge and technologies. More emphasis on project-oriented, "Pasteur quadrant" (Stokes 1997) research is appropriate for this era of convergence.
- Education for converging technologies aimed at achieving depth in a discipline in addition to attaining a basic understanding of related fields.

1.6 Conclusions and Priorities

The twentieth century saw the development of the great technologies of nanoelectronics, biotechnology, information technology, and cognitive science and the foundational knowledge and tools that enabled them. These technologies continue to advance in the twenty-first century, but the twenty-first century will be the century of converging knowledge and technologies. It will go beyond multidisciplinary research, which brings different disciplines together, to transdisciplinary research, which transforms the disciplines and creates new one. Tools will play an important role in the success of convergence. We see an opportunity to enable and accelerate progress in convergence through a new class of "convergence tools" and technologies that support the solution to problems and enhancement of human capacity. We also see opportunities to advance specific NBIC disciplines with new tools enabled by converging technologies. Scientific and educational infrastructure played a critical role in the success of twentieth century science and technologies, and new types of infrastructure will play similarly critical roles in the success of converging NBIC knowledge and technologies. New educational paradigms will be needed to inspire and prepare the next generation of scientists and engineers to realize the potential of converging knowledge and technologies. New R&D strategies will be needed to go beyond multidisciplinary research and support the kind of transdisciplinary research that is needed in the era of convergence.

1.7 R&D Impact on Society

The twentieth century witnessed great advances in the health and wealth of individuals in the United States, and the NBIC technologies played a large role in that success. Today we face an array of new challenges in providing our citizens with jobs, health, clean environment, and life security. Converging knowledge and technologies provide us with new opportunities to strengthen connections between science, technology, and the humanities, and to ensure that technology will benefit society by addressing the most urgent challenges of the twenty-first century. The development of new knowledge, disciplines, and tools at the interface of the traditional NBIC disciplines will be both a consequence and driver of convergence. Convergence has the potential to make the twenty-first century a second Renaissance, with the tight connection of science, technology, and the humanities that characterized the first Renaissance.

1.8 Examples of Achievements and Convergence Paradigm Shifts

1.8.1 A New Computing Paradigm Using the Brain Model

Contact persons: *H.-S. Philip Wong and Duygu Kuzum, Stanford University*

Introduction and Historical Background

The convergence of NBIC has led to new capabilities for brain simulation and the opportunity to create a new computing paradigm. A synapse appears to be more suitable as the basic element in understanding and simulating the brain than a neuron.

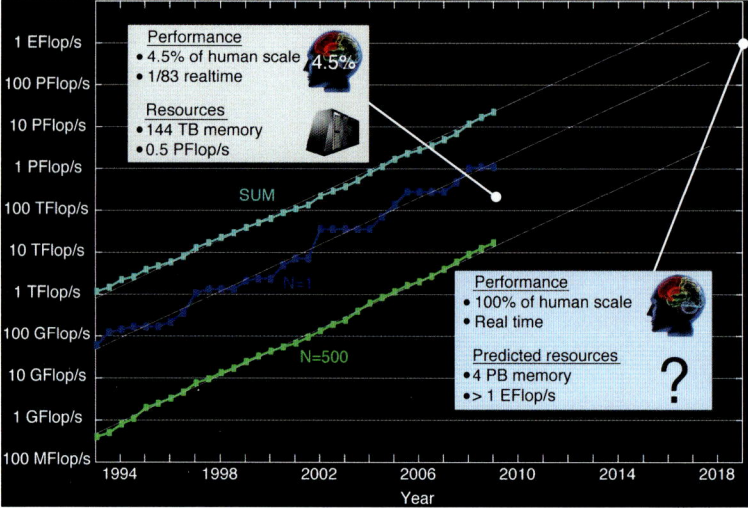

Fig. 1.8 Computation power increases over the last two decades project that human-scale computation is possible if the system design is not otherwise limited (Ananthanarayanan et al. 2009; see http://modha.org/blog/2009/11/post_3.html; source: IBM; ©ACM Digital Library; reuse cleared through Rightslink). Performance is defined for supercomputers to reach 100 % of human-scale computation, which means the computational resources to be able simulate 10^{11} neurons and 10^{15} synapses

The evolution of the microelectronics industry into a nanoelectronics industry has ushered a new era of exceptional computational capability. Figure 1.8 shows the progress that has been made in supercomputing since the early 1990s. At each time point, the green (lower) line shows the 500th fastest supercomputer, the dark blue (middle) line shows the fastest supercomputer, and the light blue (upper) line shows the summed power of the top 500 machines. These lines show a nice trend, which we have extrapolated out for 10 years. The IBM team's latest simulation results represent a model about 4.5 % the scale of the human cerebral cortex, which was run at 1/83 of real time. The machine used provided 144 TB of memory and 0.5 petaFLOPS (Ananthanarayanan et al. 2009).

Vision for a New Computing Paradigm

Turning to the future, one may project that running human-scale cortical simulations will probably require 4 PB of memory, and to run these simulations in real time will require over one exaFLOPS. If the current trends in supercomputing continue, it seems that human-scale simulations will be possible in the not too distant future, *if the system design is not otherwise limited.*

IBM projects that a human-scale simulation, running in real time, would require a dedicated nuclear power plant, whereas the power dissipation in the human central nervous system is on the order of 10 W. To give a sense of the order of magnitude, 1 MW can power a thousand homes (http://wiki.answers.com/Q/How_many_homes_can_a_megawatt_power).

The above analysis points out two key observations:

1. Continuing to do computations in the conventional way will not lead us to brain-scale simulation because it will be limited by power consumption.
2. Nanoelectronics today and its projected evolution will enable enormous computational capability.

Today's computers are based on binary logic, precise digital representation, and von Neumann architecture, and are fast and synchronous. The brain's computing elements are slow and stochastic and operate at kHz time scale. However,

1. Brain architecture and computation are massively parallel.
2. The brain is adaptive and self-learning. It modifies its hardware and computational algorithms based on learning.
3. Brain computation is extremely fault-tolerant and robust. Even 50 % noise can be tolerated.
4. On a standard computer based on the von Neumann architecture, the memory and the processor are separated by a data channel or bus between the area where data is stored and where it is worked on; in the mammalian brain, storage and computation happen at the same time and in the same physical location.
5. The brain consumes less power than a conventional incandescent light bulb.
6. The brain occupies less space than a 2-l soda bottle.
7. The brain surpasses computers when sensing and acting meet computation on complex real-world problems with massive amount of inputs and variables such as image, video, or voice analysis and recognition, autonomous navigation, and real-time ability for recognition, understanding, tagging, and decision-making.

Can one re-architect computational systems to make use of massive parallelism as the brain does (Furber 2012)? Recent developments in multicore computer architectures already have moved one step in the direction of parallelism. Is it possible to embed the storage elements within the computational fabrics in a fine-grain fashion, similar to the way that synapses (storage) and neurons (computation) are connected in the brain? Will such computing architectures be more energy-efficient than the computer architectures in use today for broad classes of applications such as searches, data mining, pattern recognition, and image understanding?

Turning to the rapid evolution of nanoelectronics, we note that electronic devices will continue to be scaled down in size with a concomitant reduction in energy (and power) consumption per device. The semiconductor industry projects that it is straightforward to scale to 10 nm feature size (Mayberry 2012), and there are efforts to push transistor scaling down to 5 nm feature size. At a 5 nm feature size, a chip of 1 cm × 1 cm × 1 cm volume will contain roughly on the order of 10^{18} devices! The use of new nanomaterials for nanoelectronics adds to the suite of capability of the conventional material of choice, silicon. New innovations in nonvolatile memory devices creates new opportunities for revolutionizing the memory hierarchy for more optimal utilization of computation resources and will further enhance system performance.

A natural question that arises is, Can we utilize the enormous capability of nanofabrication of electronic devices to perform computation that rivals the computational density, capability, and energy efficiency of the brain? If one can emulate the functions of the brain, can one then use the capability to further understand how the brain functions and thereby help advance neuroscience and our understanding about learning and cognition? In other words, can we reverse-engineer the brain using an electronic model of the synaptic and neuronal functions? With an electronic emulation of the functions of neurons and synapses, can we then ask the "what if" questions simply by tuning the electronics—something that would be rather difficult to do if we only have biological systems to work with?

Current Status

Because of the 10,000 synapses/neuron ratio, the synapse circuit dominates the implementation problem. If we want to build a computational system that is massively parallel, highly interconnected, and as compact as the brain, the most important building block is a compact nanoelectronic device emulating the functions and plasticity of biological synapses. Different from the approaches that have been investigated over the last couple of decades, which try to mimic only the connectivity and architecture of the brain, understanding and emulating the functionality of synapses is critical to achieving brain-level parallelism and efficiency. Also different from the earlier works of using software programs to implement computational algorithms, the direct use of hardware with physical connections enables a direct comparison of the energy efficiency of the computation system. The electronic version of the synapse must be very compact, consume very little energy, and have the requisite plasticity akin to the biological synapse. Synaptic plasticity is weight adaptation during the course of computation.

In 1949 Donald Hebb postulated that the connection strengths between neurons are modified based on neural activities in presynaptic and postsynaptic cells. A form of Hebbian learning called spike-timing-dependent plasticity (STDP) was discovered in about the late 1990s. Synaptic plasticity is a mechanism that regulates the experience-dependent change in connectivity between neurons, and it is believed to underlie learning and memory in the biological brain. Other forms of plasticity have also been observed in various regions of the brain (Shouval et al. 2010).

Recent works on neuromorphic or brain-inspired circuits and devices emulating synaptic plasticity fall into two categories: (1) use of conventional digital and analog circuitry to emulate the functions of the neurons and the synapses, (2) use of new materials in a two-terminal device to emulate the functions of the synapse.

Recent publications have shown two brain-inspired neuromorphic chips from IBM that are examples of implementing synaptic plasticity using conventional digital and analog circuitry. While the use of conventional circuit design techniques and device technologies have enabled the building of a moderate-scale system, the main problem with these designs is that each synapse (an 8-transistor static random-access memory cell, or 8T SRAM) occupies a significant area. If we want to build a brain-like system with massively parallel architecture and billions of synapses, this same methodology would not be practical by using these area-inefficient approaches.

An alternative research direction is focused on single-element scalable synaptic devices. A recent special issue of the *Proceedings of the IEEE* includes several review papers that summarize recent developments (e.g., Mazumder et al. 2012). In these devices, various nanomaterials can be sandwiched between two metal electrodes to form a two-terminal electronic device whose conductance can be precisely modulated by charge or flux (e.g., current) through it, or by applied voltages between the two terminals. A nanoscale silicon-based two-terminal device has been proposed as a synaptic element in neuromorphic circuits. Silver nanoparticles are incorporated into silicon medium and form conduction paths by applied bias. Implementation of STDP learning rules is demonstrated by utilizing this conductance change (Jo et al. 2010).

Atomic bridge memory work from the Ohno group (Ohno et al. 2011) at the International Center for Materials Nanoarchitectonics within the National Institute for Materials Science (NIMS) in Japan is another such approach using a different material system for a two-terminal device. Yet another approach (Fig. 1.9) uses phase change materials (PCM) to build massively parallel and compact terascale systems (Kuzum et al. 2012). The nanoelectronic synapses will have all the essential functionalities of biological synapses, and they exhibit plasticity. The energy efficiency of plastic synapses of phase change materials is on the order of hundreds of femtojoules (fJ) per synaptic event. Further reduction in energy consumption can be achieved by further down-scaling of the device from 75 nm to less than 2 nm (energy consumption is proportional to the feature size squared). Using metal oxides sandwiched between two metal electrodes (resistive random-access memory or RRAM), it is also possible to achieve plasticity with tens of fJ per synaptic event (Yu et al. 2011).

Using these two-terminal programmable resistors, various forms of STDP have been implemented. Through software emulation of the electronic synaptic functions, it has been shown by various authors that functions such as pattern recognition and associative learning can be effectively achieved (Snider 2007; Jo et al. 2010; Yu et al. 2011; Kuzum et al. 2011; Suri et al. 2011; Ohno et al. 2011; Mazumder et al. 2012) (Fig. 1.10). What remains to be demonstrated are large-scale hardware systems that combine the novel two-terminal electronic plastic synapses with neuron circuitries to perform real-time neuromorphic computing so that detailed measurements can be made of the systems' computation efficacy and energy efficiency.

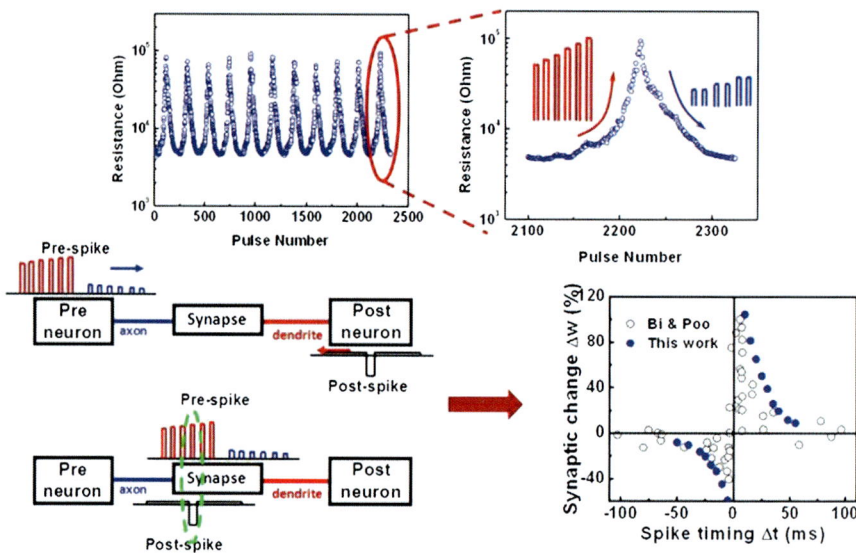

Fig. 1.9 An order-of-magnitude change in the phase change cell resistance was achieved through 100 steps for both the set and reset transitions. The repeatability of this gradual phenomenon was confirmed through many cycles, as shown in the figure on the *top left*. After confirming that gradual resistance increase and decrease can be achieved with PCM, the next step would be implementation of synaptic learning. In the spike scheme explained here, preneuron and postneuron independently generate pre-spike and post-spike without any other communication link between them. When a preneuron spikes, it sends the pre-spike, consisting of gradually increasing reset pulses and gradually decreasing set pulses, through its axon. At the synapse, pre-spike propagating through the axon meets with the post-spike propagating through the dendrite. The potential across the synapse, depending on the superposition of pre- and post-spikes as a function of relative spike timing, determines the amount of synaptic weight change. The way pre-spike is designed determines whether STDP is asymmetric or symmetric. Asymmetric and symmetric STDP can be implemented with the same scheme by simply changing the amplitude and the order of pulses in pre-spike (Adapted and used with permission from Kuzum et al. 2012, ©2012, American Chemical Society)

Impact on Other Areas of Knowledge

Brain-inspired computational systems will complement conventional digital computers and humans. We envision the following scenarios:

- New computational paradigms and architectures to extend capabilities of information technology beyond digital logic
- A new platform for real-time brain simulations, to support advances in the field of neuroscience, learning, and cognition
- Reduced need for animal experiments
- Electronic synapses for *in vivo* monitoring and stimulation of neurons or neural prosthetics

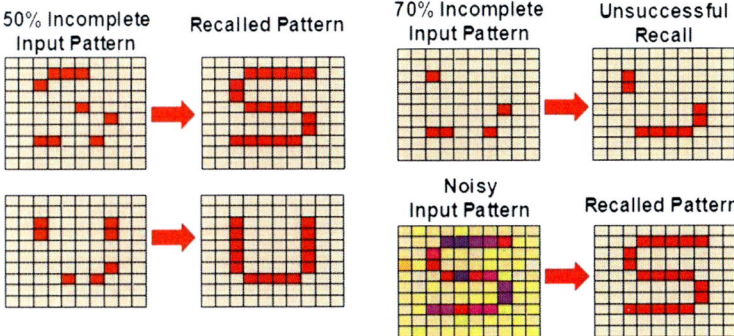

Fig. 1.10 Associative learning is central to neural computation and formation of episodic memory in the brain. In associative learning, the goal is to complete and recall a previously learned pattern from an incomplete representation. The network learns a pattern by strengthening (potentiating) the synapses between neurons, which are coactive. If an incomplete pattern is presented, the potentiated synapses can recruit the missing neurons in order to recall the original pattern. A recurrent network of 100 neurons and 10,000 synapses with asymmetric STDP is constructed; the network is stimulated with predetermined spike patterns. The network can recruit missing neurons and recall the original pattern when an incomplete pattern with up to 50 % missing neuron spikes is presented. After training, the network is stimulated with 50 % incomplete "S" or "U" patterns. The network is shown to recover the full pattern, except in the case where missing parts are more than 70 % and the stimulated parts have strong overlaps with both patterns. One of the advantages of neuromorphic computation is its immunity against noise and variation. In order to test the robustness of computation, the network is stimulated with a fuzzy input pattern with 30 % white Gaussian noise. The success rate for recalls is ~60 % for 50 % noise and more than ~85 % for 30 % noise. Device-to-device variation or cycle-to-cycle variation during programming can also affect recall performance. The resistance value of the synaptic cell may show some stochastic variation from one spike cycle to another. The cell resistance is randomly chosen from a distribution with a maximum variation in the range of 20–90 % (After Kuzum et al. 2011; ©IEEE, reuse cleared through Rightslink)

Long-Term Perspective

Brain-inspired computing that is portable, energy-efficient, and adaptable will become a reality. Initial digital implementations may emerge in 10 years. A realistic projection would see impact in a 20-year time frame considering the long time scale needed for the evolution of new technology. These systems are interactive and derive their computational power through data-driven machine learning instead of manual programming. This new computing paradigm will address applications at the intersection of sensing and computation, and incorporate learning and understanding (not just merely storage) of the data as key elements of the system. For example, machine learning can be used to understand images (not just to recognize certain geometric features). Brain-like computing will change the way we interact with machines and substantially change the landscape of the human–machine interface. Computers will no longer be limited to the digital systems we know today that need precise deterministic inputs. "Computing machines" will be able to receive dynamic and imprecise data inputs and provide probabilistic answers. These new capabilities in modeling and analysis of systems will improve economic productivity, human potential, and quality of life.

Using these nanoelectronic systems, which use unconventional devices interconnected in a massively parallel fashion (perhaps even three-dimensionally connected), it may be possible to ask the "what if" questions that biologists, psychologists, and neuroscientists have not been able to ask before, thereby complementing the efforts aimed at reconstructing the full record of neural activity across complete neural circuits (Alivisatos et al. 2012). If the properties of synapses are different from what we know today, will learning and cognition be very different? If certain functions of the neurons or the synapses are impaired, how does that affect neurological behavior? Is the connectivity itself important? Does cognition derive from the intrinsic connectivity and the intrinsic functions of the synapses? With an electronic version of the brain, one can begin to probe, change, and ask these questions that experimentalists today cannot ask. With an electronic version of the brain, we will have a new tool to understand ourselves.

1.8.2 Convergence of Knowledge and Technologies in the Semiconductor Industry

Contact persons: *Mark Lundstrom, Purdue University; Robert R. Doering, Semiconductor Research Corporation; and H.-S. Philip Wong, Stanford University*

Semiconductor microelectronics was one of the great technologies of the twentieth century. Semiconductor nanoelectronics seems destined to play an even more important role in addressing the challenges of the twenty-first century by serving as the foundation that enables continuing advances in the convergence of NBIC technologies (Fig. 1.11).

Introduction and Historical Background

In 1931, the physicist Wolfgang Pauli expressed his opinion that "one shouldn't work on semiconductors; that is a filthy mess." (Hoddeson et al. 1992, 121). Subsequently, however, the science of these interesting materials was unraveled step-by-step. While semiconductor science progressed, the invention of the vacuum tube launched the electronics era and transformed communications—creating the "golden age of radio" and producing the first digital computers. By mid-century, semiconductor science had progressed to a point where Bell Labs could mount a serious R&D effort to replace the vacuum tube with a semiconductor device to address the vacuum tube's excessive power consumption and limited reliability. The result was the invention of the germanium transistor in 1948, followed at Texas Instruments by the invention of the silicon transistor in 1954. The integrated circuit quickly followed and was patented by Texas Instruments and Fairchild

1 Convergence Platforms: Foundational Science and Technology Tools

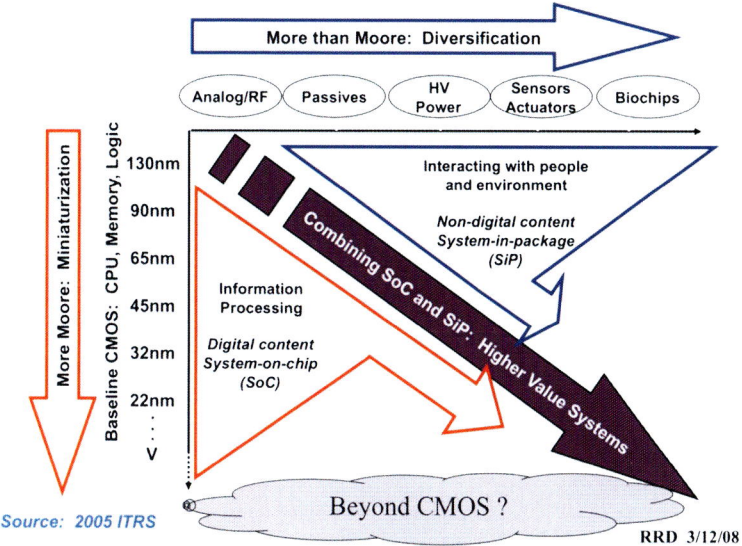

Fig. 1.11 On-chip technology convergence (SIA 2011. ©SEMATECH, used by permission, http://www.itrs.net/Links/2011ITRS/2011Chapters/2011ExecSum.pdf)

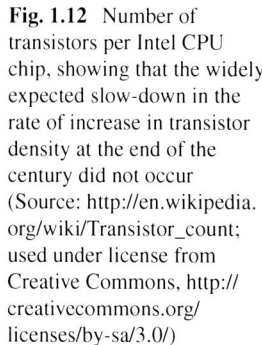

Fig. 1.12 Number of transistors per Intel CPU chip, showing that the widely expected slow-down in the rate of increase in transistor density at the end of the century did not occur (Source: http://en.wikipedia.org/wiki/Transistor_count; used under license from Creative Commons, http://creativecommons.org/licenses/by-sa/3.0/)

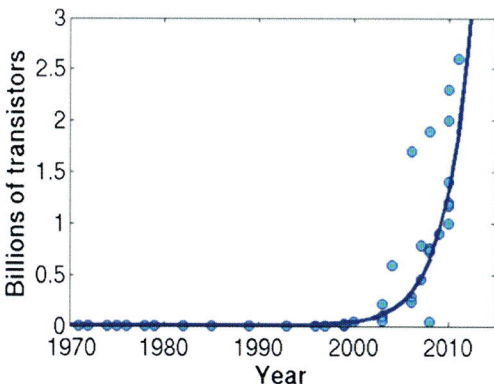

Semiconductor in 1959. Microelectronics became a viable technology, steadily progressing and addressing a larger and larger range of applications (Lojek 2007). The microelectronics revolution transformed communications and computing and in the process shaped our modern society.

By the end of the twentieth century, it was felt by many that microelectronics had run its course. It was widely thought that the technology had been pushed to its practical (and possibly fundamental) limits. Figure 1.12 shows what actually happened in the first decade of the twenty-first century. The figure shows the number of transistors on an Intel processor chip vs. time. Progress did not slow; the number of transistors per chip continued to increase according to Gordon Moore's famous "law". Microelectronics became nanoelectronics, and the resulting increase in the capabilities of electronic systems—especially personal electronics—continued to transform the world. It now seems likely that the impact of semiconductor electronics on human society will be even greater in the twenty-first century than in the twentieth century. Converging knowledge and technologies played a key role in the transformation of microelectronics to nanoelectronics. Looking to the future, semiconductor nanoelectronics will play a critical role in connecting nanotechnology, biotechnology, information technology, and cognitive sciences to address the challenges of the twenty-first century.

Microelectronics Becomes Nanoelectronics: Contributions of Converging Knowledge and Technologies

From the 1960s to 2000, semiconductor integrated circuit technology was known as "microelectronics." One definition of "nanotechnology" is a device or structure with features smaller than 100 nanometers (nm) in at least two dimensions. As the end of the century approached, gate oxides had been scaled to less than 2 nm in thickness, and channel lengths were approaching 100 nm. New knowledge and technologies played an essential role in continuing the historic pace of channel-length scaling, making the transistor a true nanodevice. The tools from nanoscience, such as tunneling and atomic force scanning probes (Binnig et al. 1982; Binnig and Quate 1986) became widely used for metrology, and new fabrication technologies such as atomic layer deposition (Ritala and Lerskela 2002) found applications when high-k dielectrics began to replace silicon dioxide (SiO_2) (Mistry et al. 2007).

Patterning nanoscale dimensions with optical illumination wavelengths of 248 and 193 nm has played a critical role in taking semiconductor electronics to the nanoscale. Sophisticated techniques such as phase shift masks (PSM), double patterning, immersion lithography, and computational lithography, which make use of numerical simulation to improve resolution and contrast, were developed, as exemplified in Fig. 1.13. The result is that at the 22 nm node, critical features are patterned with light having a wavelength almost ten times the feature size.

The trend of increasing use of a wider and wider range of new materials and technologies continues to enable progress in nanoelectronics. In addition to advances in materials and process technologies, new devices based on new physical principles are beginning to augment CMOS systems with new capabilities. One example currently being developed is so-called spin-transfer-torque memory, which uses a spin polarized electron current to switch the orientation of a magnet (Kawahara et al. 2008).

Finally, we also note that converging knowledge as well as converging technologies has played a key role in continuing the progress of nanoelectronics. The transistor models used by technology developers and circuit designers can be traced to seminal work in the 1960s (Hofstein and Heiman 1963; Sah 1964). These models were extended and enhanced into the 1990s—without any major conceptual changes.

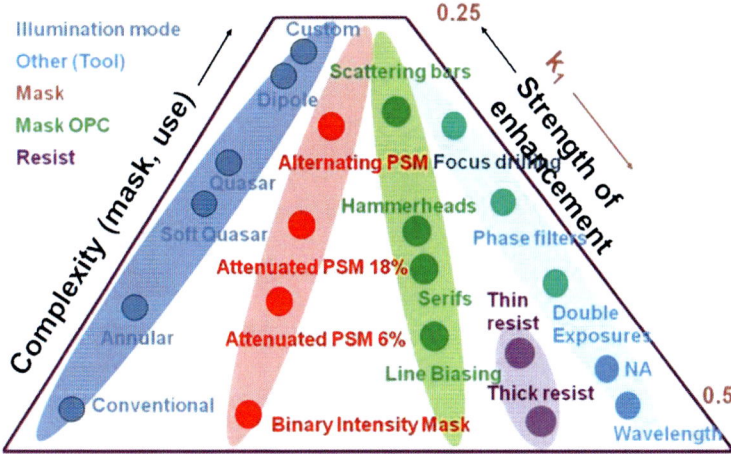

Fig. 1.13 Examples of knowledge convergence for optical lithography of integrated circuits (Source: ©ASML Holding NV, reused under ASML terms of usage). *NA* sodium, *OPC* optical proximity correction, *PSM* phase shift mask

As channel lengths shrunk below 100 nm, however, the physical assumptions in these models lost validity, and semiconductor electronics benefitted from the fundamental understanding of electronic conduction at the nanoscale that was first developed in the 1980s (Datta 1997). Computational techniques such as the non-equilibrium Green's function (NEGF) approach, which was widely used to understand conduction in molecules (Datta et al. 1997), has now become a computer-assisted design (CAD) technique used for treating quantum transport in advanced transistor design (Lake et al. 1997).

Semiconductor physics and technology was a clearly defined discipline from the 1960s through the mid-1990s, but since that time, an increasingly wide range of knowledge and technologies has been used to drive progress in semiconductor nanoelectronics.

Impact of Semiconductor Nanoelectronics on Other Areas of Knowledge, Technology, and Society

Converging knowledge and technologies have played a critical role in the progress of semiconductor technology from microelectronics to nanoelectronics. The result is an increasingly powerful technology that will play a central role in addressing almost any of the grand challenges that society faces. Nanoelectronics provides the hardware that has enabled the information age, but applications of electronics in medicine and biology and in sophisticated instrumentation for cognitive science are increasing. Enhancing human health, security, and economic well-being will require combinations of technologies, with semiconductor electronics playing an essential role in most solutions.

Outcomes

The worldwide semiconductor market is now over $300 billion annually (IDC 2012), 17 % of which is invested in R&D. The information technology industry enabled by semiconductor technology is even larger. The semiconductor industry is characterized by a rapid rate of innovation. Over the 1960–2007 timeframe, semiconductor innovation grew at 9 % per year—25 times the growth rate for the overall economy. As a whole, it accounted for 30 % of total innovation (Samuels 2012). In the United States, a unique open-innovation ecosystem plays a key role in the semiconductor industry. Central to the strategy is the concept of "precompetitive" research and development in which competing companies share the funding and intellectual property of precompetitive development in industrial consortia such as SEMATECH (http://sematech.org/), in industry-funded university research through organizations such as the Semiconductor Research Corporation (http://src.org/), and in partnerships with Federal agencies such as NSF and the National Institutes of Standards and Technology (NIST). The 30-year history of these partnerships for supporting precompetitive R&D provides a model for other industries.

Long-Term Perspective

For the past 50 years, progress in semiconductor electronics has been driven by the continued downscaling of device dimensions, increasing numbers of transistors per chip, producing increasingly powerful electronic systems on a chip. Cost-effective device scaling is increasingly challenged as gate lengths approach the 10 nm scale, but another decade of scaling appears to be possible. At that time, multi-billion-transistor chips will provide a power platform for applications that involve converging technologies. At the same time, other forms of technological progress will begin to emerge; we can expect to see a wider and wider range of devices and specialized architectures integrated on these gigascale CMOS platforms. The development of these new technologies and architectures will be driven by the requirements of the NBIC technologies and their integration to address grand challenge problems. Semiconductor nanoelectronics can be expected to play a critical role in catalyzing the converging of twenty-first century NBIC technologies.

In summary, converging knowledge and technologies played a critical role in the successful transition from microelectronics to nanoelectronics, and CKTS will continue this role in the twenty-first century. At the same time, nanoelectronics can be expected to play an expanding role in addressing society's challenges with converging technologies.

1.8.3 Cognitive Computing: Turning Big Data into Actionable Information and Valuable Knowledge

Contact person: *R. Stanley Williams, Hewlett-Packard Laboratories*

Our modern society is drowning in data. By some estimates (Knorr 2012), more than 95 % of accumulated data is never examined, and much of it is discarded to make room for the most recently gathered data. This is a tremendous waste of the expense and effort that went into collecting the data, but even more critical is the opportunity cost represented by the information and knowledge that were never created from that data. The amount of data we are presently gathering is roughly doubling every 20 months (Gantz and Reinsel 2011), but the doubling time is decreasing dramatically. The majority of this data is unstructured, i.e., it is not in the form of tables of numbers, but rather as images or data streams from various types of sensors and machines. Although there will be four billion people online by 2020, it is estimated that there will be 31 billion connected devices and another 1.3 trillion sensors of various types collecting as much as 50 zettabytes (billion terabytes) of data (Hopkins and Evelson 2011). By the end of the decade, the amount of data harvested in a year will far exceed the capacity of all storage and computing systems on earth, so only a tiny fraction will be examined, and for the most part, the tools that exist today are unsuitable to analyze the unstructured data. Although there have been many reports on how to manage big data from a business perspective (Manyika et al. 2011), the simple fact is that information technology is lagging behind the data explosion, even given the most optimistic forecasts for improvements in computers (Kogge et al 2008). We are being overwhelmed, and there is no way to catch up via evolutionary technology advancements. We must develop revolutionary new technologies to turn the tsunami of data into real-time actionable information and long-term knowledge to improve our use of resources, our health and safety, and our enterprises.

Cognitive Computing

There is a biologically inspired research agenda to build cognizant systems, i.e., software and hardware that emulate (rather than attempting to replicate or simulate) the architectures, algorithms, and processes used in brains for data processing to extract meaning orders of magnitude more rapidly and efficiently than a von Neumann computer. The goal is to build systems that can be taught rather than programmed, and that can autonomously learn how to respond to unanticipated events in an uncertain and changing world. This is an aspirational program, especially since no one actually knows in detail how brains accomplish these tasks.

However, there have been tremendous advancements in both quantitative psychology (Wolfe and Horowitz 2004) and neurophysiological research (Turrigiano and Nelson 2004); although there is still much to learn, the time is right to apply engineering discipline to cognition. One approach is to take the best existing models of cognitive processes and build experimental software and/or hardware implementations that are then applied to solve real problems. The idea is to bootstrap our understanding via a build/fail/partially succeed strategy; we may never get to the point where we understand actual brain function in detail, but we do intend to find new and vastly more efficient ways to analyze data and present the information to humans in a manner that is immediately useful. The challenge is being addressed with both top-down and bottom-up strategies by developing a software platform based on quantitative models of brain architecture and algorithms, and also hardware devices that electronically emulate the synapses and neurons in brains. The two approaches are highly complementary, and they will eventually be combined to provide extremely capable cognizant systems.

Top-Down Software Research

The introduction of multicore processor chips is dramatically increasing the scale of parallel computing, to the point that system software and algorithms are lagging far behind the new processing hardware (Kogge et al. 2008). By 2015, a one-million-core graphics processing unit (GPU) data center will require only about five racks and consume an aggregate of around 270 kW. Unleashing the power of a million cores will enable unprecedented scale and functionality for cognitive applications, but only if there are software platforms and algorithms that can utilize them. Future hardware, both mainstream (GPUs, CPUs, co-processors) and emerging nanotechnology-based systems (see below), will take this trend much further. Programming such massively parallel hardware currently requires specialists to implement even the simplest of functions, and existing high-level programming models do not scale beyond thousands of cores, much less millions. An approach to this problem is developing Cog ex Machina, a general software framework for cognitive computing (Snider et al. 2011; Snider 2012) aimed at providing a powerful, scalable programming model for deploying massively parallel cognitive algorithms without exposing parallel programming to the end user. The Cog framework is composed of a programming model and language, compiler and debugger, cognitive libraries, and runtime support, each designed to satisfy three guiding principles: performance, programmability, and portability (Fig. 1.14).

The Cog framework is a software platform for mapping cognitive algorithms to massively parallel hardware and requires a robust set of applications in order to perform useful tasks. Researchers are building a broad range of algorithms based on the most recent quantitative models of cognition into libraries on top of the Cog core (Snider 2012) so that end users can create substantial, sophisticated applications without being domain specialists or parallel programming experts. Nevertheless, the full power of the Cog core is available to specialists to build their own custom modules.

The goal of such a framework is to enable a user, as an example, to place a custom probabilistic inference module on top of a billion-node neural network for a recognition task in tens of lines of code. The platform can also be used to explore new biologically inspired algorithms, which are just now rendered practical by the

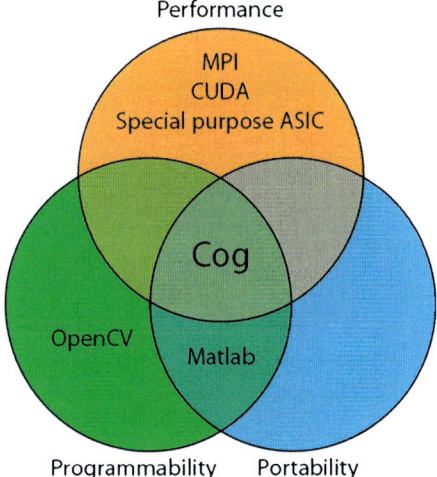

Fig. 1.14 A schematic demonstrating the guiding principles for Cog ex Machina and its position relative to competing approaches for cognitive computing (Courtesy, R. Stanley Williams, Hewlett-Packard; source, Reeke and Sporns 1993; used by permission). *MPI* message passing interface (designed for use on parallel computers), *CUDA* (parallel computing platform and programming model for boosting computing performance through use of GPUs), *ASIC* application-specific integrated circuit, *OpenCV* open computer vision library of programming functions, *Matlab* (high-level interactive technical computing language)

cost-effective availability of massively parallel hardware. Although this approach does not yet adopt a neurophysiological imitation or simulation strategy for system design, biology offers a valuable source of architecture and algorithms for minimizing energy and latency associated with nonlocal communication. Segmentation-from-motion and 3D model building from 2D images is an example of a case where this strategy may offer significant benefit to visual analytics tasks. This augments the state of the art in signal processing and machine learning with recent discoveries in computational psychology (Wolfe and Horowitz 2004).

Bottom-Up Hardware Research

The most fundamental units of brain function are dendrites, axons, and synapses. The resistive switching memory devices that have been proposed as electronic synapses are examples of the memristor (Strukov et al. 2008), a passive electronic circuit element first postulated by Leon Chua (1971). These devices are now being explored by a large number of research groups for use as synapses in biomimetic circuits with transistors. A recent demonstration of a scalable electronic circuit known as a neuristor (Pickett et al. 2013) utilized memristors with both a transient memory and a negative differential resistance that behave in a manner resembling the ion channels in axons. Neuristors are active devices that emulate the signal

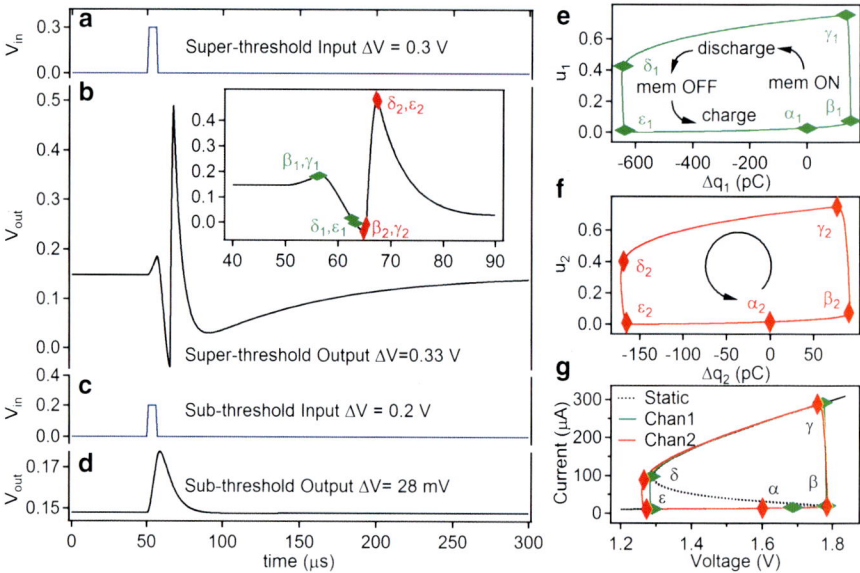

Fig. 1.15 All-or-nothing response and state variable dynamics of the neuristor (Pickett et al. 2013; ©*Nature Materials*, used under terms of non-commercial reuse). Super-threshold 0.3 V input pulse (**a**) and its corresponding spike output (**b**). A sub-threshold 0.2 V input (**c**) to the same device yields a damped output (**d**). Phase portraits of the characteristic state variables u and q for channel 1 (**e**) and channel 2 (**f**) illustrate a stable trajectory for both channels during the spike activation period of (**b**). Points labeled α through ε on the phase portraits indicate the special points associated with switching events in each channel (**g**)

processing and transmission properties of a neuron; they exhibit all-or-nothing spiking of an action potential, a bifurcation threshold to a continuous spiking regime, signal gain, and a refractory period, so they can be the basis of active circuitry that does not contain transistors (Fig. 1.15). The combination of neuristors and memristors provides the components necessary to build transistor less integrated circuits that directly emulate the signal transmission and processing capabilities of brains, are orders of magnitude faster than biological systems, and are also extremely energy-efficient. They also provide an easily accessible experimental platform for building circuits that emulate different models of brain function to study issues such as how information processing emerges from the edge of chaos (Chua et al. 2012a, b).

Key memristor attributes that will contribute to future computing platforms are (a) low switching energy ≤100 attojoule (aJ), (b) fast switching time (≤100 ps), (c) real number representation (analog) in addition to binary bits, (d) demonstrated scalability into the <10 nm range, and (e) CMOS compatibility. Currently, the roadmap for memristor introduction in the marketplace consists of an initial FLASH competitor, succeeded by DRAM and SRAM nonvolatile solutions. This critical new area requires research focused on real-world applications, and system designs based on rigorous device and circuit models validated through experimental measurements, device parameter extraction, optimization, and tailoring for the specific goal of creating embedded systems with accelerated performance and dramatically reduced power.

Potential Benefits of Cognizant Systems

The ambition is to push beyond the Cloud and current information management technologies to develop cognition-based analytics that will provide real-time analysis and automated action from structured and unstructured data inputs. In short, we want to bring awareness to the Cloud by creating cognizant systems that understand location, context, and intent. These systems need to be self-learning and self-adapting to autonomously accomplish objectives in an uncertain and changing world. They must possess the ability to plan and navigate virtual and physical spaces by building and traversing an internal representation of the world. Autonomous actions resulting from learning and inference will fulfill the goal of augmenting scarce human analysts' expertise and capacity. The most radical new capability of these cognizant systems will be visual analytics that will solve important multidimensional topological problems not possible with existing statistical or image processing techniques. An example is, given a multipetabyte five-dimensional data set of a whole-body image of a patient (represented by three spatial coordinates and two or more colors for each voxel, for instance), to develop cognizant agents that can navigate through these n-dimensional worlds and locate likely physiological problems before they are recognizable to a human.

The potential benefits span a broad range of opportunities. By seamlessly integrating the physical and virtual world, we can apply cognitive systems to transform the "internet of things" into a "swarm of cognizant objects." The number of potential use cases is large:

- Infrastructure monitoring with proactive preventive maintenance or disaster mitigation for bridges, highways, buildings, and other critical infrastructure
- Public transportation systems monitoring to maximize efficiency and to mitigate loss during a natural disaster
- Traffic management and parking that reduces the time for people to travel from point A to point B and minimize carbon footprint by analyzing the current traffic as well as the parking available on the streets and in garages
- Water and food safety and distribution that ensures the timely delivery of necessary products without waste or contamination by inspecting products before, during, and after delivery from vendors in the supply chain
- Healthcare applications for increased efficiency and patient safety; examples include personalized medical treatment that factors in a patient's genome, history, ethnicity, and aggregated patient treatment histories
- Autonomous ground and aerial sensing platforms for public safety, and mobile sensing with the ability to learn and adapt to local environments as part of emergency response teams to find a lost child or search for victims of a hurricane, tornado, or earthquake

1.8.4 Biodesign Institute at Arizona State University

Contact person: *Stuart Lindsay, Arizona State University*

Conceived as a space for the convergence of knowledge and technology, the first half of Arizona State University's Biodesign Institute's planned 800,000 square feet of shared research space was completed in 2006, housing centers that span biomedicine, nanotechnology, and bioinformatics. The institute has had a profound impact on both translational research and interdisciplinary education. The forces that have driven this initiative are best understood in the context of the recent development of Arizona State University (ASU) itself. A decade ago, the university was known mainly for the size of its student population. Michael Crow took the presidency determined to use the university as a test-bed for a radical redesign of higher education ("The New American University"; http://newamericanuniversity.asu.edu/). With the triple goals of excellence, access, and impact, ASU's Challenges Project mission statement declares, "We measure ourselves by those we include, not those we exclude." The opposite view, "More will mean worse," was how Sir Kingsley Amis expressed his disdain of broadly inclusive education in an article in *Encounter* in July 1960. Today, with 72,000 students, an acceptance rate of 89 % of qualified applicants, and a Pell-grant population that exactly matches that predicted by the economic demography of Arizona, ASU certainly has the "more." But three Nobel laureates and a 2012 crop of Fulbright Scholars that ranks it fifth in the country (in company with University of California–Berkeley and Yale University) is an important demonstration that excellence and access are not mutually exclusive.

This theme plays into the design of research at the Biodesign Institute. Research that has "impact" not only yields economic and social benefits, but it also drives public understanding of the research enterprise and draws students into the process (http://uresearch.asu.edu/home). The Biodesign Institute was the first of several such initiatives at ASU specifically designed with the convergence of knowledge and technology in mind. These goals drive the architecture, staffing, and administration of the institute. Its achievements follow from its commitments.

Architecture. Large, open lab spaces, shared by multiple investigators, are enclosed in glass, allowing the research community to see and find each other at work. Facilities are clustered around an open atrium that forms a natural interaction space. Graduate students from many disciplines share common space. The labs themselves are designed to be as flexible as possible, with mobile benches and facilities that are readily moved to accommodate new projects. The building won *R&D Magazine's* "Lab of the Year" award when it opened in 2006.

Staffing. 26 tenured (or tenure-track) faculty members work in one or more of 11 research centers at the Biodesign Institute (http://www.biodesign.asu.edu/research/research-centers). Their faculty tenure homes include the School of Life Sciences; Physics, Chemistry, and Biochemistry; Electrical, Computer and Energy Engineering; Biological and Health Systems Engineering; Sustainable Engineering; Computer Informatics and Decision Engineering; Engineering of Matter, Transport, and Energy; and Medical Bioinformatics. An additional 32 research faculty members have institute appointments. The centers are supported by 140 research staff, nearly 100 graduate students, and 40 postdoctoral fellows. Nearly 100 undergraduates are involved in Biodesign Institute research projects.

Administration. All tenure appointments reside in academic departments reporting to the ASU Provost. Graduate degrees are granted by academic departments or, in the case of the in-house Ph.D. in "Biological Design," by the Graduate College. The Institute Director reports directly to the President of the University. The Deputy Director, responsible for day-to-day operations, is also general manager of Biodesign Commercial Translation, a partnership with Arizona Technology Enterprises (http://www.azte.com/), an independent limited liability corporation that handles intellectual property (IP) and develops business based on university research.

Achievements. The institute currently brings in over $48 million of external research funding annually. About 50 invention disclosures each year result in more than 30 patent applications. Nearly 40 companies sponsor research, ranging from giants like GE Healthcare and Hoffmann-La Roche to small start-ups. Ten companies have been spun out of the Institute, of which three have been acquired by major corporations (Nanobiomics by MPI, Intrinsic Bioprobes by Thermo-Fischer, and Molecular Imaging by Agilent Technologies). Major programs that develop vaccines for the third world, personalized medicine, sustainable health, and environmental biotechnology bring important social benefits not only to Arizona but to the United States and the world.

The impact on graduate education at ASU's Biodesign Institute has been profound. Asking the average physics student to prepare a DNA sample, or asking a chemistry student to build a sensitive amplifier, is difficult in traditional departments. In an interactive interdisciplinary institute, the students teach each other, and biologists learn to write Labview programs as readily as physics students learn to make aptamer molecules. This versatility is essential in preparing the workforce of the future.

1.9 International Perspectives

The following are summaries relevant to this chapter of discussions at the international regional WTEC NBIC2 workshops held in Leuven, Belgium, September 20–21, 2012; in Seoul, Korea, October 15–16, 2012; and in Beijing, China, October 18–19, 2012. Further details of those workshops are provided in Appendix A.

1.9.1 United States–European Union NBIC2 Workshop (Leuven, Belgium)

Two panels addressed topics related to emerging and foundational NBIC tools.

Panel on Converging Tools

Facilitator and Rapporteur: *H.-S. Philip Wong, Stanford University (U.S.)*

Discussants:

Mike Adams, University of Birmingham, Unilever (UK)
Jacqueline Allan, OECD-WPB (EU)
Paolo Milani, University of Milan (Italy)
Yves Samson, CEA (France)
Mamadou Diallo, Caltech and KAIST (U.S.)
Bruce Tonn, University of Tennessee (U.S.)
George Whitesides, Harvard University (U.S.)

Panel on Human Development and Convergence

Facilitator and Rapporteur: *Laura Ballerini, University of Trieste (Italy)*

Discussants:

Mira Kalish, Tel Aviv University (Israel)
Milos Nesladek, Academy of Sciences of the Czech Republic and Haselt University (Belgium)
Francoise Roure, OECD-WPN (France)
Sylvie Rousset, CNRS (France)
Jian Cao, Northwestern University (U.S.)
Mark Lundstrom, Purdue University (U.S.)
James Olds, George Mason University (U.S.)

This group of scientists assessed the state of the art in foundational science and technology tools. The focus of the discussions was on manufacturing, education, and research infrastructure.

Convergence has been defined as "enabling technologies and knowledge systems that enable each other in the pursuit of a common goal" (Nordmann 2004). In the last decade, information technology has accelerated the globalization of the economy. Manufacturing has become a global activity, linked by tools that enable workers to communicate and transmit large amounts of information across national boundaries in real time. Increasingly, not only the big businesses are global in nature, even the small and medium-sized companies now also operate on a global scale. The notion of "shipping jobs overseas" is becoming out of date because new businesses are essentially made possible by a global work force and economic infrastructure that are optimized for specific sectors. In Europe as in the United States, the transition to a knowledge-based economy and a service-oriented economy is happening at a rapid pace.

Education as a tool to advancing human productivity and job training and as a resource for innovation will be of central importance in the decade ahead. Advances in information technology, enabled by nanotechnology, will lead to new modalities of education such as online education. In analogy with personalized medicine, we are seeing the emergence of personalized education, both in the conventional education setting and in life-long continuing education.

1 Convergence Platforms: Foundational Science and Technology Tools   45

Because of increased use of automation by all organizations, high-school education needs to be rethought to include vocational training that addresses the machine–human interface. At the Ph.D. level, converging technologies will require teaming and multidisciplinary perspectives; this will require a different approach than those presently used by most universities.

In the last decade, Europe has invested in key research infrastructures that nurture NBIC convergence. A prime example is IMEC (formerly the Interuniversity Microelectronics Centre) in Belgium, which served as the host of our meeting. With an annual budget of €600 million, IMEC has emerged as a leading research institution in NBIC technologies with strong support from a large industrial base. IMEC has become the leading institution in nanoelectronics research in the world, with state-of-the-art tools and equipment that parallel those found in the most advanced industrial R&D organizations. Researchers are able to utilize the advanced technology platform to pursue research topics beyond applications in information technology—in energy, biomedicine, and healthcare delivery, for example. The workforce is truly international and it is well-integrated with the education system in Europe.

In the last 10 years, IMEC has made strategic investments, for example, in converging neuroscience with nanotechnology. It made five faculty hires and created the NeuroElectronics Research Flanders laboratory (http://www.nerf.be/) that specifically focuses on applying the techniques of nanotechnology (such as nanostructures and nanoelectronics integration with sensors) to the study of neuroscience. In contrast, such investments at the national level are lacking in the United States.

1.9.2 United States–Korea–Japan NBIC2 Workshop (Seoul, Korea)

Panel members/discussants:

Kuang Reyol (*co-chair*), *Korea Advanced Institute of Science and Technology* (*KAIST, Korea*)
Tsuyohsi Hasegawa, National Institute for Materials Science (*Japan*)
Mark Lundstrom, Purdue University (*U.S.*)

This group discussed foundational science and technology, focusing on tools. Over the past decade, the panel members noted a significant increase in the importance of tools to analyze massive amounts of data, tools for imaging and visualization, and an increasing use of tools to deal with design complexity (e.g., integrated circuit chip design). Looking to the next decade, the panel saw a central role for tools to (1) catalyze the ability of converging technologies to benefit society and accelerate the overarching goals of NBIC2, (2) help innovation keep pace with rapidly advancing science, and (3) achieve the convergence of real and virtual spaces.

Goals for the next decade were also discussed. The development of coupled tools for computation, synthesis, and processing and characterization were highlighted (such tools would lead to so-called "virtual fabs"). The wider use of brain-inspired architectures for computing (e.g., for driverless cars) was also identified as a goal, as were new design tools that couple electronics and biology (e.g., for applications such as brain–machine interfaces), and tools for understanding the phenomena of emergence (as in the so-called flash crash of the New York Stock Exchange).

In the area of infrastructure for convergence, the panel saw a need for international infrastructure for managing big science data (e.g., storage, curation, and access).

1.9.3 United States–China–Australia–India NBIC2 Workshop (Beijing, China)

Panel members/discussants:

Chen Wang (co-chair), National Center for Nanoscience and Technology (China)
Chennupati Jagadish, (co-chair) Australian National University (Australia)
Mark Lundstrom, Purdue University (U.S.)

Panel 1 addressed foundational knowledge and tools. One issue that the panel struggled with from the very first meeting in Washington was to understand exactly how "convergence" differed from "multidisciplinary research." In this regard, the plenary talk by Prof. Jonathan H. Manton, University of Melbourne, Australia, was relevant and effectively set the stage for the panel's discussion. He labeled research beyond multidisciplinary research as "transdisciplinary research," which he defined as follows:

> Transdisciplinary research is research that does not merely borrow methods from one discipline and apply them to another. Rather it is research that, by necessity, requires the concerted efforts of a team of experts from across disciplines to create new techniques that transcend disciplinary boundaries for tackling problems which are between, across, and beyond each individual discipline.

Manton went on to quote Norbert Wiener, the founder of the science of cybernetics. Wiener wrote that it is the "boundary regions of science which offer the richest opportunities to the qualified investigator. They are at the same time the most refractory to the accepted techniques of mass attack and the division of labor" (Wiener 1948, 12). Manton noted that Wiener went on to explain,

> A proper exploration of these blank spaces on the map of science could only be made by a team of scientists, each a specialist in his own field but each possessing a thoroughly sound and trained acquaintance with the fields of his neighbors; all in the habit of working together, of knowing one another's intellectual customs, and of recognizing the significance of a colleague's new suggestion before it has taken on a full formal expression. The mathematician need not have the skill to conduct a physiological experiment, but he must have the skill to understand one, to criticize one, and to suggest one. The physiologist need not be able to prove a certain mathematical theorem, but he must be able to grasp its physiological significance and to tell the mathematician for what he should look.

As Manton pointed out, success in converging knowledge and technologies will require a different style of research; multidisciplinary teamwork provides opportunities

1 Convergence Platforms: Foundational Science and Technology Tools   47

to foster new transdisciplinary fields. For teams to be successful, the knowledge of each team member must span multiple disciplines. Great effort is required to build such a team. New educational strategies and new approaches to supporting research and development will be needed to create, nurture, and sustain such teams.

This panel's discussions expanded upon many of the points raised by the European and Korean panels. The discussion first addressed the changes in the vision over the past decade. Neuroscience went from using tools to study rats to new tools to study and interact with human brains, and as a result, cognitive science began to connect with nanotechnology, biotechnology, and information technology. Biology became "industrialized" (e.g., as for gene sequencing and as opposed to particle physics), which took biology from small-scale to large-scale science. Finally, an increasing reliance on tools to deal with complexity (e.g., design of integrated circuit processes and chips) was noted. Looking to the next decade, this panel also saw a central role for tools:

- To address issues relating to the aging society (e.g., better therapies, mobile health services, driverless vehicles, etc.)
- To transform the way we manage and treat mental illness
- To deal with complex systems and to help innovation keep pace with rapidly advancing science (including brain–machine interfaces)
- To achieve the convergence of real and virtual spaces (telepresence and teleoperation)
- Massive industrialization of nanotechnology (e.g., nanocatalysts, electrodes for fuel cells, patterning without lithography)
- Ubiquitous, networked, intelligent sensors

The panel also discussed the key advances over the past decade. The design of nanomaterials and nanofabrication of devices advanced from single component to multicomponent and even integrated devices. The combination of artificial systems and biological systems has been increasing.

Finally, another significant advance was in the development of noninvasive brain imaging, intervention, and interaction and their application to human cognition. Among the several success stories of convergence over the past decade were (1) cochlear implants, (2) smart phones, (3) adaptive optics, and (4) weather modeling.

The panel also identified six goals for the next decade:

1. Develop tools to address the challenges of an aging society
2. Develop tools to treat mental illness
3. Develop new tools to understand human cognition (brain–machine interfaces embedded in everyday life)
4. Massive industrialization of nanotechnology, e.g., nanomaterials, nano-ink-printing/green printing, nanocatalysts, nanomedicines
5. Single-cell biology and technology (from patch clamp to laser-guided protein extraction)
6. Focus on fundamental quantitative biology sciences, the mechanisms at the molecular level

The panel identified a need for three new types of research infrastructure:

1. Enhance capabilities of high-performance computing and data mining
2. Fabrication and processing facilities for hybrid CMOS + bio
3. Tools for imaging and interacting with the brain with three orders of magnitude improvement in spatial and temporal resolution

and three R&D strategies for advancing convergence:

1. Programs to grow and strengthen transdisciplinary research
2. Drive research by identifying and supporting high-impact projects
3. Jointly fund international collaboration in mission-driven research areas.

References

Abrams, Z.R., Warrier, A., Trauner, D., Zhang, X.: A signal processing analysis of Purkinje cells in vitro. F. Neural Circ. **4**, Article 13 (2010)

Alivisatos, A.P., Chun, M., Church, G.M., Greenspan, R.J., Roukes, M.L., Yuste, R.: The brain activity map project and the challenge of functional connectomics. Neuron **74**(6), 970–974 (2012). http://dx.doi.org/10.1016/j.neuron.2012.06.006

Amir, E., Maynard-Zhang, P.: Logic-based subsumption architecture. Artif. Intell. **153**, 167–237 (2004)

Ananthanarayanan, R., Esser, S.K., Simon, H.D., Modha, D.S.: The cat is out of the bag: cortical simulations with 10^9 neurons, 10^{13} synapses. In: Proceedings of the ACM/IEEE Conference on Supercomputing (Portland, OR, Nov. 14–20), pp. 1–12. ACM, New York (2009)

Bandettini, P.A.: Functional MRI: a confluence of fortunate circumstances. NeuroImage **63**, 1712 (2012) (Feb. 6, Epub)

Banghart, M., Borges, K., Isacoff, E.Y., Trauner, D., Kramer, R.H.: Light-activated ion channels for remote control of neuronal firing. Nat. Neurosci. **7**, 1381–1386 (2004)

Binnig, G., Quate, C.F.: Atomic force microscope. Phys. Rev. Lett. **56**, 930–933 (1986)

Binnig, G., Rohrer, H., Gerber, C., Weibel, E.: Surface studies by scanning tunneling microscopy. Phys. Rev. Lett. **49**, 57–61 (1982)

Boyden, E.S., Zhang, F., Bamberg, E., Nagel, G., Deisseroth, K.: Millisecond-timescale, genetically targeted optical control of neural activity. Nat. Neurosci. **8**(9), 1263–1268 (2005)

Chua, L.O.: Memristor: the missing circuit element. IEEE Trans. Circuit Theory **18**, 507 (1971)

Chua, L., Sbitnev, V., Kim, H.: Hodgkin-Huxley axon is made of memristors. Int. J Bifurcation Chaos **22**, art. #1230011 (2012a)

Chua, L., Sbitnev, V., Kim, H.: Neurons are poised near the edge of chaos. Int. J. Bifurcation Chaos **22**, art. #1250098 (2012b)

Datta, S.: Electronic Conduction in Mesoscopic Systems. Cambridge University Press, Cambridge (1997)

Datta, S., Tian, W., Hong, S., Reifenberger, R., Henderson, J.I., Kubiak, C.P.: Current-voltage characteristics of self-assembled monolayers by scanning tunneling microscopy. Phys. Rev. Lett. **79**, 2530–2533 (1997)

DHS (Department of Homeland Security): Solicitation BAA-11-02, Jan. 26, 2011, to improve security in both Federal networks and the larger Internet in 14 Technical Topic Areas. https://www.fbo.gov/index?s=opportunity&mode=form&id=40161dd972cd60642ecaaa955e247067&tab=core&_cview=1 (2011)

Fang, N., Lee, H., Sun, C., Zhang, X.: Sub-diffraction-limited optical imaging with a silver superlens. Science **308**(5721), 534–537 (2005)

Fehrentz, T., Schönberger, M., Trauner, D.: Optochemical genetics. Angew. Chem. Int. Engl. **50**, 12156–12182 (2011)

Feynman, R.P.: There's plenty of room at the bottom. Eng. Sci. **24**(2), 22–26 (1960). Available Online: http://www.its.caltech.edu/~feynman/plenty.html

FNIH (Foundation for the National Institutes of Health): Observational Medical Outcomes Partnership Annual Symposium, 28 June 2012, Bethesda, MD. Available online: http://omop.fnih.org/2012SymposiumPresentations (2012)

Foucher, J.R., Luck, D., Chassagnon, S., Offerlin-Meyer, I., Pham, B.-T.: What is needed for rTMS to become a treatment? Encephale **33**, 982–989 (2007)

Fox, M.D., Snyder, A.Z., Vincent, J.L., Corbetta, M., Essen, D.C.V., et al.: The human brain is intrinsically organized into dynamic, anticorrelated functional networks. Proc. Natl. Acad. Sci. U. S. A. **102**, 9673–9678 (2005)

Furber, S.: Low-power chips to model a billion neurons. IEEE Spectrum August. http://spectrum.ieee.org/computing/hardware/lowpower-chips-to-model-a-billion-neurons (2012)

Gantz, J., Reinsel, D.: Extracting value from chaos. IDC IVIEW (June). http://www.emc.com/leadership/programs/digital-universe.htm (2011)

Gimsa, J., Habel, B., Schreiber, U., Rienen, U.V., Strauss, U., Gimsa, U.: Choosing electrodes for deep brain stimulation experiments—electrochemical considerations. J. Neurosci. Methods **142**(2), 251–265 (2005)

Glotzer, S., Kim, S., Cummings, P.T., Deshmukh, A., Head-Gordon, M., Karniadakis, G., Petzold, L., Sagui, C., Shinozuka, M.: International Assessment of Research and Development in Simulation-Based Engineering and Science. WTEC, Inc., Baltimore (2009)

Hall, D.A., Gaster, R.S., Lin, T., Osterfeld, S.J., Han, S., Murmann, B., Wang, S.X.: GMR biosensor arrays: a system perspective. Biosens. Bioelectron. **25**, 2051–2057 (2010)

Head-Gordon, M.: Chapter 6: Software development. In: Glotzer, S., Kim, S., Cummings, P.T., Deshmukh, A., Head-Gordon, M., Karniadakis, G., Petzold, L., Sagui, C., Shinozuka, M. (eds.) International Assessment of Research and Development in Simulation-Based Engineering and Science. WTEC, Inc., Baltimore (2009)

Hoddeson, L., Braun, E., Teichmann, J., Weart, S.: Out of the Crystal Maze. Oxford University Press, London/New York (1992) (Refer to a 1931 letter to Rudolph Peierls from Wolfgang Pauli)

Hofstein, S.R., Heiman, F.P.: The silicon insulated-gate field-effect transistor. Proc. IEEE **9**, 1190–1202 (1963)

Hopkins, B., Evelson, B.: Expand Your Digital Horizon with Big Data. Forrester Research, Inc. (2011). Available at http://www.forrester.com/Expand+Your+Digital+Horizon+With+Big+Data/fulltext/-/E-RES60751?docid=60751

IDC: Press Release, April 30. Available online at: http://www.idc.com/getdoc.jsp?containerId=prUS23457712 (2012)

Jo, S.H., Chang, T., Ebong, I., Bhadviya, B.B., Mazumder, P., Lu, W.: Nanoscale memristor device as synapse in neuromorphic systems. Nano Lett. **10**(4), 1297–1301 (2010)

Johnson, K.A., Fox, N.C., Sperling, R.A., Klunk, W.E.: Brain imaging in Alzheimer disease. Cold Spring Harb. Perspect Med. **2**, a006213 (2012)

Kawahara, T., Takemura, R., Miura, K., Hayakawa, J., Ikeda, S., Lee, Y.M., Sasaki, R., Goto, Y., Ito, K., Meguro, T., Matsukura, F., Takahashi, H., Matsuoka, H., Ohno, H.: 2 Mb SPRAM (SPin-Transfer Torque RAM) with bit-by-bit bi-directional current write and parallelizing-direction current read. IEEE J. Solid-State Circ. **43**, 109–120 (2008)

Klimeck, G., McLennan, M., Brophy, S.B., Adams III, G.B., Lundstrom, M.S.: nanoHUB.org: advancing education and research in nanotechnology. Comput. Sci. Eng. **10**, 17–23 (2008)

Knorr, E., (ed.): The enterprise data explosion. InfoWorld Digital Spotlight (2012). Available at http://www.infoworld.com/d/data-explosion/download-the-enterprise-data-explosion-digital-spotlight-199696

Kogge, P., et al.: Exascale computing study: technology challenges in achieving exascale systems. DARPA IPTO Report, AFRL contract # FA8650-07-C-7724. http://users.ece.gatech.edu/mrichard/ExascaleComputingStudyReports/exascale_final_report_100208.pdf (2008)

Kuzum, D., Jeyasingh, R.G.D., Wong, H.-S.P.: Energy efficient programming of nanoelectronic synaptic devices for large-scale implementation of associative and temporal sequence learning. IEEE International Electron Devices Meeting (IEDM), December 5–7, Washington, DC, paper 30.3, 693–696. http://dx.doi.org/10.1109/IEDM.2011.6131643 doi:10.1109/IEDM.2011.6131643 (2011)

Kuzum, D., Jeyasingh, R.G.D., Lee, B., Wong, H.-S.P.: Nanoelectronic programmable synapses based on phase change materials for brain-inspired computing. Nano Lett. **12**(5), 2179–2186 (2012). http://dx.doi.org/10.1021/nl201040y

Lake, R.G., Klimeck, R.C.B., Jovanovic, D.: Single and multiband modeling of quantum electron transport through layered semiconductor devices. J. Appl. Phys. **81**, 7845–7869 (1997)

Lewin, T.: Universities reshaping education on the Web. N.Y. Times, 17 July 2012. Available online: http://www.nytimes.com/2012/07/18/education/top-universities-test-the-online-appeal-of-free.html

Liu, Z.W., Lee, H., Xiong, Y., Sun, C., Zhang, X.: Far-field optical hyperlens magnifying sub-diffraction-limited objects. Science **315**(5819), 1686–1686 (2007)

Lojek, B.: History of Semiconductor Engineering. Springer, New York (2007)

Manganas, L.N., Zhang, X., Li, Y., Hazel, R.D., Smith, D.S., Wagshul, M.E., Benveniste, H., Djuric, P.M., Enikolopov, G., Maletic-Savatic, M.: Magnetic resonance spectroscopy identifies neural progenitor cells in the live human brain. Science **318**, 980–985 (2007)

Manyika, J., Chui, M., Brown, B., Bughin, J., Dobbs, R., Roxburgh, C., Hung Byers, A.: Big data: The next frontier for innovation, competition and productivity. McKinsey Global Institute. Available online: http://www.mckinsey.com/Insights_and_Publications/MGI/Research/Technology_and_Innovation (2011)

Mardis, E.R.: Next-generation DNA sequencing methods. Annu. Rev. Genom. Hum. Genet. **9**, 387–402 (2008)

Mardis, E.R.: A decade's perspective on DNA sequencing technology. Nature **470**, 198–203 (2011)

Mayberry, M.: Peering through the technology scaling fog. In: Symposium VLSI Technology, Honolulu, Hawaii, pp. 1–4 (2012)

Mazumder, P., Kang, S.M., Waser, R.: Memristors: devices, models, and applications. Proc. IEEE **100**(6), 1911–1919 (2012). http://dx.doi.org/10.1109/JPROC.2012.2190812

McLennan, M., Kennell, R.: HUBzero: a platform for dissemination and collaboration in computational science and engineering. Comput. Sci. Eng. **12**(2), 48–52 (2010)

Merolla, P., Arthur, J., Akopyan, F., Imam, N., Manohar, R., Modha, D.S.: A digital neurosynaptic core using embedded crossbar memory with 45pJ per spike in 45nm. In: Proceedings of the IEEE Custom Integrated Circuits Conference, 19–21 Sept. 2011, pp. 1–4. http://dx.doi.org/10.1109/CICC.2011.6055294 (2011)

Metzker, M.L.: Sequencing technologies—the next generation. Nat. Rev. Genet. **11**, 31–46 (2010)

Miller, B.P.: Vulnerability assessment of open source Wireshark. Presentation at the DHS Science Technology Cyber Forensics Research & Development Technology Update and New Requirements Workshop, October 2010, Rosslyn, VA (2010)

Mistry, K., et al.: A 45 nm logic technology with high-k+ metal gate transistors, strained silicon, 9 Cu interconnect layers, 193 nm dry patterning, and 100% Pb-free packaging. IEEE International Electron Devices Meeting Technical Digest 247–250 (2007). http://dx.doi.org/10.1109/IEDM.2007.4418914

NAE (National Academy of Engineering of the National Academies): Grand Challenges for Engineering. National Academy of Sciences, Washington, DC (2008). Available online: http://www.engineeringchallenges.org

Nordmann, A.: Converging technologies – Shaping the future of European societies. European Commission Report. Available online: http://www.ntnu.no/2020/final_report_en.pdf (2004)

Ohno, T., Hasegawa, T., Tsuruoka, T., Terabe, K., Gimzewski, J.K., Aono, M.: Short-term plasticity and long-term potentiation mimicked in single inorganic synapses. Nat. Mater. **10**(8), 591–595 (2011)

Pickett, M.D., Medeiros-Ribeiro, G., Williams, R.S.: A scalable neuristor built with Mott memristors. Nat. Mater. **12**(2), 114–117 (2013). http://dx.doi.org/10.1038/nmat3510

Reeke Jr., G.N., Sporns, O.: Behaviorally based modeling and computational approaches to neuroscience. Annu. Rev. Neurosci. **16**, 597–623 (1993). http://dx.doi.org/10.1146/annurev.ne.16.030193.003121

Ritala, M., Lerskela, M.: Chapter 2: Atomic layer deposition. In: Nalwa, H.S. (ed.) Handbook of Thin Films. Academic Press, San Diego (2002)

Roco, M.C.: Chapter 24: Technology convergence. In: Bainbridge, W.S. (ed.) Leadership in Science and Technology: A Reference Handbook. Sage Publications, Thousand Oaks (2012)

Roco, M.C., Bainbridge, W.S. (eds.): Converging Technologies for Improving Human Performance: Nanotechnology, Biotechnology, Information Technology, and Cognitive Science. Kluwer Academic, Dordrecht (2003)

Roco, M.C., Mirkin, C.A., Hersam, M.C.: Nanotechnology Research Directions for Societal Needs in 2020: Retrospective and Outlook. Science Policy Report. Springer, Dordrecht (2011)

Rolls, E.T., Treves, A.: The neuronal encoding of information in the brain. Prog. Neurobiol. **95**, 448–490 (2011)

Rothberg, J., et al.: An integrated semiconductor device enabling non-optical genome sequencing. Nature **475**, 348–352 (2011). http://dx.doi.org/10.1038/nature10242

Sah, C.T.: Characteristics of the metal-oxide-semiconductor transistors. IEEE Trans. Electron Dev. **11**, 324–345 (1964)

Samuels, J.D.: Semiconductors and U.S. economic growth. Draft paper, April 1. Available online: http://www.sia-online.org/clientuploads/directory/DocumentSIA/ecoimpactsemidraft_Samuels.pdf (2012)

Seo, J., Brezzo, B., Liu, Y., Parker, B.D., Esser, S.K., Montoye, R.K., Rajendran, B., Tierno, J.A., Chang, L., Modha, D.S., Friedman, D.J.: A 45nm CMOS neuromorphic chip with a scalable architecture for learning in networks of spiking neurons. In Proceedings of the IEEE Custom Integrated Circuits Conference 2011, pp. 1–4, 19–21 (2011). http://dx.doi.org/10.1109/CICC.2011.6055293

Shouval, H.Z., Wang, S.S.-H., Wittenberg, G.M.: Spike timing dependent plasticity: a consequence of more fundamental learning rules. Front. Comput. Neurosci. **4**, 19 (2010). http://dx.doi.org/10.3389/fncom.2010.00019

SIA (Semiconductor Industry Association): The International Technology Roadmap for Semiconductors, 2011th edn. SEMATECH, Albany (2011)

Snider, G.S.: Self-organized computation with unreliable, memristive nanodevices. Nanotechnology **18**(36), 365202 (2007)

Snider, G.: Massively parallel computing on Cog ex Machina. HP Lab. Tech. Rep. **179**, 1–11 (2012). http://www.hpl.hp.com/techreports/2012/HPL-2012-179.html

Snider, G., Amerson, R., Carter, D., Abdalla, H., Qureshi, M.S., Léveillé, J., Versace, M.: From synapses to circuitry: using memristive memory to explore the electronic brain. Computer **44**(2), 21–28 (2011)

Srituravanich, W., Pan, L., Wang, Y., Sun, C., Bogy, D.B., Zhang, X.: Flying plasmonic lens in the near field for high-speed nanolithography. Nat. Nanotechnol. **3**, 733–737 (2008)

Stokes, D.E.: Pasteur's Quadrant: Basic Science and Technological Innovation. Brookings Institution Press, Washington, DC (1997)

Strukov, D.B., Snider, G.S., Stewart, D.R., Williams, R.S.: The missing memristor found. Nature **453**, 80–83 (2008)

Studer, B., Apergis-Schoute, A.M., Robbins, T.W., Clark, L.: What are the odds? The neural correlates of active choice during gambling. Front Neurosci. **6**, 46 (2012). http://dx.doi.org/10.3389/fnins.2012.00046

Suri, M., Bichler, O., Querlioz, D., Cueto, O., Perniola, L., Sousa, V., Vuillaume, D., Gamrat, C., DeSalvo, B.: Phase change memory as synapse for ultra-dense neuromorphic systems: Application to complex visual pattern extraction. In: Proceedings of the International Electron Devices Meeting (IEDM) 2011 IEEE International (Washington, DC, December 5–7), paper 4.4, pp. 4.4.1–4.4.4 (2011)

Tedeschi, G., Bonavita, S., Banerjee, T.K., Virta, A., Schiffmann, R.: Diffuse central neuronal involvement in Fabry disease: a proton MRS imaging study. Neurology **52**, 663 (1999). http://dx.doi.org/0.1212/WNL.52.8.1663

Theogarajan, L.: Strategies for restoring vision to the blind: current and emerging technologies. Neurosci. Lett. **519**, 129–133 (2012)

Turrigiano, G.G., Nelson, S.B.: Homeostatic plasticity in the developing nervous system. Nat. Rev. Neurosci. **5**, 97–107 (2004)

Valentine, J., Zhang, S., Zentgraf, T., Ulin-Avila, E., Genov, D.A., Bartal, G., Zhang, X.: Three-dimensional optical metamaterial with a negative refractive index. Nat. Lett. **455**(7211), 376–379 (2008)

Waltz, E.: Reading and writing a book with DNA. IEEE Spectrum (online) August. http://spectrum.ieee.org/biomedical/imaging/reading-and-writing-a-book-with-dna (2012)

Wang, S., Szobota, S., Wang, Y., Volgraf, M., Liu, Z., Sun, C., Trauner, D., Isacoff, E.Y., Zhang, X.: All optical interface for parallel, remote, and spatiotemporal control of neuronal activity. Nano Lett. **7**, 3859 (2007)

Wiener, N.: Cybernetics: Or Control and Communication in the Animal and the Machine, 2nd edn. The Massachusetts Institute of Technology Press, Cambridge, MA (1948)

Wolfe, J.M., Horowitz, T.S.: What attributes guide the deployment of visual attention and how do they do it? Nat. Rev. Neurosci. **5**, 495–501 (2004)

Yu, S., Wu, Y., Jeyasingh, R.G.D., Kuzum, D., Wong, H.-S.P.: An electronic synapse device based on metal oxide resistive switching memory for neuromorphic computation. IEEE Trans. Electron Devices **58**(8), 2729–2737 (2011)

Chapter 2
Convergence Platforms: Human-Scale Convergence and the Quality of Life

Donald MacGregor, Marietta Baba, Aude Oliva,
Anne Collins McLaughlin, Walt Scacchi, Brian Scassellati,
Philip Rubin, Robert M. Mason, and James R. Spohrer

The greatest benefits to human knowledge and well-being will come from convergence of all fields of science, with nanotechnology, biotechnology, information technology, and cognitive science (NBIC) being the core foundational quartet of fields drawing all the others together. But at particular points in time, one field may be having disproportionate impacts on society, even as the others are gathering strength through fundamental scientific research. The NBIC field experiencing the most rapid application changes and having the most potential for continued change in its direct effects on human lives in the coming decade appears to be *information*

Corresponding editors M.C. Roco (mroco@nsf.gov) and W.S. Bainbridge (wbainbri@nsf.gov).

D. MacGregor
MacGregor Bates, Inc., Cottage Grove, OR, USA

M. Baba
Michigan State University, East Lansing, MI, USA

A. Oliva
Massachusetts Institute of Technology, Cambridge, MA, USA

A.C. McLaughlin
North Carolina State University, Raleigh, NC, USA

W. Scacchi
University of California, Irvine, CA, USA

B. Scassellati
Yale University, New Haven, CT, USA

P. Rubin
Office of Science and Technology Policy, Washington, DC, USA

R.M. Mason
University of Washington, Seattle, WA, USA

J.R. Spohrer
IBM Global University Programs, San Francisco, CA, USA

technology, although nanoscience, biology, and cognitive science are also progressing rapidly and will contribute through their convergence. It is entirely possible that fundamental scientific advance is moving more rapidly at the nanoscale, whether reflected in biotechnology or nanotechnology, and information technology is merely going through a stage in its application to human problems where the impacts are both huge and highly visible.

Not only do vast numbers of people belong to online communities, but they also carry the equivalent of computers around with them, whether in the form of smart phones, handhelds, or tablets. In addition, information technology is changing the relations that ordinary people experience with governments and with corporations, as well as globalizing the relations between organizations both large and small. However, information technology cannot progress in isolation from the other NBIC fields, or in ignorance of its ethical, legal, and social implications. Therefore, this chapter examines the general topic of human–technology relationships, with a vantage point from information technology, but stressing convergence across multiple fields.

Electronics is the obvious meeting ground between information technology and nanotechnology, and the relationship between the two is rather mature at this point, although still offering many opportunities for progress. A few years ago, the nanoscale phenomenon called the giant magnetoresistance effect was exploited to give computer hard drives the ability to handle greater amounts of information and move it more quickly, but now solid-state hard drives are becoming common and inexpensive, handling input and output vastly faster in the absence of a disk that must be rotated, and employing electronic components with nanoscale dimensions. This chapter notes connections to biotechnology as well, but since its theme is the human dimensions of NBIC, much attention will be given to cognitive science and the social sciences.

Since the beginning of formal NBIC efforts over a decade ago, there has been an awareness that the social sciences need to be included, probably through their connections to information science and cognitive science. A different conception of how that might happen has arisen very recently, as increasing numbers of research projects enlist nonscientists in what is often called "citizen science." This significant shift has positive educational implications, because the ordinary citizens who contribute their labor learn about the science. But more than that, this shift represents a convergence of science and technology with society and thus may offer an entirely new way in which the societal implications of technology progress can be addressed to achieve progress for society by embedding science and technology more solidly in society.

Convergence platforms offer an opportunity to accelerate the scaling of the benefits of new knowledge globally and rapidly (Gawer 2009). For example, industry platforms include both technological service platforms such as smart phones for scaling "new apps" globally and rapidly (e.g., Apple iPhone, Google Android) as well as organizational service platforms such as franchises for scaling "new offerings" globally and rapidly (e.g., Starbucks). In addition, IBM and other global IT businesses are working on diverse types of "Smarter Cities Intelligent Operations

Center" platforms to scale up the benefits of new urban service innovations, across industries, globally and rapidly. These platforms depend on intelligent infrastructure of converged nano-bio-info-cogno capabilities. For example, the "digital baby" offering can be deployed to identify health problems in premature babies in hospitals where the needed platform combines nanosensors, bio-pattern databases, mobile cognitive assistance alerts, all integrated in a shared healthcare cloud-computing environment. As convergence platforms enable the rapid spread of service innovations globally, the benefits of diverse crowd-funded university-based startups can more easily be compared and evaluated. The rise of convergence platforms may foreshadow an increase in faculty and student teams working on local community-based collaborative innovation and service learning projects, with the potential to positively impact local quality of life (Baldwin and von Hippel 2011). The large global vendors may then compete to be the scale-up partner of choice for these university-based startups and open service innovations (Chesbrough 2011).

This chapter explores specific investigative and transforming approaches and overall characteristics of the human-scale platform of converging advances in science and technology. Its connections to other converging platforms (foundational tools, Earth-scale, and societal-scale) also are discussed.

2.1 Vision

2.1.1 Changes in the Vision Over the Past Decade

The past decade has seen two major shifts relevant to information technology, reinforcing each other but conceptually distinct. The first major trend is captured in such buzzwords as "Web 2.0" to designate a new system in which a billion people create online content to share with each other, rather than the original situation in which a few large companies or government agencies provided the content. Of course, this term is something of a misnomer, because the original vision of the World Wide Web enunciated over two decades ago by Tim Berners-Lee imagined that content on the Web would be created by all the users (Berners-Lee and Fischetti 1999). But from the days of Colossus and ENIAC in the 1940s, the dominant computers were the largest machines, built and operated by large bureaucracies. The development of personal computers in the 1970s was a milestone on the route to democratic computing (Freiberger and Swaine 1984), but even today much online content comes from the modern equivalents of broadcasting companies. Thus the democratization of the Internet is a gradual process, but one that has made great progress since the earliest NBIC reports.

New interaction methods are being developed at the interpersonal level, and in human–environment, human–machine, and individual–social media interactions. This is particularly evident in the case of open source software (Scacchi et al. 2010; Schweik and English 2012). The most influential early example is the Linux operating system, which runs on a number of hardware platforms, which began with a kernel released in 1991 by Linus Torvalds of Finland and has since been expanded by a large community of volunteer programmers. The software infrastructure for the World Wide Web also emerged as open-source software (Jensen and Scacchi 2005) prior to its early commercialization efforts (by Microsoft, Netscape Communications, and others); through ongoing efforts by the Mozilla Foundation, Apache Foundation, and others, much of the Web still relies on open source software. Since Linux and the Web led the way, participatory open source systems and approaches to software development have become a major methodology for creating, deploying, and sustaining participatory systems for NBIC research and development communities. Open source systems and software have transformed software-intensive industries and institutions, software development practices, and global socio-technical ecosystems.

The second major trend is a reconceptualization of computing as a service, rather than as the sale of hardware and software products, foreshadowed in the original NBIC reports (Roco and Bainbridge 2003; Roco and Montemagno 2004; Bainbridge and Roco 2006a, b), but further advanced today. Technologies are owned by service system entities, such as people, businesses, universities, and nations. Service can be defined as the application of knowledge belonging to one person for the benefit of another person. Thus, service is governed not only by software and hardware, but also by the rules of the social system in which the service provider and the customer interact. Rules are a special type of symbolic knowledge that help to govern complex systems. The co-evolution of technologies and rules is key to policymaking in the future.

2.1.2 The Vision for the Next Decade

Scientific research performed according to the open source paradigm will accelerate the development, reproduction, adaptation, and replication of participatory organizational forms. A crucial goal for the coming decade is to establish open source and adaptive models, representations, and interaction protocols that can enhance converging technology discovery, education, innovation, informatics, and commercialization within a globally self-organizing and self-regulating ecosystem.

Convergence at any scale will greatly benefit from both (1) transparent, open source models and process representations that can be expressed in both human-readable and computational forms (Scacchi et al. 2010), and (2) models and representations that can be visualized and computationally simulated to help people understand and explore the features and limitations of the models and representations (Scacchi 2012). The openness realized through open-source approaches enables both reproducibility and more rapid diffusion and transfer of concepts, techniques, and tools across organizations and disciplines. Both are critical to scientific advancement. They also encourage open access to scientific results, research methods, and data for sharing, which improves the research impact (Gargouri et al. 2010; Piwowar 2011).

Investment and implementation strategies should therefore address computational tools and techniques that embrace open source expression and collective articulation of the scientific models used to convey observable principles and practices of converging technologies. Governance methods should embrace self-organizing practices in terms of the socio-technical systems that emerge to create, continuously refine, and evolve the converging technologies at different scales of interest. Thus, open source is not merely a way of harnessing the talent and energies of computer programmers outside a formal organizational structure, but a principle for doing technical work that could be applied across all NBIC fields.

An important step toward interdisciplinary work and converging knowledge will come from technology. There are limits on our capabilities to understand numerous data points and phenomena across domains and disciplines. Current tools at our disposal are information visualizations, statistical packages, and mathematical modeling to help our minds detect patterns. However, as data grow ever larger, we need better visualizations that can be used across different fields with wildly different types and amounts of data. We need easy ways to do this, rather than individuals trying to cobble these together from other tools.

2.2 Advances in the Last Decade and Current Status

Contact: *W. S. Bainbridge, National Science Foundation*

Within the field of information technology, well-established areas have seen incremental progress, and the most rapid progress seems to have been in areas related to social computing, where a number of the developments are controversial. Considering NBIC more broadly, fields are becoming increasingly interdisciplinary, but rewards for such work are not keeping up with the demand for it. There are serious barriers that must be overcome to encourage such work. Political and economic divergence is the most significant change in the wider environment in which research and development take place. That is, even as we see great opportunities for unification within science, disunity is growing in society; thus, *convergence must combat disorganization*.

For convergence to combat disorganization, the challenge will be to ensure that all major urban regions of the world have the needed convergence platforms to rapidly scale up the service innovations that positively impact quality of life. A hopeful sign is that the adoption rate of mobile phones and smart phones in developing nations is proceeding at an incredible pace (Goggin 2012). These human-scale convergence platforms (as they become enabled with more nanosensor capabilities for perception tasks) can also be used to spread skills via massively open online courses (MOOCs) that lead to the regional upward spirals of technological infrastructure and individual skills.

However, convergence is challenged by the fact that human-relevant systems tend to have multiple levels. For example, the service system entities at all levels from people (with their smart phones) to family living structures (houses, apartments, etc.) to local universities to cities, states, and nations must be able to "keep up" and to improve their convergence platform technological infrastructure and individual skill levels in a manner that preserves equity of access to the benefits of new knowledge. This will require a successful shift in policy (rule systems) at the continental level, globally. Institutional rules may not change as rapidly as technological infrastructure or even individual skill levels. The need for better rule systems is as apparent, if not more apparent, than the need for better technological systems (Spohrer et al. 2012). To achieve better equity and increase unity, the strongest must have an incentive for helping the weakest, and the weakest must have an incentive to aspire to compete with the strongest—which requires a great deal of enlightened self-interest among all stakeholders. For example, what would inspire top MIT innovators to move to Detroit, a city in turmoil and a mere shadow of its former glory, to work on urban challenges there (Rowan 2012)? Mere investment and loans cannot be the full answer, if the debtor regions fall further behind in productivity and cannot pay back the wealthier lender nations, as illustrated by the European Union's troubled strategy, where it may be that real human talent flows are needed to allow the weakest regions to compete with the strongest regions. When top talent flows out of weak regions, instead of into weak regions, it is only a question of time before the weak region falls further and further behind. The current rule systems and policies favor the concentration of talent into single winner-take-all regions (Florida 2008).

Within many of the sciences, a considerable degree of convergence has occurred through *collaboratories* and *virtual organizations*. These are partnerships linking multiple organizations and large numbers of scientists who use information and communication technology to share data, analytical tools, and results (Noll and Scacchi 1999; Olson et al. 2008). Aspects of this can be seen in the wider society as technology compresses time. There is no lag between an event and that information flowing into the world where others are made aware of the event via a display. But within a science, rigorous quality control is essential, and to accomplish that, human experts must work in cooperation with increasingly intelligent automated systems.

The Protein Data Bank and computer tomography virtual organization are convergence examples of computer-supported cooperative scientific research, which is among the main scientific and engineering advancements of the last 10 years.

The Worldwide Protein Data Bank (http://www.wwpdb.org/) illustrates how a shared information resource that was born long ago has taken on new life through information technology. Established in 1971, it became a clearing house for information about the structure and folding properties of proteins, a topic at the intersection of biology and nanotechnology, because proteins are building-blocks of life whose size is on the nanoscale. As innovative methods made it possible to decipher more quickly and reliably the structure of proteins and comparable complex molecules, the need for computer tools to manage the information also grew. This led to such crucial, but sometimes apparently mundane, developments as revising the file structure for the data, and more obviously revolutionary forms of convergence as the effort begun in 2001 to develop collaborative centers and use the Internet to unite researchers around the world (Berman 2012).

The computer tomography virtual organization (Tapia et al. 2012) is another example, but centered on a particular high-resolution system for computer tomography using X-rays to study structures such as human bone anatomy. This unofficial community has formed around one of only three high-resolution computed tomography (HRCT) scanners in the world, that at the Center for Quantitative X-Ray Imaging at the Pennsylvania State University (http://www.cqi.psu.edu/). For example, as used by the Department of Anthropology (among other departments at Penn State), the HRCT can generate 3D internal maps of physical structures in human and animal subjects, which capability has led to a number of cross-institution and international research collaborations among physical anthropologists. It is of great value to researchers at many institutions, so it is an example of one kind of scientific collaboratory, a shared instrument. But the virtual organization that hosts the instrument also provides a community data system, sustains a virtual community of practice, and functions as a geographically distributed research center.

At the present time, it is uncertain how well telecommuting and teleconferencing can substitute for traditional collaboration in which the people share a physical location, but this option does seem promising. As an example, both the National Science Foundation (NSF) and the National Institutes of Health (NIH) have been experimenting with virtual panels to review scientific research grant proposals. As part of their funding decision process, these agencies bring together scientists and engineers who have written reviews of research proposals, so they can discuss the various projects and achieve a convergence of views through group recommendations. For example, at this point NSF has held 23 review panels on a secure "island" in the virtual world Second Life, in which each person is represented by an avatar, illustrated in Fig. 2.1 (Bohannon 2011).

All NSF panels have long used groupware called the Interactive Panel System, which provides access via Internet to research proposals and individual reviews, as well as providing a text-based communication system for writing, commenting on, and approving a written summary of the panel's discussion of each proposal. This combined virtual-groupware system reduces costs, allows convenient scheduling of meetings, and permits people who cannot travel to participate. Enhancements to facilitate transdisciplinary convergence remain to be developed.

2.3 Goals for the Next Decade

Two very different but compatible strategies suggest goals for the coming decade, based on advances in information technology in convergence with other fields. First, specific areas, of which robotics is a good example, bring together elements from different areas of science, through the engineering of new technologies, for direct application in human lives. Second, with a much more general scope, social sciences and related disciplines can be integrated with the NBIC fields, helping both to connect them and to increase the benefits they provide to society.

Fig. 2.1 One of the areas where virtual review panels are held by NSF

2.3.1 Strategy/Goal 1: Advance Information Technology in the Field of Robotics

Robotics has the potential to impact our daily lives substantially in the next decade, though perhaps not in the ways we envisioned 10 years ago. We typically have viewed the promise of robots as automated manual laborers, a vision that matches both their early capabilities (industrial automation) and fictional visions. The difficulties faced in developing these kinds of systems are well known. For example, perception is deceptively challenging, manipulation lacks flexible and compliant actuators and control algorithms, and planning requires both fine detail and extensive computational power. However, robots have the potential to offer other forms of support—cognitive, social, and behavioral. This switch is a substantial intersection for converging technology that can impact the quality of life for many individuals. Important areas of convergence in the next 20 years will be systems that offer support for the cognitively challenging society that the information revolution has produced: systems that enhance social support for individuals, that allow for a more connected and more natural experience, and that help to coach, train, and support healthy behavior and educational goals.

Robotics is seeing some of the same changes that happened to computing in previous decades. Just as the giant-room-sized computers of the 1960s became smaller, less expensive machines as they became consumer electronic devices, robotics is entering the same kind of transformation where robotics technologies are

Table 2.1 Consumer shifts in computing and robotics

	1960 Computing	2010 Computing
Cost	Institute-scale	Consumer-scale
Size	Room-sized	Desktop-sized or smaller
Training of user	Ph.D. level	None
Technology requirements	Fast, repeatable, durable	Easy to use, portable, flexible
Applications	Cryptography	Social networking
	Scientific computing	Entertainment
	Engineering design	Personalized search
	1975 Robotics	2025 Robotics
Cost	Institute-scale	Consumer-scale
Size	Room-sized	Desktop-sized or smaller
Training of user	Ph.D. level	None
Technology requirements	Fast, repeatable, durable	Easy to use, safe, flexible
Applications	Factory automation	?
	Remote sensing	
	Dangerous materials handling	

becoming more available and pervasive. Convergence in robotics will follow the trend of convergence in computing; that is, as robotic devices become consumer electronics, there will be a shift in the convergent areas toward areas of social and cognitive support for individuals. Table 2.1 sketches the possible future of consumer robotics by analogy with consumer computing.

The final cell of the table, representing 2025 consumer robotics applications, contains only a question mark, indicating that we can imagine several possibilities but not confidently predict which ones will become both feasible and popular. Just because industrial robots have been so successful, we cannot assume that robots in the home will be doing factory-based tasks. Robotics will initially move into consumer-driven niche markets, just as computers did, but which ones is exceedingly difficult to predict. We can, however, identify some key questions.

The traditional conception of a robot as a machine human dates back perhaps as early as the mythical ancient Greek giant made of bronze, Talos, who according to some accounts was manufactured by Hephaestus, the god of technology. Indeed, it may be very important for some consumer robots, or robots in educational settings, to take humanoid form, in order to interact comfortably with children and adult humans, or to perform certain tasks for which the human geometry is well-suited. However, we already have dishwashers, so the image of a robot standing at the sink washing dishes by hand is no more than a cartoon.

It may well be that robot applications will move from classical, physical-based to cognitive-based and social-based tasks. However, the form in which artificial intelligence is embodied will be a significant research and design question. For example, a security system to protect a home may be embodied in a humanoid robot, integrated into the dwelling itself, or located in the wider cloud of information services surrounding the home.

Robotics Application Areas

This raises the question of why consumers would need a mechanically embodied robot instead of a virtual agent, like Siri, the intelligent personal assistant and knowledge navigator that currently works on Apple mobile devices. However, many tasks do require physical movement and manipulation of physical objects. As the simple Roomba robot vacuum cleaner suggests, the ideal form for a particular physical-based task may not always be humanoid. When physical-based tasks are involved, actuators and sensors need to be in the mobile device, but the intelligence could be elsewhere. For example, in domestic applications, a central computer of a smart home could operate several robots of different kinds simultaneously and wirelessly; among them might be a humanoid robot or voice-operated intelligent agent that serves as the user interface for the entire system.

One application area of great potential and current research effort is *assistive robotics*, most obviously for people with problems of physical mobility. Personalized assistive devices available to the individual consumer will introduce a new range of services and demands for converging technologies. For example, convergence with biotechnology may often be required if the user has difficulty using hands to control a device like a wheelchair. Research progresses both on noninvasive brain–computer interfaces and on neural implants, for example, to operate artificial limbs. Understanding the human mind through behavioral science is needed to develop better noninvasive brain–machine interface design. Obviously, safety is a primary concern, but another is adaptability, because individual people differ widely in their needs and capabilities. In the context of a "smart" home, the entire environment could become assistive technology, including robotic components integrated into a unified system, greatly enhancing the autonomy and capabilities of the human being, so he or she would no longer be called "disabled." Achievement of these visions requires advanced forms of artificial intelligence, designed to handle a wide range of tasks reliably, and to do so in the manner most supportive of human freedom and dignity.

A second possible application area is in the *education of children*. Many people quickly respond negatively to this idea with concerns that childhood should be a natural period of life, filled with free play and exploration unconstrained by machinery. However, from an early age children are subjected to the regimentation of schools, and we typically give little thought to the harmful consequences of confinement to a classroom for half of each weekday. Robotic technology could be developed with great sensitivity to this issue, to increase rather than reduce the freedom of childhood, while providing proper discipline and helping a child to learn in the best way for that individual. We imagine a robot that can guide the child toward long-term behavioral goals; be customized to the particular needs of the child; develop and change as the child does; and engage the child as a peer, not as a parent, teacher, toy, or pet.

2.3.2 Strategy/Goal 2: Advance Information Technology by Integrating Social and Hard Sciences

Contact: *W. S. Bainbridge, National Science Foundation*

Opportunities for collaboration between the natural science and technology disciplines, on the one hand, and the social science and humanities disciplines, on the other, are increasing under the vision of "NBIC2," although there could be challenges in bringing them together. Some concern has been expressed by some experts that certain areas of the social sciences are perceived to have become moribund or marginalized, with the exception of economics, and more needs to be done to demonstrate their relevance and role in government decision-making.

Journals like *Science*, or magazines like *Scientific American* or *Science News*, hardly ever publish articles about solid advances in sociology or political science, and social psychology or anthropology are typically reported only if the particular work involves physical data. How much this state of affairs results from opposition to the social sciences, or from differences between social and physical sciences that may be responsible for delays in the progress of the former, we cannot say (Bulmer and Bulmer 1981; Fisher 1993). Revival of social sciences might be facilitated in several ways, notably: (1) providing extensive support for scientific fresh starts, often based on convergence of cognitive and information sciences, and (2) redesigning the currently problematic institutions of societal governance, with much greater guidance from social sciences (Bainbridge 2007b, c, 2008, 2009).

Expanding Citizen Science to Include Citizen Social Science

As one specific example of how convergence might revive social science, we can sketch an approach in which information technology assists in transferring the open source concept to sociological research, in the form of citizen social science. Of course, amateurs have played a role in several sciences since their historical beginning, including amateur astronomers who discover comets through their backyard telescopes and amateur archaeologists or paleontologists who hunt far and wide for specimens. One of the most famous examples was Francis Tully, an ordinary citizen who spent much of his spare time exploring the Mazon Creek area in Illinois picking up fossils. In the 1950s, he found the first specimen of the remarkable and as-yet not thoroughly understood animal named after him, the Tully Monster (*Tullimonstrum gregarium*), and took it to professional paleontologists for them to study (Johnson and Richardson 1969). While amateurs continue to contribute to science in this manner, today many major research projects enlist ordinary citizens in a systematic way, providing training and collecting data online.

A modern citizen-science project involving research on animals is *eBird*, centered on an Internet-accessible database (http://ebird.org/content/ebird/) where amateur bird watchers submit observations of particular species observed at specific times and places. As simple as this concept seems, it has led to significant research findings about the distribution and migrations of avian species, results that would have been next to impossible to achieve by any other means (Fink et al. 2010). Two related examples are *Galaxy Zoo* (http://www.galaxyzoo.org/) and Phylo (http://phylo.cs.mcgill.ca/eng/). *Galaxy Zoo* currently enlists thousands of amateurs to classify images of galaxies taken through the Hubble telescope and find anomalies for further study by professionals, using an online system to train the volunteers and combine their efforts to achieve maximum reliability (Smith et al. 2011). Phylo employs a game-based crowdsourcing approach for identification of multiple DNA sequence alignments (Kawrykow et al. 2012). Another example, closer to the social sciences but still using data about physical objects, is the *Field Expedition: Mongolia, Valley of the Khans* project (http://exploration.nationalgeographic.com/mongolia/), a crowd-sourced search for archaeological sites in aerial photographs of Mongolia that identified targets for ground-based expeditions to explore (Lin et al. 2011). More obviously relevant for NBIC is the *Foldit* project (http://fold.it/portal/) created by a collaboration between the Center for Game Science and the Department of Biochemistry at the University of Washington, which uses an online game to enlist nonspecialists in solving problems in protein folding. This is the most prominent example of citizen nanoscience (Khatib et al. 2011) and computer game-based approaches to scientific research (Scacchi 2012).

Ordinary citizens have long contributed to progress in the social sciences by answering survey questionnaires and volunteering as research subjects in laboratory experiments. But in the adjacent field of history, they have also contributed as oral history interviewers in addition to being respondents. For example, for 40 years the public library in Greenwich, Connecticut, in collaboration with the town's historical society, has collected documents and conducted 850 interviews with older residents to gain their observations about Greenwich throughout their lifetimes. In principle, every town should do this—but with enhancements in methods and concepts provided by the social sciences. Perhaps the most massive, high-tech example is the Shoah Foundation Institute at the University of Southern California (http://dornsife.usc.edu/vhi/), which at the time of the NBIC2 U.S. workshop announced completion of its project to digitally preserve video interviews with 52,000 survivors and witnesses of the European Holocaust.

With such oral history examples as inspiration and the many well-established citizen science projects as a guide, it is possible to imagine transformation of the social sciences, to make them not only more successful intellectually, but also more relevant for the lives of citizens. In the case of sociology, every person receiving a BA degree in sociology could be considered a bona fide sociologist in the context of collaborative citizen-science projects. For example, a graduate could remain a research associate of a favorite professor for the duration of the professor's career, collecting data, conducting analyses, and collaborating in publications, wherever the graduate happens to wind up in life. Organizations like the American Sociological Association or leading university departments in the field could set up research projects on specific topics, involving holders of sociology undergraduate degrees, or other reasonably well-prepared people.

As of July 9, 2012, the *Scientific American* website http://www.scientificamerican.com/citizen-science/ offered descriptions and links for fully 47 citizen science projects that were seeking volunteers, but none of these centrally involved the social sciences. One, Ancient Lives (http://ancientlives.org/), concerns ancient texts of value for historians and conceivably could be used to develop or test sociological theories, although that does not seem to be among its current goals. Another, the Health Tracking Network (http://www.healthtracking.net/index.php), could be adapted for sociological use or to chart sociologically significant phenomena like what is popularly called "mental health," but it currently is medically oriented. Additionally, the projects all seem to be old-style citizen science, in which the volunteers server as laborers rather than collaborators. Nonetheless, as online clearing houses of citizen science such as *Scientific American* emerge, there is no reason why social scientists could not submit their projects for inclusion.

Information technology would provide the data collection and team collaboration framework for "citizen social science." Thinking in terms of today's technology, a text-based forum and chat room would run constantly as members of the particular project share insights, advice, and access to other resources. A wiki, editable only by members of the team but visible to the whole world, could assemble their findings. Questionnaires, videos, and data in innumerable other formats could become part of the archive, and newly emerging analytical tools could be employed to achieve results. The range of potential topics is so great, and the opportunity to study the same phenomena in many different physical and cultural environments so attractive, that a very large number of such projects could be carried out simultaneously. Of course, proper attention would need to be paid to the ethics of research on human beings, yet the net result could be a huge transformation of the status of social science in society. Each participant would incorporate the research and its ideas in his or her own life, together accomplishing the convergence of social science and society.

2.4 Infrastructure Needs

A key infrastructure feature will be shared resources that allow scientists and engineers to accomplish great tasks, with special emphasis on those resources useful for multiple fields, because such multiple-field infrastructures are environments conducive to convergence. People think of infrastructure in terms of expensive hardware and physical installations, such as big telescopes, atom-smashers, and supercomputers. However, in the context of convergence we can identify four kinds of infrastructure that will contribute to and build on converging knowledge and technology:

1. *Physical infrastructure*, such as major production facilities, urban infrastructure, and instruments and laboratories
2. *Information infrastructure*, such as massive databases and the tools for using them
3. *Institutional infrastructure*, such as research universities, scientific associations, and government funding agencies

4. *Educational infrastructure*, for training the new generations of scientists and engineers

As the Protein Data Bank and computer tomography virtual organization illustrate, often these kinds of infrastructure are combined into one resource, a convergence of hardware and data resources within a cooperative social organization. But distinguishing the four infrastructure types avoids conceptualizing infrastructure entirely in terms of physical installations, and it facilitates thinking about many specific relevant issues, such as the following:

- Creation of laboratories, centers, and funding mechanisms should be designed to foster integrated, multidisciplinary research.
- Better scientific collaboration systems should include (1) informatics to assist in data collection, pattern detection, and data assimilations; (2) shortened publication cycles for research findings; and (3) wide availability of raw data and good procedures for their proper shared use.
- Converging technologies research should also provide a concurrent match in resources for public education in science and technology. We should consider that resources for converging technology research should be matched by resources for public education.
- Much of the future scientific infrastructure will take the form of Internet-based systems to support citizen science, employing transparent, open-source models and process representations that can be expressed in both human-readable and computational forms.
- Technology and access to information grow exponentially, but understanding within the human brain and mind cannot. Thus the gap is widening between our technological capabilities and our mental grasp of those capabilities; we need an infrastructure to help the individual and societal mind work with technology.
- Technology development for health sciences should have a home in focused Federal funding rather than falling between the areas funded by NIH and NSF. Other multidisciplinary areas may need similar innovations regarding institutional support.
- So that self-organizing communities of practice can flourish, government may need to focus on developing infrastructure and assuring free access to this infrastructure.

2.5 R&D Strategies

Converging technologies at the human scale require humans to work together, specifically humans with knowledge in the various converging domains and the ability to understand and incorporate the knowledge of those domains into their own individual work. Such interdisciplinary work will be required for converging technologies. One example is personalized healthcare. Various motivational, societal, and political forces influence the usefulness and acceptance of personalized healthcare. Years ago, the field of economics realized the need to incorporate social and behavioral science, because classical models failed to explain much of human behavior involving resources. Indeed, it was not enough for economists to merely work with psychologists; they also had to understand the theories and data in both fields to make progress. In converging technologies, we need to both encourage interdisciplinary work and create workers and scientists capable of transdisciplinary work, where a single worker or researcher can work in more than one field.

However, in research, there is not yet a widespread system that encourages interdisciplinary work. As with most work, it can be encouraged or discouraged through rewards and "punishments." Already, several Federal agencies offer incentives for interdisciplinary work through specifically solicited interdisciplinary programs; center competitions; interdisciplinary submissions to core programs; and through education, training, workshops, conferences, and symposiums. The National Science Foundation lists several of its own examples in three major categories:

1. **Solicited Interdisciplinary Programs**. Numerous NSF programs are designed explicitly to be interdisciplinary, often involving several NSF directorates. Program solicitations are developed for these programs and posted on the NSF website, e.g., that for Interdisciplinary Behavioral & Social Science Research (http://nsf.gov/pubs/2012/nsf12614/nsf12614.htm). Recent examples include Cyber-Enabled Discovery and Innovation; Water Sustainability and Climate; Collaboration in Mathematical Geosciences; Dynamics of Coupled Natural Human Systems; Macrosystems Biology; Emerging Frontiers in Research and Innovation 2010; and Decadal and Regional Climate Prediction using Earth System Models.
2. **Areas of National Importance**. NSF develops activity portfolios focusing on areas of national interest, often in collaboration with other Federal agencies. Because the challenges that we face as a society are often complex and require an integrative, collaborative approach, these areas are often interdisciplinary. Examples of interdisciplinary programs that NSF contributes resources to include Science, Engineering, and Education for Sustainability (SEES, http://www.nsf.gov/funding/pgm_summ.jsp?pims_id=504707); Networking and Information Technology Research and Development (NITRD, http://www.nitrd.gov/); and the National Nanotechnology Initiative (NNI, http://nano.gov/).
3. **Center Competitions**. Many of the centers funded by NSF bring together interdisciplinary research teams. Some examples include the Materials Research Science and Engineering Centers (MRSECs, http://www.mrsec.org/); the Science of Learning Centers (SLCs, http://www.nsf.gov/funding/pgm_summ.jsp?pims_id=5567); and the Science and Technology Centers (STCs, http://www.nsf.gov/od/oia/programs/stc/index.jsp).

NSF's solicited interdisciplinary programs operate in parallel and cooperation with the long-standing disciplinary programs, typically managed by teams of program officers drawn from the core programs. These programs' special competitions tend to last for 3–5 years and are designed to promote collaboration among selected sets of disciplines that are judged to have a good potential for collaborations at their current state of development. Areas of national importance are highlighted for similar periods of time, because a pressing societal need can be addressed by a concentrated but multidisciplinary effort. Based in one or more universities, centers play a key leadership role in addressing national areas of need, often spanning multiple disciplines.

A somewhat different approach has been taken by the National Institutes of Health in its Common Fund (http://commonfund.nih.gov/):

> The NIH Common Fund was enacted into law by Congress through the 2006 NIH Reform Act to support cross-cutting, trans-NIH programs that require participation by at least two NIH Institutes or Centers (ICs) or would otherwise benefit from strategic planning and coordination. The requirements for the Common Fund encourage collaboration across the ICs while providing the NIH with flexibility to determine priorities for Common Fund support. To date, the Common Fund has been used to support a series of short term, exceptionally high impact, trans-NIH programs known collectively as the NIH Roadmap for Medical Research [http://commonfund.nih.gov/aboutroadmap.aspx]. The Common Fund is coordinated by the Office of Strategic Coordination, one of the six offices of the Division of Program Coordination, Planning, and Strategic Initiatives (DPCPSI) within the Office of the Director.

Occasionally, several Federal agencies combine their resources in a special convergence effort. Among the most promising current examples that place human beings in the context of technology is the National Robotics Initiative (NRI; http://www.nsf.gov/pubs/2011/nsf11553/nsf11553.htm, "Synopsis of Program"):

> The goal of the National Robotics Initiative is to accelerate the development and use of robots in the United States that work beside, or cooperatively with, people. Innovative robotics research and applications emphasizing the realization of such co-robots acting in direct support of and in a symbiotic relationship with human partners is supported by multiple agencies of the federal government including the National Science Foundation (NSF), the National Aeronautics and Space Administration (NASA), the National Institutes of Health (NIH), and the U.S. Department of Agriculture (USDA). The purpose of this program is the development of this next generation of robotics, to advance the capability and usability of such systems and artifacts, and to encourage existing and new communities to focus on innovative application areas. It will address the entire life cycle from fundamental research and development to industry manufacturing and deployment. Methods for the establishment and infusion of robotics in educational curricula and research to gain a better understanding of the long term social, behavioral and economic implications of co-robots across all areas of human activity are important parts of this initiative. Collaboration between academic, industry, non-profit and other organizations is strongly encouraged to establish better linkages between fundamental science and technology development, deployment and use.

At the core of the NRI is the concept of co-robot, the design of intelligent machines that will be optimal partners with human beings. Figure 2.2, from the multiagency program solicitation, suggests the convergence of multiple disciplines in achieving this synthesis.

These examples of existing government activities reflect three somewhat different but apparently equally successful strategies: (1) creation of well-focused temporary competitions within one agency; (2) development of an over-arching convergence mechanism uniting all divisions of an agency, and (3) cooperation between agencies in a convergence area of mutual interest. All three mechanisms must be designed to emphasize areas that are technically suitable at the given point

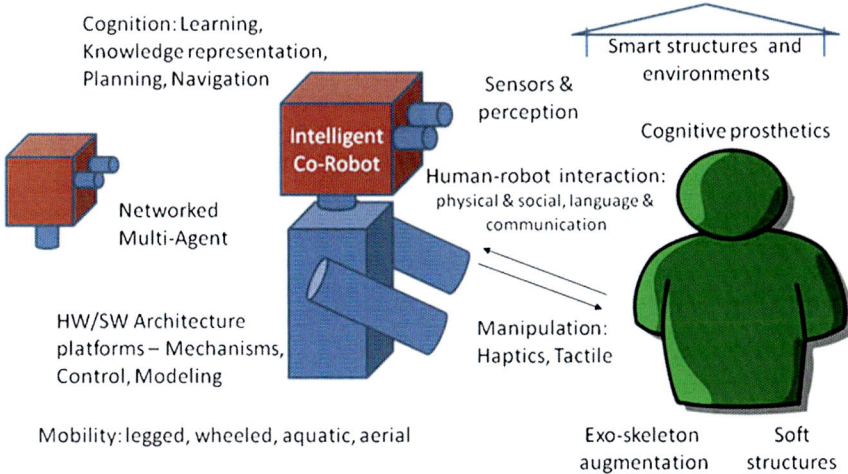

Fig. 2.2 The "co-robot" concept in the National Robotics Initiative (Open source: National Science Foundation, http://www.nsf.gov/pubs/2011/nsf11553/nsf11553.htm)

in time for rapid progress and where both the leadership and the rank-and-file employees in the agencies are prepared to work together.

It is not uncommon for universities to create multidisciplinary programs, although their functionality and longevity may often be inferior to those of departments representing single disciplines. Some examples happen on an ad hoc basis because they do not give the fields in question high priority, as some teaching institutions combine sociology and anthropology, and it is a rare research university that separates archaeology from cultural anthropology.

Reward systems differ by discipline. For example, a computer scientist gets high value from proceedings papers at conferences, while a psychologist benefits most from journal publications, and in the view of many, sociology is a book-based discipline. Even author order can be problematic: psychologists list the primary author first, mathematicians provide authors in alphabetical order, and biologists list the primary author last. In short, an interdisciplinary *curriculum vitae* is a difficult one to evaluate.

At the university level, even with the advent of interdisciplinary hires, there are still hurdles to interdisciplinary work. First, as the systems stand, there is little reward for interdisciplinary work. In the day-to-day life of the academic, it is difficult to work with researchers outside of one's discipline. Their offices or laboratories may be in different locations, requiring tedious travel to meet. Sharing data is difficult, because of a need for privacy and security in human behavioral data and the different data formats and analysis software used by different physical sciences. If researchers in two disciplines submit a grant proposal together, they may need to deal with the rules of two offices of sponsored programs that have different deadlines and requirements. It is much easier to collaborate with colleagues at the same location, with the same reward structure, and speaking the same technical and administrative languages.

Institutional Review Board approvals for human subjects research is a challenge that must be addressed for interdisciplinary teams that engage physical, natural, and social scientists. The complexity of research design around the issues of human subject protection is not always immediately appreciated by researchers whose past experience has not involved human subjects. While social scientists need to become more familiar with the language and landscape of contemporary science and technology research priorities, STEM (science, technology, engineering, and mathematics) researchers (outside medicine) also need to gain an appreciation of Federal regulations related to protection of human subjects and to become more familiar with protocols that involve these regulations if NBIC convergence is to progress.

Some solutions for these problems involve small changes to culture "at the human scale." Implementing rewards such as interdisciplinary funding from funding sources and university hiring across areas, is valuable in itself. However, success can also occur at a smaller scale by removing barriers to interdisciplinary work on a daily basis. This requires leadership in universities and government agencies, but also initiative-taking by all the people who work within these organizations.

2.6 Conclusions and Priorities

Contact: *W. S. Bainbridge, National Science Foundation*

2.6.1 Economic Effects Associated with Technological Innovation

In the economic sphere, information technology is creatively disruptive, offering marvelous new jobs for workers and profits for investors, yet also destroying or downsizing old occupations and industries. There is widespread concern that the current economic problems are more than merely the aftermath of an unusually severe downturn in the regular economic cycle, but a transition to a "new normal" with permanently higher unemployment rates (Brynjolfsson and McAfee 2011). We could even have entered a long-term decline in the prosperity of most citizens of post-industrial societies, in a "race to the bottom" as per capita incomes equalize across nations without sufficient growth to keep the rich nations rich. Clearly, we need much better-quality social science in order first to answer such disturbing questions with any confidence, then to invent the best responses to the problems we face.

Computers may be cutting-edge, but they are also double-edged swords that can cause harm as well as provide benefit. Extreme technological unemployment may come about as intelligent machines replace more and more human workers in an increasing number of sectors of the economy, which we already see as mass-market stores install automatic check-out machines, bookselling moves away from local bookstores to online services, and journalistic careers vanish as newspaper subscriptions plummet and people read blogs instead of magazines. Information technology facilitates off-shoring of jobs, for example, of local telephone help operators in Indiana who are replaced by workers in India.

A standard concept in the economics of innovation is creative destruction, the principle that innovations often destroy old industries and jobs, but the net result is creation of more and better-paying jobs in new industries (McKnight and Kuehn 2012). The usual criticism of this postulate is that it does not take account of the suffering of people who lose employment in old industries, especially those unable to get jobs in the new industries. But if we think of creative destruction in terms of a production function in which old industries are the input and new industries are the output, we cannot in fact specify exactly what the function is, especially when innovations of different kinds may have different consequences. Some scholars suggest that innovation has a declining utility function, such that in earlier stages of human history the new jobs gained from creative destruction outnumbered the old jobs lost, but later on in history, the jobs lost outnumber the jobs gained. They postulate we may have passed the inflection point, after which innovation is more destructive than creative.

The current situation with information technology may illustrate a different point, conceptualizing innovation as a staged process, in which creation and destruction occur at different times, following a complex coupling function. For example, the two could proceed in surges like two sine waves of the same frequency but out of phase with each other, or even of different frequencies and thus causing a third kind of wave like a beat frequency that occasionally causes either creation or destruction to predominate without that fact marking a long-term trend. This metaphor suggests that the current unusually severe economic distress could be partially caused by an extreme but temporary dissonance in the rates of progress in different fields of science and technology, which convergence could harmonize.

2.6.2 Social Effects Associated with Technological Innovation

While today we commonly think of scientific and technological progress as a gradual improvement in knowledge and abilities, convergence can unleash sudden advances with profound implications that can be realized quickly. A classical view of scientific and technological progress stressed decisive innovations that have great consequences, but too much emphasis on day-to-day management may currently blind science and technology leaders to the very real possibility of technical revolutions. Ninety years ago, William F. Ogburn (1922; cf. Bimber 1994) outlined a four-step "technological determinism" model of human history that gave special prominence to rare, decisive innovations:

1. *Invention*: the appearance of a new technological form that does not depend upon the genius of any one inventor, because major inventions are made roughly simultaneously by several different people
2. *Accumulation*: the process in which new inventions are added to the general store of culture more rapidly than they are forgotten or rendered obsolete
3. *Diffusion*: the transfer of knowledge, skills, and technology from one geographic location to another, and from one domain of application to another
4. *Adjustment*: the process by which nontechnical aspects of a culture adapt to invention, usually happening with greater difficulty for the more significant innovations

Ogburn was especially worried about *cultural lag*, a tendency of society's institutions to resist change, thus hindering adjustment. On the positive side, he noted that accumulation of inventions could feed back into the invention process via diffusion, as new inventions combined old ones in new ways. Thus, Ogburn was an especially early theorist of convergence. A rather different episodic theory of scientific progress was offered by Thomas Kuhn (1957, 1962), who wrote about historical paradigm shifts in which a new generation of scientists brought with them a radically different conception of their field than that possessed by their elders. It is possible that convergence represents a *metaparadigm*—a paradigm about the combination of existing paradigms across fields, leading eventually to a wholesale reconceptualization that transcends many of the older paradigms.

Information technology relates to these ideas in three ways. First, it offers tools for rapid diffusion of innovations around the globe and across fields of science and technology. Second, it is an example of disjunctive progress, just as the invention of integrated solid-state circuits half a century ago made possible a true revolution across many areas of human accomplishment. Third, we may be experiencing severe cultural lag as a consequence of this technological revolution, and we have hardly begun to decide which traditional societal institutions need significant adjustment, let alone in which direction. Thus, the overarching themes of this report—(1) improved economic productivity, (2) increased human potential, and (3) sustainable life security—are themselves areas for research and public debate.

2.6.3 Relationship of NBIC Convergence to Economic and Social Aspects of Technological Change

Economic competiveness can be analyzed on several levels: individual, community, corporate, national, and global. For individuals, a key dimension of competitiveness is education, not only the formal schooling early in life, but life-long learning to keep up-to-date in terms of work skills and to enhance the quality of life outside of work. Communities, such as cities and states or provinces, compete for industries against other communities, but at any given time only a few can be centers of technical innovation, while other geographical areas may rely upon extractive industries, agriculture, or commerce—leaving many geographic regions at a competitive disadvantage unless new institutions can be developed that compensate for their disadvantages. Corporate competitiveness is harsh but often highly rewarded in new industries, while innovation in well-established companies can be especially difficult. On the global level, the question may become how to emphasize cooperation and mutually beneficial division of labor across nations, or else the world economy may become "a race to the bottom" in which the nation with the lowest labor costs yet sufficient technical competence undercuts the prices of all other nations for goods and services.

Increasing human capacity today often means giving people technologies that effectively increase the scope of their memories and their capacity to process more information, for example, through the Internet, personal computers, and mobile devices. In the past, human physical capacity was increased through motor-driven machines that ranged from water pumps to drill presses to automobiles. We can ask whether we might stand at the threshold of another phase of technology-assisted human development, that of increasing human capacity not at the physical or mental levels, but at the social level of communities, corporations, and families.

Security and sustainability cannot be achieved through resistance to change, but only through *wise management of innovation*. This is not to suggest that we would continue a technocratic, typically hierarchical, top–down policy-setting and management approach in which people who consider themselves knowledgeable help the resisters see the benefits of change. Resistance provides information that may often be valuable and sometimes correctly advises against a particular course of action, or for a better one that technocratic managers may not have noticed. If viewed in this way, resistance provides data for a scientific and reflective observation about what underlies the lack of enthusiasm or even active resistance. Today's concepts of management may be supplemented by emerging management forms that can arise as a result of the increased connectivity among people possessing a range of perspectives and interests.

Wise management may often mean defining environments in new ways, such as redesign of the workplace to balance the human needs of the workers with increased productivity. Smart homes may be both more comfortable and more cost-effective than big homes, integrating sensors, machines, and computation into unified habitats. Information-integrated transportation systems can be more satisfying than exhausting traffic jams or waiting lines, avoiding both accidents and unnecessary costs in time and money. It is easy to name problems and suggest superficial solutions, but it is extremely difficult to preserve the beneficial aspects of customary life while enhancing human capabilities in profound ways. A proper balance requires very significant research and development efforts, bringing together the methods and insights of all relevant fields, in technology-enhanced human convergence.

2.7 R&D Impact on Society

To properly assess the impact of NBIC convergence on society, we need to develop better models and approaches for prospectively gauging social impacts of new technologies. A difficulty here is assessing impacts that are beyond economic and quantitative assessment. We may want to consider looking toward governance concepts that are less oriented than currently toward utilitarian concepts like risk/benefit analysis. This is true for at least three reasons. First, we know from social science research that economic models do not adequately account for how broad populations of people evaluate prospects put before them. Second, as NBIC emerges in the applied world, it will interact with other cultures, many of which are based on traditionalism that would reject utilitarian thinking in favor of more holistic or morality-based ideas. Third, priorities for research and development in converging technologies should take account of broad social needs, as opposed to allowing commercialization to be heavily driven by market forces.

The impact depends on the state of society, as well as upon the nature and application focus of the technologies. Today, the world faces unprecedented aging of its population. The stakeholders, workers, and even the scientists will be increasingly older. In consequence, the focus of much medical work will shift from curing diseases to managing chronic conditions and degeneration. Aging is not a disease; evolution shaped humans to live long enough to complete the task of being parents, given the long duration of human childhood—one of the prices for our high intelligence—and not much longer. At present, we lack a consensus about priorities concerning how much society is willing to invest in adding a few months to the life of an elderly person versus investing in universal good nutrition and healthcare for children. We also lack a clear understanding of what new "nano-bio" methods might increase not only the average life span, but the years during which a person can be productive.

One idea considered by the Quality of Life working group at the U.S. NBIC2 conference was the antagonistic pleiotropy hypothesis (Williams 1957), postulating that the expression of a gene results in multiple competing effects, some beneficial but others detrimental to the organism. Some genes responsible for increased fitness in the younger, fertile organism contribute to decreased fitness later in life. This means declining forces of natural selection. Antagonistic pleiotropy can conceivably be fought by up-regulating the repair mechanisms we already have, which decay with time. Using molecules derived from the large inventory of substances generally recognized as safe might open pathways to use genomic information, perhaps in the form of dietary supplements. Genomics focused on finding repair mechanisms, rather than disease precursors, would contribute to progress in this area, as would attacking aging holistically as a complex system of disorders requiring convergence of the relevant sciences and technologies.

One of the attractive features of the antagonistic pleiotropy hypothesis, quite apart from the ultimate determination of its correctness within the science of biology, is that it can be applied by analogy to features of social evolution. Characteristics of societal institutions that served us well earlier in history may be dysfunctional now, and we may seek other analogies from across the spectrum of convergent sciences to suggest cures for the maladies of an aging civilization. For example, comparison of winner-take-all (WTA) and improve-weakest-link (IWL) policies are needed. Shifting the competitive framework to accelerate learning from regional competitions and experiments has great potential for improving quality of life. Improvements in innovativeness, equity, sustainability, and resiliency will result from converging technologies. A shift may be required in the competitive framework to balance WTA and IWL policies to accelerate learning in and between geographic regions so as to improve quality of life for all.

A cornerstone of this session is quality of life. Yet, quality-of-life research has, over several decades, moved from "standard-of-living" types of (economic) measures in the direction of concepts such as "happiness." We need to develop ways to identify how the concept "quality of life" can be brought into the risk-based decision processes associated with new theories of risk governance. This may require higher synergies between funding agencies, academia, and industry: the funding agency defines parameters of the problems, and industry and academia receive funding to solve them. That could imply creating generally higher rewards for scientist-practitioners. In education, convergence R&D can promote learning based on creative synthesis rather than learning based on facts and problem decomposition, with increased emphasis on philosophy and values.

A crucially important area where convergence has failed to take place concerns the ethical, legal, and social implications (ELSI) of information technology. Both nanotechnology and biotechnology have well-established ethical codes, and one of the original sources of government NBIC efforts was the concentrated consideration of the social implications of nanotechnology. Two areas of clear public concern having implications for the economy as well as for human well-being are information privacy and intellectual property rights. A third area not often discussed but of real significance is the possibly excessive regimentation of daily life by information technologies. With guidance from social science, cognitive science, and public debate, these are three very high priorities for the next few years:

1. *Information Privacy*. Legitimate concerns focus on government surveillance of personal communications, justified by a national-security ideology that includes data mining, Internet scraping, computer vision through ubiquitous cameras, and requirements that many kinds of records in such areas as health and education be stored in central databases. At the opposite extreme, movements like WikiLeaks publicize government secrets, and irresponsible individuals can post online harmful, false, or defamatory information about other individuals in a manner that makes it impossible to prove who is responsible and difficult to remove the material from public view. Whether or not there exist technological solutions for these problems, a fresh look at public policies is needed.
2. *Intellectual Property Rights*. There has been much recent focus on online file-sharing practices by ordinary citizens that violate existing copyright laws, and on actions by legislatures to increase the penalties for such violations. The legal structure is the result of centuries of accidental historical accretion, leading to many anomalies. For example, why is the duration of a patent so much shorter than the duration of a copyright? Or, why can engineers patent their inventions,

but scientists cannot patent their discoveries? A social movement in Europe represented by the Pirate Parties International, originating in Sweden but currently popular in Germany, advocates radical reduction or even elimination of intellectual property laws. Whatever the right course may be, information technology has changed the realities surrounding intellectual property rights, so fresh thinking should be a high priority.
3. *Computer Regimentation.* Computer technology permits imposition of harsh control over every detail of worker behavior, for example in white-collar jobs where everything must be done through computers, imprisoning the worker by rigid and poorly designed software. As an illustration, medical professionals have complained about new government regulations promoting electronic health records, indicating that requirements were too aggressive, not all laboratories have the technology to send clinical lab tests into electronic health record technology as structured data, and that the new requirements will force providers to engage in too much manual data entry (Fiegl 2012a, b). There is a similar concern about increasingly dogmatic indoctrination of students in schools, which often require computer-based lessons and tests that include artificial intelligences serving both as teachers and graders, structured within an increasingly rigid ontology of permissible ideas. The relative dearth of public discussion of such possible threats leaves open the question of how real they are, but innovators developing new information technologies for work and school need to be cognizant of the dangers of inappropriate application.

Even as information technology presents such dangers, in convergence with the cognitive and social sciences, it can help us navigate past the dangers to a bright new future for humanity. New online forms of public decision-making can be developed, involving all sectors of society impacted by technological changes, in a manner that combines expert analysis with the consent of the governed. This could not only prevent potential negative consequences of improper application of information technology but also unleash human creativity to overcome cultural lag while safeguarding the well-being of the general public. Expanding on the success of the virtual organization scientific collaboratories developed over the past decade, and building through increasing involvement of the general public in citizen science, the ultimate convergence could be achieved, uniting science with society.

2.8 Examples of Achievements and Convergence Paradigm Shifts

2.8.1 Case Study: Arts, Humanities, and Culture

Contact: *W. S. Bainbridge, National Science Foundation*

Human artistic creativity almost invariably uses tools, so throughout history, technology advances have facilitated the development of new artforms. While the first novels were written in the ancient world, novel-writing did not become a popular artform until the development of the printing press and widespread distribution of manufactured goods. Today, as so many forms of art are moving to digital media, we forget how technically demanding color and sound motion pictures were, using three microscopically thin photographic layers on the film to record the colors, and one or more separate optical sound tracks. Yet this exceedingly complex technology was completely mature by 1939, as demonstrated by *Gone With the Wind* and *The Wizard of Oz* that date from that year.

Today, the relatively new digital media are already influencing artistic creativity in at least two ways: (1) by enabling new artforms that were either impossible with earlier technologies, or so difficult they could not be widely adopted, and (2) by establishing new forms of distribution that not only reduce costs and facilitate speedy diffusion of new artforms, but also call into question traditional norms concerning intellectual property. Convergence of all the NBIC fields has the potential for a third, related transformation: (3) inclusion in many arts of themes derived from progress in science and technology, whether in the plots of stories or in other artistic dimensions such as images and background metaphors.

A new medium that illustrates all three of these factors is massively multi-player online role-playing games (MMOs), which are actually far more than mere "games" but total works of art that include music, drama, and visual arts—even in many cases allowing the player's avatar to dance. Players can team up with each other despite being thousands of miles apart and belonging to different cultures. *Second Life* includes the Google Translate system that facilitates communication across languages, and *Final Fantasy XI* includes an extensive phrase book for communication between English, French, German, and Japanese players. *World of Warcraft* reached a peak subscriber population of 12,000,000. *EVE Online* has "only" 400,000 subscribers, yet for them it is a realistic economy based on exploitation of virtual natural resources, manufacturing of virtual goods, and economic exchange (Bainbridge 2011; Plumer 2012). Already, MMOs have shown promise for education, and now innovative mobile educational games are being developed (Steinkuehler 2007; Scacchi et al. 2008; Scacchi 2010), as well as enabling advances in the arts and humanities, social sciences, and broader scientific research (Scacchi 2012).

The ease of distributing MMOs over the Internet has facilitated the rapid emergence of major companies in China and South Korea, often using a free-to-play but pay-to-win economic strategy very different from the Western subscription model that is in rapid decline. Many MMOs, and smaller games for mobile devices as well, are more-or-less direct copies of existing successful ones, so there have been many lawsuits, and intellectual property protections seem to be breaking down. The Chinese MMO *Perfect World* reverse-engineered *World of Warcraft*, but filled it with Chinese culture, so no laws were broken. The company operating *Perfect World* has bought *Star Trek Online*, thereby taking possession of a major product of American culture. Among the most innovative features of some MMOs, including *Star Trek Online*, is the freedom they give players to create their own missions, which itself erodes intellectual property protections, because the players lose control over their own creations and have no property rights, at the same time that some players create game missions that violate franchise copyrights. A new convergent discipline is emerging that might be called *virtual socio-legal studies*, that examines the radical implications of MMOs and other revolutionary communication developments (Lastowka 2010).

Many MMOs have begun to feature culture based on technological convergence. For example, nanotechnology is the central metaphor of the Norwegian MMO, *Anarchy Online*, in which players may simulate building human-sized robots from nanobot components, and use fictional nanomedicine to improve their avatars' functions. *Fallen Earth* is a survivalist MMO set in the future around the Grand Canyon, where a convergence of biotechnology and information technology has caused the collapse of civilization but also offers the possibility of rebuilding it. Of course, many NBIC convergence ideas have appeared in other artforms. For example, Neal Stephenson's (1994) novel, *The Diamond Age*, conceptualizes nanotechnology in terms of information technology, while Kathleen Ann Goonan's tetralogy (1994, 1997, 2000, 2002) conceptualizes nanotechnology in terms of biotechnology in convergence with music.

Research and policy reconsiderations are desperately needed in the area of intellectual property rights. Intellectual property norms have varied across cultures and throughout history, so the copyright regulations of the past two centuries may become outmoded as future culture evolves away from traditional assumptions. The sharing of data files such as movies and music tracks has become trivially easy via online computers, so why should people not be free to share any artworks they wish? From the perspective of copyright law, Sophocles and Shakespeare were plagiarists, because they freely borrowed from other authors. Yet, they might argue that culture belongs to the entire society, and dramatists are adequately rewarded with either honor or money by being directly involved in theater productions. Under current U.S. law, the movie version of *The Wizard of Oz* will not enter the public domain until perhaps 2034, but the novel the movie was based on is already in the public domain, having been published in 1900, and both are easy to share online.

Information technology is blurring the practical distinctions among three conceptually and legally different forms of art: (1) folk art that is associated with an historical or ethnic culture but not owned by individuals, (2) classical high art that may be owned by its creators for a brief period but then passes into the cherished cultural heritage of humanity, and (3) popular art that is commercially created and distributed by mass media companies but loses value when the stylistic fad and associated celebrities pass from the scene. Whether or not this conceptualization is adequate, the new communications technologies can be the basis for a new science of art, as researchers develop new tools and consider their impact.

The humanities can be conceptualized either as scholarship oriented toward the arts, or as humanistic scholarship about non-artistic aspects of human culture. Both dimensions of the humanities link to the social sciences and are an interesting focus for cognitive science research. Developing a new convergence science of the arts need not inhibit aesthetic creativity, and indeed, the merging technology can nourish artistic innovation. Likely areas of innovation include:

- Comprehensive entry points to a particular cultural complex, especially useful for students, such as The Perseus Project gateway to classical civilization (Crane 2012)
- Digital preservation projects, which need not only to archive cultural works and provide ways of searching and comparing them, but also to manage the intellectual property context for each one (Caroli et al. 2012)
- New methodologies to transform the humanities into sciences, for example, quantitative analysis of changing artistic patterns, adapting the recommender system technology currently used to advertise products the user is predicted to like (Bainbridge 2007a)
- Adaptation of traditional theories and methods from cultural anthropology to post-modern culture, for example, participant observation ethnography of MMOs (Nardi 2010)

2.8.2 Case Study: Digital Government 2.0

Contact: *W. S. Bainbridge, National Science Foundation*

The form and structure of government has varied greatly across the time dimension of history and the cultural dimensions around the globe. Thus, there is reason to doubt that our current institutions of government are the best ones possible, or the ones best adapted to the unique conditions of the twenty-first century. However, there is every reason to preserve our current institutions, as they are, until we have solid evidence that improvements are both needed and feasible. Many current research projects are exploring alternative possibilities, based on convergence of social science and information technology, to offer feasibility assessments and design innovations for consideration by the citizenry.

The concept of *digital government* originated in the 1990s, initially with a goal of improving dissemination of government information between agencies and to the general public, but increasingly also seeking ways in which ordinary citizens could participate more actively in government decision-making. Thus, "Digital Government 1.0" emphasized information flow in one direction, from government to citizens, whereas "Digital Government 1.5" increased input from the citizens. Now, "Digital Government 2.0" seeks to go further, also developing ways in which citizens can use information technology for self-governance, especially on the local level, without necessarily any involvement by formal governmental agencies.

Soon after U.S. Federal Government agencies had set up websites and started using the Internet extensively for both internal and external communication, the revolutionary potential of the technologies began to dawn on sophisticated users. Already in 2002, a workshop to develop a research agenda in this area noted such innovative possibilities: "The fundamental restructuring of government from bureaucratic structures joined through oversight bureaucracies and Congress to greater use of horizontal arrangements using, at times, less formal governance mechanisms, market mechanisms, and temporary configurations signals an emergent change in the structure of the state and policymaking capacity" (Fountain 2002, 5).

An excellent example of recent Digital Government 1.5 research is a study at Cornell University aimed at achieving wider public involvement in the rule-making process of conventional government agencies (Farina et al. 2011). Often the legislation passed by Congress outlines steps to be taken but does not specify the details, so a complex technical process of drafting specific rules must follow before the legislation can be fully enacted. The goal often may be regulation of technology, in such areas as the environment, transportation, workplace safety, or now even intellectual property rights in the information sphere. Government agencies are under a legal obligation to include stakeholder groups and citizens in the process of evaluation of proposed rules, but exactly how to do this both fairly and efficiently has been a difficult challenge. It is one thing to set up an online forum where people may post comments, but quite another to do so in a manner that objectively classifies the diverse contributions to achieve a high-quality consensus. Current research systematically charts participation trends and identifies problems of message content, with the hope that future research will develop methods based on computerized natural language processing to cluster the inputs and measure their quality.

Several very recent projects have begun developing communication tools that citizens can use in their local communities to undertake joint projects and perform tasks that otherwise government might need to do at greater cost and lesser flexibility, or that lie outside the traditional domain of government responsibilities. Janne Lindqvist at Rutgers University and Winter Mason at Stevens Institute of Technology are developing a system called Myrmex to facilitate local community crowdsourcing of physical tasks, like package delivery, optimizing such variables as efficiency, privacy, usability, and motivation for people to participate. John Carroll at Pennsylvania State University is studying time banking, a form of community-based volunteering in which participants provide and receive services, to develop new design principles in human-centered computing. Loren Terveen at the University of Minnesota is developing algorithms and interaction mechanisms to improve the functioning of social media technologies through which members of a community can assist each other with very personal goals, such as parenting and bicycling for good health. Such Digital Government 2.0 research often faces surprisingly difficult intellectual challenges, and makes unexpectedly profound discoveries, despite the apparently humble and "down-to-earth" nature of the applications in ordinary people's lives.

Digital Government 2.0 also includes research projects to examine or prepare for major changes in political processes. Many researchers have focused on the role of new Internet-based communication technologies in the so-called "Arab Spring," although their research has shown that the Internet is not necessarily a powerful force that operates only to promote democracy (Howard 2010; cf. Howard and Jones 2004). Rolf Wigand at the University of Arkansas and Merlyna Lim at Arizona State University have begun to study factors leading to success or failure in collective online action, using the global female Muslim blogosphere as their case study. Philip Howard at the University of Washington studied online communications around the historic 2011 Tunisian election.

Other researchers are examining the changing dynamics of American politics in the Internet era. Jason Thatcher at Clemson University is developing techniques for understanding the impacts of the social web on political polarization, in the context of Congressional elections. Geraldine Gay at Cornell University and Francesca Poletta at University of California, Irvine, are developing computational tools to support citizens in framing political issues in the ways most conducive for reasonable public deliberation, as opposed to mere partisanship. In two separate projects using different methods to achieve convergence between social and information sciences, Robert Mason at the University of Washington and John Jost at New York University are developing new research techniques, theories, and design principles related to online social movements, using the case study of the "Occupy Wall Street" movement.

Digital Government 2.0 research has as yet produced very few publications and is among the very newest areas of science and engineering. Thus, there is a great need for contributions from all the many disciplines that might help establish this new field of convergence.

2.8.3 Case Study: Post-industrial Society

Contact: *W. S. Bainbridge, National Science Foundation*

While nanotechnology and biotechnology give new life to manufacturing industries, for decades it has been clear that the economy has been shifting toward services, in which innovation may require very different approaches, including new ways of managing information (Bell 1973; Spohrer et al. 2012). As noted earlier, a new discipline called *services science* has been emerging, beginning first with reconceptualization of computing as a service provided to the user by companies specializing in advanced information processing, rather than primarily thinking in terms of hardware possessed by the user.

If the economy continues to evolve away from industrial production to provision of services, there will be great challenges for the social and information sciences, and we may not yet be clear about what the most difficult problems or greatest opportunities will be. Communication between users and the organizations providing the services will be essential, and the users may initially lack a full technical appreciation of the requirements for using the service or of the benefits from using it competently. This line of thought implies that tools and procedures for collaboration between people possessing different kinds and levels of expertise will be important, and this is a classical convergence issue. One solution may be professional *convergence consultants*, independent from the major service-providing corporations, able to help the corporation's customers gain the best possible service in the context of their own special needs.

Other issues may arise at the societal level. Outside the specific area of information technology services, there is some concern that service industries are less profitable than manufacturing ones, having less to invest in research and development, and less to pay employees who therefore will have lower salaries. To the extent that many kinds of services are optional, even luxuries, they may exaggerate the magnitude of economic recessions, even during the normal economic cycle. This is somewhat true already for manufactured durable goods, as people continue to drive their old cars during economic recessions rather than buy new ones, and factories put off installing or do not install new equipment. But services that are not essential for survival may suffer almost complete loss of business during economic hard times, thus exaggerating the depth of a recession and even prolonging it (Keynes 1936).

The NBIC2 workshop in Seoul, Korea, contributed a number of ideas that relate to the economic shift from physical production to services, suggesting three related dimensions of this transformation:

1. The aging of the population shifts the balance of both the labor force and human needs.
2. Many challenges will be presented, in moving from an economy based on physical production, to one emphasizing entertainment and psychological well-being.
3. Numerous factors combine in complex ways to require fundamental redesign of the healthcare system.

The fundamental goal becomes enhancing the quality of human life, in a context that is ubiquitous in three senses: (1) society is global rather than local, (2) information technology connects people wherever they happen to be, and (3) achievement of the most important goals requires broad convergence across multiple domains of science and technology rather than narrow specialization.

The world is evolving beyond industrial society, precisely because it has been so successful in providing food, housing, and other material goods required for life. Especially in the context of an aging population, health priorities shift to improved management of incurable chronic conditions, then beyond physical well-being altogether to mental health. Scientific progress in understanding the human brain clearly has an important role to play. Psychiatric disorder can be treated, based on the development of new methods, by finding the root cause of problems such as alcohol abuse, mental diseases, and impaired physical functions. To the extent that the causes are physical, whether chemical or structural, we can reasonably hope that effective new treatments can rely upon convergence technology. But mental health is more than simply cure of "mental illness" following a medical paradigm. It is also well-being in terms of social relationships, engagement in pleasurable activities, and a sense of intellectual and spiritual growth.

During such a massive societal change away from industrial production to services, the entertainment industries become more important, both as a fraction of the economy and in the functions they perform for human beings (Scacchi 2012). An entertainment-driven society can spoil human beings and distract them from important responsibilities they have for human progress, a complex set of problems that science needs to tackle. One clear example is how much Japan has contributed to worldwide entertainment, through popular videogames, manga, and cartoons, even as it has faced industrial stagnation and all the problems attendant to an aging population. Korea has become a major center for the computer gaming industry, illustrating how different nations around the globe can play creative roles for the benefit of all nations. A key research question becomes how to elevate the quality of entertainment so that it becomes art, not merely diverting people but also ennobling their experience of life.

2.8.4 Mimicking Avatar: Uploading Your Experience

Contact: *J.-W. Lee, Department of Convergence Nanoscience, Hanyang University, Korea*

Preserving one's memories or even living eternally has been among the most interesting subjects in popular culture. In 2009, *Avatar* became the first motion picture ever to gross two billion dollars, by using advanced computer graphics to depict the transfer of human consciousness into an alien body (Bainbridge 2011: 190–194). The scene was a densely forested habitable moon in another solar system, Pandora, where humans developed software to transfer the soul of a physically disabled human into a cloned body of a member of the Na'vi—a humanoid but very alien species. In so doing, humans created Na'vi–human hybrids with human sprits, called *Avatars*. They can be controlled remotely by genetically matched humans, communicating with each other via a neural network. The background for the creation of Avatars is the following. In the middle of the twenty-second century, humans try to mine a valuable mineral, "unobtanium," on Pandora in order to cope with the exhaustion of natural resources on earth. However, humans face great difficulty in obtaining unobtanium, because Pandora's atmosphere is poisonous to humans. The implication is that advanced avatar technology could be used here on Earth, for other equally radical purposes.

For decades, science fiction writers have speculated about the possibility of transferring human memories and personalities to artificial platforms based on information technology (Clarke 1956; Brunner 1975), and *Avatar* expanded this concept through bio–info convergence. However, some fantasies of the past can become realities of the future, and technically competent visionaries have begun to write serious nonfiction books on this radical possibility (Moravec 1988; Kurzweil 1999). In February 2011, Dmitry Itskov, a Russian media billionaire, announced the "2045 Initiative" to disembody our conscious minds and upload them into holograms (http://2045.com). This is intended to achieve cybernetic immortality and the artificial body as conceptualized in *Avatar*. A research team of leading Russian scientists will be engaged in R&D on humanoid robots, modeling of the brain and consciousness, and so on. They have sketched the following technology roadmap:

2015~2020 (Avatar A); A robotic copy of a human body is remotely controlled via brain–computer interface (BCI). This robot could work in hazardous environments and for rescue operations.

2020~2025 (Avatar B); An avatar in which a human brain is transplanted at the end of the human's life. Achieving this technology could result in another IT revolution by the materialization of hybrid bioelectronics systems.

2030~2035 (Avatar C); An avatar with an artificial brain in which a human personality is transferred at the end of one's life. This really expands human capability by restoring and modifying one's brain at will.

2040~2045 (Avatar D); A hologram-like avatar. This is the time when a humanoid robot exceeds the capability of ordinary humans, called singularity, as predicted by a futurist Ray Kurzweil. It is forecasted that humans will coexist with humanoid robots at that time.

Over the past century, scientists and engineers have tried to make machines that would be able to think, learn, or behave like a human being. The key benchmark in the history of *artificial intelligence* (AI) was a workshop held at Dartmouth College in 1956 (McCarthy et al. 1955). In 1970, *Life* magazine quoted one of the organizers, Marvin Minsky, as saying, "In from three to eight years we will have a machine with the general intelligence of an average human being" (Darrach 1970, 58). Later, Minsky said he was misquoted, but it appears that progress was slower than he and many other Dartmouth participants had expected (Crevier 1993; McCorduck 2004). One reason may be that for them *intelligence* meant the ability to solve complex but well-defined puzzles, such as the technical design challenges faced by engineers, rather than the ability to behave exactly as a human being does (Simon 1996). However, progress has constantly been achieved in AI, and already 15 years ago, affective computing was developing means for allowing machines to understand human emotions (Picard 1997).

AI was one of the fields that converged to create cognitive science, and a full NBIC convergence can accelerate progress. A major challenge in duplicating the human mind is the fact that nobody fully grasps how our brains work. Nevertheless, in the past few years, there have been real advances in several fields of research in the building of an AI system. For example, neuroscientists have gained considerable knowledge about how learning occurs at the level of a single synapse. Computer scientists have modeled and simulated the neural mechanisms to obtain the behavior of neurons using supercomputers. Engineers have investigated nanoelectronics with some promising results such as more computation power, higher density, and less power consumption.

What kinds of AI systems will we introduce in the future? A first approach to get thinking machines is to develop the technology that can let us control a machine simply by thinking. This is called either *brain–machine interface* (BMI) or *brain–computer interaction* (BCI). BCI has had reasonable success in moving parts of the human body like arms, legs, and wrists. BCI technology could be further extended into stimulating brain and muscles simultaneously to restore the movement of physically disabled people. Nevertheless, real enhancement of the physically disabled still has a long way to go in that researchers all over the world are still struggling to find the basic principles of neural activities in the motor cortex, a part of the brain controlling movement.

In the long-term future, we could directly feed a much denser stream of information to our retinas by optical implants, including contact lenses, and to nanosystems implanted on our cerebral cortexes. An app will tell you what to do and will guide your work. As illustrated in *Avatar*, brain waves from thinking thus would control all the machines and even yield person-to-person communication via a shared neural network.

After 2020, we expect to have AI robots substantially better in performance than present humanoid robots. However, if we use conventional CMOS integrated circuits for them, the power consumption may surpass 100 kW, which is almost one-third of that of the IBM Watson computer that beat a human expert in the game Jeopardy (Markoff 2011). The human brain runs on about 20 W, while the power for the robot as smart as the human brain (100 petaFLOPS) requires 100 MW—a small nuclear power plant if you use conventional CMOS utilizing von Neumann-based software. Thus, energy efficiency fundamentally limits our ability to realize AI systems unless we do not use CMOS for the AI hardware. Furthermore, other big technological obstacles in today's CMOS integrated circuits are unacceptable variations in properties, size, noise, defects, and density of devices. Up to now, no solution to overcome these barriers has been found, regardless of the form in the name of beyond-CMOS.

In addition, computer-based computation is very inefficient at human tasks such as adaptability, pattern recognition, and error tolerance. This is because the human brain processes information in parallel mode even 10,000,000 times slower than today's desktop computer. Thus, the R&D priority for the materialization of AI systems could be given to brain-inspired neuromorphic devices. Mimicking the function of human brain through hardware development could be the first prerequisite to the successful buildup of AI systems. Fortunately, recent advancements in nanoelectronics including phase change memory (PCM) and metal oxide resistive switching (random-access) memory (RRAM) and other memories have allowed renewed hope in emulating the human brain, because they offer much higher memory density and lower power consumption, which are essential for artificial synapses in neuromorphic hardware. Nevertheless, the development of brain-inspired neuromorphic devices is still in the early stage of research.

From this perspective, mapping the entire brain of an individual would be required for complete preservation of the brain, and perhaps even to understand the mechanisms of brain functions. This is called the *connectome*, a comprehensive map of neurons to fully catch up with learning, recognition, and reasoning mechanisms (Kasthuri and Lichtman 2010). However, there is an insurmountable technological difficulty associated with mapping the entire human brain, since our brain consists of perhaps 10^{11} neurons and 10^{14}–10^{15} synapses. More than 1 zettabyte (10^{21} bytes) of random access memory (RAM) would be needed to store all the mapping information. To put that figure in perspective, IBM's Watson computer—the winner of Jeopardy—contains only 16 terabytes (1.6×10^{13}) of RAM, and the year 2012's total digital data produced from all over the world was 2.8 zettabytes.

An alternative viewpoint argues that the information defining an individual human's mind with reasonable fidelity is much, much smaller, for example encompassing perhaps only 50,000 episodic memories of life events, each of which consists of a very small net of connections to memories of concepts, which themselves number only in the tens of thousands (Bainbridge 2002, 2003). Perhaps the large number of neurons in the human brain is merely evolution's way of compensating by means of massively parallel processing for how slowly each neuron reacts compared to computer components. Thus, while neurons are the fundamental units in the brain, perhaps much larger modules encompassing many neurons actually represent concepts at a far grosser scale, and only they would need to be emulated by a computer. Or, it may be that nanoscale processes inside each neuron are crucial, and the previous paragraph even underestimates the difficulty of the challenge. These debates are among many that can be resolved only through convergence across many fields of science.

2.8.5 Human–Robot Interaction: An Emerging Field Dependent Upon CKTS

Refer to: WTEC *Study on R&D in Human–Robot Interaction* (http://www.wtec.org/reports.htm)

In the last decade or so, robotics research has become increasingly focused on understanding and defining the dynamic interactions between human and robot, to help make robots easier for people to use in a wider variety of situations. Many emerging applications of robotic systems are being based on models of human intelligence and behavior. In general, robotic applications trend today toward *proximate* interactions of humans with robots, and "*toward peer or mentor roles*" (Goodrich and Schultz 2007, 234–235).

These trends have pushed the development of a new field, that of human–robot interaction (HRI), where clinical and rehabilitative medicine, biomedical engineering, social psychology, neuroscience, cognitive science, human factors research, artificial intelligence, organizational behavior, anthropology, linguistics, and even standards-setting governance, all have become essential to advancing the ability of robots to meet human needs.

As the U.S. Office of Naval Research puts it, HRI aims to "develop the underlying principles and technology that will enable autonomous vehicles and robots to work with people as capable partners."[1] Prestigious university, industrial, and national laboratories across the United States, Europe, and East Asia are investigating and refining myriad applications for HRI, including the ones listed below. Each has different human–robot interaction paradigms and interface modalities, depending on the robot's role:

- Robotic assistance for seniors and persons recovering from injuries or with disabilities to provide physical, cognitive, safety, occupational and physical

---

[1] http://www.onr.navy.mil/en/Media-Center/Fact-Sheets/Human-Robotic-Interaction.aspx

therapy, and/or social support, where it is critical for the human to feel safe, comfortable, and "in control" when interacting with the robot
- Robotic devices for microsurgery and telesurgery where the communication between surgeon and robot must be seamless and intuitive and the mechanical capabilities must be intricate
- Robots that provide teaching, interaction, and therapy for individuals with autism spectrum disorders or trauma, who may respond better to mechanical devices than to social interaction
- Service robots that perform innumerable, mostly precise, automated tasks in manufacturing, mining, inventory management, agriculture, even as receptionists
- Robots for entertainment and education, as dance partners, tour guides, storytellers, even pets
- Unmanned intelligent or autonomous space, air, naval, underwater, and ground vehicles for military, academic, private, and public use under a variety of complex operating conditions
- Robot-assisted search and rescue operations in which human–robot interactions must be capable of considerable complexity because of the inherently unstructured nature of such work
- Robot-assisted space exploration with specific challenges due to extreme operating conditions. Interactions include both time-lag and proximal interactions such as the robot assisting a human to transport equipment, perform physical tasks, and/or provide sensing and information

2.9 International Perspectives

The following are summaries relevant to this chapter of discussions at the international regional WTEC NBIC2 workshops held in Leuven, Belgium, September 20–21, 2012; in Seoul, Korea, October 15–16, 2012; and in Beijing, China, October 18–19, 2012. Further details of those workshops are provided in Appendix A.

2.9.1 United States–European Union NBIC2 Workshop (Leuven, Belgium)

Participants: *Elements of all three EU working groups' deliberations had relevance for this topic, as presented in the final plenary session. Participants in the Leuven workshop are listed in Appendix B.*

This workshop's participants identified three major interrelated themes where convergence could achieve great continued progress: (1) productivity, (2) human capacity, and (3) education. Increased productivity can create many new jobs, even as some older jobs become obsolete, so long as human capacity is also increased, and transforming education is one of the best ways to accomplish this.

NBIC-enhanced automation, carefully designed standardization across fields, and more efficient use of time and materials will be essential for future productivity. Techniques such as neuromorphic engineering and transformational communication media can employ machines to increase human capabilities, thereby improving both ourselves and the tasks we can perform. Essential for success will be changes in education at all levels—vocational, continuing education, undergraduate, and graduate education. This is not just a zero-sum game, but progress with a positive sum in which the entire world can participate, through knowledge-sharing, creative imagination, and new approaches to economic challenges, prominently including social justice and issues of negative externality defined in terms of the limited resources of the planet.

Nearly limitless productivity growth can be achieved through convergence of a wide range of new technologies, from nanomaterials, to robotics, to knowledge-based manufacturing that enables personalized production. Immersive technologies, based on sensors and adaptive human–machine interaction systems, can render robots more autonomous and human-like, even as the human experience of work can be improved on the basis of anthropomorphic interactions between psychology and cognitive science on the one hand, and robotics and machine learning on the other hand. Education must empower people to act—democratizing the ways to play and to experiment, while permitting appropriate standardization—in the context of a rethinking of high-school education to include vocational training in machine–human interfaces because of the changes in the job market. Ph.D. training could be reconceptualized as a team effort, or always involving more than one field, to escape the inhibiting constraints of assuming higher degrees are based on one lone person in one narrow field.

2.9.2 United States–Korea–Japan NBIC2 Workshop (Seoul, Korea)

Panel members/discussants:

Wonjong Yoo (Co-Moderator), SKKU Advanced Institute of Nanotechnology (Korea)
Takeshi Kawano (Co-Moderator), Toyohashi University of Technology (Japan)
H.-S. Philip Wong (Co-Moderator), Stanford University (U.S.)

Others:

Sung Ha Park, Sungkyunkwan University (Korea)
Jiyoung Kim, Kookmin University (Korea)
Mitsuo Kawato, ATR Brain Information Communication Research Laboratory (Japan)
Kazunobu Tanaka, Japan Science and Technology Agency (JST, Japan)
Myung-Ae Chung, Electronics and Telecommunications Research Institute (ETRI, Korea)

S. Kawamura, JST (*Japan*)
Young-Jae Lim, ETRI (*Korea*)
Yong-Joo Kim, Korea Electrotechnology Research Institute (*KERI, Korea*)
Changhwan Choi, Hangyang University (*Korea*)
Sanghee Sun, Korea Institute of Science and Technology (*KIST, Korea*)
Young Jik Lee, ETRI (*Korea*)

This group emphasized the ubiquitous nature of communications, globally, and the shift from a world economy based on industrial production to one giving greater emphasis to services such as healthcare and entertainment. It identified some specific convergence trends, including collaboration between electronic and biological research, nanobiotechnology, and the increasing role for social science as the economy shifts further toward services. The volatility of new industries poses challenges for people pursuing careers in science and engineering, even as research and development become global enterprises. The economically advanced nations must provide adequate healthcare and other services for their aging populations but must look beyond merely physical survival to mental and emotional well-being. Some forms of mental disorder may be cured or at least managed by new methods resulting from nano-bio-cogno convergence, but it will also be crucial to support improved mental functioning through life-long education and forms of entertainment designed to be substantively beneficial rather than merely diverting.

2.9.3 United States–China–Australia–India NBIC2 Workshop (Beijing, China)

***Panel members/discussants*:**

Shushan Cai (*Co-Moderator*), *Tsinghua University* (*China*)
Tanya Monro (*Co-Moderator*), *University of Adelaide* (*Australia*)
H.-S. Philip Wong (*Co-Moderator*), *Stanford University* (*U.S.*)

***Others*:**

Jonathan Manton, *University of Melbourne* (*Australia*)
Tianzi Jiang, *Academy of Sciences* (*China*)
Tingshao Zhu, *Academy of Sciences* (*China*)
Chen Chen (*China*)

This group emphasized issues directly or indirectly connected to health. In recent years, convergence between fields has begun but is incomplete. Medical patients have greater knowledge about their conditions, and there exists increased public concern about the environmental impact of new technologies, coupled with a general social acceptance of technological change. In the near future, we can expect improved mental health and general quality of life improvements from convergence. Mental health will benefit from far more comprehensive understanding of the causes of problems, for example, based on data mining of information about large numbers of cases and more advanced diagnostic methods, leading to improved treatments. Personalized medicine can especially advance through ubiquitous diagnostics, using sensors and other information technologies to collect data about peoples' individual health as they go about their daily activities. New assistive medical devices may replace and augment functions. Outside the domain of health, the quality of life can benefit from improved social well-being, including social management and appropriate intervention, in a safe environment with secure and efficient access to food, water, and other necessities. We can create an NBIC-technologically literate society by bringing together clinicians, scientists, and engineers to create a common language and train the next generation to be real-world problem solvers, through transdisciplinary research and education.

References

Bainbridge, W.S.: A question of immortality. Analog **122**(5), 40–49 (2002)
Bainbridge, W.S.: Massive questionnaires for personality capture. Soc. Sci. Comput. Rev. **21**(3), 267–280 (2003)
Bainbridge, W.S.: Expanding the use of the internet in religious research. Rev. Relig. Res. **49**(1), 7–20 (2007a)
Bainbridge, W.S.: Governing nanotechnology: social, ethical and political issues. In: Bhushan, B. (ed.) Handbook of Nanotechnology, 2nd edn, pp. 1823–1839. Springer, Berlin (2007b)
Bainbridge, W.S.: Nanoconvergence. Prentice-Hall, Upper Saddle River (2007c)
Bainbridge, W.S.: The convergence of sociology and computer science. In: Hartman, H. (ed.) Integrating the Sciences and Society: Challenges, Practices, and Potentials, pp. 257–278. JAI, Bingley (2008)
Bainbridge, W.S.: Converging technologies. In: Kyrre, J., Olsen, B., Pedersen, S.A., Hendricks, V.F. (eds.) A Companion to the Philosophy of Technology, pp. 508–510. Wiley-Blackwell, Malden (2009)
Bainbridge, W.S.: The Virtual Future: Science-Fiction Gameworlds. Springer, London (2011)
Bainbridge, W.S., Roco, M.C. (eds.): Managing Nano-Bio-Info-Cogno Innovations: Converging Technologies in Society. Springer, Berlin (2006a)
Bainbridge, W.S., Roco, M.C. (eds.): Progress in Convergence: Technologies for Human Well-Being. New York Academy of Sciences, New York (2006b)
Baldwin, C., von Hippel, E.: Modeling a paradigm shift: from producer innovation to user and open collaborative innovation. Organ. Sci. **22**(6), 1399–1417 (2011)
Bell, D.: The Coming of Post-industrial Society. Basic Books, New York (1973)
Berman, H.M.: The protein data bank. In: Bainbridge, W.S. (ed.) Leadership in Science and Technology, pp. 661–667. Sage, Thousand Oaks (2012)
Berners-Lee, T., Fischetti, M.: Weaving the Web. Orion Business, Britain (1999)
Bimber, B.: Three faces of technological determinism. In: Smith, M.R., Marx, L. (eds.) Does Technology Drive History? pp. 79–100. MIT Press, Cambridge (1994)
Bohannon, J.: Meeting for peer review at a resort that's virtually free. Science **331**, 27 (2011)
Brunner, J.: The Shockwave Rider. Harper & Row, New York (1975)
Brynjolfsson, E., McAfee, A.: Race Against the Machine. Digital Frontier Press, Lexington (2011)
Bulmer, M., Bulmer, J.: Philanthropy and social science in the 1920s: Beardsley ruml and the Laura Spellman Rockefeller memorial, 1922–29. Minerva **19**, 347–407 (1981)
Caroli, C., Scipione, G., Rrapi, E., Trotta, G.: ARROW: Accessible Registries of Rights Information and Orphan Works towards Europeana, D-Lib Magazine **18**. http://www.dlib.org/dlib/january12/caroli/01caroli.html (2012)

Chesbrough, H.W.: Open Services Innovation: Rethinking Your Business to Grow and Compete in a New Era. Jossey-Bass, San Francisco (2011)
Clarke, A.C.: The City and the Stars. Harcourt, Brace, New York (1956)
Crane, G.: The perseus project. In: Bainbridge, W.S. (ed.) Leadership in Science and Technology, pp. 644–652. Sage, Thousand Oaks (2012)
Crevier, D.: AI: The Tumultuous Search for Artificial Intelligence. Basic Books, New York (1993)
Darrach, B.: Meet Shaky, the first electronic person. Life: 56–68, 20 November 1970
Farina, C.R., Miller, P., Newhart, M.J., Cardie, C., Cosley, D., Vernon, R.: Rulemaking in 140 characters or less: social networking and public participation in rulemaking. Pace Law Rev. **31**(1) (2011)
Fiegl, C.: First look at the next stage of meaningful use. American Medical News, March 5 (2012a)
Fiegl, C.: Stage 2 meaningful use rules sharply criticized by physicians. American Medical News, May 13 (2012b)
Fink, D., Hochachka, W.M., Zuckerberg, B., Winkler, D.W., Shaby, B., Munson, M.A., Hooker, G., Riedewald, M., Sheldon, D., Kelling, S.: Spatiotemporal exploratory models for broad-scale survey data. Ecol. Appl. **20**(8), 2131–2147 (2010)
Fisher, D.: Fundamental Development of the Social Sciences: Rockefeller Philanthropy and the United States Social Science Research Council. University of Michigan Press, Ann Arbor (1993)
Florida, R.L.: Who's Your City? Basic Books, New York (2008)
Fountain, J.E.: Developing a Basic Research Program for Digital Government: Information, Organizations and Governance. Report of a National Workshop Held in May 2002, National Center for Digital Government, Kennedy School of Government, Harvard University (2002)
Freiberger, P., Swaine, M.: Fire in the Valley: The Making of the Personal Computer. Osborne/McGraw-Hill, Berkeley (1984)
Gargouri, Y., Hajjem, C., Lariviere, V., Gingras, Y., Brody, T., Carr, L., Harnad, S.: Self-selected or mandated, open access increases citation impact for higher quality research. PLoS ONE **5**(10), e13636 (2010). arXiv:1001.0361v2. http://dx.doi.org/10.1371/journal.pone.0013636
Gawer, A.: Platform dynamics and strategies: from products to services. In: Gawer, A. (ed.) Platforms, Markets, and Innovation, pp. 45–76. Edward Elgar, Cheltenham (2009)
Goggin, G.: Cell Phone Culture: Mobile Technology in Everyday Life. Routledge, New York (2012)
Goodrich, M.A., Schultz, A.C.: Human–robot interaction: a survey. Found. Trends Hum. Comput. Interact. **1**(3), 203–275 (2007). http://dx.doi.org/10.1561/1100000005
Goonan, K.A.: Queen City Jazz. Tor, New York (1994)
Goonan, K.A.: Mississippi Blues. Tor, New York (1997)
Goonan, K.A.: Crescent City Rhapsody. Eos, New York (2000)
Goonan, K.A.: Light Music. Eos, New York (2002)
Howard, P.N.: The Digital Origins of Dictatorship and Democracy: Information Technology and Political Islam. Oxford University Press, New York (2010)
Howard, P.N., Jones, S.: Society Online: The Internet in Context. Sage, Thousand Oaks (2004)
Jensen, C., Scacchi, W.: Process modeling across the web information infrastructure. Softw. Process Improv. Pract. **10**(3), 255–272 (2005)
Johnson, R.G., Richardson Jr., E.S.: Pennsylvanian invertebrates of the mazon creek area, Illinois: the morphology and affinities of *tullimonstrum*. Fieldiana Geol. **12**(8), 119–149 (1969). Chicago: Field Museum of Natural History
Kasthuri, N., Lichtman, J.W.: Neurocartography. Neuropsychopharmacology **35**(1), 342–343 (2010)
Kawrykow, A., Roumanis, G., Kam, A., Kwak, D., Leung, C., et al.: Phylo: a citizen science approach for improving multiple sequence alignment. PLoS ONE **7**(3), e31362 (2012). http://dx.doi.org/10.1371/journal.pone.0031362
Keynes, J.M.: The General Theory of Employment, Interest and Money. Harcourt, Brace, New York (1936)
Khatib, F., DiMaio, F., Foldit Contenders Group, Foldit Void Crushers Group, Cooper, S., Kazmierczyk, M., Gilski, M., Krzywda, S., Zabranska, H., Pichova, I., Thompson, J., Popović,

Z., Jaskolski, M., Baker, D.: Crystal structure of a monomeric retroviral protease solved by protein folding game players. Nat. Struct. Mol. Biol. **18**, 1175–1177 (2011)

Kuhn, T.S.: The Copernican Revolution: Planetary Astronomy in the Development of Western Thought. Harvard University Press, Cambridge, MA (1957)

Kuhn, T.S.: The Structure of Scientific Revolutions. University of Chicago Press, Chicago (1962)

Kurzweil, R.: The Age of Spiritual Machines. Viking, New York (1999)

Lastowka, G.: Virtual Justice: The New Laws of Online Worlds. Yale University Press, New Haven (2010)

Lin, A.Y.M., Novo, A., Har-Noy, S., Ricklin, N., Stamatiou, K.: Combining GeoEye-1 satellite remote sensing, UAV aerial imaging, and geophysical surveys in anomaly detection applied to archaeology. IEEE J. Sel. Top. Appl. Earth Obs. Remote Sens. **4**(4), 870–876 (2011)

Markoff, J.: Computer Wins on "Jeopardy!". New York Times, February 17: A1 (2011)

McCarthy, J., Minsky, M.L., Rochester, N., Shannon, C.E.: A proposal for the Dartmouth summer research project on artificial intelligence. http://www-formal.stanford.edu/jmc/history/dartmouth/dartmouth.html (1955)

McCorduck, P.: Machines Who Think. A. K. Peters, Natick (2004)

McKnight, L.W., Kuehn, A.: Creative destruction. In: Bainbridge, W.S. (ed.) Leadership in Science and Technology, pp. 105–113. Sage, Thousand Oaks (2012)

Moravec, H.P.: Mind children: The Future of Robot and Human Intelligence. Harvard University Press, Cambridge, MA (1988)

Nardi, B.: My Life as a Night Elf Priest: An Anthropological Account of World of Warcraft. University of Michigan Press, Ann Arbor (2010)

Noll, J., Scacchi, W.: Supporting software development in virtual enterprises. J. Digi. Inform. **1**(4) (1999). http://journals.tdl.org/jodi/index.php/jodi/article/view/13/12

Ogburn, W.F.: Social Change with Respect to Culture and Original Nature. B.W. Huebsch, New York (1922)

Olson, G.M., Zimmerman, A., Bos, N. (eds.): Scientific Collaboration on the Internet. MIT Press, Cambridge, MA (2008)

Picard, R.W.: Affective Computing. MIT Press, Cambridge, MA (1997)

Piwowar, H.A.: Who shares? Who doesn't? Factors associated with openly archiving raw research data. PLoS ONE **6**(7), e18657 (2011). http://dx.doi.org/10.1371/journal.pone.0018657

Plumer, B.: The economics of video games, Washington Post. http://www.washingtonpost.com/blogs/ezra-klein/wp/2012/09/28/the-economics-of-video-games/ (2012). Accessed 28 September 2012

Roco, M.C., Bainbridge, W.S. (eds.): Converging Technologies for Improving Human Performance. Kluwer, Dordrecht (2003)

Roco, M.C., Montemagno, C.D. (eds.): The Coevolution of Human Potential and Converging Technologies. New York Academy of Sciences, New York (2004)

Rowan, D.: Open university: Joi Ito plans a radical reinvention of MIT's Media Lab. Wired Magazine UK. http://www.wired.co.uk/magazine/archive/2012/11/features/open-university?page=all (2012). Accessed 15 November 2012

Scacchi, W.: Game-based virtual worlds as decentralized virtual activity systems. In: Bainbridge, W.S. (ed.) Online Worlds: Convergence of the Real and the Virtual, pp. 225–236. Springer, New York (2010)

Scacchi, W. (ed.): The Future of Research in Computer Games and Virtual Worlds. Institute for Software Research, University of California, Irvine, Irvine (2012)

Scacchi, W., Nideffer, R., Adams, J.: Collaborative game environments for informal science education: DinoQuest and DinoQuest Online. In IEEE Conf. Collaboration Technology and Systems (CTS 2008), pp. 229–236. Irvine (extended version) (2008)

Scacchi, W., et al.: Towards a science of open source systems. Final Report from the CCC 2010 Workshop on the Future of Research in Free/Open Source Software. http://foss2010.isr.uci.edu/sites/foss2010.isr.uci.edu/files/CCC-FOSS-FinalReport-29Nov10.pdf (2010)

Schweik, C.M., English, R.C.: Internet Success: A Study of Open-Source Software Commons. MIT Press, Cambridge, MA (2012)

Simon, H.A.: The Sciences of the Artificial. MIT Press, Cambridge, MA (1996)

Smith, A.M., Lynn, S., Sullivan, M., Lintott, C.J., Nugent, P.E., Botyanszki, J., Kasliwal, M., Quimby, R., Bamford, S.P., Fortson, L.F., Schawinski, K., Hook, I., Blake, S., Podsiadlowski, P., Joensson, J., Gal-Yam, A., Arcavi, I., Howell, D.A., Bloom, J.S., Jacobsen, J., Kulkarni, S.R., Law, N.M., Ofek, E.O., Walters, R.: Galaxy zoo supernovae. Mon. Not. R. Astron. Soc. **412**, 1309–1319 (2011)

Spohrer, J., Piciocchi, P., Bassano, C.: Three frameworks for service research: exploring multilevel governance in nested, networked systems. Serv. Sci. **4**(2), 147–160 (2012)

Steinkuehler, C.: Massively multiplayer online games and education: an outline of research. In: Proceedings of the 2007 Conference on Computer-Supported Collaborative Learning, pp. 675–685. ACM, New York (2007)

Stephenson, N.: The Diamond Age. Bantam, New York (1994)

Tapia, A., Ocker, R., Rosson, M.B., Blodgett, B., Ryan, T.: Computer tomography virtual organization. In: Bainbridge, W.S. (ed.) Leadership in Science and Technology, pp. 602–610. Sage, Thousand Oaks (2012)

Williams, G.C.: Pleiotropy, natural selection, and the evolution of senescence. Evolution **11**, 398–411 (1957)

Chapter 3
Convergence Platforms: Earth-Scale Systems

Bruce Tonn, Mamadou Diallo, Nora Savage, Norman Scott,
Pedro Alvarez, Alexander MacDonald, David Feldman,
Chuck Liarakos, and Michael Hochella

Earth-scale systems have many forms, and all are dynamically interrelated. *Environmental* Earth-scale systems include the global atmosphere, geological systems, Earth–Sun and space interactions, Earth electric and magnetic fields, nitrogen cycles, climate patterns, ocean currents, regional biodiversity and regional freshwater systems, minerals, water, and energy resource distributions on Earth (Fig. 3.1). The global production and consumption of energy can be considered as a separate

---

Corresponding editors M.C. Roco (mroco@nsf.gov) and W.S. Bainbridge (wbainbri@nsf.gov).

B. Tonn
University of Tennessee, Knoxville, TN, USA

M. Diallo
California Institute of Technology, Pasadena, CA, USA

Korea Advanced Institute of Science and Technology, Daejeon, Korea

N. Savage
U.S. Environmental Protection Agency, Washington, DC, USA

N. Scott
Cornell University, Ithaca, NY, USA

P. Alvarez
Rice University, Houston, TX, USA

A. MacDonald
National Aeronautics and Space Administration, Washington, DC, USA

D. Feldman
University of California, Irvine, Irvine, CA, USA

C. Liarakos
National Science Foundation, Arlington, VA, USA

M. Hochella
Virginia Polytechnic Institute and State University, Blacksburg, VA, USA

M.C. Roco et al. (eds.), *Convergence of Knowledge, Technology and Society: Beyond Convergence of Nano-Bio-Info-Cognitive Technologies*, Science Policy Reports, DOI 10.1007/978-3-319-02204-8_3, © Springer International Publishing Switzerland 2013

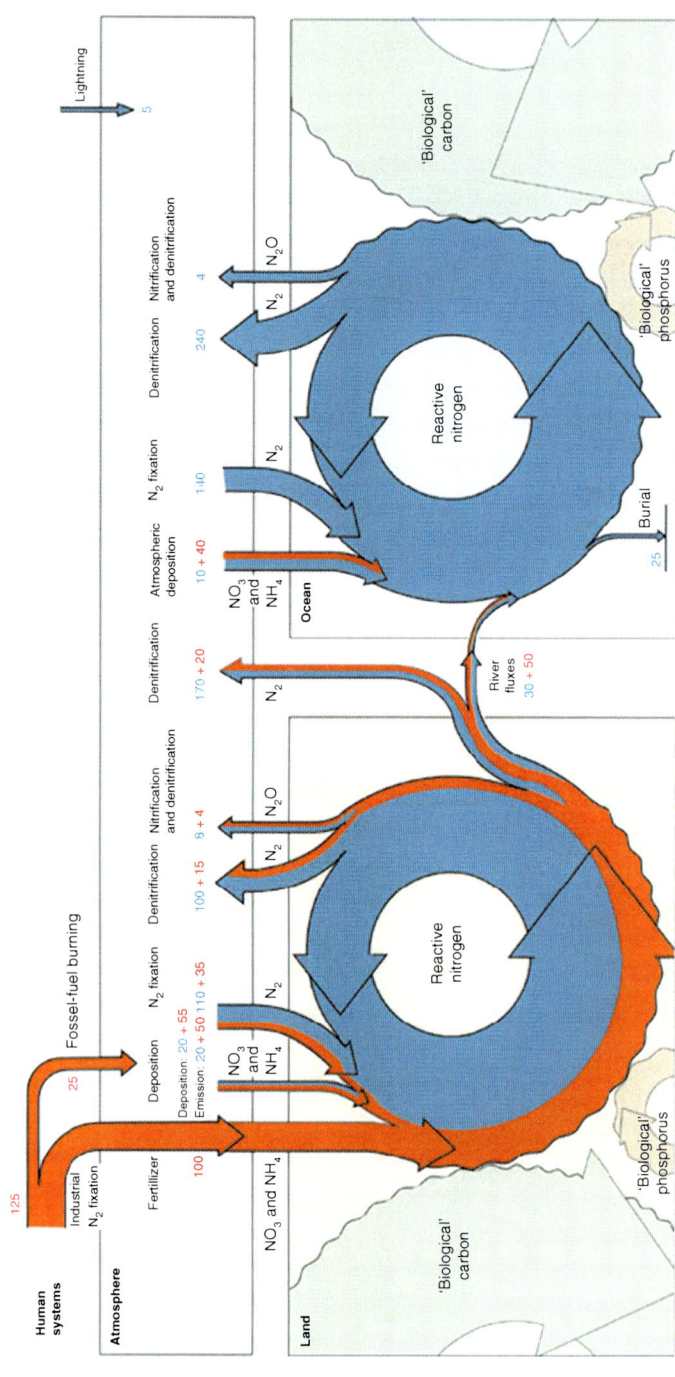

Fig. 3.1 Depiction of the global nitrogen cycle on land and in the ocean (Gruber and Galloway 2008; © Nature Publishing Group, used by permission)

Earth-scale system that is intimately tied to the global environment. Various *man-made technological* systems have the requisite scale and scope to be considered Earth-scale, from the global telecommunications system to massive infrastructure systems that serve major metropolitan areas and span nations and continents. For the purposes of this discussion, humanity's exploration and exploitation of space also falls within the purview of this chapter.

Maintaining the health of Earth-scale systems is a constant and seemingly growing challenge. The global environment continues to suffer from a plethora of changes and risks, from astronomical events and climate change to human activities and widespread species extinction. Pollution, land use change, and over-exploitation are combining to diminish the benefits that Earth's ecosystems can provide to humanity (Burkhard et al. 2010). Water is in short supply worldwide, and the availability of potable water is decreasing at the same time as population and per-capita usage is increasing (IWMI 2007; UN-Water 2013). By 2025, water scarcity will have spread further; India and China will continue to be the largest countries facing water stress, and a number of countries in Africa and the Middle East will face extreme water scarcity (Grail Research 2009). Poor water quality and food imbalances adversely impact human health across the globe (Pimentel and Pimentel 2006). The expansion of the world economy is accelerating the consumption of nonrenewable resources such that peak consumption of oil (Deffeyes 2010), natural gas, coal (Maggio and Cacciola 2012), rare earth minerals (Cherry 2011), and even phosphates may all be reached within this century (Clabby 2012). Dealing with these and other Earth-scale system problems is complicated by many factors, including aging populations and rapid urbanization. Also complicating this task is that most Earth-scale systems are intrinsically dynamic, nonlinear, quite sensitive to natural and human perturbations, and interrelated in very complicated fashions.

This chapter explores the relationships between Earth-scale systems and converging technologies, which are broadly defined to include the "NBIC" technologies: nanotechnologies, biotechnologies, information technologies, and cognitive technologies (Roco and Bainbridge 2003), plus derivative advanced technologies (e.g., many that compose the global energy system) that together may be termed "NBIC2" technologies or, broadly, converging knowledge and technology for society (CKTS). Earth-scale systems can encompass converging technologies (e.g., the global energy system encompasses various components built on nanotechnology advances); can be studied using converging technologies (e.g., simulating the Earth's climate with supercomputer-based global climate models); and/or can be managed using a convergence of technologies (e.g., through geo-engineering). In this chapter, we consider Earth-scale systems from five viewpoints: knowledge systems, monitoring systems, communication systems, management systems, and Earth-scale and other contributing technologies:

- *Knowledge Systems*: These systems are related to how we gain knowledge about Earth-scale systems. Thus, falling within this rubric are specific methods of investigation, data resources, and theoretical frameworks for conceptualizing Earth-scale systems (e.g., as complex, nonlinear systems). It can be argued that a global knowledge network ties all this together.

- *Monitoring Systems*: These systems collect and process data about the Earth-scale systems. These systems also facilitate the measurement of Earth-scale systems. Components of these systems include satellites, sensors, telescopes, gauges, Internet-based traffic information systems, experiments, and social science surveys.[1]
- *Communication Systems*: These systems provide the physical infrastructure for communications involving the knowledge networks and monitoring systems. These systems encompass human–system interfaces needed to support Earth-scale science and policymaking. These systems also facilitate communication with the public.
- *Management Systems and Tools*: This perspective encompasses those things needed to manage Earth-scale systems. Thus, included here are systems to control Earth-scale systems (e.g., geo-engineering of the global environmental system; Kintisch 2010; Victor et al. 2009); technologies to ameliorate negative aspects of Earth-scale systems (e.g., scrubbers on fossil fuel power plant smoke stacks); and institutions (including organizations, incentives, international agreements) needed to manage Earth systems.
- *Earth-scale and other contributing technological systems*: The Earth-scale technological systems that receive the most attention in this chapter include the global energy production and consumption system; the worldwide telecommunications system; major urban and national infrastructure systems (e.g., roads, bridges, ports, canals, electricity transmission systems, waste and storm water systems, subways, railroads, airports); and humanity's efforts in space. An important contributing technology that links to other chapters of this report is robotics.

Using this framework, this chapter puts forth a vision of Earth-scale systems and converging technologies, documents advances made in the last 10 years, proposes goals for the next decade, discusses infrastructure and R&D strategies to achieve these goals, and sets out priorities. Relationships with other convergence platforms discussed elsewhere in this report are also addressed. The chapter concludes with a set of case studies and examples.

3.1 Vision

3.1.1 Changes in the Vision Over the Past Decade

The last 10 years have witnessed the historically unprecedented growth in science and exponential technology change. If anything, the vision for science and technology's contributions to solving difficult societal problems has become stronger

---

[1] One example in the field of climate science and sustainability is SNOTEL, a weather-related automated or telemetered monitoring system of weather stations that serves as a snowfall depth monitoring network (established by the USGS; see Trimble et al. 1997). Another is NSF's NEON observatory designed to gather and provide ecological data on the impacts of climate change, land use change, and invasive species on natural resources and biodiversity.

during this time period as the problems have become worse and as political systems around the world have struggled to find institutional solutions to the problems.

This vision is especially strong with respect to Earth-scale environmental, energy, and infrastructure systems. Technological solutions to simultaneously solve the world's energy problems while reducing greenhouse gas emissions are given exceptional credence. This is due in part to spectacular technological change that has impacted the daily lives of almost every human on Earth. Advancements in science—in nanotechnology and biotechnology in particular—are seen as playing major roles in the energy arena. A best-case scenario is one where cost-effective scientific and technological solutions are found to Earth-scale systems problems but do not have serious political, social, or economic externalities.

Our vision of science and technology has matured over the last decade to where is it widely understood that no matter how much we wish this scenario to become reality, it is not likely that this scenario will unfold. It is increasingly clear that every technological solution has drawbacks and will require rational consideration of tradeoffs. For example, environmental/energy solutions such as harnessing wind and biomass raise concerns about environmental aesthetics (Johansson and Laike 2007) and food prices (Timilsina et al. 2012), respectively, and may have unanticipated environmental impacts (e.g., on bird populations or soil erosion). Intensive monitoring of Earth-scale systems and human use of and interaction with these systems may raise serious concerns about privacy. Various scenarios of technological convergence may solve energy, climatic, and other environmental problems while requiring significant changes in our conceptions of employment and the roles of markets, government, and the nonprofit world (Tonn and Stiefel 2013). Complicating matters further are the realizations that all Earth-scale systems are dynamically connected and our data and knowledge about the systems are limited.

The acceleration of technological change, in part fueled by advances in NBIC technologies, ironically has increased impatience in finding and implementing scientific and technological solutions. Even though at some level people must understand that changing any Earth-scale system is much more difficult than creating the next version of the iPhone, the pace of technological change witnessed by people in their lives cannot help but influence their expectations for swifter technological changes to solve Earth-scale system problems (Gleick 1999; Davis and Meyer 1998).

One can argue that balancing impatience is humanity's embracing of the concept of sustainability. Over the past 10 years, there has been a sea change in thinking about the future of the Earth, from grassroots nonprofit organizations (e.g., see ICLEI Local Governments for Sustainability, http://www.iclei.org) to corporate boardrooms (e.g., the International Institute for Sustainable Development, http://www.iisd.org/), to global institutions (e.g., the United Nations Division for Sustainable Development, http://www.un.org/esa/dsd/ and the World Bank[2]), where

---

[2] See the Sustainable Development Portal of the World Bank: http://web.worldbank.org/WBSITE/EXTERNAL/TOPICS/EXTSDNET/0,,menuPK:4812133~pagePK:64885066~piPK:4812134~theSitePK:5929282,00.htm

the connective thread is sustainability. The sea change involves the now almost universal acceptance of the concept of (global) sustainability as a fundamental human societal value.[3] This means that people and governments the world over are truly concerned about Earth-scale systems and are dedicated to finding long-term solutions, which may help to temper impatience with forthcoming solutions.

3.1.2 The Vision for the Next Decade

Knowledge Systems

One trend that can be expected to accelerate in the next 10 years is the globalization of Earth-scale systems research, especially research on Earth-scale environmental systems. Research capabilities are rapidly improving in Asia, South America, and Africa, to complement those in Western countries. Global climate modeling will continue to increase in spatial resolution, facilitating more effective regional modeling and climate policymaking. New data describing natural and human systems (e.g., data describing the environmental impacts of products over the course of their life cycles) will foster the development of more powerful analytical frameworks and the integration of systems analysis across theoretical disciplines. Frameworks will be developed to more readily identify opportunities to synthesize NBIC technologies for convergence on Earth-scale system platforms to produce synergistic benefits for humanity (i.e., CKTS).

One can also imagine that the number of boundary organizations will increase tremendously, following the lead of the Intergovernmental Panel on Climate Change (IPCC) internationally and various very successful efforts in the United States and other countries. Advances in information and cognitive technologies, many noted below, will facilitate improvements in online group collaboration and access to data, models, and information. As the world's societies continue to flatten (Friedman 2005), collaborative public policymaking will become the de facto standard approach to deal with Earth-scale systems.

Monitoring Systems

Monitoring of the environment, and energy, telecommunication, and infrastructure systems will increase by several orders of magnitude as the cost of sensors, cameras, tracking devices, and other equipment drops precipitously, along with the costs of transmitting and storing such data. Real-time data coverage of air and water quality and transportation systems will continue to increase. Technology will allow the

---

[3] Because of this, prospects for renewable energy, for example, have gone from *if* they will ever be widely implemented to *when* they will be implemented and how best they can be integrated into existing electricity grids.

closer monitoring of agricultural lands and crops, livestock, and water supplies. Ecosystem monitoring will improve through expanded satellite coverage and the tagging of key indicator species with tracking technology. Increasingly, people will wish to monitor not only their own health in real time but also their surroundings for toxic substances (Bostrom 2003).

Communication Systems

One can confidently assume that the capabilities of the globe's communications systems will continue to increase. Computing speeds and telecommunication bandwidths will continue to increase (Kurzweil 2005). Mobile computing platforms will become ubiquitous and more functional. More data, information, models, resources, and applications will become available to researchers and the public. Advances in information technology and nanotechnology will underlie these achievements. However, advances in cognitive technologies, in combination with advances in intelligent software, will be needed to assist users in accessing and using these resources. Software will seamlessly cascade models, aggregate content from social networking databases, customize presentation of complex information to non-experts (e.g., to overcome common heuristics and biases; Kahneman et al. 1982), and present visualizations of complex phenomena. A new generation of sophisticated systems will be developed to facilitate R&D collaboration as well as policymaking with respect to Earth-scale environmental systems.

Management Systems

One can envision that over the next 10 years, management of Earth-scale technological systems will become much more sophisticated. For example, to increase reliability and incorporate increasing contributions from renewable energy sources, electrical grids will become "smart."[4] In this same vein, intelligent transportation systems will be deployed to improve the efficiency and safety of highways and transit systems.[5]

During this period of time, scientists and policymakers will more seriously consider options for managing the global environment, especially if early signals are confirmed that the global climate is changing more rapidly than forecast. Innovative and creative experiments will be designed to test various schemes for carbon sequestration and climate cooling. More intensive management of regional water systems, natural amenities (e.g., forests, wetlands, fisheries), and hot spots of biodiversity can be expected. All these improvements in management of Earth-scale systems

---

[4] See the U.S. Department of Energy Smart Grid website http://energy.gov/oe/technology-development/smart-grid

[5] See the U.S. Department of Transportation Intelligent Transportation Systems website http://www.its.dot.gov/

will be supported by more robust databases created from a combination of more ubiquitous monitoring systems and pervasive global communications systems. Computer modeling and simulation of Earth systems will continue to improve as computers become ever more powerful and as more data are available to build and validate the models.

Earth-Scale and Other Technological Systems

One can envision substantial advances driven by NBIC technologies in the global energy sector over the next 10 years. Advances in nanotechnologies will dramatically increase the efficiencies of lights and photovoltaic cells and decrease the costs of hydrogen fuel cells. More broadly, materials research will lead to advances in composites for blades for wind turbines and car bodies, among many applications that impact the energy sector (e.g., high-temperature superconducting materials or carbon nanotubes for high-voltage transmission lines). Advances in biotechnologies will benefit the production of hydrogen and other fuels (e.g., from algae) and reduce the costs of cellulosic ethanol and other energy crops. A convergence of technologies will improve the capabilities to store energy from renewable sources in batteries and other media. A convergence of information and cognitive technologies will help individuals reduce energy use in their homes and vehicles.

The immense investments needed to build and maintain the world's infrastructure will spur an explosion of research and new technology development. Improvements will be made in systems designed to capture carbon during the production of cement. Innovations will continue in the areas of drinking water purification, road building materials, railroad technology, bridge building materials, and waste water management.

One major trend anticipated over the next decade is the globalization of space. The number of countries with space programs will continue to increase, as will their financial commitments to space programs (Devezas et al. 2012). Opportunities for international collaboration and competition will increase.

Advances in space systems will follow several lines of investigation. Commercial space activity will benefit from continued improvement in materials used to construct spacecraft on the one hand, and demands for space-based manufacturing of advanced nanomaterials and other materials on the other hand. Advanced robotics can be expected to play a major role in the latter. Nanotechnology and information technologies will converge to allow the development of more reliable and functional unmanned space exploration technologies. Nano-, bio-, information, and cognitive technologies will further converge on the problem of manned space exploration, with the design goal of allowing humans to live in space or on other planets without the need for resupply from Earth. Systems will be designed to provide power in the form of microwaves to the Earth from space and to deploy materials to shield portions of the Earth from the sun in order to cool the Earth's atmosphere, but these will probably not be built during the next 10 years.

3.2 Advances in the Last Decade and Current Status

3.2.1 *Knowledge Systems*

Knowledge systems that benefit research on Earth-scale systems are numerous, most of which have been supported by advancements in information technology. Over the past 10 years, advances have been made in:

- Complex and nonlinear systems
- Theoretical frameworks for understanding Earth-scale environmental problems (e.g., resiliency theory, ecological footprints)
- Emergence of Earth systems science as an organized discipline
- Data mining
- Machine learning, from genetic algorithms to neural nets
- Computer visualization, from power walls to 3D modeling
- High-performance computing
- Global climate modeling
- Crowd-sourced research supported by distributed computing data analysis[6]

There have also been advances in the overall knowledge network (Cash et al. 2003; Sarewitz and Pielke 2007; Jacobs et al. 2005). A prime example is the emergence of the Intergovernmental Panel on Climate Change as a well-respected global knowledge network that has significant influence on the management of the global environment. Boundary organizations, which are usually universities, have also arisen during the past 10 years (Guston 2001; Andrews et al. 2008). These organizations are seen as neutral parties and facilitate and supply technical information to public policy decision-making processes that address Earth-scale and regional-scale environmental and energy system issues.

Within the United States there are several examples of advances in knowledge networks that could be cited, such as those that connect public-supported land grant colleges, local irrigation district managers, and county extension agents who, among other things, transform highly technical knowledge about, say, drought, climate variability, and crop and livestock conditions into information useful to farmers, ranchers, local governments, and homemakers (Cash 2001). Another example is the Regional Integrated Sciences and Assessments (RISAs) of the U.S. National Oceanic and Atmospheric Administration (NOAA), which seek to facilitate communication among various disciplines, within specific regions, regarding the possible impacts of climate change on regional resources and economies (NRC 2008). Again, the RISAs connect generators of knowledge with users such as members of the public or policy community who need this information to manage drought, alleviate flood damage, water their crops, and even manage fire hazard risks. These "networks" share another feature: they mediate between disciplines, translate science into useable forms, and integrate user needs into knowledge-generators' activities.

---

[6] See http://en.wikipedia.org/wiki/List_of_distributed_computing_projects

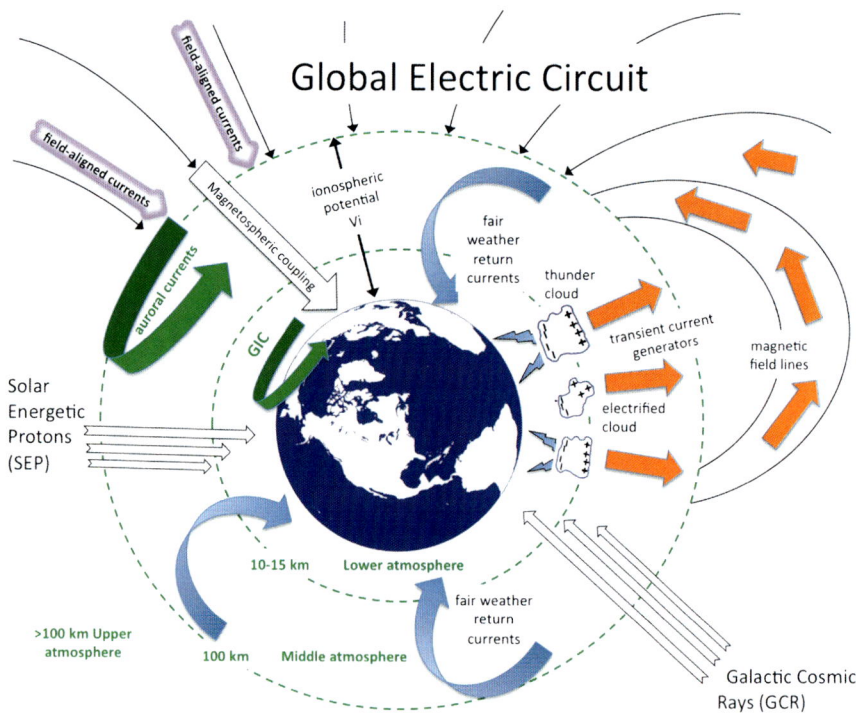

Fig. 3.2 Global electric circuit and its effects on Earth (Source: Jeffrey Forbes, University of Colorado Boulder, used by permission; available online: http://nsf.gov/news/news_images.jsp?cntn_id=121842&org=NSF)

3.2.2 Monitoring Systems

Many advances in Earth-scale systems research can be traced to substantial improvements in data gathering and management that support both statistical analysis and systems modeling. Real-time and longitudinal monitoring of a host of Earth-scale system indicators, many of them environmental, has improved tremendously. These indicators include:

- Air quality
- Water quality and supply
- Global air and ocean water temperatures
- Traffic and congestion
- Land use changes
- Global electric (illustrated in Fig. 3.2) and magnetic circuits
- Sun–Earth interactions

3.2.3 Communication Systems

The information revolution has continued unabated and maybe even accelerated over the past 10 years. Moore's law continued to hold. Bandwidth and access to computer networks continued to expand. Hundreds of millions of people worldwide now have mobile devices and access to tens of thousands of applications. Computing resources residing in the "cloud" further facilitate mobile applications and collaborative computing. Social networking has burst onto the scene. Some believe that now only there are only 4° of separation among most of the world's population because of the ubiquity of cell phones and social networking sites. Access to Earth-scale systems data and information has increased.

3.2.4 Management Systems and Tools

Access to more and higher-quality data more readily has contributed to the improvement of management systems. Global positioning system technology is positively contributing to the management of the transportation infrastructure and to the tracking of key aquatic and terrestrial species (e.g., Clark et al. 2006). Improvements in modeling and simulation are allowing more informative and reliable predictive management that incorporates information about location and timing of extreme weather events, potential brownouts and blackouts of electricity systems, and disruption of global telecommunication systems from solar activity.

3.2.5 Earth-Scale and Other Contributing Technological Systems

Wind and solar energy are two industries driving the creation of new jobs, pushed by significant amounts of government and corporate R&D funding for renewable energy (see Fig. 3.3).

Biofuels are becoming more viable, such as in Brazil, although concerns over tradeoffs between fuel and food are still very legitimate. Advances in materials science have led to steady improvements in the energy efficiency of lights, furnaces, air conditioners, insulation, windows, hot water heaters, computers, and consumer electronics. Similar stories describe advances in the efficiency of vehicles characterized by more efficient motors and transmissions and lightweighting of materials.

During the last 10 years, advancements and investments in space have been comparatively limited in Western Europe and North America, while there has been significant growth in space programs and associated technologies in the developing world, and particularly in China (which launched its first manned orbital mission in

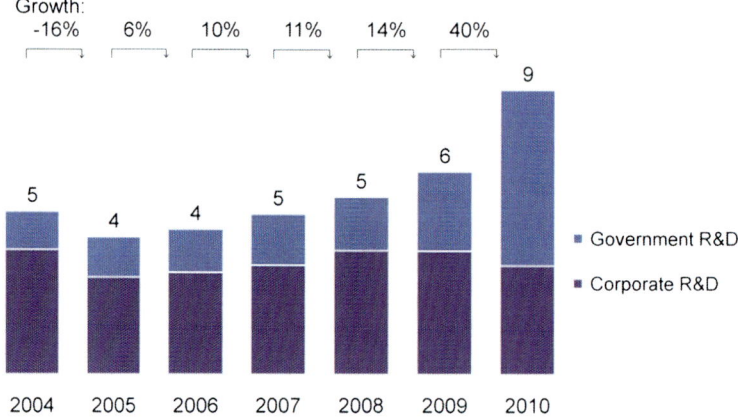

Fig. 3.3 R&D investment in renewable energy, 2004–2010, in $ billions (United Nations Environment Programme and Bloomberg New Energy Finance 2011, p. 30, Fig. 28; ©UNEP, used by permission)

2003 and plans a permanently manned space station by 2020[7]), India, and Korea. There is only one major operational manned platform currently orbiting the Earth. The number of probes sent to explore other astrological bodies has been relatively few, although there have been significant successes such as the U.S. robotic explorers on Mars.

Private sector investment in space has grown significantly over the past 10 years, spurred in part by the Ansari X-prize ($10 million for the first non-government organization to launch a reusable manned spacecraft into space twice within 2 weeks), which was awarded to the Scaled Composites Tier One team in 2004, but stimulated many other contestants and the development of a variety of novel launching and recovery technologies.[8] Meanwhile, NASA further stimulated innovation in the private launch industry by sponsoring the Commercial Orbital Transportation Services (COTS) program beginning in 2006,[9] which in 2012 yielded a successful resupply mission to the International Space Station (see http://www.spacex.com/). The program is now being extended to include crew as well as cargo resupply missions (see http://www.nasa.gov/offices/c3po/home/c3po_goal_objectives.html). A number of privately funded companies are planning commercial space tourism flights (e.g., Virgin Galactic, further developing the Scaled Composites technologies, http://www.virgingalactic.com/), and even mining of near-Earth-approaching asteroids (see http://planetaryresources.com). These activities have underlined the importance of the general S&T infrastructure

---

[7] See http://en.wikipedia.org/wiki/Chinese_space_program

[8] http://en.wikipedia.org/wiki/Ansari_X_Prize

[9] http://www.nasa.gov/offices/c3po/home/cots_project.html

Fig. 3.4 Infrastructure assessment "grades" in 2009 for the United States (American Society of Civil Engineers 2009, p. 2, used by permission)

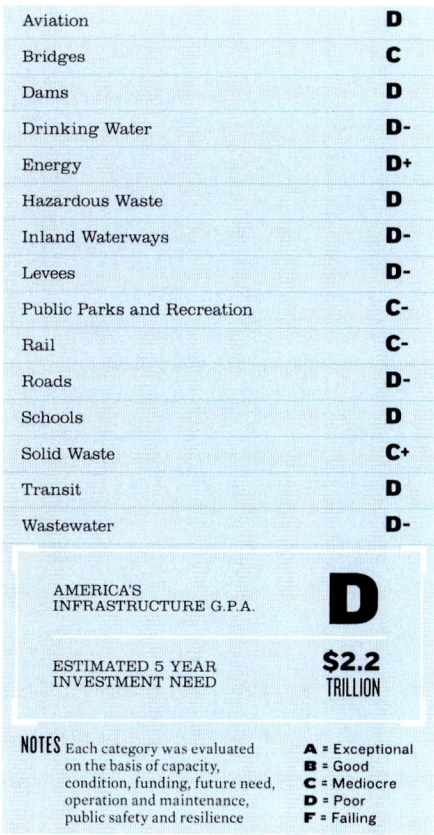

in the United States and the importance of involving various sectors of the society in a holistic approach.

The picture with respect to infrastructure is mixed. Previously, less developed countries such as China and India have embarked upon ambitious programs to build new roads, bridges, high-speed rail lines, dams, subways, and ports. China's electricity production and transmission infrastructure increased in size tremendously this past decade. Conversely, infrastructure investments in the developed world have lagged behind, resulting in systems that are in dire need of repair and replacement. Figure 3.4 presents a recent alarming scorecard for the infrastructure of the United States.

Contributions of NBIC technologies to improving infrastructure materials and infrastructure management systems have been modest during this period of time. More sophisticated information systems run transportation and other infrastructure systems. Green chemistry has shown some promise to reduce the negative externalities of cement production. This limited set of achievements is not surprising because NBIC investments have generally not focused on infrastructure issues.

3.3 Goals for the Next Decade

The convergence of NBIC technologies and knowledge systems could help to achieve the following types of goals related to meeting Earth-scale challenges:

- Build a knowledge platform to provide seamless access to integrated data and models supporting Earth-scale systems science.
- Increase access to these resources for policymakers and citizens by emphasizing translational research devoted to interpretation of NBIC technology capabilities to broader publics.
- Improve user interfaces and visualization techniques to empower researchers and others to gain unprecedented insights into environmental data and model outputs. Expand outreach and engage potential customers in the design and implementation of these new tools.
- Develop and implement real-time indicators of global environmental systems health. The concept "health" should include not just baseline conditions but also approaches to improve understanding of natural variability and causes of observed events.
- Reduce species extinction rates (or slow their rate of increase). Technological solutions include more efficient land use for food, human settlements, recreation, mineral extraction, and other resource extraction.
- Develop frameworks and tools to promote policy-oriented foresight with respect to Earth-scale systems and technological change.
- Commercialize a suite of NBIC2-related technologies to support the achievement of state-level "Renewables Portfolio Standards" or "Alternative Energy Portfolio Standards" goals and a national goal of 25 % of electricity generated by renewable resources by 2025.[10]
- Commercialize a suite of NBIC2-related technologies to significantly reduce emissions of greenhouse gasses (GHG) from production of materials for infrastructure projects (e.g., cement).
- Commercialize a suite of NBIC2-related technologies to significantly reduce the construction and operations and maintenance costs of infrastructure systems.
- Commercialize a suite of NBIC2-related technologies to implement large-scale and cost-efficient nanomaterials manufacturing processes.
- Demonstrate added value of converging technologies in the context of megacities.
- Utilize NBIC2-related technologies to achieve breakthroughs to significantly decrease the cost of space exploration.
- Identify NBIC2-related, high-value-added opportunities for space-based manufacturing.

---

[10] According to the U.S. Energy Information Administration, absent extensions of Federal subsidies for renewable energy generation, the projected share of U.S. electricity generation coming from renewable fuels (including conventional hydropower) will be only 16 % by 2040, although this figure is sensitive to natural gas prices and the relative costs of alternative generation; see http://www.eia.gov/forecasts/aeo/er/early_elecgen.cfm

- Develop a global risk assessment and warning system for potentially hazardous astronomical events (such as Sun activity, meteorites, etc.).
- Integrate these goals with other goals of high national and international importance, such as the UN Millennium Goals (http://www.un.org/millenniumgoals/).

3.4 Infrastructure Needs

The research infrastructure needed to support the vision and goals set out above could have the following types of components:

- Cloud computing platforms to support seamless integration of and access to Earth-scale systems models and to support distributed searches for solutions to R&D problems (e.g., combinations of materials for composite applications)
- Software repositories to facilitate reuse and repurposing of modeling and other types of software commonly used by Earth-scale system researchers
- Shared user facilities for sensor fabrication and for advanced materials fabrication
- A global network to provide integrated real-time access to sensor data streams
- A repository or warehouse of nanomaterials available to NBIC researchers at low cost
- Converging technologies "skunkworks" centers (highly autonomous and focused on pioneering R&D), with associated discretion to pursue innovative projects

3.5 R&D Strategies

Numerous kinds of R&D strategies can contribute to achieving the vision and goals set out above:

- Support of research to develop frameworks to understand and measure Earth-scale systems resiliency and, conversely, sources of system vulnerability.
- Support for small teams of researchers (e.g., who can use the above-mentioned user facilities).
- Support of use-inspired basic research. This includes identification of strategies that could better integrate scientific findings into decision-making, as well as identification of cultural and organizational barriers to the use of NBIC-inspired basic research.
- Replication of the Intergovernmental Panel on Climate Change (IPCC) and Intergovernmental Platform on Biodiversity and Ecosystem Services models to foster global collaboration in Earth-scale systems research.
- Continued support for innovative approaches to multidisciplinary education.
- Support projects that explicitly explore the convergence of NBIC technologies on Earth-scale system platforms (e.g., the global energy system).
- Support for research into developing frameworks for thinking about technology convergence.

- Support for research and processes to eliminate the "valley of death" that separates laboratory advances from real-world applications. This includes identifying the work contexts and circumstances of scientists and decision-makers and possible institutional impediments to innovation.
- Adoption of practices to improve university and national laboratory research and industrial R&D. This should include investigation of technology transfer impediments such as patenting, licensing, and other comparable policies and rules that tend to inhibit adoption of technologies.
- Support for foundational research to catalyze private-sector development of commercially viable space businesses.

3.6 Conclusions and Priorities

Meeting the challenges involved with improving Earth-scale systems requires advances in knowledge, monitoring, communications, and management systems, combined with better approaches for constructing the built environment. Presently, one can argue that ameliorating and adapting to global climate change is the top priority. This is because the risks of global climate change are potentially immense to humanity and other species that inhabit the Earth. Reducing GHG emissions through renewable energy, energy efficiency, and other technological advancements can have many additional economic benefits, for example, by increasing economic productivity and decreasing vulnerability of economies to energy price shocks. The convergence of technologies that could achieve these goals could also enhance self-sustainability and therefore could empower the achievement of human potential. Insight-producing knowledge systems built upon the foundations of ubiquitous monitoring and effective communication systems are needed as the backbone of these activities. Advances in space can contribute to monitoring and communications systems as well as to future global energy systems.

These goals and priorities are interrelated with many themes addressed in other chapters of this report. For example, global-scale efforts to improve energy, transportation, and water systems need to be integrated with local sustainability initiatives (Chap. 9). Efforts to reduce industrial waste can be combined with the additive manufacturing methods discussed in Chap. 7. The use of boundary organizations to facilitate collaborative decision-making about Earth-scale systems can be a shared goal to improve human quality of life (Chap. 2) and improve governance of NBIC technologies (Chap. 10).

It is also important to consider goals beyond the next 10 years. Here are four proposed initiatives that bundle together several converging knowledge and technology aspects:

- *Global Data & Information Infrastructure* ("GDDI"). Tremendous added value would be created by seamless integration of global data across spatial scales, contexts, and time (e.g., *from brains to space*). Contributing to this initiative would be wearable sensing devices that individuals would wear on their wrists

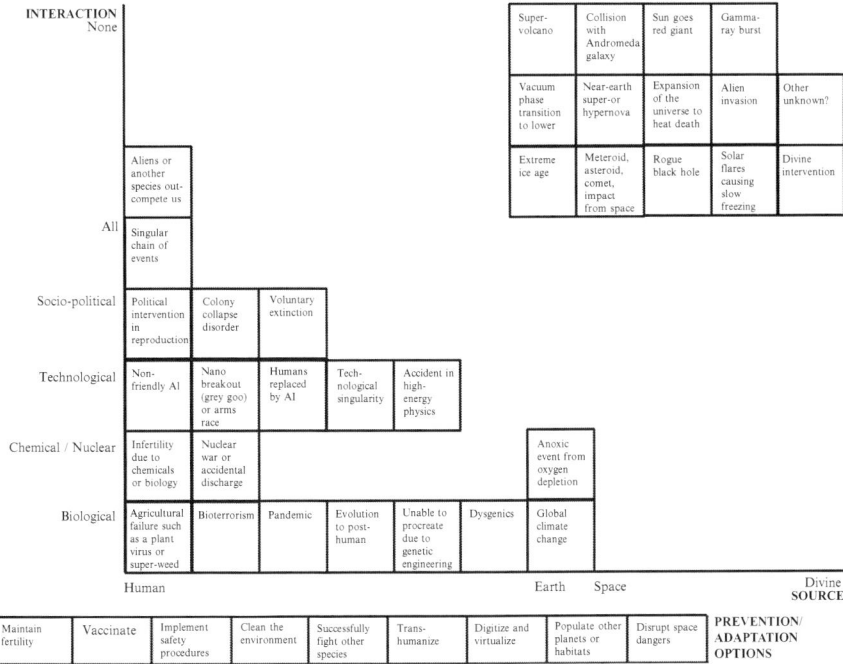

Fig. 3.5 Existential risks and prevention/adaptation options (Tonn and Stiefel 2011, courtesy of B. Tonn)

and that could collect a range of environmental data, transmit the data to the cloud for aggregation in real time, and then display, also in real time, key global health indicators.

- *Global Catastrophic and Existential Risk Assessment* ("GCERA"). Convergence of human knowledge is needed to understand and estimate these types of long-term risks. A convergence of data and modeling tools would be needed to support this initiative, along with theoretical advances in the estimation of existential risks and an IPCC-like organization to manage the assessment processes (e.g., see Fig. 3.5, which presents a set of existential risks by source and interaction, along with a range of potential prevention and adaptation options).
- *Global Megacities Initiative* ("GMCI"). This initiative focuses on comprehensive systems analyses and integrated intelligent system solutions to use resources more efficiently and improve the quality of life of residents of megacities. To achieve these goals, these types of systems would need to be intelligently integrated: smart grids and systems that generate and store locally produced energy; intelligent transportation systems; systems that manage drinking and waste water systems; smart homes; and systems that may manage local, decentralized manufacturing.
- *Global Materials Initiative* ("GMI"). The organizing concept of this initiative is the transformation of megacity and/or industrial ecosystems to be composed

of materials that are infinitely recyclable, reusable, and renewable ("IR^3"). A convergence of knowledge and technology would need to focus on creating these materials (see Chap. 9). "Nano-tagging" of metals and components can make materials tracking, recycling, and reuse more efficient. One could also consider co-designing these eco-systems to sequester carbon in the built environment.

The following are some additional visionary convergence ideas related to Earth-scale systems:

- Improve distribution of human activity on the Earth to improve overall efficiency in using natural resources, increasing productivity, and protecting health.
- Create a global electricity grid (e.g., see http://www.geni.org/energy/assets/swf/Dymax_grid_flows.html), and develop a global system of deep geothermal facilities and a space-based microwave system to power this grid. Design and build an operational space elevator to support the latter.
- Improve the efficiency, effectiveness, and safety of transportation systems around the world through use of high-speed rail lines, driverless vehicles, and other emerging solutions.
- Develop methods and approaches to dematerialize Earth-scale infrastructures (e.g., remove dams, develop permeable highways).
- Meet a substantial fraction of the world's demand for fresh water with energy-efficient and environmentally benign desalination technologies.
- Design and build comprehension support software that would work collaboratively with users to facilitate learning about complicated, massive "systems," from human behavior in megacities to actions within a cancer cell or human brain, to the architecture of billion-transistor chips and multimillion-component smart grids.
- Design Web X.0 to provide real-time feedback between human activities, Earth-scale systems status, and decision-making across scales of activities and systems operations.

With respect to Earth-scale systems, one can argue that convergence can take on many forms. Here are five forms that emerged from various NBIC2 workshop discussions:

- *Kurzweilian*. A large number of exponentially changing technologies converge on a platform (e.g., megacities' culture, social fabric, economics).
- *Added value*. Convergence components come together in a system that offers more than the sum of its parts (e.g., see the "GDDI" proposal above; global environmental systems monitoring and management).
- *Synergistic*. Convergence components come together to create a new, tightly integrated system where none of the components can be removed without destroying the system (e.g., IR^3, user-friendly communication systems, wearable computing).
- *Transdisciplinary*. Disparate knowledge converges to produce new knowledge (e.g., see the "GCERA" proposal above; systems knowledge, comprehension support systems).

- *Consilience.* Ideas, concepts, methods, and/or tools developed in one area of science are used to help explain phenomena in other areas, creating intellectual threads tying the areas together.

Lastly, with respect to Earth-systems-focused converging knowledge and technologies R&D, these challenges must be met:

- *Synthesis.* Determine how best to synthesize NBIC technologies to meet Earth-systems-scale challenges to produce added value.
- *Synergies.* Determine how best to promote the synergistic benefits of converging technologies.
- *Technological foresight.* Determine how best to assess nontechnological barriers, social and environmental impacts, and unintended consequences.
- *Governance.* Determine how best to promote synthesis and synergies and deal with issues raised through foresight exercises.

3.7 R&D Impact on Society

Protection of Earth-scale systems is essential for providing the security that, globally, all individuals will be provided with clean air and water, high-quality and reliable food, and aesthetically pleasing environments. Provided the security that essential needs will be met, humans are then free to achieve their potentials. Certainly, the construction, operation and maintenance of Earth-scale systems are a source of jobs. These systems are among the most sophisticated, extensive, and expensive that humankind has ever produced. Countries with innovative cultures and a strong base in science, technology, engineering, and mathematics will be the most competitive in designing and building these systems in the future.

3.8 Examples of Achievements and Convergence Paradigm Shifts

3.8.1 Water Management in Three World Regions: Translating Science

Contact person: *David L. Feldman, University of California, Irvine*

This section discusses three examples from the water sector that illustrate convergence among Earth-environmental-scale systems to achieve sustainability. They are from Brazil, Nigeria, and the Southeastern United States (Beller-Simms et al. 2008). Each one illustrates the challenges that might be addressed by converging knowledge and technology for Earth-scale systems under varying levels of national development. For each case, we consider four component subsystems: monitoring, communication, management, and cross-cutting interactions.

Fig. 3.6 Brazil's Ceará State (open source: http://www.ceara.gov.br)

State of Ceará, Brazil: Climate Information and Empowered Decision-Making

In Brazil as in many rapidly developing countries, climate and weather information are collected by national- or state-level agencies and translated to various users. In 1992, in response to a long, severe drought, the water management agency COGERH of the state of Ceará (Fig. 3.6) established a new system of multilevel water management for climate information. While COGERH continued to *monitor* weather data and *communicate* it to users (i.e., farmers and urban users), efforts were made to better translate information and ensure cross-cutting innovation to solve weather- and climate-related problems. Three specific reforms were pursued, as described below.

First, COGERH simplified reservoir modeling to enhance local users' knowledge about river basins and the risks shared by various groups as a result of drought. Local ways of conceiving of water flows were combined with sophisticated computer models.

Second, the state of Ceará enacted a new law for water management that created local watershed users' commissions as well as a state-level Water Resources Council

(Lemos and Oliveira 2004; Formiga-Johnsson and Kemper 2005; Pfaff et al. 1999). The state-level council, in turn, formed an interdisciplinary group comprised of social and physical scientists and local water users to better engage stakeholders in how to use and manage climate information (Lemos and Oliveira 2005). Meanwhile, local users' commissions negotiated water allocation among different users directly instead of relying on nationally determined allocations as had been done in the past. Both of these efforts encouraged a "bottom-up" approach to use of climate information for locally directed water allocation (Lemos and Oliveira 2004).

Third, at the river-basin level, the state water council trained users to use climate information to make more adaptive decisions in response to rainfall variability, and social science was better incorporated into decision-making to optimize use of climate forecast tools in specific management contexts. Evaluations have concluded that these reforms have helped build social capital in local communities with regard to information use and democratized decision-making, and have helped citizens better understand the different ways in which seasonal forecasting works, does not work, and could be improved (Lemos et al. 2002; Lemos 2003; Lemos and Oliveira 2004; Taddei 2005; Pfaff et al. 1999). While use of seasonal climate knowledge is limited so far and many logistical problems remain (e.g., continued ignoring of local knowledge and experience in some regions, and lack of available seed distribution and other economic incentives), there is great potential for use of well-translated seasonal forecasts to improve water management in the region.

Northern Nigeria: Adaptive Management and Local Initiative

Since the early 2000s, Nigeria has pursued an ambitious effort to reverse environmental degradation and the loss of rural livelihoods in an 84,000 km^2 region in the Hadejia-Jama'are-Komadugu-Yobe river basin (Fig. 3.7). Between 1970 and 1992, Nigeria's federal government built two projects to provide irrigation and flood control: the Tiga and Challawa Gorge Dams. From the very start, the projects suffered from unexpected adverse impacts, including slow flows in the Hadejia River, high turbidity in the Challawa, large deposits of silt behind the dams leading to greater downstream floods, and infestation of Typha grass, a hard-to-remove herbaceous plant notorious for clogging streams and irrigation channels. These impacts led to losses of farmlands and grazing lands and severe losses of local fisheries, all of which revealed the need for state-of-the-art converging knowledge and technologies.

Historically, decisions regarding water *monitoring*, *communication*, and *management* in the Hadejia region have been shared by several federal, state, and local agencies, which has led to rival plans for river basin development and management, and fragmentation of information. Agencies' plans were poorly coordinated: the same agencies charged with regulating water use are also large water users; thus, conflicts of interest were common. In 2002, a stakeholders' forum in the basin was convened with the support of the International Union on Conservation of Nature (IUCN), several Nigerian ministries interested in improving agriculture and water

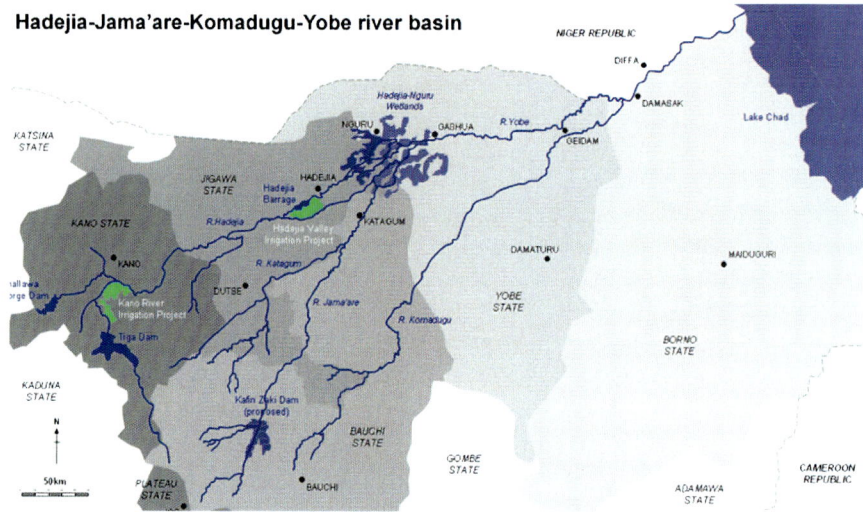

Fig. 3.7 Hadejia-Jam'are-Komadugu-Yobe river basin, Nigeria (Chiroma et al. 2010; courtesy M. Chiroma)

policy, and the United Kingdom's Department for International Development, which has an abiding interest in sustainable development (Barbier 2002).

Called the Joint Wetlands Livelihood (JWL) project, the forum instituted community-level improvements in how to use water *information* and introduced pilot projects to demonstrate best-management practices to restore livelihoods. A coordinating committee is brokered by high-level directors and permanent secretaries within the ministries. This committee fosters exchange of ideas between local farmers and government officials. In turn, local forums comprised of farmers, women's groups, and others advise the JWL and provide community-level training.

Within the watershed, farm and village groups are regularly convened in a series of role-playing tabletop "games": farmers serve as students and simulate solutions to local watershed problems by using information generated by central government climate and water data platforms. Over 2 days, participants engage in four distinct sessions that:

1. Allow them to brainstorm methods they think would work to maintain income and production while using less water
2. Prioritize these methods by a system of voting, so that farmers and other users agree on what works best. Their votes then become the basis for by-laws and agreements by farmers so that they can try these methods when they return home
3. Divide into two groups, one of which discusses the role of the watershed authority and the other of which discusses other formal institutions
4. Provide a final review that evaluates future actions (Lankford 2005)

Over the past decade, the JWL has mobilized local laborers to clear typha grass blocking waterways, restored some dry-lands farming and fisheries in the region, and helped build *local capacity* to manage freshwater problems without resort to large, highly engineered waterworks. Federal ministries support these efforts and provide technical support and translational science to communities, while villagers applaud these small-scale ventures because they participate in their implementation. Instead of grandiose public works projects, the region now increasingly relies on "gravity flow" irrigation, small dams and irrigation works, and microscale investments in the current floodplain economy (Muhammad et al. 2010).

Southeast U.S. Climate Consortium, Regional Integrated Sciences Assessment: Building Capacity

The Southeast Climate Consortium (SECC) is an association of academic researchers from universities in Alabama, Georgia, and Florida. Like other U.S. National Oceanic and Atmospheric Administration (NOAA) Regional Integrated Sciences Assessments, the primary purpose of SECC is to develop the capacity of local stakeholders—decision-makers, farmers, ranchers, and forestry managers—to better utilize climate information for managing water-related issues. Unlike other capacity-building approaches, however, SECC has since its formation in the 1990s relied upon top-down provision of information focused on this region's multibillion-dollar agricultural sector (Jagtap et al. 2002).

Early in its existence, SECC researchers recognized the potential to use knowledge of the impact of the El Niño Southern Oscillation on local climate to provide guidance to farmers, ranchers, and forestry sector stakeholders on yields and changes to risk (e.g., frost, floods, drought). Through a series of needs and vulnerability assessments, SECC researchers determined that the potential for producers to benefit from seasonal forecasts depends on factors that include their flexibility and willingness to adapt farming operations to the forecast, and the effectiveness of the communication process, and not merely on documenting the effects of climate variability and providing better forecasts (Jones et al. 2000; Hildebrand et al. 1999).

Extension specialists and faculty are members of the SECC research team. SECC engages agricultural stakeholders through *communication and outreach* efforts, including video conferences, one-on-one meetings with extension agents and producers, training workshops designed to instill confidence in climate decision tool use and to identify opportunities for their application, and traditional extension activities such as commodity meetings and field days (Fraisse et al. 2005). SECC is also able to draw upon the trust that the cooperative extension's service to the agricultural community has built up in the region through its various online tools such as AgClimate (e.g., Fraisse et al. 2006). Direct engagement with stakeholders has provided feedback to improve the design of these tools and to enhance climate forecast communication (Breuer et al. 2007).

Current activities of SECC are focused upon improving understanding of seasonal climate variability and climate predictability at local to regional scales across the South; characterizing contributions of climate variability to risks in management of agricultural, forestry, and water resources; developing decision aids based on the use of climate forecasts to help decision-makers identify management options to reduce risk; design and implement appropriate vehicles for disseminating climate and decision support information (e.g., an Internet-based learning and decision support system); and develop partnerships to build socially equitable extension and outreach programs (Southeast Climate Consortium, http://www.seclimate.org/objectives.php).

In conclusion, SECC appears to have been successful in integrating new information with established interaction networks (i.e., agricultural producers and extension agents). Its own evaluations suggest that benefits from producers' use of seasonal forecasts depends on several factors: the flexibility and willingness to adapt farming operations in response to forecasts; the effectiveness of forecast communication; sustained interactions with agricultural producers in collaboration with extension agents; and direct engagement with stakeholders that provides feedback on how to improve climate forecast communication.

3.8.2 Unifying Earth Databases: Data Observation Network for Earth (or Spaceship Earth Mission Control Center)

Contact person: *Alexander MacDonald, NASA*

A convergence initiative is necessary for planetary sustainability. NASA has proposed the Spaceship Earth Mission Control Center (SEMCON) in order to create an immersive data visualization environment for collaborative interdisciplinary research, modeling, and education focused on understanding the entire Earth system environment, the dynamic human impact on it, and decision options for effecting long-term planetary sustainability. Experts from a wide variety of scientific and research disciplines will be involved for good understanding of options for long-term planetary sustainability and economic development, including climate scientists, oceanographers, modelers, energy experts, nanotechnologists, systems biologists, information scientists, data visualization experts, economists, cognitive scientists, behavioral sociologists, political scientists, policy analysts, and others. The initiative is intended to become a kind of Manhattan Project for planetary sustainability, suited to a public–private partnership.

3.8.3 Global Nano-Geobiochemistry

Contact person: *Michael F. Hochella, Jr., Virginia Polytechnic Institute and State University*

Global nano-geobiochemistry is a relatively new NBIC field of study that is becoming somewhat established even though it has only appeared in the last few years. It is a classic example of a convergent platform of an Earth-scale system, only made possible due to knowledge, monitoring, communication, and management system components, as defined earlier in this chapter, that have been developing and maturing for decades. With each of these systems now reaching a critical level of maturity in support of understanding Earth's nanochemistry, geochemistry, and biochemistry, convergence allows for global nano-geobiochemistry to efficiently develop and provides opportunities for broad, global-scale thinking that was not possible just a few years ago. Below, we provide two examples of the power of a global nano-geobiochemistry convergent platform based on the NBIC model. These examples, both dealing with long-term Earth sustainability, show that broad implications and predictions in complex systems can be achieved.

Iron Fertilization of the Oceans

Here is an example that most strongly captures the essence of global nano-geobiochemistry (Hochella et al. 2008, 2012, and many references therein) in a complex, profound, and very serious high-stakes scenario in which the human race is deeply engaged. Over the last century, Earth mean surface temperature has increased by 0.8 °C, and the rate of temperature increase is likely accelerating. This warming is a direct observable, and is therefore indisputable. A general consensus of the world's climate science academic communities puts the probability of this increase stemming from human-induced global climate change at about 90 %, via the massive and still rapidly accelerating burning of carbon-based fossil fuels, as well as deforestation. Global warming is most often named by Earth scientists as by far and away the most important environmental issue of our times, and the number one Earth sustainability issue literally for centuries to come.

Global nano-geobiochemistry is central to this unfortunate situation, as global temperatures to a significant degree involve phytoplankton production levels in the open oceans, a process that provides a connection between nanoparticles, oceans, and global atmospheric and hydrospheric chemistry. The foundational scientific discussion that directly follows here, describing this situation, leads to possible actions that have at least some potential for ameliorating what could otherwise be a global, long-term, very difficult or dire situation. This possible action to consider is iron fertilization of the oceans.

Phytoplankton, as highly abundant chlorophyll-containing autotrophs populating the sunlit portions of nearly all oceans and continental surface waters, accounts for about half of the photosynthetic activity on Earth and are therefore one of Earth's single most abundant consumers of atmospheric CO_2. We also know that ocean phytoplankton populations are often nutrient-limited, primarily due to limited iron availability. This is where nanoparticles come in, because potential input of iron (Fe) within nanoparticles to the oceans has been predicted to be capable of far exceeding the natural inputs of dissolved iron from rivers. This is due to the fact that ferric (oxidized) iron, which is by far the dominant iron species in the Earth surface's highly oxidizing environment, is exceptionally insoluble in water.

Recently, NBIC platform development has allowed us to begin to estimate the global production of naturally occurring nanomaterials and their global movement,

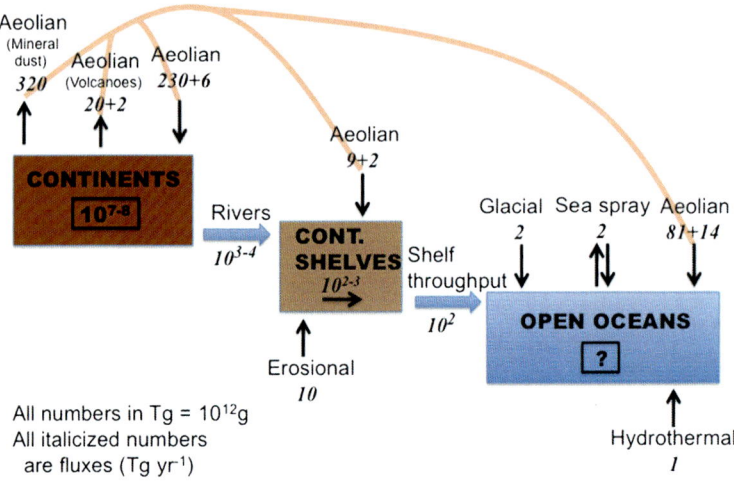

Fig. 3.8 The global budget for naturally occurring inorganic nanoparticles. All numbers are in units of teragrams (Tg = 10^{12} g = 1 million metric tons). All italicized numbers are fluxes (Tg yr^{-1}), and the number in the continent box is as estimated reservoir size. The inorganic nanoparticle reservoir size in the open oceans is not known. Most of the aeolian (i.e., produced by actions of the wind) fluxes are listed as two components: for the volcanic input to the atmosphere, 20 Tg is due to SO$_2$ aerosol formation, and 2 Tg is due to mineral ash. For the three aeolian inputs to the continents, continental shelves, and the open oceans, the first number is due to the 320 Tg continental mineral dust output, and the second number is due to the 22 Tg volcanic output (figure courtesy of Michael Hochella)

which is critical to understanding the supply of iron-containing nanoparticles to the world's oceans (Fig. 3.8). Although the primary production of iron-containing nanoparticles in the oceans is far from being quantified due to lack of sufficient information about such a massive and remote system, we can say that continental soils are otherwise the most prolific producers of inorganic nanoparticles on Earth, generating an existing reservoir of 10^{7-8} Tg (teragrams, equal to one million metric tons).

Clays are by far the most abundant naturally occurring inorganic nanomaterial in soils, many of which have significant Fe concentrations, but iron oxides and other iron-containing nanoscale minerals are also present. Rivers, and to a lesser extent, glaciers bring between 0.1 and 0.01 % of this nanomaterial reservoir to the continental edges on an annual basis, but only about 1.5 % of this makes it past continental margins and shelves to the deep oceans, due to aggregation and settling as a result of the very high ionic strength of seawater, which limits the repulsive forms that would normally prevent the particles from agglomerating. However, together with the airborne nanomaterial input via atmospheric mineral dust generated from windblown, arid, and devegetated lands (and this input rivals that from rivers), as well as much smaller but very important inputs from melting glaciers, ice sheets, and/or icebergs and hydrothermal inputs on ocean floors typically at tectonic spreading centers, the world's oceans receive a vital supply of iron-containing reactive/catalytic nanomaterial.

These minute mineral grains are bioavailable to marine phytoplankton via a number of mechanisms, including reductive dissolution processes, direct ingestion, and probably other mechanisms. It has even been shown recently for the first time that, at least for dissimilatory iron-reducing bacteria, the rate of ferrous iron release from ferric nanoparticles is dependent on the size, shape, and aggregation state of the nanoparticles (Bose et al. 2009), very typical of nanoscale behavior observed in laboratories as well as in nature.

Being aware of and understanding this background is essential to considering, let alone attempting, iron fertilization of the oceans on a large scale. The idea is to increase the biological population and therefore productivity of phytoplankton in the oceans to take up more CO_2, presumably with practical and logistical advantages over increasing land-based green plants that could accomplish the same thing. The idea has been around for well over two decades. Over that time, international teams of scientists have completed more than ten trials in the open oceans, often in the Southern Ocean that encircles Antarctica where the waters are relatively nutrient-rich except for iron. Many more experiments have taken place in laboratories. These ocean trials and experiments demonstrate that phytoplankton blooms can be stimulated by the addition of small amounts of iron compounds, typically submicron grains of iron sulfate. CO_2 consumption is increased, resulting in more biomass through photosynthesis, which is then deposited in the deep ocean as the organisms die. The chemical leverage is tremendous: each atom of iron results in the capture of roughly 13,000 atoms of carbon, due to the fact that iron is a micronutrient, but obviously still essential for growth.

However, using what amounts to geoengineering of the oceans is exceptionally controversial, ultimately and primarily based on the cognition portions of the NBIC platform. Specifically, the skill and art of taking on a highly complex situation comes into play, trying to solve problems without making new ones, and in the end, hopefully making the correct overall decision for the long term. Higher-level cognitive analysis of these types of situations suggests that when a vast, complex system is intentionally perturbed for desired benefits, unintended consequences and deficits are invariably created. This kind of scenario has been seen over and over again in attempted human engineering of the Earth's physical and ecological environments. Ocean and laboratory fertilization tests have already revealed a potentially very serious problem. It has been seen that iron stimulation can change the balance of phytoplanktonic species populations. Even worse, if any of the organisms most favorably promoted are not wanted in higher percentages within larger populations, there is a very serious problem. This is the case with a phytoplanktonic organism like *Pseudo-nitzschia*. This organism produces domoic acid, a potent neurotoxin. As consumers take in *Pseudo-nitzschia*, domoic acid invariably moves up the food chain. This neurotoxin is known to harm fish, birds, sea mammals, as well as humans who eat contaminated seafood.

In such a grand scheme, there will invariably be unintended consequences, as well as potentially great benefits. Nevertheless, at least assessment of the ocean iron fertilization scheme has been vetted at a high level thanks to an informed NBIC platform assessment, with this in turn greatly informed by our knowledge of global nano-geobiochemistry. Decisions in the future will continue to depend on updated knowledge, monitoring, communication, and management system components as described earlier in this chapter. Only one certainty regarding this situation is that whatever the action or inaction is in the future, we will be playing an exceptionally high-stakes game.

Environmental Implications of Nanotechnology

Critical areas of our understanding and preservation of Earth deal with our leading industries and their impact on the planet. One need look no further than the relationship between the energy industry based on fossil fuels, by far our greatest source of energy, and global climate change, as described earlier. When it comes to another massive and exceptionally rapidly growing industry, nanotechnology, we are much farther from understanding short- and long-term global implications, but due to NBIC convergent techniques, we are definitely on our way to a level of understanding that is already providing us with critical guideposts (e.g., Wiesner et al. 2009, 2011 and references therein). We have started by gaining some understanding of the mobility, transformations, and ultimate fate of nanomaterials in highly complex, natural, environmental contexts. We know that we need at least this before we have any chance of developing intelligent regulation. Certainly, as nanotechnology quickly ascends industrial growth projections, and it is already well on its way to a multi-trillion-dollar industry, it is very clear that the number, chemical and structural variation, and quantities of manufactured nanomaterials will grow exponentially. Therefore, it is inevitable that Earth's collection of ecosystems, and organisms within those ecosystems, will be exposed to various manufactured nanomaterials that will likely have transformed along their passage through the Earth system with which they have been in contact.

Yet, most interestingly, as these manufactured nanomaterials enter Earth's natural and man-made environments, they will be, and are, in the severe minority. Naturally occurring nanomaterials are continuously forming within and distributed throughout the Earth system, in soils, surface and ground water on the continents, the oceans (as discussed in more detail in the above section on iron fertilization of the oceans), in the atmosphere, as well as directly and indirectly by biological agents, most importantly by microorganisms (nanomaterials are both organic and inorganic, the latter including metals, metal oxides, and even metal sulfides). Our estimates of the production of naturally occurring nanomaterials in all Earth compartments vastly exceeds the amounts of manufactured nanomaterials produced today, and most likely into the foreseeable future (Fig. 3.9; Auffan et al. 2009; Hochella et al. 2012). The variation of the composition, atomic structure, and size/shape of both naturally occurring and manufactured nanomaterials is staggering, but there is significant overlap between these two groups, e.g., various metals, metal oxides and sulfides, and carbon-based nanomaterials.

Besides engineered and naturally occurring nanomaterials, there is a third class to consider, that of incidental nanomaterials. These nanomaterials are formed, either

Fig. 3.9 A comparison between the annual production of inorganic nanomaterials by industry and nature (figure courtesy of Michael Hochella)

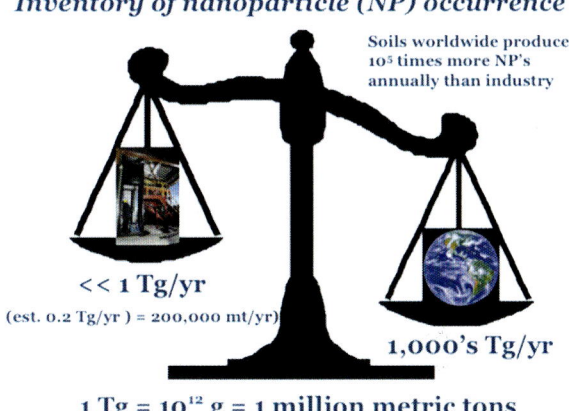

directly or indirectly, as a result of human activity, but unintentionally. Examples include nanocarbonaceous soot from diesel engines, metals sulfides from the waste streams associated with mining activity, and nanosilver in wastewater treatment plants.

The distinction between naturally occurring, engineered, and incidental nanoparticles in the environment will always be confounded by natural and variable modification of all materials in complex environments, including of course, nanomaterials (see below), but also because there is considerable overlap between these three classes of materials. On the other hand, many of these nanomaterials will have telltale signs of their origin, whether characterized by direct or circumstantial evidence based on chemistry, distribution, or specific location. It is also interesting to note that the European Commission on Nanomaterials, in October 2011, provided the following definition of a nanomaterial as "a natural, incidental or manufactured material containing particles, in an unbound state or as an aggregate or as an agglomerate and where, for 50 % or more of the particles in the number/size distribution, one or more external dimensions is in the size range 1–100 nm" (http://ec.europa.eu/nanotechnology/policies_en.html).

All nanoparticles, regardless of their origin, will transform in the environment as a result of growth or dissolution, chemical exchange, aggregation or agglomeration, recrystallization, biologic interactions, light exposure, redox reactions, and so on. Nanoparticles, depending on their type and the geo-environment in which they find themselves, may persist for long periods of time, transform rapidly, or anything in between. At each stage, their interaction with and potential toxicity to all things biologic will vary over time relative to each species encountered.

With so much dynamic complexity, one might conclude, prematurely, that we have not made progress in our recent quest for understanding the environmental impact of nanotechnology. Nothing could be further from the truth. In the last decade, we have made dramatic gains in our understanding of nanoparticle formation, transformation, and distribution in natural and man-made environments. No doubt countless details will be filled in over the next several decades.

Due to transformative thinking resulting from NBIC platforms assembled from advanced Earth-system knowledge, monitoring, communication, and management components, we can draw useful assessments, even at this point:

1. Early Earth, although geochemically primitive relative to today, would have been well suited for producing a vast array of nanomaterials. Therefore, first life on Earth, and all evolution since, has occurred in the presence of a ubiquitous array of naturally occurring nanomaterials.
2. Very recently, humans have added vanishingly small to significant amounts of incidental and manufactured nanomaterials to the Earth system, on local, regional, and global scales.
3. Direct and indirect introduction of nanomaterials due to human activity will result in the potential of transformations and transport into and between various Earth compartments (soil, water, and organisms, including potential trophic transfer, atmosphere, and oceans).
4. The behavior of certain states of each type of nanoparticle, with respect to certain organism types, in the right environment, will result in neutral, beneficial, and toxic effects.

We now have the tools and basic understanding, which was not available until recently, to begin to monitor and study such effects. This will likely allow us to take regulation and policy actions earlier rather than later, which has most often not been the case in the past with the rapid growth of new industries and technology revolutions. Such is the benefit of NBIC platforms.

3.8.4 Impact of Convergence on NASA's Earth Science Missions

Contact person: *Michael A. Meador, NASA Glenn Research Center*

The Earth is a complex, ever-changing planet. A more complete understanding of how the Earth is changing, the internal and external factors that are driving and influencing this change, and how changes in climate and the environment, and the availability and distribution of resources, impact the Earth's ability to sustain life are critical questions that must be addressed for the future of life on the planet. In 2007, the National Research Council completed the first ever decadal survey identifying future needs and opportunities in space-based Earth Science. In its report the NRC panel states, "Understanding the complex, changing planet on which we live, how it supports life and how human activities affect its ability to do so in the future is one of the greatest intellectual challenges facing humanity. It is also one of the most important challenges for society as it seeks to achieve prosperity, health, and sustainability" (NRC 2007, 19). Addressing this critical challenge will require significant developments in nanotechnology and biotechnology, information science, and cognitive and behavioral sciences in an integrated approach that draws upon the convergence of these technologies.

Fig. 3.10 The Orbiting Carbon Observatory-2, scheduled for launch in 2015, will perform precise, time-dependent measurement of atmospheric CO_2 levels. (Artist rendition of the OCS-2 by John Howard, JPL: http://oco.jpl.nasa.gov/, open source)

Advances in sensors and instrumentation, through the use of nanoscale materials and novel device architectures and fabrication techniques, are needed to enhance the accuracy and expand the capability of Earth-observing satellites (NRC 2012). Development of long-life, high-power, multibeam, and multiwavelength (0.3–2 μm) lasers (both continuous-wave and pulsed) will enable more accurate detection of atmospheric CO_2 and other gasses, measurement of aerosol and cloud properties and composition, and Doppler velocimetry-based wind measurement. Advances in optical detectors and hyperspectral imaging (from visible to short-wavelength infrared) will lead to improved global resource mapping and measurement of ocean activity, including 3D profiles. For example, significant enhancements have been achieved in the spatial resolution of LIDAR (light detection and ranging) *through use of quantum-entanglement-based* detection schemes (Dowling 2009). While quantum LIDAR has been developed for terrestrial ranging and imaging applications, it could be applied to Earth-observing satellites, such as NASA's Orbiting Carbon Observatory-2 (Fig. 3.10, above) and Earth Observatory, which recorded the image shown in Fig. 3.11.

Improvements in the design of instrument electronics, through the use of nanoelectronic materials and devices, are needed in order to minimize power, mass, and volume requirements and enable increased instrumentation in space-based Earth observation systems. Recent developments, such as the use of self-assembly to develop high-density carbon nanotube transistors, reported by IBM scientists (Park et al. 2012), and carbon-nanotube-based nonvolatile memory, developed by Nantero, suggest possible approaches to meeting this need.

The current Earth observation system of satellites and ground-based sensors generates over 4 terabytes of information per day (NASA 2010, 31). This is expected to

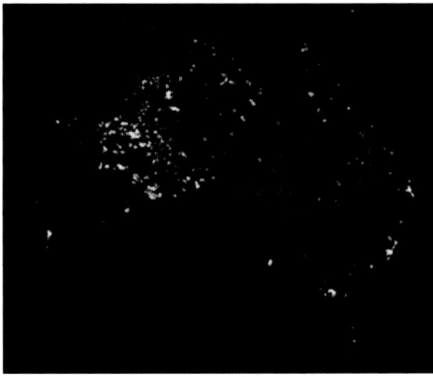

Fig. 3.11 Nighttime image of Australia captured by the NASA-NOAA Suomi NPP (National Polar-orbiting Partnership) satellite. Bright spots are sources of light from the surface, including city lighting, wildfires, and lighting on oil drilling platforms. (Photo credit: NASA Earth Observatory/NOAA NGDC, © NASA, http://www.nasa.gov/mission_pages/NPP/news/aus-fires.html, used under open-source policy)

grow significantly as new satellites are added. The ability to process this data into useful information is critical for the development of accurate and timely representations of the state of the Earth and its inhabitants, as well as for robust models to predict changes in climate and the ecology and natural resource distribution. This will require not only radical advancements in computational hardware, including alternatives to CMOS technology, but also the development of new software architectures, e.g., cognitive and hybrid computing, that can efficiently mine and process large data sets. Some of these needs could be met by adapting technologies and approaches developed under the Human Genome Initiative, which also faced challenges with processing of large amounts of data, as well as those currently under development in the Big Data Research and Development Initiative.

Solving these problems will also enable the development of autonomous satellites with decision-making capabilities to alter their mission, based upon observed events on Earth, such as forest fires or volcanic eruptions, to collect data on those specific events. These autonomous satellites could be used in the construction of a sensor web, a distributed global network of small sensor satellites that could be used for more in-depth mapping of resources, monitoring changes in climate and geology, and monitoring disasters.

3.8.5 Global Climate Change Coordination in the United States

Contact Person: *Geoffrey Holdridge, World Technology Evaluation Center*

For several decades there has been a concerted international interdisciplinary research effort to document, understand, and suggest remedial approaches to global climate change. The U.S. Government's part of this effort is coordinated through the

U.S. Global Change Research Program (USGCRP, http://www.globalchange.gov/) and through the Subcommittee on Global Change Research of the National Science and Technology Council. USGCRP periodically issues a comprehensive report summarizing the science of climate change and the impacts of climate change on the United States, now and in the future. It integrates the results of the U.S. program with related research from around the world and discusses climate-related impacts for various societal and environmental sectors and regions across the nation.

The 2009 USGCRP report, *Global Climate Change Impacts in the United States*, analyzes climate change impacts by sector, including water resources, energy supply and use, transportation, agriculture, ecosystems, human health, and society, and reviews ongoing and proposed efforts to remediate any adverse impacts of global climate change in each of those sectors. It also proposes an agenda for climate impacts science that includes recommendations to (1) expand our understanding of climate change impacts (including research on ecosystems and on economic systems, human health, and the built environment); (2) refine our ability to project climate change, including extreme events, at local scales; (3) expand our capacity to provide decision-makers and the public with relevant information on climate change and its impacts; (4) improve our understanding of thresholds likely to lead to abrupt changes in climate or ecosystems; (5) improve our understanding of the most effective ways to reduce the rate and magnitude of climate change, as well as unintended consequences of such activities; and (6) enhance our understanding of how society can adapt to climate change. In January 2013 an updated National Climate Assessment (NCA) report was posted as a draft for public comment (see http://ncadac.globalchange.gov/).

3.8.6 *Spaceship Earth Mission Control Center: A Convergence Initiative for Planetary Sustainability*

Contact person: *Alexander MacDonald, NASA*

If we are all passengers on "Spaceship Earth", then we should consider the potential need for a sort of "Spaceship Earth Mission Control Center"—a scientific facility that could serve as a focal point for the coordinated collection of global-scale earth science and socio-economic information and for the assessment of our options for ensuring long-term planetary sustainability. We have established physical data-immersive architectures to help our leaders come up to speed on our most difficult technical challenges—from the National Military Command Center in the Pentagon, to NASA's Mission Control Center. These are places where our national decision-makers can focus and immerse themselves in the state of affairs of national defense or spacecraft operations. But where does a newly elected President, or other world-leader, visit in order to understand the current state of affairs with regard to planetary sustainability? The construction of such a facility would be an important driver of technological convergence and would be a powerful symbol of our commitment to address the challenge of planetary sustainability today and to serve as the Earth's caretakers on behalf of future generation.

Although the term "Spaceship Earth mission control center" implies a governance role, the concept as envisioned would have no operational environmental or industrial policy responsibilities. Instead, the emphasis would be on multidisciplinary science—on data collection, data visualization, and systems modeling within a co-located research environment inspired by NASA's Mission Control Center in Houston. Whereas the Mission Control Center in Houston tracks the health and status of the technological subsystems of NASA's spacecraft—e.g., power, propulsion, thermal, communications, and attitude control—the Spaceship Earth Mission Control Center would track the health and status of the Earth's ecological subsystems—e.g., water, air, soil, flora, fauna, human action. Integrating these planetary subsystems within an overall model would require the participation of world-class experts not only from the established fields of earth and human sciences (physics, chemistry, biology, geology, oceanography, economics, sociology, psychology) but from researchers working within the emerging fields of technology that present both potential risks and solutions for sustainability—nanotechnology, biotechnology, information technology, and cognitive technologies. Although full convergence within the proposed framework would be a matter of decades rather than years, the convergence of information technologies and cognitive technologies for improved decision-making could show practical results within the decade.

The successful realization of a Spaceship Earth Mission Control Center would depend primarily on integrating the current state of the art in information technologies and stimulating new convergence technologies through the creation of unified Earth databases and an integrated data observation network for Earth. The convergence of these information technologies with the cognitive sciences of decision-making could potentially create a new practical capability for long-run planetary planning. This would in turn motivate the further development, over the coming decades, of new information collection capabilities critical to gathering planetary-scale environmental data at high resolution, including everything from low-cost remote sensing satellite constellations to integrated cellular-scale biosensor networks. In order to rise to the technological and environmental challenges of the twenty-first century, let alone the challenges of the twenty-second century and beyond, the multidisciplinary approach embodied by converging technologies is going to be vital to providing effective situational awareness and modeling capabilities for planetary sustainability.

3.9 International Perspective

The following are summaries relevant to this chapter of discussions at the international regional WTEC NBIC2 workshops held in Leuven, Belgium, September 20–21, 2012; in Seoul, Korea, October 15–16, 2012; and in Beijing, China, October 18–19, 2012. Further details of those workshops are provided in Appendix A.

3.9.1 United States–European Union NBIC2 Workshop (Leuven, Belgium)

This section is a distillation of several separate breakout session discussions in Leuven that pertained to Earth-scale systems, based on plenary presentations and discussions and breakout session notes. There was no single session or group that addressed this topic. Please see Appendix B for names of all attendees at the Leuven NBIC2 workshop. The sessions taken together were extremely informative in terms of conveying European perspectives, particularly the European commitment to considering the social and environmental implications of converging technologies.

Formally titled "EU–U.S. Workshop on Converging Nano-Bio-Info-Cognitive Science and Technology (S&T) for Responsible Innovation and Society," this workshop addressed a wide range of topics related to research, governance, and innovation. The breakout group charged with discussing NBIC2 concepts and directions within the rubric of sustainability focused a good deal of attention on megacities as a platform for technology convergence. Megacities offer a rich array of challenges and opportunities for converging technology solutions. For example, with respect to transport, advanced information technologies can foster more intelligent management of traffic and mass transit systems, and nanotechnologies can lead to the lightweighting of vehicles. A combination of nano-, bio-, and info technologies can lead to substantial improvements in energy efficiency (e.g., for heating and lighting systems in buildings) and increases in the production of renewable energy in megacities (e.g., rooftop photovoltaics, wind, and biofuels). Nanotechnology and information technologies will underpin advancements in smart grids and energy storage.

NBIC2 technologies also promise reduction and reuse of waste in megacities. New materials could be designed to be infinitely repurposed. Intelligent decentralized manufacturing (e.g., 3D printing) has the potential to greatly reduce waste in the production process. New material separation technologies based on advances in nanotechnology could lead to the cost-effective mining of urban waste depositories. The NBIC2 workgroup participants anticipate that megacities around the world will pursue large-scale, integrated infrastructure systems solutions in order to create urban environments more attractive to high-end companies and workers and to compete globally for jobs.

3.9.2 United States–Korea–Japan NBIC2 Workshop (Seoul, Korea)

Panel members/discussants:

Hee Chan Cho (*Co-Moderator*), *Seoul National University* (*Korea*)
Bruce Tonn (*Co-Moderator*), *University of Tennessee* (*U.S.*)

***Others*:**

Jiwan Ahn, Korean Institute of Geoscience and Mineral Resources (KIGAM, Korea)
Masafumi Ata, National Institute of Advanced Industrial Science and Technology (AIST, Japan)
Mamadou Diallo, Caltech (U.S.) and Korea Advanced Institute of Science and Technology (KAIST)
Choon Han, Kwanhwoon University (Korea)
Wanseok Kim, Electronics and Telecommunications Research Institute (ETRI, Korea)
Takashi Kohyame, Hokkaido University (Japan)
Kazuto Matsubae, Tokyo University (Japan)
Shinya Nakamoto, Japan Science and Technology Agency (JST, Japan)

Due to the composition of the participants in this breakout group, technical discussions focused mainly on materials issues. Ideas that were discussed include the following:

- Improvements in desalinization techniques
- Urban mining for valuable metals
- Improvements in oil field imaging techniques
- Materials with increased surface areas good for filtration applications
- Improvements in extraction of metals from ores
- New materials to lightweight automobiles
- Scrubbing CO_2 from stacks at cement plants
- Tracking global material flows through the use of nano-tags on metals and components

The group also took the opportunity to discuss a range of associated policy and social science issues, including:

- Trust must developed by the public in scientists to recommend acceptable solutions.
- Trust must also be given to politicians to decide on best solutions for societies.
- Social systems need to be developed to foster recycling and waste reduction.

Lastly, the group identified these important education and professional issues:

- Kids in Japan, Korea, and the United States are becoming much less interested in S&T.
- Young women in these countries need encouragement to pursue S&T education and need more professional opportunities in these areas.
- In Japan & Korea, S&T careers of young women are often derailed because of marriage and children; could government policies help reintegrate women into the workplace, especially to work in converging technology areas?
- Can graduate education be designed to train students to work in the converging technology area?
- There is so much to learn; schools need to train specialists just as matter of course; however, maybe S&T generalists can be trained to manage converging technology research.

3 Convergence Platforms: Earth-Scale Systems

- More experiments are needed in designing programs and even universities that revolve around the solution of grand challenges (e.g., water, energy).
- Students are pushing for sustainability; if professors do not follow, they become dinosaurs.
- Maybe the solution is to find a balance where a few students can be trained to manage cross-disciplinary converging technologies research; specialists then need an open attitude to collaborate with others.

3.9.3 United States–China–Australia–India NBIC2 Workshop (Beijing, China)

Panel members/discussants:

Xingyu Jian (co-moderator), National Center for Nanoscience & Technology (China)
Bruce Tonn (co-moderator), University of Tennessee (U.S.)

Others:

Mamadou Diallo, Caltech (U.S.) and Korea Advanced Institute of Science and Technology
Craig Johnson (Australia)
Ian Lowe, Griffith University (Australia)
Lianmao Peng, Peking University (China)

The breakout group discussions related to Earth-scale systems were productive and wide-ranging. The results of these discussions are presented below. To begin, the group identified three major changes in vision over the past 10-years:

- There is now a strong realization that climate change is major threat to food production, among many threats, and is also a major health threat.
- There is acknowledgement that sustainability is now an important driver behind Earth-scale environmental system decision-making, but more needs to be done to act on this trend.
- It is now generally accepted that there are various difficult and complex tradeoffs between societal and cultural challenges versus direct technological solutions to the problems.

With respect to knowledge systems, the group offered these thoughts:

- There is a need for tools to better understand contributions of multiple factors to climate change (e.g., biological contributors).
- Researchers should not ignore behavioral change in lieu of only technological solutions. Therefore, there is a need to study system solutions—those that combine behavioral change and technological solutions—where all tools are considered that can be used to influence behavior, including education and financial incentives.

- There is a need for more effective data and analytical tool sharing among researchers, agencies, companies, etc.
- A robust computational system needs to be developed to understand upscale and downscale phenomena (e.g., global-regional-subregional climate projections).
- Australia has coral reefs and tropical rainforests; focus on these resources as key indicators of climate change. Expand this approach internationally.
- There is a need to greatly increase understanding of global environmental systems as complex nonlinear systems with dramatic phase–change thresholds.
- There is a need for more modeling and computational capabilities, resources, access, and generic modeling platforms (such as already programmed in physics, some other phenomena).
- There is a need for more cross-disciplinary, physically collocated research infrastructure.
- There is a need to leverage national R&D priorities to aggregate up to rigorous creation of knowledge for humanity.

With respect to monitoring systems, the group offered these thoughts:

- More concrete data are needed, (e.g., through a hundreds of times more dense global system of sensors) to reveal, for example, patterns of volatile organic compounds (VOCs, e.g., organic aerosols) with respect to climate change.
- Therefore, there is a need for low-cost (nanowires, nanoparticles) sensors and complex models to fuse the sensor data to feed into global climate models.
- A challenge is the low concentration of some types of particles in the atmosphere, which poses a good application for nanotechnology.
- Track key marker viruses in species to better understand species vulnerability with respect to climate change. Could climate change lead to accelerated virus mutations? If so, virus tracking could be an early warning system.
- Improve space-based monitoring of epidemiological trends, conditions, and disease outbreaks.
- Improve data/sensor coverage in the Southern Hemisphere.
- More integrated environmental observatories are needed.

The group offered these thoughts with respect to communication systems:

- Link aggregated sensor data and real-time feedback to people via their cell phones with regard to climate change. This is an example of crowd-sourced data collection and communication. Maybe even give real-time feedback on the sustainability or lack thereof of personal behaviors.
- Support improvements in video conferencing to substitute for travel.

Management systems and tools also received some attention:

- Reduce traffic congestion through accelerated implementation of intelligent transportation systems.
- Accelerate research and implementation of nanocomposites to lightweight vehicles and trains to decrease pollution.

- Change the transport task: challenge all assumptions, substitute information technology for transport, reconceptualize urban design.
- Support the rigorous and inclusive development of technology roadmaps.

These thoughts were offered concerning Earth-scale and other contributing technological systems:

- In China, major advancements in various areas of infrastructure have been made (e.g., roads, healthcare), but gaps still exist.
- Artificial photosynthesis energy storage via chemical approaches needs more attention.
- Plans should move ahead to design and build an operational space elevator.
- Create a net-zero carbon emissions cycle for transportation to help ameliorate climate change. Focus on biofuels to substantially substitute for fossil fuels.
- Use CO_2 wastes to make carbon nanotubes for storing carbon in the built environment.

At times, the discussants made points that overlap with other chapters in this volume:

- Deal with aging populations using innovative technologies and management techniques. (Chap. 5)
- Develop simple diagnostic treatment technologies that can be deployed in rural/small clinics in China and other similar contexts. (Chap. 5)
- Even in the developing world, technology seems to make people busier, but not necessarily in a good way. This trend should be reversed. (Chap. 5)
- Design better products made from wood. Use genomic information from wood to make wood inherently stronger. Then maybe one could go back to using wood for wind turbines and airplane wings, for example. (Chap. 9)
- Develop more imaginative ways of producing healthier foods that are resilient to climate change and water shortages. (Chap. 9)
- Social engagement with regard to NBIC2 technologies needs to be greatly enhanced. (Chap. 10)
- Rational governance works with regard to technology development but needs to be faster. One should consider using feedback from governance processes to guide NBIC2 technology development. When targets are set by governments, technology can evolve to meet the targets. (Chap. 10)
- More efficient innovation from NBIC2 academic discoveries and advances needs to be facilitated. (Chap. 10)
- Improvements are needed in evaluating NBIC2 technology impacts on society. (Chap. 10)
- Improvements are needed in global governance and decision-making in setting global goals, processes, collaborations, and application of NBIC2 to deal with Earth-scale environmental issues. (Chap. 10)
- Design international collaborations and contributions with respect to Earth-scale systems according to capacities and needs and local/regional benefits. (Chap. 10)

The group offered these three as the top-priority issues associated with Earth-scale systems:

- Major improvements are needed in modeling of complex Earth-scale environmental systems/processes (e.g., climate change).
- Knowledge generation can be greatly accelerated through the worldwide sharing of and seamless access to important Earth-scale system R&D data, tools, and models. Responsibilities to achieve this goal need to be shared, internationally, maybe through replicating the IPCC model for various Earth-scale systems and processes (e.g., biodiversity).
- There is a need to develop more powerful and sophisticated intelligent systems to more effectively manage Earth-scale systems and communicate with all global citizens in real time about the status of the Earth.

References

ASCE (American Society of Civil Engineers): 2009 Report Card for America's Infrastructure. ASCE, Reston/Washington, DC (2009). Available online: http://www.asce.org/reportcard

Andrews, C., Jonas, H.C., Mantell, N., Solomon, R.: Deliberating on statewide energy targets. J. Plan. Educ. Res. **28**(1), 6–20 (2008)

Auffan, M., Rose, J., Bottero, J.Y., Lowry, G.V., Jolivet, J.P., Wiesner, M.R.: Towards a definition of inorganic nanoparticles from an environmental, health and safety perspective. Nat. Nanotechnol. **4**, 634–641 (2009)

Barbier, E.B.: Upstream dams and downstream water allocation: The case of the Hadejia-Jama'are floodplain, Northern Nigeria, Paper prepared for the Environmental Policy Forum. Center for Environmental Science and Policy, Institute for International Studies, Stanford University (2002)

Beller-Simms, N., Ingram, H., Feldman, D., Mantua, N., Jacobs, K.L., Waple, A. (eds.): Decision-Support Experiments and Evaluations Using Seasonal to Interannual Forecasts and Observational Data: A Focus on Water Resources. National Oceanic and Atmospheric Administration, Asheville (2008). U.S. Climate Change Science Program Synthesis and Assessment Product 5.3

Bose, S., Hochella Jr., M.F., Gorby, Y.A., Kennedy, D.W., McGready, D.E., Madden, A., Lower, B.H.: Bioreduction of hematite nanoparticles by Shewanella oneidensis MR-1. Geochim. Cosmochim. Acta **73**, 962–976 (2009)

Bostrom, A.: Future risk communication. Futures **35**(6), 553–573 (2003)

Breuer, N., Cabrera, V.E., Ingram, K.T., Broad, K., Hildebrand, P.E.: AgClimate: a case study in participatory decision support system development. Clim. Chang. **87**(3–4), 385–403 (2007)

Burkhard, B., Petrosillo, I., Costanza, R.: Ecosystem services – bridging ecology, economy and social sciences. Ecol. Complex. **7**(3), 257–259 (2010)

Cash, D.W.: In order to aid in diffusing useful and practical information: agricultural extension and boundary organizations. Sci. Technol. Hum. Values **26**(4), 431–453 (2001)

Cash, D.W., Clark, W.C., Alcock, F., Dickson, N.M., Eckley, N., Guston, D.H., Jäger, J., Mitchell, R.B.H.: Knowledge systems for sustainable development. Proc. Natl. Acad. Sci. U. S. A. **100**(14), 8086–8091 (2003)

Cherry, S.: The upside and downside to rare earth metals shortages. IEEE Spectrum This Week in Technology. Online: http://spectrum.ieee.org/podcast/energy/renewables/the-upside-and-downside-to-rare-earth-metals-shortages (2011). Accessed 3 September 2012

Clabby, C.: Does peak phosphorus loom? Am. Sci. **98**(4), 291 (2012). Available online: http://www.americanscientist.org/issues/pub/does-peak-phosphorus-loom

Clark, P.E., et al.: An advanced, low-cost, GPS-based animal tracking system. Rangeland. Ecol. Manage. **59**, 334–340 (2006)

Davis, S., Meyer, C.: Blur: The Speed of Change in the Connected Economy. Warner Books, New York (1998)

Deffeyes, K.: When Oil Peaked. Hill & Wang, New York (2010). See also http://www.princeton.edu/hubbert/the-peak.html. Accessed September 1, 2012

Devezas, T., de Melo, F., Gregori, M., et al.: The struggle for space: past and future of the space race. Technol. Forecast. Soc. Chang. **79**(5), 963–985 (2012)

Dowling, J.P.: Quantum LIDAR: Remote Sensing at the Ultimate Limit. Report AFRL-RI-RS-TR-2009-180. AFRL Information Directorate, Rome. Available online: http://www.dtic.mil/cgi-bin/GetTRDoc?AD=ADA502521 (2009)

Formiga-Johnsson, R.M., Kemper K.E.: Institutional and policy analysis of river basin management – The Jaguaribe River Basin, Ceará, Brazil. World Bank Policy research working paper 3649. World Bank, Washington, DC. Available online: http://go.worldbank.org/06H9KDDFH0 (2005)

Fraisse, C., Bellow, J., Breuer, N., Cabrera, V., Jones, J., Ingram, K., Hoogenboom, G., Paz, J.: Strategic plan for the Southeast Climate Consortium Extension Program. Southeast Climate Consortium technical report series. University of Florida, Gainesville. Available online: http://secc.coaps.fsu.edu/pdfpubs/SECC05-002.pdf (2005)

Fraisse, C.W., Breuer, N.E., Zierden, D., Bellow, J.G., Paz, J., Cabrera, V.E., GarciayGarcia, A., Ingram, K.T., Hatch, U., Hoogenboom, G., Jones, J.W., O'Brien, J.J.: AgClimate: a climate forecast information system for agricultural risk management in the southeastern USA. Comput. Electron. Agric. **53**(1), 13–27 (2006)

Friedman, T.: The World is Flat: A Brief History of the Twenty-First Century. Farrar, Straus, and Giroux, New York (2005)

Gleick, J.: Faster: The Acceleration of Just About Everything. Pantheon, New York (1999)

Grail Research: Water, The India Story (presentation). Available online: http://www.grailresearch.com/pdf/ContenPodsPdf/Water-The_India_Story.pdf (2009)

Gruber, N., Galloway, J.: An earth-system perspective of the global nitrogen cycle. Nature **451**, 293–296 (2008). http://dx.doi.org/10.1038/nature06592

Guston, D.: Boundary organizations in environmental policy and science: An introduction. Sci. Technol. Hum. Values **26**(4), 399–408 (2001)

Hildebrand, P.E., Caudle, A., Cabrera, V., Downs, M., Langholtz, M., Mugisha, A., Sandals, R., Shriar, A., Veach, K.: Potential Use of Long Range Climate Forecasts by Agricultural Extension in Florida. University of Florida, Gainesville (1999)

Hochella Jr., M.F., Aruguete, D., Kim, B., Madden, A.S.: Naturally occurring inorganic nanoparticles: general assessment and a global budget for one of Earth's last unexplored geochemical components. In: Barnard, A.S., Guo, H. (eds.) Nature's Nanostructures, pp. 1–42. Pan Stanford Publishing, Singapore (2012)

Hochella Jr., M.F., Lower, S.K., Maurice, P.A., Penn, R.L., Sahai, N., Sparks, D.I., Twining, B.S.: Nanominerals, mineral nanoparticles, and Earth systems. Science **319**, 1631–1635 (2008)

IWMI: Comprehensive assessment of water management in agriculture. In: Water for Food, Water for Life: A Comprehensive Assessment of Water Management in Agriculture. Earthscan, and Colombo: International Water Management Institute, London. Available online: http://www.iwmi.cgiar.org/assessment/ (2007)

Jacobs, K.L., Garfin, G.M., Lenart, M.: More than just talk: connecting science and decision-making. Environment **47**(9), 6–22 (2005)

Jagtap, S.S., Jones, J.W., Hildebrand, P., Letson, D., O'Brien, J.J., Podestá, G., Zierden, D., Zazueta, F.: Responding to stakeholders' demands for climate information: from research to applications in Florida. Agric. Syst. **74**(3), 415–430 (2002)

Johansson, M., Laike, T.: Intention to respond to local wind turbines: the role of attitudes and visual perception. Wind Energy **10**(5), 435–451 (2007)

Jones, J.W., Hansen, J.W., Royce, F.S., Messina, C.D.: Potential benefits of climate forecast to agriculture. Agric. Ecosyst. Environ. **82**(1–3), 169–184 (2000)

Kahneman, D., Slovic, P., Tversky, A.: Judgment Under Uncertainty: Heuristics and Biases. Cambridge University Press, New York (1982)

Kintisch, E.: Hack the Planet: Science's Best Hope or Worst Nightmare for Averting Climate Catastrophe. Wiley, Hoboken (2010)

Kurzweil, R.: The Singularity is Near: When Humans Transcend Biology. Viking Press, New York (2005)

Lankford, B.: Facilitation of Water Sharing Arrangements in the Hadejia-Jama'are Komadugu Yobe Basin with the River Basin Game Dialogue Tool. Overseas Development Group (ODG), School of Development Studies, Norwich (2005). Final Report

Lemos, M.C.: A tale of two policies: the politics of climate forecasting and drought relief in Ceará, Brazil. Policy Sci. **36**(2), 101–123 (2003)

Lemos, M.C., Oliveira, J.L.F.: Can water reform survive politics? Institutional change and river basin management in Ceará, Northeast Brazil. World Dev. **32**(12), 2121–2137 (2004)

Lemos, M.C., Oliveira, J.L.F.: Water reform across the state/society divide: the case of Ceará, Brazil. Int. J. Water Resour. Dev. **21**(1), 93–07 (2005)

Lemos, M.C., Finan, T.J., Fox, R.W., Nelson, D.R., Tucker, J.: The use of seasonal climate forecasting in policymaking: lessons from Northeast Brazil. Clim. Chang. **55**(4), 479–507 (2002)

Maggio, G., Cacciola, G.: When will oil, natural gas, and coal peak? Fuel **98**, 111–123 (2012)

Muhammad, J., Chiroma, M.J., Kazaure, Y.D., Karaye, Y.B., Gashua, A.J.: Water management issues in the Hadejia-Jama'are-Komadugu-Yobe basin: DFID-JWL and stakeholders experience in information sharing, reaching consensus and physical interventions. (Working paper). Online at http://www.iwmi.cgiar.org/research_impacts/Research_Themes/BasinWaterManagement/RIPARWIN/PDFs/14%20Muhammad%20Chiroma%20SS%20FINAL%20EDIT.pdf (2010)

NASA: 2010 Science Plan for NASA's Science Mission Directorate. National Aeronautics and Space Administration, Washington, DC (2010). http://science.nasa.gov/media/medialibrary/2010/08/30/2010SciencePlan_TAGGED.pdf

NRC (National Research Council): Earth Science and Applications from Space: National Imperatives for the Next Decade and Beyond. The National Academies Press, Washington, DC (2007)

NRC (National Research Council): In: Ingram, H.M., Stern, P.C. (eds.) Research and Networks for Decision Support in the NOAA Sectoral Applications Research Program. The National Academies Press, Washington, DC (2008)

NRC (National Research Council): Science instruments, observatories, and sensor systems (TA08). In: Appendix K in NASA Space Technology Roadmaps and Priorities: Restoring NASA's Technological Edge and Paving the Way for A New Era in Space. The National Academies Press, Washington, DC (2012)

Park, H., Afzali, A., Han, S.-J., Tulevski, G.S., Franklin, A.D., Tersoff, J., Hannon, J.B., Haensch, W.: High-density integration of carbon nanotubes via chemical self-assembly. Nat. Nanotech. **7**, 787–791 (2012). http://dx.doi.org/10.1038/nnano.2012.189

Pfaff, A., Broad, K., Glantz, M.: Who benefits from climate forecasts? Nature **397**(6721), 645–646 (1999)

Pimentel, D., Pimentel, M.: Global environmental resources versus world population growth. Ecol. Econ. **59**(2), 195–198 (2006)

Roco, M.C., Bainbridge, W.S. (eds.): Converging Technologies for Improving Human Performance: Nanotechnology, Biotechnology, Information Technology, and Cognitive Science. Kluwer Academic, Dordrecht (2003)

Sarewitz, D., Pielke Jr., R.A.: The neglected heart of science policy: reconciling supply of and demand for science. Environ. Sci. Policy **10**(1), 5–16 (2007)

Taddei, R.: Of clouds and streams, prophets and profits: the political semiotics of climate and water in the Brazilian northeast. PhD Dissertation, Columbia University. Published electronically at http://bdtd.ibict.br (2005)

Timilsina, G., Beghin, J.C., van der Mensbrugghe, D., Meve, S.: The impacts of biofuels targets on land-use change and food supply: a global CGE assessment. Agric. Econ. **43**(3), 315–332 (2012)

Tonn, B., Stiefel, D.: Willow Pond: A decentralized low-carbon future scenario. Futures, http://dx.doi.org/10.1016/j.futures.2013.10.001 (2013)

Tonn, B., Stiefel, D.: Evaluating methods for estimating existential risks. Presentation at Society for Risk Analysis Conference, Charleston, SC, Dec. 6. Risk Analysis published online 28 March 2013, http://dx.doi.org/10.1111/risa.12039 (2011)

Trimble, P.J., Santee, E.R., Neidrauer, C.J.: Including the effects of solar activity for more efficient water management: an application of neural networks. In: Proceedings, Second International Workshop on Artificial Intelligence Applications in Solar-Terrestrial Physics, Sweden. Available online: http://my.sfwmd.gov/portal/page/portal/xrepository/sfwmd_repository_pdf/final_dec3.pdf (1997)

United Nations Environment Programme, Bloomberg New Energy Finance: Global trends in renewable energy investment 2011: Analysis of trends and issues in the financing of renewable energy. UNEP Collaborating Centre for Climate & Sustainable Energy Finance, Frankfurt School of Finance & Management, Frankfurt, Germany, http://www.unep.org/pdf/BNEF_global_trends_in_renewable_energy_investment_2011_report.pdf (2011)

UN-Water: UN Year of International Water Cooperation, Facts and Figures. http://www.unwater.org/water-cooperation-2013/water-cooperation/facts-and-figures/en/ (2013)

USGCRP (U.S. Global Climate Change Research Program): Global Climate Change Impacts in the United States. Cambridge University Press, Cambridge (2009)

Victor, D., Morgan, M.G., Apt, J., Steinbrunner, J., Ricke, K.: The geoengineering option. Foreign Affairs. **88**(2), 64–76 March/April (2009)

Wiesner, M.R., Lowry, G.V., Casman, E., Bertsch, P., Matson, C., Di Giulio, R.T., Liu, J., Hochella Jr., M.F.: Meditations on the ubiquity and mutability of nano-sized materials in the environment. ACS Nano. **5**, 8466–8470 (2011)

Wiesner, M.R., Lowry, G.V., Jones, K.L., Hochella Jr., M.F., Di Giulio, R.T., Casman, E., Bernhardt, E.S.: Decreasing uncertainties in assessing environmental exposure, risk and ecological implications of nanomaterials. Environ. Sci. Technol. **43**, 6458–6462 (2009)

Chapter 4
Methods to Improve and Expedite Convergence

Mihail C. Roco, George Whitesides, Jim Murday, Placid M. Ferreira, Giorgio Ascoli, Chin Hua Kong, Clayton Teague, Roop Mahajan, David Rejeski, Eli Yablonovitch, Jian Cao, and Mark Suchman

4.1 Vision

4.1.1 Defining Convergence

Convergence includes all relevant areas of human and machine capability that enable each other to allow society to answer questions and to resolve problems that isolated capabilities cannot, as well as to create *new* competencies, knowledge, technologies, and products on that basis. Convergence of knowledge, technology and society (CKTS) also includes societal dimensions (such as socio-ethical, institutional, and governance aspects), and has been called in this study "beyond NBIC" (nanotechnology, biotechnology, information technology, and cognitive sciences) or "NBIC2." General CKTS convergence platforms are the foundational and emerging (NBIC) tools, human-scale platform, Earth-scale platform, and societal-scale platform (Fig. 4.1a).

Corresponding editors M.C. Roco (mroco@nsf.gov) and W.S. Bainbridge (wbainbri@nsf.gov).

M.C. Roco
National Science Foundation, Arlington, VA, USA

G. Whitesides
Harvard University, Cambridge, MA, USA

J. Murday
University of Southern California, Los Angeles, CA, USA

P.M. Ferreira
University of Illinois at Urbana, Champaign, IL, USA

G. Ascoli
George Mason University, Fairfax, VA, USA

C.H. Kong
Indiana University, Bloomington, PA, USA

The convergence process in this report is defined as the escalating and transformative interaction of seemingly different disciplines, technologies, and communities to (a) achieve mutual compatibility, synergism, and integration, (b) create added value (generate new things, with faster outcomes) to meet shared goals, and (c) branch out to new things. The convergence process is evolutionary and non-uniform. It requires a preliminary degree of development in each domain; it begins with achieving reciprocal compatibility such as in communication and knowledge exchange, and it leads to changes in the system in terms of its assembly, functions, and outcomes. The initial interaction between fields of study may be either coincidental or deliberate. The convergence process typically is followed by a process of divergence (branching and growing out) between newly created knowledge and its technological components. This convergence–divergence cycle is a typical, coherent process in science and technology (Roco 2002) that ultimately leads to novel systems with unanticipated applications and benefits (Fig. 4.1b).

The human activity system evolves during the convergence process. In order to better understand and characterize the system, we have defined four essential general convergence platforms of human activity that are driven by societal values and needs and that ultimately lead to progress in human development (Fig. 4.1a). Creativity and innovation are enhanced by the circuit of information and ideas between various platforms of the system. As exchanges happen faster and between larger domains within the platforms, the foundation for creativity and innovation broadens.

This study expands the concepts of convergence and divergence megatrends in science and engineering that were applied in 2001 to NBIC convergence (Roco and Bainbridge 2003; Fig. 4.2) to include the more general concept of convergence defined above, which is broadly applied to unprecedented, interconnected advancements in knowledge, technology, and social systems. The result of CKTS thus broadly conceived is expected to include numerous new applications with significant added value to society.

C. Teague
Independent Nanotechnology Consultant, Gaithersburg, MD, USA

R. Mahajan
Virginia Polytechnic Institute and State University, Blacksburg, VA, USA

D. Rejeski
Woodrow Wilson International Center for Scholars, Washington, DC, USA

E. Yablonovitch
University of California, Berkeley, CA, USA

J. Cao
Northwestern University, Evanston, IL, USA

M. Suchman
Brown University, Providence, RI, USA

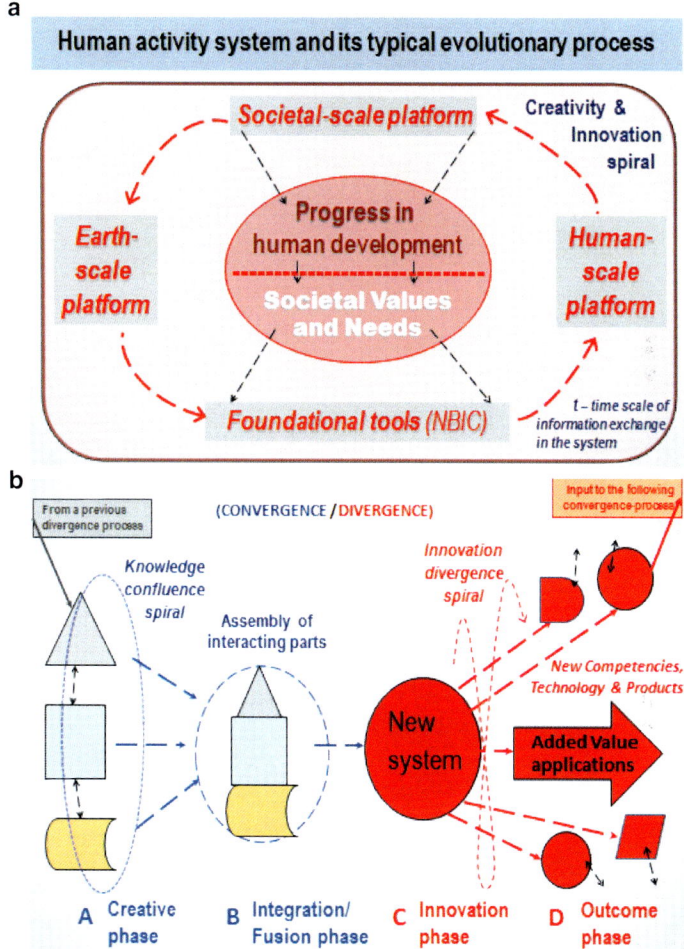

Fig. 4.1 Human activity system (a) and its typical convergence–divergence evolutionary process (b) (based on Roco 2002; both images courtesy of M. C. Roco). a The four general convergence platforms of the evolutionary human activity system: foundational tools, human-scale, Earth-scale, and societal-scale. It includes the major human activities involved in and impacted by creativity/innovation inputs into the system. This system evolves from A to D during the convergence–divergence process outlined in (b). b Schematic of the convergence (A, B)–divergence (C, D) process, initially formulated for megatrends in science and engineering (S&E) (based on Roco 2002). This concept is extended here to other knowledge, technology, and societal domains. It includes (A) a creative phase dominated by synergism between multidisciplinary components, (B) integration/fusion into a new system, (C) an innovation phase leading to new competencies and products, and (D) an outcome phase leading to new competencies, technologies, and products

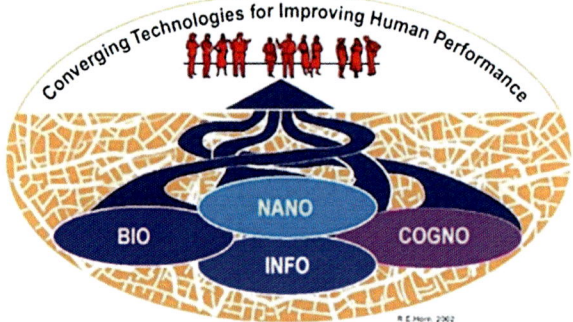

Fig. 4.2 As envisioned in 2001, convergence of foundational and emerging technologies would have the effect of changing the societal fabric towards a new structure that would improve human potential: *"The integration and synergy of nanotechnology, biotechnology, information technology, and cognitive science originate from the nanoscale, where the building blocks of matter are established. This picture symbolizes the confluence of technologies that offers the promise of improving human lives in many ways, and the realignment of traditional disciplinary boundaries that will be needed to realize this potential. New and more direct pathways towards human goals are envisioned in working habits, in economic activity, and in the humanities."* (Roco and Bainbridge 2003; figure by R.E. Horn)

There are important and potentially transformative benefits in proactively advancing systemic convergence of knowledge and technology. Such convergence will allow us to better understand socio-economic processes, stimulate creativity in knowledge-generation activities, increase the speed of development and innovation in technology, and create valuable new tools and products in various science and technology domains (Sharp and Langer 2011; Roco 2012). Together these kinds of developments offer new prospects not only for improving economic productivity and human potential but also for understanding and solving the immensely complex problems that the world faces today. As a result, the participants in this study see an urgent need to thoughtfully but deliberately expedite the convergence process. This chapter discusses approaches for systematically promoting compatibility, synergism, and integration in the CKTS processes that have already begun.

4.1.2 Changes in the Vision Over the Past Decade

Since the first NBIC conference in 2001, convergence has been reported mostly in just two or three domains of NBIC, in citation index papers, funded projects, and multiple-technology organizations motivated by the benefits of convergence. Reinforced by seed projects, use of all four domains has been reported, although somewhat sporadically. However, there has been a growing appreciation in scientific and academic communities worldwide that converging technologies in general, and synergies among "nano, bio, info, and cogno" areas of science, engineering, and technology specifically, are likely to create important advances toward societal gain (Bainbridge and Roco 2006; h+ 2010). Meanwhile, significant progress has been made in most regions of the world toward achieving the 20 major goals of NBIC convergence identified in 2003 (Roco and Bainbridge 2003); these goals are listed in Appendix D.

Several other noteworthy trends in NBIC convergence over the past decade are described below:

- *Overall, the R&D focus for converging technologies publications has remained reactive* (or "coincidental") to various opportunities of collaboration, rather than being driven by a holistic, systematic, proactive approach towards promoting convergence. Reactive convergence is observable over the last 10 years mainly in terms of paired collaborative activities between nanotechnology and biotechnology and between information technology and cognitive sciences.
- The pervasive availability and growing sophistication of *converging knowledge and technology has been revolutionizing the way science, engineering, commerce, and government are practiced.* Colossal changes have taken place in computing, handling of "big data," materials processing, digital sensing, biomedical advances, robotics, space communication, brain research, and other fields, building upon the NBIC foundational tools. In terms of the changes in commerce and government, for example, witness the growth in webinars and virtual meetings, and in global communications services companies such as Skype (http://beta.skype.com/en/) and Cisco WebEx (http://www.webex.com/). The nanoscale has enabled the devices that have enabled such trends and is also providing opportunities for measurement and control of biologically important processes, thereby opening up opportunities to marry information technologies with biology, medicine, and cognition (Murday et al. 2009).
- *There is increasing recognition that investments in innovative technologies must engage society* in terms of focus on social sciences, entrepreneurial education (e.g., Roco and Bainbridge 2006; Centers for Nanotechnology in Society founded in 2006), and overall human development (e.g., the UN Millennium Project studies: http://www.millennium-project.org/). Such trends help to facilitate better-informed decisions on benefit/risk management, legislation, and regulation. In parallel, there is a growing belief that the NBIC science and engineering convergence will also enable the evolution of social sciences (which reflect the study of human behavior and cognition) from phenomenological observation toward a much stronger basis in the fundamental laws of chemistry and physics.
- *R&D activities have become more participatory and group-oriented in the process of developing NBIC foundational tools.* The last decade witnessed appearance on an ad hoc basis of integrated S&T platforms driven by users; these include smart phone platforms, formation of associated communities of interest or affiliation, collective creation of large databases, and "cloud"-based computing that provides distributed and shared access to information and tools. The decade also witnessed the integration of converging technologies into social science—a reflection of the multipolar research world. There is now a self-

regulating convergence of communities facilitated by social networks. New data collection tools utilize multiple technologies, create big data, and indirectly create opportunities for integration. In addition, more social scrutiny of emerging technologies has been observed since 2000.

- *Converging trends between academia and society have become increasingly geared to industrial and societal needs.* In the last decade, with large deficits constraining governmental budgets, partnerships between industry and government have increasingly been touted as mechanisms to support university research (NRC 2012). Further, there has been growing attention paid to fostering mechanisms that will facilitate the transition of university-based science and engineering discovery into functional innovative technology to address societal concerns.
- *Convergence concepts have become progressively more focused on sustainable development.* For example, the increasing evidence for global climate change (NRC 2009b; NRC 2009c; see also http://globalissues.org) and security and economic concerns over petroleum resources (NRC 2009a) have focused global attention on renewable energy. It is recognized that secure petroleum resources and renewable energy resources require an integrated international effort to foster both a multidisciplinary science approach to create the necessary technological innovations and the necessary societal behavioral transformations.

4.1.3 The Vision for the Next Decade

In the next decade proactive, holistic systemic convergence will be cultivated in various domains of knowledge, technology, and society. A science of convergence is likely to emerge. Convergence methods will become ubiquitous in decision analysis and planning processes. Different fields of S&T will increasingly feed into each other and provide new opportunities for synergy, creating a virtual "spiral of innovation evolving over time" (or a "Lagrangian path of ideas") within the broader converging knowledge and technology platforms, including synergies between theoretical and applied sciences and social and behavioral sciences. From the science of convergence is expected to emerge better methods of promoting and managing convergence, better insights about which methods are best for which convergence contexts (e.g., academic or manufacturing contexts), and how to collect and evaluate data that could indicate how well convergence methods work in various contexts.

Convergence in the investigation of natural phenomena (analogous to the convergence in the forces of matter) *will accelerate*, and will underpin the technological effectiveness of society.

Higher-level convergence languages based on new concepts, relationships, and methods are envisioned to be developed to allow integration of components from the language levels for particular topics; that will facilitate integration into and among essential platforms, and in turn that will facilitate innovation (confluence of streams), improved manufacturing (hybrid NBIC methods, mind-cyber-physical systems), and advances in all emerging technologies. For example, the unified theory of physicochemical forces provides a higher-level language than the language for investigating individual mechanical forces, and concepts of converging technologies have a higher degree of generality than concepts for a specific technology such as biotechnology. Using higher-level convergence languages will enhance comprehension of surrounding complex systems and understanding of how knowledge is generated over time and crosses multiple fields using logic models (patterns). Cultural anthropologists will help make this transition. A database and an expert system could be established to select the methods of convergence to accelerate unprecedented discovery and innovation.

Important drivers will be the emerging technologies informed by a holistic view of society. For example, converging knowledge and technology will support personalized medicine (rooted in the integration of medicine, electronics, robotics, bionics, and data handling), leading to an increase in average life expectancy. CKTS will support personalized online education—for formal and informal lifelong learning—that exploits advances in neuroscience, communication, psychology, and understanding of learning.

There will be an increased focus on communities defining their own needs for using CKTS to address major issues affecting them, on an as-needed/as-requested basis, relying on self-regulation (through professional societies, nongovernmental organizations [NGOs], etc.) in a manner that is less paternalistic than that of current systems.

Implementing convergence methods will create a continuing focus on discovery, innovation, and responsible development, as well as on economic benefit. Scientific convergence and technological change are inevitable, and the public is expected to demand more social convergence, such as fusion of foods, music, and fashion, providing new experiences. Convergence will drive innovation in R&D organizations around the world and will become a source of competitive economic advantage in the global economy.

Sustainable development on Earth over the long term will depend on using converging knowledge and technology to develop new observatories and new methods to simulate Earth's dynamic systems using new sensing capabilities, large data, and so forth.

Changes in governance (organizational structures, decision analysis, and measures to reach decisions and solve conflicts) *will be needed to enable societal benefits rooted in CKTS developments* such as those noted above. As a result, government organizations and regulations will need to be updated to allow the convergence processes to work better and provide increased benefits to society.

4.2 Advances in the Last Decade and Current Status

The National Science Foundation (NSF) and U.S. Department of Commerce (DOC) organized the initial NBIC workshop in 2001 (Roco and Bainbridge 2003). Science and engineering professional literature addressing NBIC convergence first appeared after that workshop but has averaged only about five papers per year where all four domains were addressed, as determined by a simple keyword search of the ISI

Thompson Web of Science data base – *nano** + *bio** + *info** + *cognit**. In the same time frame, papers on nanoscale science and engineering continued to experience quasi-exponential growth. While the literature addressing convergence of all four NBIC areas has not grown appreciably, there has been significant growth in the number of papers addressing the convergence of two or three of the NBIC areas.

A number of U.S. universities, as well as institutions in other nations, have begun efforts to foster convergence (MIT 2011; Sharp and Langer 2011; also see Table 4.1 below and Fig. 8.1 in Chap. 8 of this report). The need for integration in industry had been recognized earlier (Ericson 1960).

As shown in Table 4.1, targeted NBIC efforts have been initiated by the European Union,[1] Japan, Russia,[2] and China,[3] among others, and attention has been given to this issue by the independent International Risk Governance Council (Renn and Roco 2006; IRGC 2009).[4] Several national and EU science and technology (S&T) multiyear plans use components of the convergence approach within their predetermined political context and requirements; Table 4.2 gives several examples.

Figure 4.3 shows 10-year trends in numbers of publications in NBIC areas. Work in all four NBIC domains began to be cited after 2001 when NSF and DOC organized the first NBIC workshop and the U.S. National Nanotechnology Initiative was begun. The rate of increase in publications with at least some convergence component has been roughly doubling each 3 years since 2001. Figure 4.4 shows the increase in the number of NBIC awards at NSF having at least two components in each award. The search was performed by using keywords in the abstracts published on the NSF website (http://www.nsf.gov). NSF's rate of increase of awards in NBIC domains is significantly higher than the overall budget increases for all research areas. Since 2010, the NBIC awards represent about 5 % of the total NSF awards.

A mechanism for improving convergence is increasing the level and speed of interactions in the creative phase (A in Fig. 4.1b). Support for interdisciplinary interactivity among scientists from different traditional disciplines appears to be increasing. Recent research describes creativity as being spurred by real physical environments where people and ideas converge and synergies arise (e.g., in the books *Imagine* by Jonah Lehrer (2012) and *inGenius: A Crash Course on Creativity* by Tina Seelig 2012), In a number of large companies such as 3M and IBM, scientists are rotated from one department to the next, regardless of their training, to spur

---

[1] An example is the project Knowledge Politics and New Converging Technologies; A Social Science Perspective, funded by the European Commission under Priority 7 of the 6th Framework Programme; http://converging-technologies.org/

[2] For example, the Kurchatov Institute Centre of Nano-, Bio-, Info-, and Cognitive Sciences and Technologies in Moscow, http://www.kiae.ru/e/nbic.html. The Kurchatov Institute is Russia's National Research Center

[3] For example, see "NBIC–New Opportunity for China," the National Natural Science Foundation of China, http://www.nsfc.gov.cn/Portal0/InfoModule_584/37502.htm

[4] The IRGC White Paper on Nanotechnology Risk Governance (Renn and Roco 2006) notes that in the fourth generation of nanotechnology products and production process (after 2014/2020), "NBIC convergence will play an increased role" and addresses some of those issues, including their sociological components.

Table 4.1 Examples of U.S. and Other Programs with Seed NBIC/CKTS Approaches

Program name and website	Home institution
Biodesign institute: http://biodesign.asu.edu/research/research-centers	Arizona State University, Tempe, AZ (U.S.)
The Wisconsin Institute for Discovery: http://wid.wisc.edu/about/	University of Wisconsin, Madison (U.S.)
Center for Nanotechnology in Society (CNS-ASU): http://cns.asu.edu	Arizona State University, Tempe, AZ (U.S.)
Center for Nanotechnology in Society (CNS-UCSB): http://cns.ucsb.edu/	University of California–Santa Barbara (U.S.)
Nanoscale Informal Science Education Network (NISEnet): http://nisenet.org	Boston Museum of Science, Boston, MA (U.S.)
Converged Technologies for Security, Safety, and Resilience (CTSSR): http://www.it.vt.edu/organization/ctssr/	Virginia Polytechnic Institute and State University, Blacksburg, VA (U.S.)
Centre for Mobile and Converging Technologies: http://www.leedsmet.ac.uk/aet/computing/centre-for-mobile-and-converging-technologies.htm	Leeds Metropolitan University, UK
Kurchatov Centre of Converging of Nano-, Bio-, Information and Cognitive Sciences and Technologies: http://www.kiae.ru/e/nbic.html	Moscow, Russia
Tsukuba Innovation Arena: http://tia-nano.jp/en/about/index.html	Japan
Inter-university MicroElectronics Center (nano-bio-IT-cogno): http://www2.imec.be/be_en/home.html	Belgium
MINATEC (nano-bio-info): http://www.minatec.org/en	Grenoble, France
"NBIC–New Opporrunity for China," China National Natural Science Foundation: http://nstc.gov.cn/Portal0/InfoModule_584/37502.htm	China
CRPC(Convergence Research Policy-Development Centre) (2012-)	Korea
Knowledge NBIC Project, http://www.converging-technologies.org/	Zeppelin University, Germany
Centech, http://www.centech.de	Munster, Germany
Interdisciplinary Center for Technology Analysis and Forecasting (ICTAF): http://ictaf.tau.ac.il/index.asp?lang=eng	Tel Aviv University, Israel
Centre for Converging Technologies: http://www.uniraj.ac.in/cct/	University of Rajasthan, Jaipur, India

Table 4.2 National and EU S&T multiyear plans with approaches related to CKTS

Program name and website	Economy
National Nanotechnology Initiative (NNI), http://nano.gov	The United States, 2000
National Information Technology R&D, http://www.nitrd.gov	The United States, 2000
Horizon2020 http://ec.europa.eu/research/horizon2020/index_en.cfm?pg=h2020-documents	European Union, 2013
HighTech 2020 Strategy for Germany, http://www.hightech-strategie.de/de/390.php	Germany, 2010
Japan's Science and Technology Basic Policy Report, http://www8.cao.go.jp/cstp/english/basic/4th-BasicPolicy.pdf	Japan, 2010
Korea S&T plans	Korea
China S&T plans	China

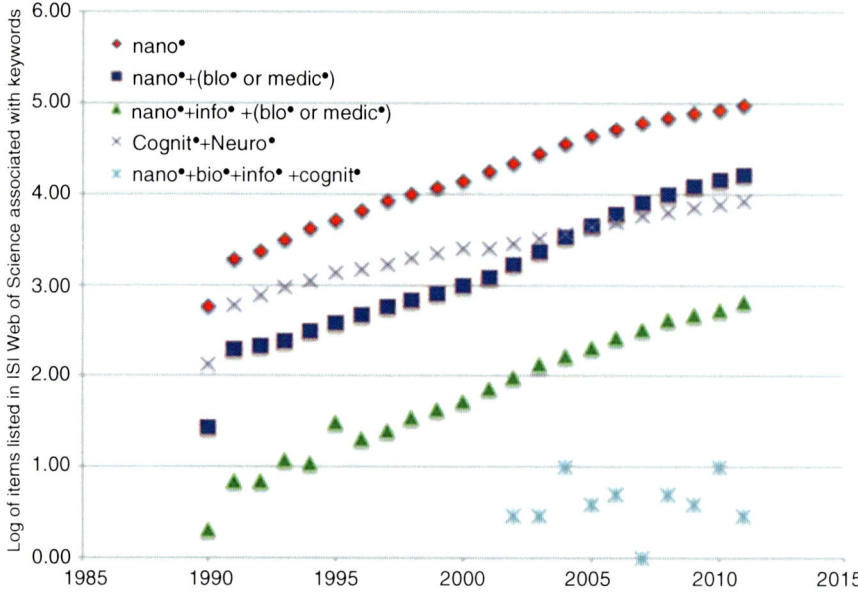

Fig. 4.3 Number of publications in various domains of NBIC, 1990–2011 (Courtesy of J. Murday)

synergistic and creative thinking. In several universities such as Cornell University and MIT, Ph.D. students are rotated through various laboratories during the course of their studies. Creation of multidisciplinary campus spaces for faculty and students is a reality in an increasing number of campuses such as University of California–Los Angeles and Seoul National University. Such practices need to be expanded and systematized. One could imagine routinely rotating professors from department to department. Leading engineering design centers may provide models for problem solving using a systems approach.

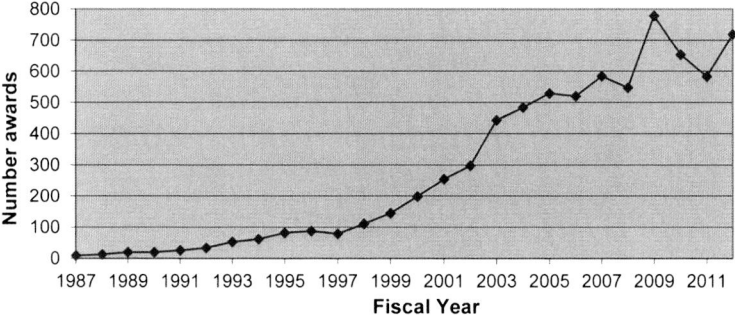

Fig. 4.4 Funding per various domains of NBIC (at least two NBIC domains in each award) by NSF, 1987–2012 (Courtesy of M. C. Roco)

Another mechanism to stimulate convergence is evidenced in the increasing collaboration between academic, industry, and government sectors. As industry has reduced its investment in internal basic research, it has looked to the academic community for research discovery. While that has mostly been on an ad hoc basis, in one specific instance, the semiconductor industry has formed an extended partnership with Federal and state governments to co-fund use-inspired basic research pertinent to its interests. For example, the Semiconductor Research Corporation's (SRC) Nanoelectronics Research Initiative (http://src.org/program/nri/) is partially funded by industry and is supplemented by funding from NSF and the National Institute of Standards and Technology (NIST). Another SRC effort, the Focus Center Research Program (http://www.src.org/program/fcrp/), is funded by roughly 50/50 cost sharing with government.

Overall, in the past 10 years, areas of NBIC convergence have been integrated within various U.S. Government R&D programs mostly by pairs, in a reactive mode, generally without a holistic system view. At the same time, the emerging technologies have developed both independently and jointly to a level that now more readily enables structured convergence. Examples in the United States of program developments for national priorities are the Information Technology Research for National Priorities program (ITR) and its successors (such as NITRD, http://www.nitrd.gov), the National Nanotechnology Initiative (NNI; http://nano.gov), and several large database initiatives such as ones in the medical, communication, and defense fields.[5] In the same period, the 2001 NBIC study (Roco and Bainbridge 2003) and other studies also planted seeds for convergence. Simulations of reality are increasingly providing a method of evaluating possible integration efforts. International organizations such as the International Risk Governance Council

---

[5] Examples of U.S. Big Data initiatives in 2012 are at DOD http://www.DefenseInnovationMarketplace.mil; at DARPA http://www.darpa.mil/NewsEvents/Releases/2012/03/29.aspx: at DOE http://science.energy.gov/news/; at HHS/NIH http://www.nih.gov/news/health/mar2012/nhgri-29.htm; at NSF http://nsf.gov/news/news_summ.jsp?cntn_id=123607; and at USGS http://powellcenter.usgs.gov

(since 2005) and various international scientific communities have recognized the need for transformative interdisciplinary synergism. Despite their importance, systems approaches and corresponding methods of convergence, and connections between the transformative tools, nomenclature, standards, legislation, and policies of CKTS, have not received sufficient attention in scientific and public discourse.

Despite the fact that these kinds of inroads have helped to partially advance convergence in the past 10 years, they are insufficient to enable the synergies and momentum that are critical if the potential benefits of convergence are to be realized. The relatively large differences between the rates of discoveries and economical outputs can be observed in most of the world economies, with some specific characteristics in each country. In the last decade, U.S. policy leaders began to realize that the 60-year dominance of the United States in science and technology is eroding, and national interests require a new perspective. In 2005 Thomas Friedman published his book *The World is Flat*, pointing out that the world was rapidly becoming a level playing field in terms of commerce, where all competitors have equal opportunity. In 2007 the U.S. National Academy of Sciences published *Rising above the Gathering Storm: Energizing and Employing America for a Brighter Economic Future* (NRC 2007b). It observed that without a renewed effort to bolster the foundations of U.S. competitiveness, we can expect to lose our privileged global position. The report was revised in 2010 to *Rising Above the Gathering Storm, Revisited: Rapidly Approaching Category 5* (NRC 2010b), which asserted that in the face of so many daunting near-term challenges, U.S. government and industry were letting the crucial strategic issues of U.S. competitiveness slip below the surface. Transformative technology, particularly in emerging areas, was identified as a critical component in retaining competitiveness. Systematic actions are needed to extract the full value out of convergence trends going into the future. Friedman and Mandelbaum (2011) identify one of the main challenges for the United States is to "act collectively for the common good."

4.3 Goals for the Next Decade

The main purpose in promoting a proactive and systematic focus on converging knowledge and technology is to increase creativity, innovation, and productivity in society, to enable people to solve both local and global problems that otherwise seem to defy solution. To achieve value-added goals, it will be necessary to create a portfolio of convergence methods or an expert system platform for selecting methods of integration and domains of application. A number of 10-year goals for reaching synergism and integration in CKTS are described below. They include methods applicable to the entire convergence–divergence cycle—a systems-based approach (see Sects. 4.3.1 and 4.3.3)—and other methods more specific to one of the four phases of the knowledge development cycle: the creative, integration/fusion, innovation, and outcomes phases (see Fig. 4.1 and Sects. 4.3.2, 4.3.3, 4.3.4, 4.3.5, 4.3.6, 4.3.7, 4.3.8, and 4.3.9).

4.3.1 Improve Systems-Based Convergence Methods

Improved systems-based methods for achieving CKTS will encompass the following kinds of approaches:

- *Support system science approaches*, including specific tools such as neural networking and logic model approaches to underline system interactions.
- *Support evolutionary approaches* for system convergence (see Sect. 4.8.1, Erik Goodman).
- *Develop knowledge mapping and network visualization techniques* for identifying large patterns in the knowledge, technology, and societal systems, as well as related work going on in other disciplinary communities and other hierarchical levels that might benefit from collaboration.
- Unify large sets of multi-topic databases and expert systems for the converging platforms, coupled with informatics, and linked computer simulations, including different phenomena, processes, scales, applications, and social events.
- Promote more integrated and interactive education programs (e.g., Khan Academy for large groups; the Integrative Graduate Education and Research Traineeship [IGERT], as described in Sect. 8.8.8 of this volume; and the vertical [from K–16], horizontal [disciplinary] and global [international] approach for education as described in Sect. 8.8.1 of this volume).
- Apply CKTS fractal patterns (similar patterns of contents and presentation but used in different contexts) in various areas of R&D, applications, and education.
- Combine organizations' technical pushes and societal pulls for R&D planning and setting organization structures.
- Use S&T policy and directed investment to reach societal goals.
- Combine deductive (e.g., nanoscale to macroscale analyses; apply direct design; application of topical-based science and technology to society) and inductive (e.g., Earth- to human-scale, apply reverse design; societal goal-driven science and technology) approaches to long-term decision analysis.
- Apply scenarios development and research for uncovering the potential open issues that may be "invisible" at the beginning of societal planning.
- Encourage international sharing of models for exploiting R&D investments, based on mutual interests.
- Move to open publications on converging methods.

4.3.2 Create and Implement Higher-Level (Multidomain, Convergent) Languages

By convergence language we mean the common concepts, network relationships, methods, and nomenclature used in a multidomain of science, technology, and society. Languages evolve over time. Each scientific discipline and technology area has a specific language. An effective convergence process across a CKTS platform will require a more comprehensive and faster exchange of language for communication and synergism among its disciplines, areas of relevance, and stakeholders. This would allow for better integration of the components and faster spirals of creativity and innovation in order to support successful communication and synergism across disciplines and cultures. Establishing of multidomain convergent (or higher-level) languages is using knowledge, technology, and cultural integrators such as unifying theories. The approach finds what is common and essential in multiple domains, and bottom-up and horizontally establishes languages and rules that are suitable to all those involved.

An example of an existing *convergence language is using music* to help bridge cultural divides at the human and societal scales. Throughout human history—probably before word-driven language—music has been used to create a common mood, bringing people together during important social events, from tribal meetings to funerals and modern military parades; it is a universal means of communication. Music education can contribute to development of the human personality, be used as therapy in difficult life situations, and be a means of communication and mutual understanding between different cultures in problem-solving processes.

Another example of a convergence language is one that will support understanding, reasoning, and decision analysis with respect to *unstructured data, patterns, and methods*. Such language is in development and will allow generalizations across apparently unrelated fields, novel solutions, and new mathematical concepts.

Unifying physical, chemical, and biological concepts would lead to a convergent scientific language for those scientific domains.

Multidomain benchmarking is a form of higher-level language relevant to both knowledge and technology areas. It can be used in both creative and divergent phases of convergence. It may help to identify where to focus efforts, steer the interest and efforts to broader goals, spotlight the key areas for creativity and innovation, and better communicate across fields.

An example of a process to establish a convergent language is *using shared databases* to connect computer simulations and evaluation methods for the respective convergence platforms such as NBIC tools or Earth-scale. This would facilitate interactions and broad principles of optimization.

An approach to establishing a convergence platform and its suitable convergence language is to identify the *knowledge and technology integrators* describing the essential features of the platform. A higher level of generality of a convergence platform and its language are reached when the respective integrators are applicable for larger domains and with faster information exchanges. Three successive levels of convergence have been reached, by advancing nanotechnology in 2000 (Roco et al. 2000), foundational tools in 2002 (Roco and Bainbridge 2003), and CKTS (this study; see Fig. 1 in Overview and Recommendations):

1. *Nanotechnology*—Integrate knowledge and disciplines for all sectors of the material world from the nanoscale.
2. *NBIC*—Integrate foundational and emerging technologies from elemental features (atoms, bits, DNA, and synapses) and integrate them into larger-scale systems.

3. *CKTS*—Integrate essential convergence platforms in human activities, including knowledge, technology, human behavior, and society, driven by societal values and needs.

4.3.3 Apply a Holistic Deductive Approach to Decision-Making

Human activities are interdependent through an evolutionary system (Fig. 4.1a). When such a system is specified for a given goal, the most suitable approach in decision analysis and establishing partnerships is the deductive approach, where all system components and their causal evolution are considered. The results are different from coincidental collaborations or interdisciplinary work, where the collaborating parts and approaches are defined *a priori*.

The applicability of the deductive approach is a function of how well the hierarchical structure of the respective system is known and whether the bottom-up, top-down, interdisciplinary horizontal, and longer-range links and processes are well understood.

4.3.4 Improve the Creative Phase of the Convergence–Divergence Cycle

Several methods to improve the creative phase of the convergence–divergence cycle are:

- Open collaboration.
- Multidisciplinary education: Apply CKTS fractal patterns in education (similar patterns of content and presentation but used in different components of the convergence platform) in various areas of R&D, applications, and education. For example, presenting nanoscale principles in chemistry, mechanical, and social sciences courses has created a foundation for collaboration and synergism in these several fields.
- Connect basic research projects based in different disciplines.
- Team science.
- Use social media and game theory to connect communities and increase the synergism of ideas.
- Explore concepts of new funding mechanisms conducive to convergent S&T ideas, e.g., processes for idea incubation prior to formal proposal formulation. Examples in the United States are the Integrated NSF Support Promoting Interdisciplinary Research and Education (INSPIRE) programs, the goal-oriented solicitations of the Defense Advanced Research Projects Agency (DARPA), and science-engineering-medical collaborative programs at the National Institutes of Health.
- Better understand psychological processes in creativity and human capabilities (for illustration of the role of intelligence, see Neisser et al. 1996).

4.3.5 Improve the Integration/Fusion Phase of the Convergence–Divergence Cycle

Several methods are:

- Use broad-based S&T platforms that offer more integration opportunities.

  – An example of the kinds of opportunities opened up through convergence–divergence processes is the cell phone platform, which has been underestimated throughout its history. It is rapidly becoming one of the most important technological platforms in mankind's history, with 1.6–1.8 billion sold per year, over 5.6 billion subscriptions, and devices owned by >70 % of the human race. Convergence of multiple disciplines and technology sectors yielded the cell system. It depends on high-frequency communications technologies and packet switching protocols; materials science and nanoelectronics for logic units, data storage, touch screens, antennas, etc.; and cognitive science and human–computer interface technologies—all of which converged to create the "smart phone" about a decade ago. In turn, the platform has yielded a cascade of outcomes across science and society. There are thousands of applications scarcely imagined 10 years ago, from social networks to ability to control swarms of very inexpensive miniaturized satellites, and many other examples. These impacts in turn have profound "cascade" implications for and secondary impacts on areas as diverse as national security, education, and cognitive science. In the area of healthcare, widely available "smart" portable devices are beginning to enable round-the-clock medical monitoring, and identification and interpretation, of data related to critical body functioning; these kinds of devices and their associated applications will accelerate highly personalized medicine, where nanoelectronics, sensors, cognition, and information technology are integrated. Soon they are likely to be connected through video devices in glasses, watches, and other portable devices in daily activities. Although such personalized information would serve initially for diagnostic and preventative purposes, it is widely acknowledged that such use of this platform is an important step toward improved quality in healthcare.

- Use multidisciplinary design methods.
- Make various manufacturing methods and phases of production compatible and synergistic by integration of knowledge and technologies.
- Develop advanced mind–cyber-physical systems.

4.3.6 Improve the Innovation Phase of the Convergence–Divergence Cycle

Several methods are:

- Promote open science and innovation, and participatory design and governance (Sect. 10.8).

- Support the field of futures studies, including use of trend analysis, Delphi, scenario writing techniques, and virtual reality methods.
- Use specialized software providing alternatives in decision-making (NSF 2012).

To conceptualize the influence of convergence on innovation, we have defined an index of innovation rate (I). The three-dimensional innovation divergence spiral (Fig. 4.1b) has a projection in the convergence platform plane (Fig. 4.1a) that is characterized by a time scale "t" of information exchange in the respective convergence platform. The time scale of the convergence–divergence process (from the beginning of the creative phase "A" to outcomes "D" in Fig. 4.1b) is "T". The index of innovation rate is estimated to be in direct proportion with the size S of the convergence platform from where information is collected (the domain circumscribed by the projection of innovation spiral; that is, the number of disciplines or application areas intersected by the circumferential spiral, in Fig. 4.1a), the speed of information exchange supporting innovation in that platform (S / t), the speed of the convergence–divergence cycle (1/T), and the divergence angle in realizing the outcomes (~ O/T).

$$I \sim k(S)(S/t)(1/T)(O/T) \sim kS^2O/(TTt) \sim k'S^2O/T^3 \quad (4.1)$$

where:

t ~ T because they are the time scales on the normal and axial projections of the displacement on the innovation spiral; O is a measure of the outcomes; and k and k' are coefficients of proportionality.

This qualitative correlation shows that the innovation rate increases with the size of the convergence domain ($I \sim S^2$) and is significantly affected by the time scale of the convergence-divergence cycle ($I \sim 1/T^3$). This correlation has similarities with the Metcalfe's Law in information research (Shapiro and Varian 1999) that says that the number of possible cross-connections in a network grows as the square of the number of computers in the network increases, and the community value of a network grows as the square of the number of its users increase. Metcalfe's Law is often cited as an explanation for the rapid growth of communication technology and the Internet. The (O/TT) term in the equation is in agreement with the empirical exponential growth model for science and technology (for illustration see Kurzweil 1999).

4.3.7 Improve the Outcomes Phase of the Convergence–Divergence Cycle

Several methods and evaluation considerations are:

- Reevaluate and take steps to realign criteria for success in society by adding—besides economic output (such as GDP)—accumulations in knowledge and technology, preparation for the future, and changes in education and infrastructure

- Identify and pursue high-priority common values and goals
- Cultivate user-driven approaches such as use-inspired basic research
- Create top–down governance focused on convergence for specific goals
- Sponsor self-regulating convergence in self-reliant communities (with constraints for sustainable technologies and practices)
- Explore downstream activities as integrators (provide funding for application areas)
- Focus on computer-assisted policy design methods
- Pursue methods for distributed and hybrid manufacturing
- Investigate medical-cognitive advances; develop governance science

From one perspective that shaped the working group's discussion at the U.S. NBIC2 workshop, policymaking progress will depend on better tools, including the inclusive rules themselves but also computer-assisted policy design methods. The equivalent of a computer-aided design tool for modeling the world's service systems is much needed. The world's service ecology is increasingly composed of major nested, networked service system entities. Modeling cities and their top universities is essential, because universities are the knowledge engines creating the next generation of workers, knowledge, and regional economic development. Flows of people, information, capital, technology, energy, water, waste, and other resources or products are also important to understand.

Convergence results also may be expressed in terms of increases in number of patents from converging technologies, number of new jobs created nationally, percent increase in GDP, as well as the emergence of new industries and academic disciplines.

4.3.8 Vision-Inspired Basic Research and Grand Challenges

Using forecasting, early signs of change, scenario setting, and other approaches, it is possible to establish a credible vision for what is desired in the longer term for a knowledge and technology field. Then, a recommended approach is to work backwards from the vision to investigate intermediate research steps and approaches. This approach was used in researching and writing the two nanotechnology research directions reports (Roco et al. 1999; Roco et al. 2011).

A typical knowledge and technology convergence–divergence process (Fig. 4.1b) has four phases (creative assembling, system integration for known uses, innovation, and outcomes that lead to emerging new uses); the research approach has to adapt to the corresponding phase of the process. The CKTS methods can provide connections between the long-term knowledge and technology vision and basic research activities in each phase. An additional "Vision-Inspired Basic Research" quadrangle is proposed (Fig. 4.5) that expands the existing four domains—particularly the Pasteur quadrangle—of the Stokes diagram of basic scientific research paradigms, to address new and emerging areas of research and applications ("new use" section of Fig. 4.5).

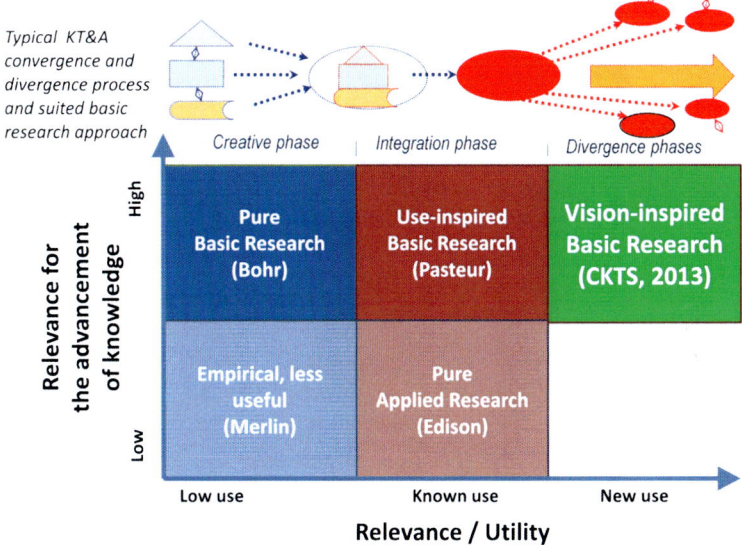

Fig. 4.5 Schematic for "Vision-Inspired Basic Research" quadrangle in the modified Stokes diagram: three basic research approaches (pure, use-inspired, and vision-inspired) are suggested as a function of the phase (Creative, Integration, or Divergence) in the knowledge, technology, and applications (KT&A) convergence–divergence process (Figure is a modification of that by Stokes (1997), courtesy of M. C. Roco)

To efficiently and responsibly achieve the benefits of research in new areas, the convergence processes need to be used to identify the vision and its corresponding basic research strategic areas, changing priorities periodically as interdependencies change, old goals are achieved, and new ones come within reach. The "vision-inspired" domain is extending the Stokes diagram to basic research in emerging areas where their solution and use are not known *a priori*. Different basic research approaches are needed as a function of the phase in the knowledge and technology convergence–divergence process (Fig. 4.1b), that is, the Bohr approach for the creative phase (A), the Pasteur approach for the integration/fusion phase (B), and vision-inspired basic research for the divergent phase (C and D).

4.3.9 *Change the Knowledge and Technology Culture*

An overall goal of CKTS is supporting a *culture* that is guided by principles of interdependence and connectivity. Two approaches for changing the culture are discussed below: open science and anticipatory technology assessment.

Open Science

A key approach for the creative and innovation phase is "open science" consisting of virtual organization collaboratories and online citizen science. Already, scientists are beginning to talk about open science as the approach in which specialists, nonspecialist scientists, and nonscientists collaborate fruitfully by means of new methods of communication and novel research methodologies (Woelfle et al. 2011). Two important recent information technology developments are expected to converge in the near future, with marvelous benefits for the unification of the sciences towards open science:

- The evolution of scientific data archives into digital libraries in the 1990s and into *virtual organization collaboratories* in the 2000s. Prominent examples include The Protein Data Bank (http://rcsb.org/pdb/; Berman 2012), The Computer Tomography Virtual Organization (Tapia et al. 2012), and a computer-based archives network shared by astronomers (Djorgovski 2012).
- The emergence of *online citizen science* over the past decade, which began by using ordinary people as volunteer data collection and classification workers, but which is beginning to include people from a variety of technical backgrounds as volunteer collaborators. An example that has been used as an innovative information technology approach to unite biology with nanoscience is the Foldit project (http://fold.it/portal/info/science), in which nonscientist volunteers have been able to solve difficult problems in protein folding through an online game (Khatib et al. 2011). Another example is the 2010 Oregon Citizens' Initiative (Gastil and Knobloch 2010).

Several future convergence scenarios come readily to mind; the following five suggest the range of possibilities, each organized around a new or existing virtual organization collaboratory:

- A professional scientist in one discipline volunteers as a citizen scientist for a project in a different discipline, thereby gaining greater awareness of that other science and growing into a professional collaborator with the scientists managing the project or with others in his/her field.
- A serious student in high school becomes a citizen volunteer for projects in two or more distinct sciences, gaining increased motivation to become a professional scientist and the experience to succeed as a double-major in college.
- A member of the staff of a college or university migrates from one project to another, after contributing significantly as a citizen scientist in each, and then becomes the official coordinator of citizen science projects at his or her educational institution on the basis of this extensive personal experience.
- Members of two high-quality scientific teams in adjacent fields volunteer as citizen scientists to contribute labor to the other's success as a conscious step in development of a permanent collaboration between the teams.
- A scientist in one field becomes a citizen science volunteer for a project in his or her own field but conducted in a different nation and using a different language, as preparation for increased international collaborations.

Anticipatory, Participatory, and Adaptive Technology Assessments and Decision-Making

The idea of anticipatory technology evaluation for CKTS fits within a larger national and global movement toward sustainable chemical, material, and product development and use. As an example, work on public responses to nanotechnology has included fine-grained comparative work with other contemporary merging/converging technologies like neurology and synthetic biology and with past technologies and risk controversies such as biotechnology and nuclear power. A key aim of risk perception and mental models research is to create an empirical basis for linking risk and benefit perception to risk and benefit communication. Anticipatory technology evaluation will require a reflexive approach to CKTS development. That is, it will be necessary to look critically at technology itself in terms of impacts as well as benefits.

4.4 Infrastructure Needs

Changes expected in the next decade include accelerated developments in emerging and converging technologies, increased needs for sustainable development in a more crowded world, more rapid globalization of commerce, and aging of the world's overall population. This section discusses several S&T infrastructure elements that are needed to adapt to these kinds of changes.

To effectively achieve integration and synergism in CKTS, an infrastructure is needed that can provide diverse networks and tools for communication and convergence across knowledge and technology domains. The primary objectives are to develop new convergence methods by exploiting expertise that already exists to build a specialized expert network infrastructure. The infrastructure is expected to provide platforms for:

- Creating specialized networks and tools for communication, with seed funding for converging of people and communities. Horizontal information systems in university, industry, and government organizations will have an increased focus on building knowledge versus conveying information.
- Developing methods for advancing convergence in planning, decision analysis, investment policy, and other areas.
- Building collaboration spaces that are free from geographical boundaries, organizational constraints, and single-domain restrictions. Examples of biology-centered convergence in the United States are the Arizona State University's Biodesign Institute (see Chap. 1, Sect. 1.8.4), Stanford's Bio-X Institute, Harvard's Wyss Institute, and the University of Wisconsin–Madison's Wisconsin Institute for Discovery (a public–private partnership).
- Drawing on international forecasts of science and technology trends. Forecasts of essential aspects in the world can be garnered by mining information from social network and news services such as Twitter, Facebook, Youtube, CNN, etc.

- Developing information systems for finding experts from different domains across the world, providing a communication space by adopting functionality from existing professional and social applications.
- Enabling the sharing of resources and exchange of knowledge by designing a knowledge profiling service with well-defined resources such as data, software, source code, documentation with standardized formats, publications, laboratories, etc.
- Creating collaborative groups and projects for solving real-world issues.
- Providing tools or services for analyzing and visualizing trends, networks, project progress, discussion topics, etc.

Convergence can be supported by rejuvenation of university laboratories and creating CKTS curricula early in the educational continuum. A number of researcher network systems have been introduced to expose high-quality institutional data via websites and semantically structured RDF (resource description framework) data. Several well-known systems are VIVO (the "Facebook-for-researchers", http://vivoweb.org); Harvard Catalyst Profiles (http://profiles.catalyst.harvard.edu); Stanford CAP (http://med.stanford.edu/profiles); Academia.edu (the "Linked-In for academics"); and Elsevier Scival Experts (http://www.scival.com/experts). These systems are institution-based, and only the members of the system are able to access the information. International Researcher Network (IRN) visualization (http://nrn.cns.iu.edu) has provided a single interface that allows experts to find other experts across these systems. Since these systems are served as a profiling service where members can find expert members in the system, they lack the opportunities created by open-communication functionality and resource-sharing. Collaboration has to be done through email outside the systems. HUBzero® (http://hubzero.org/) has taken this web-based researcher network concept further by providing an online space for sharing resources and collaborating; however, it is still limited to members only, and each hub is only available for a specified expertise group. For example, nanoHUB (http://nanohub.org/) is for nano researchers and the c3Bio hub (http://c3bio.org/) is for researchers involved in energy and carbon efficiencies of biofuel production. Mendeley (http://www.mendeley.com/) has provided a space for experts from different disciplines to collaborate. (A visualization of the cross-disciplinary collaboration is shown at http://cns.iu.edu/research/2011_Mendeley_Binary_Battle_entry.pdf.)

Software, computer, and visualization equipment are needed for improved communication, information, and virtual reality environments to allow hybrid cluster facilities for manufacturing.

4.5 R&D Strategies

The following strategies are suggested:

- *Restructure "broader impact" requirements for projects to include convergence aspects* such as manufacturability, planning, investment policies, education, and

decision analysis. Establish positive closed-feedback loops across multiple technologies, medicine, education, and manufacturing, regulated appropriately.
- *Expand the "science for science" programs to include a focus on "science for governance"*, methods to improve and expedite convergence, and institutionalize them to do so. The need may be illustrated by how different the approaches are toward grant proposals in different government agencies. Researchers who have spent the majority of their careers dealing with one agency should deal with various agencies that have completely different research philosophies. Plans for interdisciplinary centers should ensure that communications in calls for proposals are explicit about expectations for prior research and that evaluations of proposals are conducted by panels constituted of members who have multi- and interdisciplinary backgrounds.
- *Promote mechanisms to engender more effective stakeholder involvement in improving education and research to support CKTS.* One strategy is to support research grants that encourage universities to develop new programs and courses that address convergence-driven research that is holistic and personalized in concept. In plans for any new cross-disciplinary efforts for converging technologies, capturing and leveraging stakeholders' enthusiasm and excitement should be a high priority.
- *Provide open access to science and technology journals* with the goal of widely disseminating best practices to promote accelerated advancement of converging technologies. Current efforts toward open access are being restricted by the need to resolve intellectual property issues, to compensate "publishers" for the effort and expertise to review and provide the resources to archive the publications, and to provide appropriate rewards systems for authors and their institutions. These difficulties about open access are being compounded by the increasing prices of journals (Sample 2012). Arguments have been put forth (Ross n.d.) that open-access publications are beginning to garner prestige equivalent to conventional publications, and that taxpayers have already paid for the research results.
- *Extend NNI lessons learned for promoting interdisciplinary research.* There are some significant lessons learned from the National Nanotechnology Initiative in the efforts by participating agencies to promote interdisciplinary research among their multidisciplinary centers and programs. Nanotechnology development has been similar to CKTS in that researchers from all the physical, chemical, biological, and medical sciences participate, but CKTS broadens this expectation to include the geological, geospatial, social, neurological, environmental, and educational regimes.
- *Convene a group of agency and academic experts to identify and design appropriate centers around national research priorities.* Converging technologies imply cross-disciplinary efforts, and R&D centers provide a mechanism to locally house expertise derived from multiple disciplines. With the growing sophistication of information technology, virtual centers can certainly play a role, but unscheduled, informal meetings between people are often the more likely occasions for nurturing insightful, innovative ideas. This effort must explicitly engage the challenges of converging technologies.
- Foster open innovation environments and innovation mechanisms for CKTS (Kim et al. 2008).

- Foster anticipatory and participatory governance integrating the four basic functions of CKTS (transformative, inclusive, responsible, and visionary) as a guiding vision for the next 10 years. Foster diversity in the pipeline of future researchers; this will promote innovation by providing differing perspectives. Devote attention to enabling participation by broader representation in the student population, including lower-income students (Osterman 2008) and students from segments of society that are underrepresented due to gender (NRC 2010a; Beede et al. 2010) and/or minority status (Frehill et al. 2008).
- *Define realistic "grand challenges"* in problem-driven research for proper selection of converging methods. Examples are the "mapping brain activity" project (Alivisatos et al. 2012) and the technology challenge project (Mills and Ottino 2012).

4.6 Conclusions and Priorities

The main priorities as they relate to the methods of convergence are as follows:

- *Proactively promote holistic and balanced convergence of knowledge, technology, and social developments* with a strong focus on problem-solving, transdisciplinary S&T platforms, stakeholder involvement, and use of emerging resources. Emerging resources include multidomain databases; cloud information, computing, and manufacturing; open-access publications; open innovation mechanisms and environments; and virtual reality methods.
- *Added-value decision-making and transformation of scientific knowledge* are based on the convergence–divergence process (see below) in science, technology, and applications. Methods are needed to improve and expedite each of the four phases of the process.
- *Develop a systematic approach for selecting the methods of convergence for synergy and integration,* with a focus on emerging technologies in society and enhancement of individual human quality of life in terms of both mental satisfaction (pursuit of happiness) and improved physical condition.
- Identify the subset of problems likely to benefit most from the convergence approach. Address priority large-scale imminent problems of fundamental concern to communities across the globe, Define specific, realistic "grand challenges" in problem-driven research, such as brain-inspired neuromorphic devices, digital and distributed manufacturing, Earth-scale interventions, and personalized health and education.
- *Establish methods for convergence suitable for a variety of self-reliant communities* (sized from a thousand to millions of people) for sustainable development over time. Define specific, realistic "grand challenges" in problem-driven research proposed by various communities.
- *Establish broad user "platforms" for interdisciplinary collaboration* to support communication, inputs from various communities, international partnerships, and sharing of resources and expertise, e.g., portal websites serving diverse S&T communities.

- *Fully involve social scientists* and decision analysis procedures in CKTS projects.
- Incorporate dynamic feedback loops across multiple technologies in performance evaluation.
- Establish a common language for scientists, engineers, industry, regulators, and the general public, and develop compatible standards for different fields.
- Seek alternatives and modifications to the current peer-review system in R&D project selection in order to encourage converging technologies R&D, and consider a set-aside of funding for exploratory CKTS projects.
- Focus early in the innovation cycle on identifying potential risks and hazards. Support international collaboration to enhance value and reduce risk in high-return R&D projects for CKTS.

4.7 R&D Impact on Society

Accelerating the beneficial use of CKTS in society to solve high-interest problems will deepen human insights into natural phenomena, offer significant new opportunities to increase quality of life, and bring an infusion of excitement into society concerning S&T similar to that seen for the Apollo Mission.

The United Nations goals for human development provide a good reference for guiding convergence methodologies (UN 2012a, b). The methods of convergence need to enable the following kinds of outcomes:

- Affordable access of the world's citizens to new knowledge, technologies, and products. Such access is anticipated through:
  - Increased and open access to the latest advances in S&T publications and databases in all disciplinary areas, including platforms for interdisciplinary collaboration, host communities, and shared resources and expertise.
  - Innovation for advanced manufacturing/jobs/competitiveness such as that provided in personalized ("mass customization" or "do-it-yourself") manufacturing and the availability of manufacturing information as a service. These exemplify the integration of information, communications, materials, and economics in an evolving new paradigm for manufacturing. Downstream activities such as *design* and *manufacturing* can serve as important means of integrating diverse knowledge bases. Realizing artifacts and services of benefit to society necessarily requires the consideration of multiple facets: economics, technology, sustainability and the environment, societal effects, etc. A main goal is increasing worker efficiency.
  - Scalable models for distributed manufacturing: The emerging idea here is the integration of small distributed facilities to obtain scaled-up manufacturing. Crowd sourcing is one emerging idea related to this trend.
  - Information-technology-inspired communication and manufacturing systems (cores, buses, peripherals, operating systems, networks, etc.) following architecture similar to that employed in computer systems.

- *Significant support for increased innovation* through inclusion of multiple science paths, application areas, and societal targets.
- *Sustainable and equitable use of natural resources* (energy, water, sustainable food and nutrition security, climate, oceans, forests, biodiversity) and management of wastes.
- Universal access to high-quality healthcare and education.
- *Enhanced individual quality of life* through personalized medicine, education for the entire life span, and other measures.
- Use of new governance paradigms enabled by CKTS.

Sustainable development is a challenge for the world because current economic development models (i.e., development based on increases in global population and industrial output of all economies) will accelerate resource depletion and global warming. Piecemeal progress in different geographic areas, supported by advances in single areas of technology, could be counterproductive. This challenge can be met successfully only by treating the world as a single complex system, from the perspective of unified science based on convergence. A comprehensive approach is required to address the economic, social, and environmental dimensions of sustainable development (UN 2012a, b). Methods of convergence have particularities as a function of goals, such as those listed in Sect. 4.6.

Much of the CKTS discussion concerns human beings working together to achieve convergence of the sciences, but it is also important for the sciences to work together to achieve convergence of humanity. This is obvious in the international context, and the CKTS discussions and workshops have been carried out explicitly as a global activity. Within each nation, convergence of people having a variety of native talents must also be achieved. There will be many challenges.

The world has already advanced several decades into a new form of economy, building upon the original Industrial Revolution and worthy of being called the Second Industrial Revolution, in which machines and physical products are still important but where information systems are paramount. At the beginning of this revolution, a number of theorists proposed very powerful ideas about this future, but we have largely failed to conduct the empirical scientific research needed to evaluate their ideas and to understand the new world we are rapidly entering. Whether the emerging world society is called *technological, technetronic, informatics,* or *post-industrial*, it places a high priority on the ability to handle complex systems of information (Ellul 1964; Brzezinski 1970; Bell 1973; Castells 1996). What, then, will be the place for people who lack the abilities or skills to do so?

Often, this question has been framed in terms of the *digital divide*, originally defined in terms of lack of affordable access to computers by economically disadvantaged people, but that also may apply to people having easy access to Internet but lacking the ability to use it effectively (DiMaggio et al. 2001; Attewell 2001). Universal access to quality education will reduce the number of people lacking essential skills, although very deep analysis will be required to design educational improvements that are really effective (Ray and Mickelson 1993) for all kinds of students. Ethical considerations must be addressed for providing equal opportunities for all kinds of individuals and communities to participate fully in the progress that CKTS is expected to bring.

4.8 Examples of Achievements and Convergence Paradigm Shifts

4.8.1 *Evolutionary Approach for Convergence*

Contact person: *Erik Goodman, Michigan State University*

The BEACON Center (Bio/computational Evolution in Action CONsortium) for the Study of Evolution in Action (http://beacon-center.org/) is a science and technology center funded in 2010 by NSF. It teams evolutionary biologists conducting experiments in the lab and field with computer scientists and engineers experimenting with evolution of artificial organisms in the digital domain, where the organisms compete for resources and reproduce themselves with a specified mutation rate. Experiments are carried out that cast light on the mechanisms operating in the other domains.

4.8.2 *Open Learning Systems for Convergence*

Contact person: *Clayton Teague, National Nanotechnology Consultant*

The new online Internet learning efforts by the Khan Academy, MIT OpenCourseWare, Udacity, EdX (a joint venture of Harvard and MIT), Coursera (Stanford, University of Michigan, and University of Pennsylvania), 2Tor (UNC Chapel Hill and Georgetown University), and others are challenging conventional approaches to learning systems for students. These kinds of approaches to delivering high-quality lectures to individual students are being combined with such efforts as those by the Knewton Company to obtain real-time feedback from a student's response in terms of his or her speed, accuracy, delays, keystrokes, click-streams, and drop-offs. In the future, one could envision a tutor watching a student via a camera and having the tutor respond with the intelligence of say a "Watson"-equivalent computer to the student's responses, such as depicted in the novel *The Diamond Age* by Neal Stephenson (1995).

4.8.3 Convergence in Regional Partnerships: Oregon Nanoscience and Microtechnologies Institute (ONAMI)

Contact person: Skip Rung, ONAMI

Knowledge, innovation, and entrepreneurship converge in regional partnerships such as ONAMI to build better citizen futures and serve a diversity of stakeholders. ONAMI was formed (and funded by the State of Oregon) in 2003 to locally respond to the implications—both benefits and challenges—of accelerated technological advancement and global integration. The gifts of freedom, prosperity, and global peace (comparatively speaking) have raised the stakes for communities, regions, and nations to enable their citizens and institutions to create and deliver economic value commensurate with their consumption, as determined by a global marketplace that increasingly erodes the economic potency of political borders.

The current "state of play" calls for critical and apparently non-substitutable collaborative roles for (a) researchers/inventors, (b) entrepreneurs/early-stage investors, (c) large manufacturing and "customer-facing" businesses, and (d) governance at all levels in order to sustain and reasonably distribute the productivity and opportunity growth that has led to human flourishing in the leading economies—and to do so without undermining the human responsibility and initiative that made the productivity and opportunity possible in the first place.

Two important examples of these changing interdependent roles are (1) increasing outsourcing of early-stage R&D to universities and startups (which requires both early-stage capital and large-company customers and scale-up partners) and (2) "horizontalization" of large-company roles, e.g., even large original design manufacturers (ODMs) such as Apple and Hewlett-Packard have become customer-facing operations, relying on contract manufacturing firms for production and even supply chain operations. These developments are mature in the electronics and pharmaceutical industries and are moving to other sectors such as energy and materials.

Just as nanotechnology is a convergence of previously distinct science, engineering, and medical disciplines, the laws of the human ecosystem (economics) are becoming a convergence of knowledge and knowledge workers, innovation, and productivity enhancement in major industries and markets, and entrepreneurship and risky leadership and investment in disruptive advancements that at first threaten, but ultimately transform, large private and public organizations.

ONAMI's chosen approach is project-centric. It funds collaborative developments involving research institution assets (intellectual property [IP], talent, and facilities), entrepreneurial ventures (with experienced management teams), a regional industrial base in materials and device manufacturing, and public economic development functions that are rapidly evolving (from corporate recruiting from and assistance to "mom & pop" operations to cluster development, "gazelle" company, and "economic gardening").

Fig. 4.6 The ONAMI gap fund process (Courtesy of S. Rung)

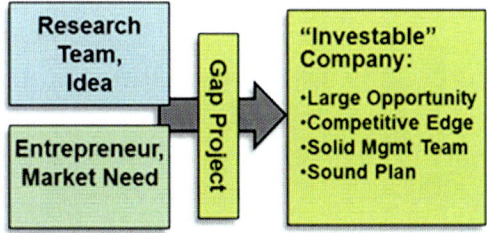

The crucial inputs to the process are technology IP, a startup business plan, and a project team to advance them both. The "gap project"[6] consists of up to $250,000 in 3–4 installments of milestone-tranched grant funding and rigorous periodic technical and business review. The desired output is an "investable company" (Fig. 4.6), as defined by terms acceptable to "super angels," angel groups, venture capital firms, and (typically, later) private equity/capital management firms.

In the simplest terms, what new technology startups need is a good chance for the kind of breakthrough economic performance that only significant innovations can achieve and that require a transition in orientation from "technology push" to "market pull" as evidenced by successful sales to and partnerships with significant (i.e., large) customers. With venture capital increasingly scarce (because of poor performance) and demanding ("capital efficiency"), the expectations for deal quality are bracing. Business community and local government assistance (in many forms, including limited partner investments) can provide the necessary edge.

ONAMI's results with this process have exceeded expectations (see (Fig. 4.7). As of 2012, approximately $6 million has been committed and $5 million disbursed, resulting in over $112 million in leveraged funds ($96 million in private capital, $16 million in SBIR/STTR and other government grants) to ONAMI gap fund portfolio companies (Fig. 4.6). All of the companies are early-stage—two thirds are pre-revenue—but most of them are candidates for making world-impacting differences in medical devices, cancer treatments, semiconductor fabrication, energy generation and storage, solid state lighting, flat-panel display manufacturing, and water resource management (see http://www.onami.us/commercialization/current_gap_projects).

Not every gap project has been successful, but the experience is consistent with the original intuition that deep collaboration between technical and business development talent (both of these guided by outstanding leadership and mentoring) is a critical factor in success. The "investment" criteria therefore remain unchanged: build a solid business plan oriented around customer/strategic partner needs, a clear

---

[6] In general terms, the "gap" is the normally underfunded period in the evolution of a new technology development between the time it is funded at the R&D level and the time, when applied in a startup business, that it successfully attracts adequate commercial investment funding and/or becomes a profitable enterprise.

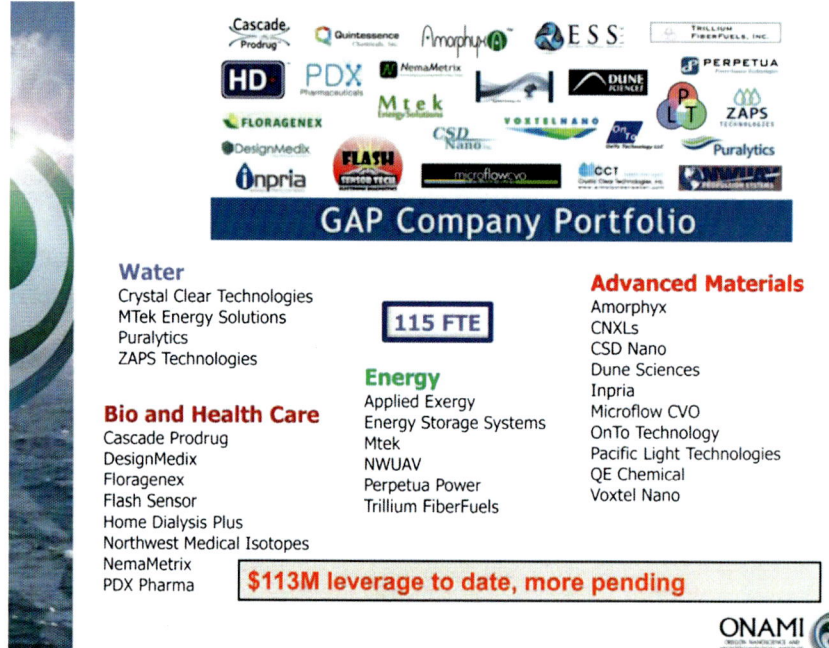

Fig. 4.7 ONAMI's gap fund grants have enabled 25+ companies to raise over $112 million in leveraged private and Federal funding (http://www.onami.us/PDFs/innovation-economic-development-9-14-2012.pdf; courtesy of S. Rung)

definition of the "gap" to be crossed, and quantified success criteria for in-project milestones and end-of-project goals.

4.8.4 ICTAS as an Exemplar of an Organization Built on the Foundation of Converging Technologies

Contact person: *Roop Mahajan, Virginia Polytechnic Institute and State University (Virginia Tech)*

Converging interdisciplinary research as a mode of discovery and learning is a powerful paradigm. It allows us to break down and restructure our knowledge about a subject in order to gain new insights into its nature and to use our imaginations to invent and innovate. A report from the National Research Council of the National Academies, *Facilitating Interdisciplinary Research* (NRC 2004), considers interdisciplinary research to be "one of the most productive and inspiring of human pursuits—one that provides a format for conversations and connections that lead to new knowledge" (p. 16).

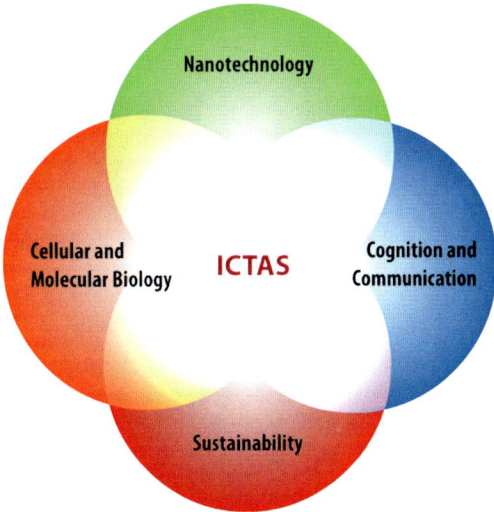

Fig. 4.8 ICTAS view of transformative technologies (Courtesy of R. Mahajan)

Perhaps nowhere is the potential of interdisciplinary research more promising than at the intersection of the four converging NBIC technologies. As noted by Roco and Bainbridge (2003), this set of powerful technologies is poised to unleash a new understanding of matter at the atomic scale and of the complex working of the human brain, creating opportunities for new industries and jobs and enhanced human capabilities. While each of the original NBIC technologies is powerful, with the potential for huge impact, it is the interfaces between them that are most transformative and exciting. However, in harnessing these technologies, attention must be paid to unintended consequences, as well as to the short- and long-term implications of these technologies in a range of areas that include the environment, quality of life, and human dignity (Fisher et al. 2006).

Motivated by these considerations, the Institute for Critical Technology and Applied Science (ICTAS) at Virginia Tech formulated a vision to be a premier institute to advance transformative, interdisciplinary research for a sustainable future. To this end, ICTAS decided to focus its research on tapping the potential of the confluence of the NBIC technologies, anchored by the principles of sustainability (Fig. 4.8), where sustainability is defined as a metadiscipline that integrates economic, social, and environmental processes to meet the needs of the present without compromising the ability of future generations to meet their needs.

The following research thrust areas selected for investment and advancement are either a subset of or at an intersection of these technologies: nanoscale science and engineering, nano–bio interface, cognition and communication, sustainable energy, sustainable water, renewable materials, and national security. Considering the nature of our times when technology progresses at breakneck speed, another thrust area, "emerging [disruptive] technologies," was added as a focus. Each of these thrust areas is populated by interdisciplinary research teams ranging from a roster of 12 or more faculty members to smaller teams of four to six researchers. A minimum of two doctoral students per faculty member/researcher is the norm.

In the nano–bio thrust area, for example, research is carried out under the umbrella of six research teams engaged in investigating cellular interactions with nanostructured surfaces; the seamless union of computational and experimental models to drive the next generation of advances in tissue engineering and in systems biology; multiscale bio-engineered devices and systems; nanoscale assembly of biological building blocks to understand healing mechanisms; regenerative medicine in animals; and targeted delivery of multifunctional nanoparticles for imaging, therapeutic, and immunological applications.

It is estimated that the sphere of influence of ICTAS activities is at least 60 % on the Virginia Tech campus, where five of the eight major colleges are represented on the ICTAS Stakeholder Board: the College of Engineering, College of Science, College of Agriculture and Life Science, College of Natural Resources, and Regional College of Veterinary Medicine. ICTAS interacts with all of these colleges substantially (and to some extent with the remaining three colleges) and supports their faculty in the strategic areas of ICTAS (NBIC-based) research through facilities, seed grants, equipment, matching funds, start-up funds, and large proposal development.

Activity focus areas for the next years are included in the ICTAS strategic plan for 2012–2018; ICTAS holds frequent meetings with thrust leaders, who are leading scientists/engineers in their fields, to brainstorm where the technology/science is headed.[7] Projects are then selected on the basis of members' unique capabilities and a goal of being among the top three in the selected field. For example, Sustainable Agriculture was added as a thrust area to pursue in the next few years. "Black swan" seminars are held to help identify new areas of research to seed and grow. An example is the Virginia Tech project, the "Humanoid Hospital and Hospital Room of the Future."

Supported by over 180,000 square feet of state-of-the-art collaborative laboratories spread over three buildings in Blacksburg and 6,700 square feet of rental space in the Virginia Tech Research Center at Arlington, ICTAS has grown significantly since 2006 in its contribution to the research and learning enterprises of Virginia Tech. Research expenditures over the first 5 years grew tenfold to a level of over $80 million/year in FY 2011. This has been made possible by the collaborative work of over 300 faculty members[8] from a wide spectrum of disciplines spanning engineering; the physical, life, and social sciences; and the humanities.

Looking ahead, NBIC will continue to serve as platform technologies for ICTAS, at least for the next decade. However, these technologies will give rise to or will be supplemented by cutting-edge developments in various other technology fields, i.e., emerging technologies. In a *Wall Street Journal* op-ed, "The Coming Tech-led

---

[7] Thrust leaders include Mike Hochella (nanoscale science and engineering), Marc Edwards (sustainable water), Jeff Reed (cognitive communications), and Naren Ramakrishnan (discovery analytics).

[8] This number does not include faculty supported in earlier years or through provision of laboratory space, equipment, etc.

Boom" (January 30, 2012), Mark Mills and Julio Ottino named three grand technologies—big data, smart manufacturing, and the wireless revolution—poised to transform the twenty-first century as much as telephony and electricity did in the twentieth century. Other novel technologies, so-called "black swans," may not yet have been recognized but may still carry an extreme impact. Best-selling author Nassim Nicholas Taleb in his book *The Black Swan* (2010) defines a black swan as an event that has three characteristics; it is an outlier; it carries an extreme impact; it has retrospective predictability. He further makes a claim that our world is dominated by black swans. To identify future disruptive technologies, ICTAS holds a monthly "Black Swan" seminar series that generally focuses on a broad field of inquiry and is guided by the question, "What technology/innovation idea will transform your field in seven years?" These seminars are open to all who want to innovate and stay ahead of the times.

Built on a strong foundation of the converging technologies, ICTAS is well-positioned to integrate emerging technologies and sustainable practices to bring positive transformation in the lives of people locally and globally.

4.8.5 *Knowledge and Technology Convergence and Decision Analysis*

Contact persons: *Matthew E. Bates and Igor Linkov, U.S. Army Corps of Engineers, Engineer Research and Development Center*

Developments in transportation, information technology, and the physical, life, and social sciences have led to discoveries, knowledge, products, and capabilities that have revolutionized science and changed how we live our lives. Although independent progress in these fields has been invaluable, the increasingly complex nature of our world and its management requires additional fusion of information and intentioned synchronization between fields and across organizational, temporal, spatial, and social scales to develop converging technologies and knowledge capable of solving the most pressing economic, technical, and social problems facing humanity now and in the future.

Previous attempts to understand the convergence of disparate fields for social benefit have been largely qualitative and not well integrated across management levels. This NBIC2 study addresses the mechanisms of convergence where decision analysis can be adapted to align with convergence. For example, in previous studies, bottom-up approaches provided available technical information to organizational decision-makers as raw data or, at best, as visualized through the use of statistical tools and dashboards. Without additional decision tools integrating information across scales and domains and quantifying tradeoffs from stakeholders and top leadership, the ensuing decisions were limited to ad hoc treatment of the provided data based on management intuition, expert opinion, and often misleading oversimplifications (Hammond et al. 1998).

Similarly, top-down approaches implemented by governments and leading scientific organizations as road maps and vision statements have been helpful in guiding various communities towards information fusion at the highest levels, but they have lacked concrete and prescriptive means to guide the daily technology-development decisions through which convergence ultimately takes place. This lack of practical decision relevance in projects has been noted on several occasions by the scientific community, with calls for the adoption of quantitative decision-analytical frameworks that fuse relevant data and preferences from all levels of the management hierarchy and social structure and use transparent and replicable decision tools with this information to guide decisions from a holistic perspective (NRC 1996; NRC 2009d; OECD 2003; The Presidential/Congressional Commission on Risk Assessment and Risk Management 1997; NSET 2011).

Decision analysis is a science with tools relevant for supporting a broad range of economic, social, and technical decisions at the organizational, national, and international levels. It aids knowledge and technology convergence for social or other benefit through the fusion of preferences and information across temporal, spatial, and organizational scales and domains. Decision analysis has developed over the past half-century drawing on ideas from operations research and the cognitive and management sciences to transparently and quantitatively identify the underlying values guiding decisions, map those values to data for comparing alternatives in a specific decision context, integrate all relevant preferences and information available from any level of the management hierarchy, and fuse this information to evaluate decision alternatives or potential product designs (Raiffa and Schlaifer 2000; Keeney and Raiffa 1993). These methods transcend the directional hierarchy of typical approaches by combining technical information and needs (bottom-up) strategies with broad organizational, national, or international objectives and tradeoffs (top-down) strategies to guide the creation and selection of alternatives from a holistic perspective.

Of particular interest are the Multi Criteria Decision Analysis (MCDA) and Value of Information (VoI) analysis tools. Based on available scientific information, MCDA identifies feasible alternatives (typically enumerated by experts) and decision criteria (typically from decision-makers), assess the performance (via experts) of each alternative relative to those criteria, and elicits or explores relative priorities (from stakeholders and decision-makers) among the continuum of incommensurable criteria (e.g., characteristics that cannot be reduced to single units, as in a cost-benefit analysis) (Linkov and Moberg 2012; Tervonen and Figueira 2008). VoI analysis extends MCDA by further identifying uncertainties whose resolution has a good chance of changing the decision or that most undermines confidence in the results. This leads decision-makers to guide knowledge and technology convergence for social benefit by quantitatively demonstrating how differences in the design of integrated products or technologies affect long-term, holistic social outcomes (Howard 1996; Linkov et al. 2011, 2012a) (Fig. 4.9).

The ad hoc nature of many current technological decisions is worrisome because knowledge convergence and the management of emerging technologies are filled with complicated, contentious, and risky decisions. For example, a manufacturer's

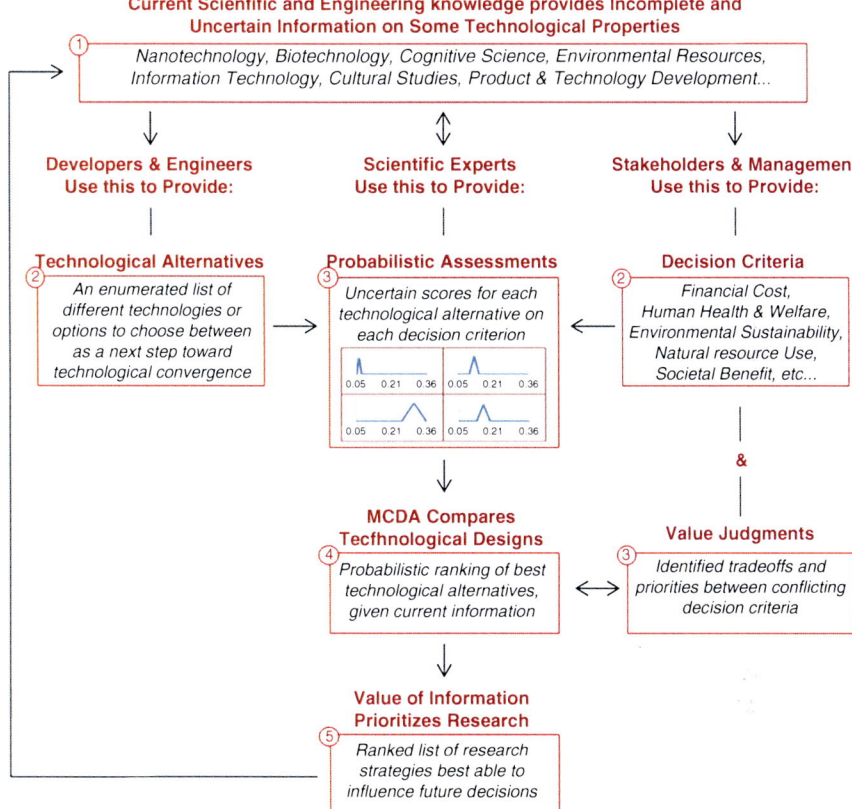

Fig. 4.9 Example of a five-step decision analysis process to integrate quantitative and qualitative information across organizational, temporal, and spatial scales to develop alternatives, evaluate decisions, and prioritize research from a holistic perspective encompassing multiple decision criteria (Bates et al. 2012; ©*NanoToday*/Elsevier, permission for reuse through RightsLink)

choice to use a specific type of nanomaterial in a new biomedical device may involve significant uncertainty about life-cycle material properties, dozens of potential process inputs and consequences, unknown human health and environmental risks, multiple stakeholder groups, competing objectives, and dynamic nonlinear interdependencies that test the limits of unaided human judgment. In situations like these, without structured decision processes that incorporate computational exploration of uncertainties and holistic trade-offs, human decision processes can often lead to erroneous, inefficient, or suboptimal outcomes that decision-makers may later come to regret (Linkov Cormier et al. 2012b).

Formal decision analysis is useful for guiding knowledge and technology convergence because it is entirely transparent and quantitative (Keeney and Raiffa 1993). With all data, values, perspectives, and relevant information clearly specified in a governing framework and applied to the data of each unique decision, all parties now and in the future can be made aware of and deliberate the pros and cons of alternative pathways towards convergence. This also enables scenario and sensitivity analysis, where impacts on individual decisions can be assessed based on possible differences in perspective or data (Karvetski et al. 2011; Montibeller et al. 2006). Also, by eliciting and incorporating information from all relevant organizational, political, and social levels, applications of decision analysis transcend many limitations of traditional bottom–up and top–down approaches.

Applied at the organizational level, quantitative decision frameworks can help businesses and organizations efficiently navigate technological decisions and maintain consistent direction toward their organizational goals. At the national or international level, instead of letting organizations of scientists and engineers independently stumble towards convergence, governments and scientific authorities can aid the process by clearly specifying the objectives of technological progress.

Recognizing that convergence is not an end in itself but rather a means toward more sustainable and equitable social outcomes, decision frameworks can be used to clearly communicate and quantify the goals of the desired future. For example, governments and scientific academies can systematically explore their decision preferences, reveal their decision criteria (perhaps drawing from the three common economic, environmental, and societal pillars of sustainability), and quantify their values and the tradeoffs between criteria to provide practical and decision-relevant guidelines for individual scientific organizations to follow. These guiding frameworks can then be applied to evaluate the long-term net expected social or other benefits of specific variations of converging technologies being developed, drawing from knowledge and components in different scientific and engineering fields.

Decision analysis tools explicitly take into account data and preferences across many scales and domains and quantitatively relate this information to outcomes on multiple management criteria, helping decision-makers transparently and consistently create better alternatives, identify preferable decisions, and guide knowledge and technology convergence for holistic social benefit (Keeney 1992).

4.8.6 Achieving Continuous Operation of Huge, Complex, and Ever-Changing Systems in the Nanotechnology Era

Contact persons: *Mario Tokoro, Professor, Faculty of Science and Technology, Keio University, and President and CEO, Sony Computer Science Laboratories, Japan*

In the nanotechnology era, every system becomes smart, compact, and small, and at the same time huge in functionality, complex in structure, and ever-changing to meet the requirements of users. These systems are often connected to each other to form much more complex ones, and the boundary of a system becomes vague. Our daily lives are supported by such systems, from telecommunication to traffic and distribution, from manufacturing to financing, and from healthcare to defense. The prevention of a failure, and isolating a failure from propagation when it occurs, becomes crucial to achieve continuous operation of such systems.

Open systems dependability is a methodology of achieving continuous operation of huge, complex, and ever-changing systems. It sees such a system as an "open system" (as these words are used in scientific language), in which failures cannot be completely prevented, and therefore, minimizing the damage, identifying the causes, and achieving accountability have the highest priority.

The *DEOS process* (dependability engineering for open systems; http://www.dependable-os.net/) is defined as one that integrates the continuous development process (forever changing nature) and the operational processes that overlap with the continuous development process. It is applicable not only to ICT (information, communication, and technology) systems but also to a wider range of systems, including mechanical and cyber-physical ones.

4.8.7 Scenario Development Approach

Contact person: *Cynthia Selin, Arizona State University*

The future of emerging technologies is not preordained and can therefore not be predicted. There are critical uncertainties surrounding both the technological pathways and the societal implications of scientific discoveries. The development of emerging technologies depends on choices made today, choices that occur throughout society—in the boardroom, within the laboratory, in the legislature, and in shopping malls. There are numerous complex, interrelated variables that impinge upon what those technologies will ultimately look like in 10 years' time. Future-oriented methods like scenario planning provide a means to structure key uncertainties driving the co-evolution of technology and society. These critical uncertainties range from the health of the economy, to regulatory frameworks, to public opinion, to the actual technical performance of many of nanotechnology's projected products. Anticipation and foresight, as opposed to predictive science, provide means to appreciate and analyze uncertainty in such a way as to maximize the positive outcomes and minimize the negative outcomes of new technologies. The value of scenario development in particular is to rehearse potential futures to identify untapped markets, unintended consequences, and unforeseen opportunities.

4.9 International Perspectives

The following are summaries relevant to this chapter of discussions at the international regional WTEC NBIC2 workshops held in Leuven, Belgium, September 20–21, 2012; in Seoul, Korea, October 15–16, 2012; and in Beijing, China, October 18–19, 2012. Further details of those workshops are provided in Appendix W.

4.9.1 United States–European Union NBIC2 Workshop (Leuven, Belgium)

Principal discussants:

Christos Tokamanis, European Commission (EU)
Françoise Roure, OECD (France)
Alfred Nordmann, Darmstadt Technical University (Germany)
Sylvie Rousset, Université Denis Diderot (France)
George Whitesides, Harvard University (U.S.)
Bruce Tonn, University of Tennessee (U.S.)
Jo De Boeck, IMEC (EU)
Barbara Harthorn, Center for Nanotechnology in Society, Univ. of California–Santa Barbara (U.S.)
Maurizio Salvi, European Commission (EU)
Mark Lundstrom, Purdue University (U.S.)
Mira Kalish, Tel Aviv University (Israel)
Todd Kuiken, Woodrow Wilson International Center for Scholars (U.S.)

This group of scientists found consensus around three important themes: (1) human development, (2) sustainability and human development, and (3) co-evolution of human development and technology.

Prof. Alfred Nordmann and colleagues (under the subtitle "Magic Moment in History—Sound Science Policy Concept") presented the evolution of the concept, suggested a definition, and spoke about methods of convergence. Converging technologies are enabling technologies and knowledge systems that enable each other in the pursuit of a common goal—through neither emergence, nor planning, but through agenda-setting that achieves robustness through a political process. There are two ways of considering "converging technologies": as a technological push and as a policy initiative. The main challenges are the following:

- Defining methods of convergence and essential technology platforms based on developing mechanisms for matching diverse research capabilities to societal needs.
    - Two core ideas here were (1) an emerging trend to use personalized and molecular information to enhance both medical treatment and cognition, and (2) individualized long-term education.

- Examining and assessing human–technology co-evolution, i.e., placing S&T research in the context of social science and humanities research (since all technologies structure human–human and human–nature interactions).
    - The core idea here was using convergent technologies to enhance human sustainability. It includes using recycled water and sea water as sources of clean water, nutrients, and materials; urban planning for megacities—maximizing positive effects; and renewable materials.

- Providing a knowledge generation-to-governance framework, i.e., allowing for inclusive agenda-setting processes.
    - The core idea here was the accelerating co-evolution between human development and technology by using the societal convergence approach. The image of robots as taking abilities and jobs away from people is changing: it is better to view robots as assisting and being complimentary to people; people are the robot creators and controllers.

An example of a convergence network in France was presented by Sylvie Rousset. Since 2005 in France, the Centers of Competence in Nanosciences for the Ile-de-France region (C'Nano IdF) has been established at the initiative of the National Research Centre (CNRS), the French Atomic Energy and Alternative Energies Commission (CEA), and the Ministry of Higher Education and Research (MESR). This network of 2,700 researchers federates the largest European cluster in nanoscience and technology, coming from multidisciplinary fields that range from natural sciences to the human and social sciences (including physics, chemistry, engineering, medicine, biology, toxicology, law, philosophy, sociology, and economics). The existence of large technology facilities with technology platforms and a synchrotron plays a key role in the emergence of highly innovative interdisciplinary projects. Another example is the Interuniversity MicroElectronics Center (IMEC) with bio-nano-cogno programs in Leuven, Belgium.

The time has come for convergence between the USA-NSF NBIC initiative (2001 workshop) and the EU, owing to the numerous accomplishments in industry, academia, medicine, and government (see the EU report, *Converging Technologies—Shaping the Future of European Societies* (Nordmann 2004). The formation of a meeting point platform to combine and bridge basic research (bottom-up) and the application-driven research (top-down) in the converged micro and macro environment will pave the way to ensure overall progress.

4.9.2 United States–Korea–Japan NBIC2 Workshop (Seoul, Korea)

Panel members/discussants:

Jo-Won Lee (co-chair), Hanyang University (Korea)
Mihail Roco (co-chair), National Science Foundation (U.S.)
Takahiro Fujita (co-chair), NIMS (Japan)

Others:

Hanjo Lim, Ajou University (Korea)
Y. Eugene Pak, Seoul National University (Korea)
Jian Cao, Northwestern University (U.S.)
Jim Murday, University of Southern California (U.S.)

Considering the enrichment of human life without our present concerns such as sickness, climate change, and running out of natural resources, NBIC would be a key technology to ease our concerns and may be a solution provider. Increasing scientific and technological complexity with all the necessary core capabilities cannot be sustained in one discipline. This vision has not been changed for the last 10–12 years since the first NBIC report was published (Roco and Bainbridge 2003), even if many goals are only just getting under way. NBIC, robotics, virtual reality, avatar approach, human–machine interfaces, and other CKTS approaches will become dominant in the next decade.

An example of a multidisciplinary platform is the International Center for Materials Nanoarchitectonics (MANA) in Japan. A general method to enhance convergence is creating environments to stimulate encounters with disparate disciplines.

Equipment is needed for communication and virtual reality to allow converging platforms and networks to operate.

Urgent global problems in the twenty-first century—population growth, shortage of resources and energy, greenhouse effects, deterioration of biodiversity, shortage of food and water, and the aging population—are given conditions in converging society planning. New technologies in the twenty-first century should be designed for meeting global issues and social wishes. A specific method of convergence for new technologies is inverse design based on science for design. Examples of visible social wishes for converging society are green innovation, life innovation, and recovery and reconstruction from a disaster, as illustrated, for example, *in the Japanese 4th S&T Basic Plan (2011–2015)*. Potential/invisible social wishes, such as having an active aged society, are important (Yoshikawa 2012). The interaction between society and scientific community needs to be a closed loop, combining application of science to society with society tasks driving science and technology. A better science of science is needed.

An approach to advance convergence in a domain of activity is through the mission of dedicated organizations. An example for environmental aspects is South Korea's Korea Environmental Industry and Technology Institute (KEITI), whose focus areas are shown in Fig. 4.10.

Over the past decade, the R&D approach for converging technology has been generally "reactive" in achieving a collaborative approach (compared to the NBIC2 holistic, systematic approach). From "Priority R&D Areas"–driven research we are moving to "Social-Issue Targets"–oriented research (i.e., from "seeds push" to "needs pull" science and technology). "Scientific challenges" make big breakthroughs for real applications, judging from history. "Social-issue targets" should be well translated to the scientific way of thinking and scientific approaches.

All 20 illustrations cited in the first NBIC report have seen significant progress (see Appendix D). The vision for the next 10 years includes a systematic convergence (proactive and holistic) of various domains of knowledge, technology, and society. Creating a database with an expert system to select the method of

Fig. 4.10 The functions of the public organization KEITI with a role in environmental convergence in South Korea (Courtesy of Seung-Joon Yoon)

convergence to accelerate unprecedented discoveries and innovation is a priority. Important drivers will be robotics, bionics, and road-mapping using a holistic view of human society. A higher-level convergence language needs to be developed that will facilitate integration into essential platforms and facilitate innovation (confluence of streams), improved manufacturing (hybrid methods, cyber-physical systems), and advances in all emerging technologies. Personalized, online, life-long education, combined with advances in neuroscience, communication, psychology, and understanding of learning, will be needed for the human-scale convergence platform. Observatories and simulation for Earth's dynamic systems, new sensing, large data, etc., will be needed for Earth-scale convergence platforms. The focus will be on communities and on addressing major issues affecting those, using converging technologies. Self-regulating convergence among communities is facilitated by organizations such as professional societies and NGOs. Changes in governance will be needed, especially for changes that depend on a holistic view. There is a planned converging technologies organization to be created in the Office of the Prime Minister in Korea.

Several emerging topics and priorities emerged in the discussions:

- Self-reliant communities for sustainable development
- "Platforms" for interdisciplinary collaboration that include host communities and shared resources and expertise, e.g., portal websites serving diverse S&T communities
- Define specific, realistic "grand challenges" in problem-driven research, such as brain-inspired neuromorphic devices, digital and distributed manufacturing, Earth-scale interventions, and personalized health and education
- Enhancement of human quality of life enabled by individual "pursuit of happiness"
- International collaboration to enhance the value and reduce the risk in this high-return area

4.9.3 United States–China–Australia–India NBIC2 Workshop (Beijing, China)

Plenary speaker: Chunli Bai, Chinese Academy of Sciences (CAS)

Panel members/discussants:

Dongyi Chen (co-chair), University of Electronic Science and Technology (China)
Ron Johnston (co-chair), Australian Centre for Innovation
Mike Roco (co-chair), National Science Foundation (U.S.)

Others:

Xing Zhu, Peking University (China)
Jian Cao, Northwestern University (U.S.)
Ke Xu, Suzhou Institute of Nano-tech and Nano-bionics (China)
James Murday, University of Southern California (U.S.)
Calum Drummond, CSIRO (Australia)

Convergence of important frontier technologies, such as those represented by NBIC, will be crucial to the ways in which we innovate, and eventually to the success of human society (Chunli Bai, plenary presentation on September 17, 2012). In China, important steps have already been made in tearing down the traditional boundaries of individual scientific disciplines. For example, the CAS has set up entirely new research entities to explore convergent technologies. In Shenzhen, Chongqing, and several cities, China is organizing research institutes that combine information science, advanced manufacturing, and biomedicine.

Developing compatible standards for different fields is an important tool for convergence. This may allow for establishing a common language for scientists, engineers, industry, regulators, and the public.

An approach to integration is supporting personalized, online education, combined with advances in neuroscience, communication, psychology, and understanding of learning. Another approach is to focus on communities and major issues affecting them using converging technologies and a self-regulating approach. Changes in governance (structures and measures to reach decisions and solve conflicts) will be needed for achieving societal benefits. Updating government organizations and regulations will be required.

Multidisciplinary teams are inadequate for the challenges of this century unless they are also transdisciplinary; knowledge of each team member should span multiple disciplines. There is a need for continual refinement based on both top-down strategies and bottom-up findings.

An illustration for a converging center is Suzhou Industry Park (China), co-located with the National Nano-tech Innovation Park, both focused on nanotechnology and nanobiology. It includes one institute of the Chinese Academy of Sciences, more than 20 universities, about 50 engineering research centers, and more than 1,000 spin-off high-tech companies.

Several emerging topics and priorities are:

- Implement investment policy in multisector, multidisciplinary centers and networks
- Develop standards for various fields in convergence platforms
- Expand and institutionalize the programs "science for science" to "science for governance"
- Explore mechanisms to engender more effective stakeholder contributions towards improvements in personalized education
- Open access to all S&T journals, to promote accelerated advancement of converging technologies, resolution of IP issues, and exploration of the rewards system vs. the advantages of open access.

References

Alivisatos, P., Chun, M., Church, G.M., Greenspan, R.J., Roukes, M.L., Yuste, R.: The brain activity map project and the challenge of functional connectomics. Neuron **74** (2012)

Attewell, P.: The first and second digital divides. Sociol. Educ. **74**(3), 252–259 (2001)

Bainbridge, W.S., Roco, M.C. (eds.): Managing Nano-Bio-Info-Cogno Innovations: Converging Technologies in Society. Springer, Dordrecht (2006)

Beede, D., Julian, T., Langdon, D., McKittrick, G., Khan, B., Doms, M.: Women in STEM: a gender gap to innovation. Washington, DC: U.S. Department of Commerce, Economics and Statistics Administration, ESA Issue Brief #04-11. Executive summary online: http://www.esa.doc.gov/Reports/women-stem-gender-gap-innovation (2010)

Bell, D.: The Coming of Post-Industrial Society. Basic Books, New York (1973)

Berman, H.: The protein data bank. In: Bainbridge, W.S. (ed.) Leadership in Science and Technology, pp. 661–667. Sage, Thousand Oaks (2012)

Brzezinski, Z.: Between Two Ages: America's Role in the Technetronic Era. Viking, New York (1970)

Castells, M.: The Rise of the Network Society. Blackwell, Malden (1996)

DiMaggio, P., Hargittai, E., Neuman, W.R., Robinson, J.P.: Social implications of the Internet. Annu. Rev. Sociol. **27**, 307–336 (2001)

Djorgovski, S.G.: Data-intensive astronomy. In: Bainbridge, W.S. (ed.) Leadership in Science and Technology, pp. 611–618. Sage, Thousand Oaks (2012)

Ellul, J.: The Technological Society. Knopf, New York (1964)

Ericson, R.F.: The growing demand for synoptic minds in industry. J. Acad. Manag. **3**(1), 25–40 (1960)

Fisher, E., Mahajan, R.L., Mitcham, C.: Midstream modulation of technology: governance from within. Bull. Sci. Technol. Soc. **26**(6), 485–496 (2006)

Frehill, L.M., DiFabio, N.M., Hill, S.T.: Confronting the "new" American dilemma—Underrepresented minorities in engineering: A data-based look at diversity. National Action Council for Minorities in Engineering, White Plains. Available online: http://www.nacme.org/user/docs/NACME%2008%20ResearchReport.pdf (2008)

Friedman, T.L., Mandelbaum, M.: That Used to Be Us: How America Fell Behind in the World It Invented and How We Can Come Back. Farrar, Strauss, and Giroux, New York (2011)

Gastil, J., Knobloch, K.: Evaluation report to the Oregon State Legislature on the 2010 Oregon Citizens' Initiative Review. Oregon State Legislature documents and University of Washington. Available online: http://www.la1.psu.edu/cas/jgastil/CIR/OregonLegislativeReportCIR.pdf (2010)

h + (magazine online): "Nano-Bio-Info-Cogno: Paradigm for the future." February 12. http://hplusmagazine.com/2010/02/12/nano-bio-info-cogno-paradigm-future/ (2010)

Hammond, J.S., Keeney, R.L., Raiffa, H.: The hidden traps in decision making. Harv. Bus. Rev. **76**, 47–58 (1998)

Howard, R.A.: Information value theory. IEEE Trans. Syst. Man Cybern. **1**, 22–26 (1996)

IRGC: Appropriate risk governance strategies for nanotechnology applications in food and cosmetics, consider products with the highest risk perception. Report available on http://www.irgc.org (2009)

Karvetski, C., Lambert, J., Linkov, I.: Integration of decision analysis and scenario planning: application to coastal engineering and climate change. IEEE Trans. Syst. Man Cybern. Syst. Hum. **41**, 63–73 (2011)

Keeney, R.L.: Value-Focused Thinking: A Path to Creative Decision Making. Harvard University Press, Cambridge (1992)

Keeney, R.L., Raiffa, H.: Decisions with Multiple Objectives: Preferences and Value Tradeoffs. Cambridge University Press, Cambridge (1993)

Khatib, F., DiMaio, F., Foldit Contenders Group, Foldit Void Crushers Group, Cooper, S., Kazmierczyk, M., Gilski, M., Krzywda, S., Zabranska, H., Pichova, I., Thompson, J., Popovic, Z., Jaskoski, M., Baker, D.: Crystal structure of a monomeric retroviral protease solved by protein folding game players. Nat. Struct. Mol. Biol. **18**, 1175–1177 (2011)

Kim, S.K., et al.: Open Innovation: Theory, Practices and Policy Implications. STEPI, Washington, DC (2008). Research Report

Kurzweil, R.: The Age of Spiritual Machines: When Computers Exceed Human Intelligence. Viking, Penguin Group, New York (1999)

Lehrer, J.: Imagine: How Creativity Works. Houghton Mifflin Harcourt Publishing Company, New York (2012)

Linkov, I., Moberg, E.: Multi-Criteria Decision Analysis: Environmental Applications and Case Studies. CRC Press, Boca Raton (2012)

Linkov, I., Bates, M.E., Canis, L.J., Seager, T.P., Keisler, J.M.: A decision direct approach for prioritizing research into the impact of nanomaterials on the environment and human health. Nat. Nanotechnol. **6**, 784–787 (2011)

Linkov, I., Bates, M.E., Trump, B.D., Seager, T.P., Chappell, M.A., Keisler J.M.: For nanotechnology decisions, use decision analysis. Nano Today http://dx.doi.org/10.1016/j.nantod.2012.10.002 (2012)

Linkov, I., Cormier, S., Gold, J., Satterstrom, F.K., Bridges, T.: Using our brains to develop better policy. Risk Anal. **32**, 374–380 (2012b)

Mills, M.P., Ottino, J.M.: The coming tech-led boom. The Wall Street Journal, 30 January 2012

MIT (Massachusetts Institute of Technology): The third revolution: The convergence of the life sciences, physical sciences, and engineering (white paper). MIT, Washington, DC (Washington, DC, office). Available online: http://dc.mit.edu/sites/dc.mit.edu/files/MIT%20White%20Paper%20on%20Convergence.pdf (2011)

Montibeller, G., Gummer, H., Tumidei, D.: Combining scenario planning and multi-criteria decision analysis in practice. J. Multi-Criteria Decis. Anal. **14**, 5–20 (2006)

Murday, J.S., Siegel, R.W., Stein, J., Wright, J.F.: Translational nanomedicine: status assessment and opportunities. Nanomedicine **5**(3), 251–273 (2009)

Neisser, U., Boodoo, G., Bouchard Jr., T.J., Boykin, A.W., Brody, N., Ceci, S.J., Halpern, D.F., Perloff, J.C.L.R., Sternberg, R.J., Urbina, S.: Intelligence: knowns and unknowns. Am. Psychol. **51**(2), 77–101 (1996)

Nordmann, A.: Converging technologies—Shaping the future of European societies. A report from the High-Level Expert Group on "Foresighting the new technology wave." Belgium: Luxembourg: Office for Official Publications of the European Communities. Available online: http://ec.europa.eu/research/social-sciences/pdf/ntw-report-alfred-nordmann_en.pdf (2004)

NRC (National Research Council): Understanding Risk: Informing Decisions in a Democratic Society. The National Academies Press, Washington, DC (1996)

NRC (National Research Council): Facilitating Interdisciplinary Research. The National Academies Press, Washington, DC (2004)

NRC (National Research Council): Is America Off the Flat Earth? The National Academies Press, Washington, DC (2007a)

NRC (National Research Council): Rising Above the Gathering Storm: Energizing and Employing America for a Brighter Economic Future. The National Academies Press, Washington, DC (2007b). The initial report release was in 2005, with the final, edited book issued in 2007

NRC (National Research Council): America's Energy Future: Technology and Transformation. The National Academies Press, Washington, DC (2009a)

NRC (National Research Council): Informing Decisions in a Changing Climate. The National Academies Press, Washington, DC (2009b)

NRC (National Research Council): Restructuring Federal Climate Research to Meet the Challenges of Climate Change. The National Academies Press, Washington, DC (2009c)

NRC (National Research Council): Science and Decisions: Advancing Risk Assessment. The National Academies Press, Washington, DC (2009d)

NRC (National Research Council): Gender Differences at Critical Transitions in the Careers of Science, Engineering, and Mathematics Faculty. The National Academies Press, Washington, DC (2010a)

NRC (National Research Council): Rising Above the Gathering Storm, Revisited: Rapidly Approaching Category 5. The National Academies Press, Washington, DC (2010b)

NRC (National Research Council): Research universities and the future of America: Ten breakthrough actions vital to our nation's prosperity and security. The National Academies Press, Washington, DC (2012)

NSTC Subcommittee of the Committee on Technology of the National Science and Technology Council: National Nanotechnology Initiative environmental, health, and safety research strategy. NSET, Washington, DC. http://nano.gov/sites/default/files/pub_resource/nni_2011_ehs_research_strategy.pdf (2011). Accessed 20 Nov 2012

NSF: Software infrastructure for sustained innovation. NSF 13-525. Available on http://www.nsf.gov (2012)

OECD (Organization for Economic Co-operation and Development). 2003. Emerging risks in the 21st century: An agenda for action. OECD, Paris. http://www.oecd.org/dataoecd/20/23/37944611.pdf. Accessed 20 November 2012

Osterman, P.: College for all? The labor market for college-educated workers. Center for American Progress, Washington, DC. http://www.americanprogress.org/issues/2008/08/college_for_all.html (2008)

Presidential/Congressional Commission on Risk Assessment and Risk Management: Framework for Environmental Health Risk Management. Final report, Volume 1. RiskWorld, Washington, DC. Available online http://www.riskworld.com/nreports/nr7me001.htm (1997). Accessed 20 Nov 2012

Raiffa, H., Schlaifer, R.O.: Applied Statistical Decision Theory. Wiley, New York (2000)

Ray, C.A., Mickelson, R.A.: Restructuring students for restructured work: the economy, school reform, and non-college-bound youths. Sociol. Educ. **66**(1), 1–20 (1993)

Renn, O., Roco, M. (eds.): Governance of nanotechnology, Geneva, Switzerland: IRGC. Available online: http://www.irgc.org/IMG/pdf/IRGC_white_paper_2_PDF_final_version-2.pdf (2006)

Roco, M.C.: Coherence and divergence of megatrends in science and engineering. J. Nanopart. Res. **4**, 9–19 (2002)

Roco, M.C.: Technology convergence. In: Bainbridge, W.S. (ed.) Leadership in Science and Technology, pp. 210–219. Sage, Thousand Oaks (2012). Ch. 24

Roco, M.C., Williams, R.S., Alivisatos, P. (eds): Nanotechnology research directions: Vision for the next decade. IWGN Workshop Report 1999. National Science and Technology Council, Washington, DC. Available online: http://www.wtec.org/loyola/nano/IWGN.Research.Directions/. Also, Dordrecht, New York: Springer (2000). (1999)

Roco, M.C., Bainbridge, W.S. (eds.): Converging technologies for improving human performance: Nanotechnology, biotechnology, information technology, and cognitive science. Kluwer Academic, Dordrecht (now Springer) (2003)

Roco, M.C., Bainbridge, W.S. (eds): Nanotechnology—Societal implications: Maximising benefits for humanity. Vol. 1. Springer, Dordrecht (2006)
Roco, M.C., Mirkin, C.A., Hersam, M.C.: Nanotechnology research directions for societal needs in 2020: retrospective and outlook. Springer, Dordrecht/New York (2011). Available online: http://nano.gov/sites/default/files/pub_resource/wtec_nano2_report.pdf.
Ross, G.: An interview with Lawrence Lessig. American Scientist (online): http://www.americanscientist.org/bookshelf/pub/lawrence-lessig (n.d.)
Sample, I.: Harvard University says it can't afford journal publishers' prices. The Guardian 24 April 2012
Seelig, T.: inGenius: A Crash Course on Creativity. Harper Collins, New York (2012)
Shapiro, C., Varian, H.R.: Information Rules: A Strategic Guide to the Information Economy. Harvard Business School Press, Boston (1999)
Sharp, P.A., Langer, R.: Promoting convergence in biomedical science. Science **222**(6042), 527 (2011)
Stephenson, N.: The Diamond Age: Or, a Young Lady's Illustrated Primer. Bantam Bell, New York (1995)
Stokes, D.E.: Pasteur's Quadrant: Basic Science and Technological Innovation. Brookings Institution Press, Washington, DC (1997)
Taleb, N.N.: The Black Swan: The Impact of the Highly Improbable, 2nd edn. Random House, New York (2010)
Tapia, A., Ocker, R., Rosson, M., Blodgett, B., Ryan, T.: The computer tomography virtual organization. In: Bainbridge, W.S. (ed.) Leadership in Science and Technology, pp. 602–610. Sage, Thousand Oaks (2012)
Tervonen, T., Figueira, J.R.: A survey on stochastic multicriteria acceptability analysis methods. J. Multi-Criteria Decis. Anal. **15**, 1099–1360 (2008)
UN (United Nations) General Assembly: Future we want: Outcome Document Resolution A/RES/66/288 on the Outcomes of the Rio + 20 UN Conference on Sustainable Development in Rio de Janeiro, Brazil, 20–22 June 2012; issued September 11, 2012. Available online: http://sustainabledevelopment.un.org/futurewewant.html. United Nations, New York (2012a)
UN (United Nations) System Task Team on the Post-2015 UN Development Agenda: Realizing the future we want for all. Report to the Secretary-General, New York (2012b)
Woelfle, M., Olliaro, P., Todd, M.H.: Open science is a research accelerator. Nat. Chem. **3**, 745–748 (2011)
Yoshikawa, H.: Design Methodology for Research and Development Strategy Realizing a Sustainable Society. CRDS-JST, Tokyo (2012)

Chapter 5
Implications: Human Health and Physical Potential

Robert G. Urban, Piotr Grodzinski, and Amanda Arnold

Societies all over the world benefit from medical advances made possible by the interplay of science and technology. Seldom are these advances simply the straightforward result of traditional improvements in manufacturing or automation or increases in scale. Instead, these improvements emanate in large part from the application of newly developing knowledge towards solving technological challenges. Convergence is the next-generation iteration of that process, resulting in new solutions and enhanced return on investment. This chapter deals with the role of convergence and its application in advancing health and wellness for the world.

5.1 Vision

5.1.1 Changes in the Vision over the Past Decade

The pre-convergence process in science and technology advancements was dependent on coincidence ("passive convergence"). In fact, many of today's keystone medical tools were derived by happenstance, and their histories often illustrate the fundamentals of progress in action (as illustrated in Sect. 5.2.1).

With contributions from James Heath, Ramy Arnaout, Larry Bell, Clem Bezold, Jake Dunagan, and Robert Langer.

Corresponding editors M.C. Roco (mroco@nsf.gov) and W.S. Bainbridge (wbainbri@nsf.gov).

R.G. Urban
Johnson & Johnson Boston Innovation Center, Cambridge, MA, USA

P. Grodzinski
National Cancer Institute, Bethesda, MD, USA

A. Arnold
MIT, Washington, DC, USA

Organizational models have played an important role in the probability of productive convergence. Highly productive and innovative organizations like Bell Labs, Genentech, Hewlett Packard (leading to Agilent), and MIT's Media Lab tend to share a few common characteristics. The first is a collection of highly accomplished experts gathered from a range of only marginally related fields. The second is an operational and cultural "openness" to considering what could be possible or impactful. The third is a determined group-based drive and expectation to deliver revolutionary rather than incremental outcomes. Transforming medicine and human potential will depend on these types of attributes, and much can be learned from organizations that operate on these principles.

5.1.2 The Vision for the Next Decade

The practice of medicine is moving from a passive coincidence model to an active convergence model. At the center of this transformation is the intersection of physiology, molecular profiling, and information technology. Today's hallmarks of wellness and disease have their origin in measurements of vital signs such as pulse, temperature, blood pressure, and respiration, each with its specific method of measurement. The vital signs of tomorrow will be based on concentrations of molecular analytes detected in various body fluids and in exhaled breath and will be measured using minimally invasive imaging or on-body sensors. The resulting molecular profiles, often simultaneously measured via multiplexed interrogation of numerous analytes, will more precisely distinguish wellness from a disease state, efficacy from toxicity, and regression from progression. This form of "precision medicine" is centered on the consideration of the individual, is deeply reliant on a molecular understanding of a disease, and is coupled with emerging analytical tools competent in measuring the resulting biochemistries (NRC 2011). We expect that these new models will improve the health of patients by transforming medicine from a reactive into a proactive practice and result in efficiencies that reduce the overall cost of healthcare (Fig. 5.1).

5.2 Advances in the Last Decade and Current Status

5.2.1 Advances in Converging Technologies

Coincidence in medical advances of the past was a dominant trend. Blood pressure measurement (sphygmomanometry), fundamental to the field of cardiology, is an example. It primarily began with Stephen Hales, who in 1773 proved that the mammalian circulatory system operates on the basis of pressure when reporting his observation, "that by attaching a long length of glass capillary tubing to the artery of horse, the blood rose in the tube 8 ft and 3 in. perpendicular above the level of the

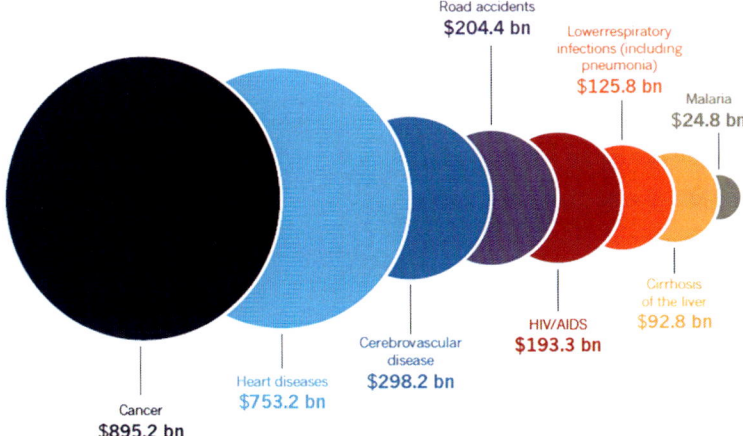

Fig. 5.1 Worldwide economic impact of treating different diseases (O'Callaghan 2011; ©Nature Publishing Group 2011; permission to reuse through Rightslink)

left ventricle of the heart" (Booth 1977). It took until the middle of the nineteenth century for the physician-physicist Jean Louis Marie Poiseuille to further advance the field by inventing the mercury-based manometer. This instrument enabled accurate arterial pressure measurements for the first time. But his instrument and those noninvasive instruments that followed are only half of the story. Considered a "vital sign," blood pressure measurement has proven invaluable in the clinical management of trauma. However, it is its use in detecting hypertension that it resulted in saving millions of lives.

In the late 1800s, the astute pathologist Rudolf Virchow first reported his observation of arterial occlusion (Caplan 2000) to be coincident with hypertension. Subsequent chemical analysis of these occluding substances—yellow lipid-like materials—revealed their principle component to be a molecule termed "cholesterol." Over the ensuing decades of the twentieth century, the cholesterol biosynthetic pathways were characterized and critical enzymes were identified, with HMG-CoA reductase being particularly interesting, because it was druggable with statins. In the 1970s the first clinical trials of statins began, and after a few starts and stops associated with ensuring safety and efficacy, the widespread use of these drugs in patients with elevated blood lipid levels has translated into a tremendous reduction in cardiovascular system-associated death and morbidity (Tobert 2003). Cardiovascular system-associated deaths per 100,000 in the United States have dropped from ~600 in the 1950s to ~200 today (Fig. 5.2); similar dramatic reductions have been seen everywhere these medicines are given.

When invented, the tools to measure blood pressure were designed to help us understand physiology and the mechanics of how our bodies worked, but unexpected extrapolation and convergence resulted. The subsequent linkage of coronary occlusion, lipid metabolism, and hypertension (i.e., high blood pressure) would have been impossible to predict, yet their role in health or disease is indelible and

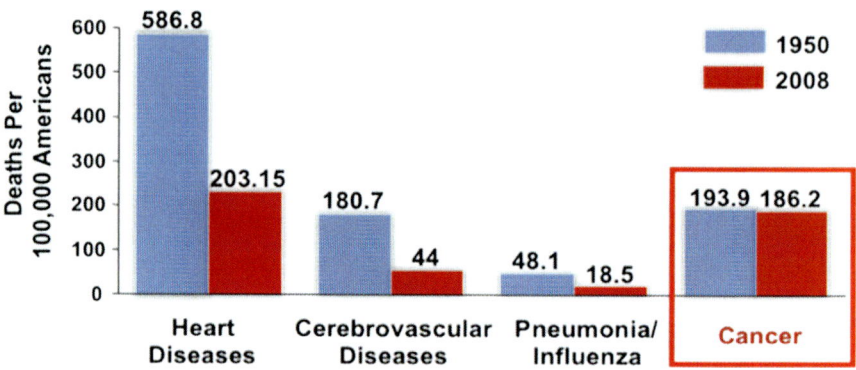

Fig. 5.2 Evolution of death rates in the last 60 years for different diseases. (Source for the 1950 and 2008 age-adjusted death rates is NCHS 2008, http://www.cdc.gov/nchs/hus/contents2011.htm#024, Available for public use)

illustrative. The development of a reliable and accurate analytical method (sphygmomanometry) by experts in one field and its deployment in measuring a disease process by experts in a completely different field (primary care and/or cardiology) developed by coincidence. Events and individuals separated in time and space were connected through sharing of information and the development and dissemination of analytical methods.

Medicine today is supported by a dizzying array of tests. Detailed molecular testing began in earnest with blood transfusions, followed by its use in tissue transplantation in which "testing" patients to find those with sufficiently similar "types" was a fundamental requirement for treatment success (Murray and Holden 1954). These precursor tests were soon followed by a tsunami of diverse medical tests to catalog and index a remarkable number of analytes and/or anatomical details, with some now serving to confirm wellness while others are used to screen for evidence of disease.

Convergence is a primary driver in this aspect of medicine. In recent years, especially given advances resulting from the effort to map the human genome, the potential for new tests have emerged that can accurately predict the efficacy of certain medicines and the potential toxicities among subsets of the population corresponding to distinct genetic markers. While there is still much work to do, many medicines in the regulatory approval process today are co-developed with tests that strive to define the optimal patient population (Hamburg and Collins 2010). In many cases these therapies for targeted populations represent only a small fraction of the total patient pool, but in these precise subsets, profound benefits of treatment are apparent.

In addition to this new era of medicine, patients and their physicians today are presented with remarkable, albeit in many cases overwhelming and disjointed, amounts of medical information. Driving much of this are information systems and hand-held devices that are now ubiquitous agents in society. In medicine, these resources are changing the relationships between symptoms and diagnosis and doctor and patient. Concurrent with these changes in information systems and related access is a rapid change in the physical form of the medical record itself. Rapidly vanishing are the individualized paper-based medical records of the past. These are being replaced with electronic medical filing systems connected to networks. While standardization and interoperability are still challenging issues, a primary goal of this effort is to learn more about modeling health for populations by sharing instantly the results collected in one medical setting with other medical professionals and then comparing results with millions of anonymous patient files with similar attributes (Grossman et al. 2011).

5.3 Goals for the Next Decade

Given the advances over last 10 years, the practice of medicine will change in remarkable ways over the next decade, and convergence, along with mounting economic pressures, will drive many of these changes. It is clear that a focus on a few targeted advances could have broad catalytic implications for healthcare. These actionable goals are outlined below.

5.3.1 Goal 1: Advance Cancer Detection and Treatment with Reduced Side Effects

Why, Why Now, What Are the Strategy and the Drivers?

Since its earliest descriptions, the experience and the treatment of cancer have been synonymous with suffering. Although new treatments are being developed, improvement in reducing disease-associated death has been slow (Fig. 5.2). As we have learned, the ability to detect aggressive cancers "early," meaning before they have metastasized and so can be removed surgically, is the most effective method of achieving treatment success. However, identifying reliable and predictive markers of disease has been difficult. The first major success was derived from the work of George Papanicolaou who in the 1920s reported observations of "abnormal cells" derived from cervical scraping, but it was not until the 1950s that the "Pap Smear" began to make its way into standard of care (Elgert and Gill 2009). When coupled with a minimally invasive surgical technique, this early detection and subsequent treatment has dramatically reduced death and morbidity associated with cervical cancer. Mammography, colonoscopy, and various blood tests are important contributors in detecting cancers early, but minimally invasive approaches that can clearly differentiate aggressive from benign tumors, or metastatic from pre-metastatic lesions, are generally lacking.

The early successes in diagnosing and treating some cancers unfortunately do not translate to all types of cancer. Figure 5.3 shows typical time of disease diagnosis for

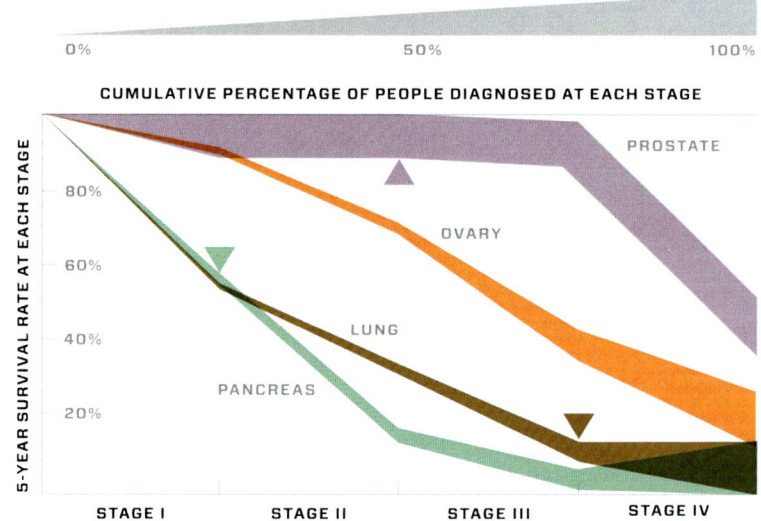

Fig. 5.3 Time of diagnosis for different types of cancer varies with advancement (stage) of disease (Goetz 2008; © *Wired* Condé Nast 2009, used by permission)

prostate, ovarian, lung, and pancreatic cancers. It is clear that the majority of pancreatic cancers are both aggressive and diagnosed very late. This combination means that less than 10 % of diagnosed cases qualify for surgery—the only path to a lasting cure—and the resulting 5-year survival rate is less than 15 % (Goetz 2008). Significant improvements in detecting early stage disease are anticipated as the human proteome is more deeply explored over the next decade and panels of biomarkers emerge that can provide increasingly detailed diagnostic guidance. Implementation of these diagnostic assays into clinical tools will rely on the convergence of analytical tools, mathematical and computational approaches, surface and materials chemistry and engineering, and nanotechnology-based imaging and measuring tools.

Cancer treatments typically involve a combination of surgery, chemotherapy, and radiation therapy. The collateral toxicity of such treatments and the tendency of certain tumors to develop a resistance to chemotherapy are often primary constraints in achieving meaningful remission. New improvements related to the use of nanoparticle-based drug delivery may provide new solutions (Ferrari 2005; Sinha et al. 2006; Wang et al. 2012). Most of the recent developments in nanotherapeutics are focused on delivery of existing, already-FDA-approved chemotherapeutics (Farrell et al. 2010, 2011).

Nanotherapeutic formulations of these existing drugs can yield multiple advantages. First, nanoparticle circulation times can be made to be many times longer than small molecules, and so the nanoparticles have more time to find the tumor before they are passed from the body. Second, surfaces of nanoparticles can be modified to recognize the tumor, thus increasing retention time near the diseased site (Hrkach et al. 2012). Third, a single nanoparticle may hold a thousand or more copies of the drug molecule. The net result is that the drug is selectively delivered to the tumor at a high dose, while the overall amount of drug that is given to the patient is dramatically reduced. The potential, which is already being realized for certain cancers (Heath and Davis 2008), is to maximize the tumor-killing capacity of the drug while minimizing the toxic side effects of that drug, with the goal of quickly and fully removing the tumor before it evolves a resistance to the therapy (Yoo et al. 2011).

The resulting design modifications of nanocarriers will lead to further improvement of therapeutic effect through endocytosis-mediated uptake of nanoparticles to bypass efflux pump mechanisms of multidrug resistance (Farrell et al. 2010; Fox et al. 2011). The goal of these improvements in treatment selectivity, as well as patient stratification, is *cancer therapies with no side effects*. If these advances are made in cancer treatment, they will have parallel and high-impact contributions in many other medical conditions as well.

5.3.2 Goal 2: Improve Health Data Analysis and Delivery for Real-Time Health Monitoring Towards Wellness

Why, Why Now, What Are the Strategy and the Drivers?

Today in medicine the average consumer depends on the internet, physician, and pharmacist for regular health monitoring and maintenance. However, a new kind of wellness opportunity is emerging through the use of digital health aids. This includes the introduction of a new stage of in-home or on-body medical and monitoring sensors coupled with medical expertise, as well as treatment distribution and information systems. At an early stage, such aids may act like health coaches, as integrated systems capable of providing dynamic feedback as well as appropriate or corrective action to support personalized wellness regimes. Versions of these aids already exist in the "app" world for smart phones. While these are useful as monitoring partners for goals that include exercise, weight loss, and medication regimes, the apps still for the large part depend on the information the user manually enters.

Moving forward, the convergence for medical information technologies poses unique opportunities to advance human wellness on an automated level. For example, smart homes, originally conceived as a way to provide convenience, improve security, and save energy, are now increasingly envisioned as a means to improve the quality of life for individuals with disabilities or decreased function due to aging (Ding et al. 2011; Chan et al. 2009).

Potential functions in smart homes include health monitoring, assistance with daily living activities, and illness and injury prevention. Relevant technologies might include interactive displays or video communication with caregivers to provide reminders of daily schedules, tracking of medication usage and physical therapy needs, care and even feeding of pets, and smart appliance reminders when staples such as milk or bread are getting low. GPS technologies can be used for small tasks like tracking the location of eyeglasses, e-readers, or house keys, or more aggressively for potentially tracking the location of an individual with mild dementia and even triggering interactions with caregivers or emergency personnel.

A key factor in the effectiveness of eHealth/telehome-care lies in the design of human-network connections that can be personalized and specialized to the needs, privacy requirements, and habits of the user (Kadouche et al. 2009). Figure 5.4 outlines the factors included in developing such an in-home monitoring approach. These home monitoring technologies are not only useful in cases of monitoring the health and wellbeing of individuals at risk in the home. Such system technologies could also enable a new understanding of the interactions between wellness and the "exposome," or the impact on health outcomes of an individual's or even of a community's environment over time.

This potential contribution of smart monitoring to understanding the contributions of exposomes to wellness and disease is evidenced in a recent NAS report, which calls for the development of an "Information Commons and Knowledge Network" to measure patient outcomes over time using the phenotypic and molecular data on individual patients, and including information about social and physical environments: "[D]ata added to the Information Commons should not be limited to molecular parameters as they are currently understood: patient-related data on environmental, behavioral, and socioeconomic factors will need to be considered as well in a thorough description of disease features" (NRC 2011, 37). The goal of this effort is not only to maximize treatments for individual patients, but also to develop a new taxonomy of disease using the macro-level data gathered to better understand health trends in populations.

The effort to understand health at the macro level is reflected in a companion effort to gain a better understanding of health at the cellular level. *In both cases, the eventual goal is to provide automated digital responses that do not depend on manual data entry, but on the behavior and biology of the user.* For instance, the National Science Foundation Science and Technology Center Integrative Partnerships program called EBICS (the Emergent Behaviors of Integrated Cellular Systems) is led and managed through a partnership between multiple universities and hospitals.[1] Among a long list of EBICS projects is an effort related to diabetes, to create an injected biosensor that can detect the level of glucose in the body and provide the readout visible on the skin using fluorescent markers. This kind of biomachine would lessen the pain of monitoring experienced by diabetes patients, and increase overall health outcomes.

---

[1] Membership includes Massachusetts Institute of Technology (lead), Georgia Institute of Technology, University of Illinois at Urbana-Champaign, City College of New York, Morehouse College, University of California-Merced, Brigham and Women's Hospital, Emory University, Princeton University, Tufts University, and University of Georgia.

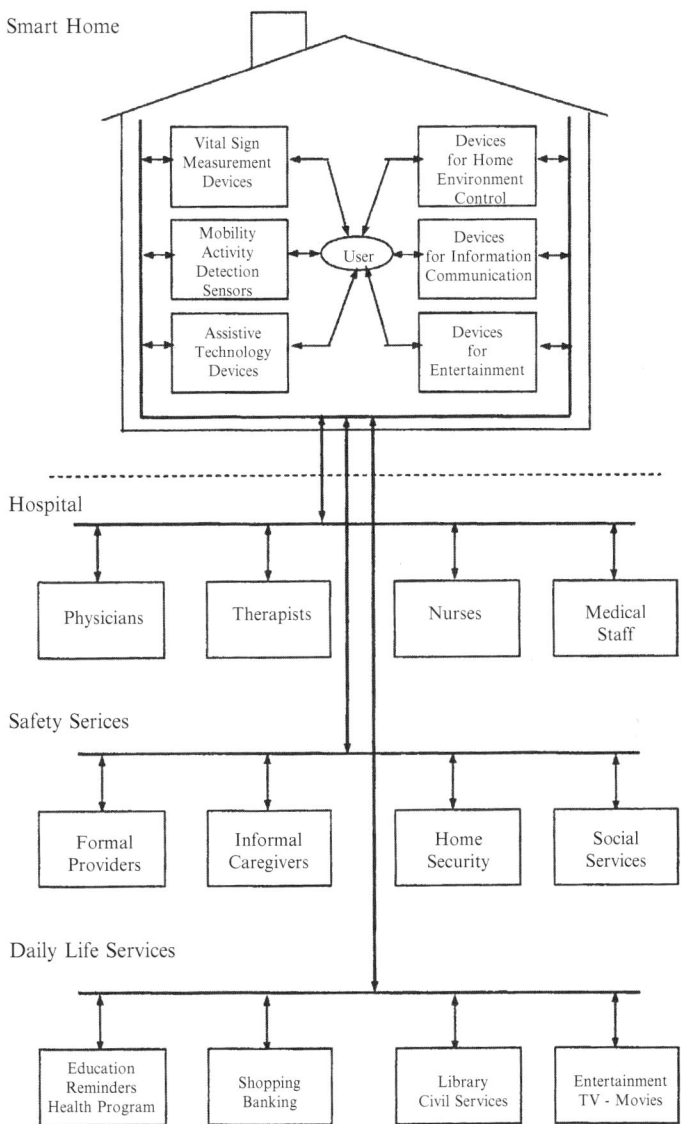

Fig. 5.4 Key organization in a smart home (Chan et al. 2009; ©*Maturitas* 2009; permission for reuse through Rightslink)

As described above, the development of next-generation digital system aids that will function both outside and within the patient will take into account all possible levels of data and will significantly change the way we "participate" in the healthcare system. Of course, as with all scientific developments that change the way we work and function, the ethical, legal and social implications should be investigated along the way.

5.3.3 Goal 3: Promote Regenerative Medicine and Advanced Prosthetics, a Revolution in Treatment

Why, Why Now, What Are the Strategy and the Drivers?

In the twentieth century, medicine was changed forever by the discovery and dissemination of antibiotics and vaccines. Injuries and infections that would have otherwise proven to be maiming or fatal were shown to be treatable or even preventable. Moving forward, the hallmarks of the twenty-first century's transformative moments in medicine will likely involve tissue regeneration and repair. Many of our most grand medical challenges arise at the level of tissue dysfunction. Some of these are the result of trauma, while others are the result of genetic abnormalities or degenerative processes. To regain or restore function requires introducing cells with regenerative capacity or transplanting preformed tissue structures, or in some cases, medical devices like pacemakers, whole joints, and prosthetics.

The art and science of tissue transplantation has made tremendous progress since its origins in the 1950s; millions of people are alive today as result of artificial organ therapy (Khademhosseini et al. 2009). Advanced surgical techniques coupled with immunosuppressive drug regimens have continued to push survival rates higher, yet access to transplantable tissues remains sharply below demand. Over 110,000 Americans are on the lists for organ replacement, and more than 87,000 of these patients need kidneys. But only about 17,000 Americans get kidneys each year, while several die waiting. Increasing the percentage of organs that are available for donation would be an important contribution, but progress is also well underway to "manufacture" patient-specific fully functioning tissues and organs in the laboratory.

In simple terms, regenerative medicine can be grouped into three functional categories:

1. Replacement, which reflects mechanical exchange
2. Repair, which typically uses exogenous cells added to damaged tissues
3. Regeneration, which relies on the mobilization of endogenous pools of stem cells

Progress toward "making" patient-specific tissues will be driven by advances in autologous stem cell biology and advanced material sciences—a convergence of "replacement" and "repair".

The subject of daily media attention and endless controversy, stem cells have the remarkable ability to differentiate into diverse cell types and are the precursors of all the tissues in our bodies. During development, each tissue builds its own custom scaffolding from materials called the extracellular matrix (ECM). Recently, it has been discovered that stem cells "read" these 3D architectures and microenvironments to determine which differentiation program to activate. This new understanding opens the door for a convergence of biological engineering and nanotechnology. Using 3D nanoprinting technologies, work is underway to create tissue-specific structures out of ECM onto which patient-derived stem cells can be seeded. Once those cells are fully encased and differentiated, whole and autologous replacement organs or structures can be potentially realized.

With these advances, one can imagine "curing" diabetes by replacing a compromised pancreas, as could other diseases' negative impact be lessened by replacing organs such as the heart, liver, or kidney with a custom-manufactured version, which would remove the days of waiting (or dying) on organ transplant lists. A greater understanding of how to compensate for compromised organs would also lessen issues associated with organ rejection.

These interventions may seem costly on a per-patient basis; however, they would counter the current cumulative cost of chronic disease treatment, in addition to the loss of economic productivity while patients are disabled by organ under-function or failure. A focused "convergence push" in regenerative medicine over the next decade could deliver these important medical advances and set the stage for an earlier interventional and regenerative medicine of the future, a medicine in which endogenous regenerative cellular capacities are "evoked" in patients before late-stage tissue failure predominates.

Related to biologically based regenerative medicine is the field of advanced prosthetics. Using advanced actuators, sensors, lightweight materials, and power systems, the field of limb prosthetics is making great strides toward achieving the goal of producing a prosthetic that would be so human-like that a patient or an observer could not tell the difference from an actual human limb. The march toward highly effective brain–prosthetic interfaces for brain–prosthetic communication and the fuller understanding of brain processing of information is also occurring. The current limit in neural control of prosthetics lies not within the limb itself, but in the brain—in our incomplete understanding of not only how to get signals out but how to send them. Monitoring signals from the brain with EEG, MEG, and fMRI has progressed at surprising rates, with great progress in interpreting these signals also progressing well. It is expected that we will be able to reduce to practice the interpretation of brain signals (output) using sensors placed either in the brain or on the skull, and to develop successful signal (input) interfaces.

5.3.4 Goal 4: Harness the Human Immune System as a Steady-State Monitor of Health and Disease and Tool for Next-Generation Vaccines

Why, Why Now, What Are the Strategy and the Drivers?

It is hard to overstate the importance of the immune system. It plays key roles in infection, cancer, and autoimmune disease, conditions that together account each year for a quarter of deaths in the United States (Murphy et al. 2012). It is also responsible for the protective effects of vaccines, which, aside from the provision of clean water and good nutrition, constitute the single most effective public health measure in history (Rosen 1993).

It is also hard to overstate the immune system's complexity. It comprises numerous cell subsets—including B cells, which make antibodies, and T cells, which destroy virus-infected and cancerous cells—as well as a host of molecular players. The cast of characters involved has grown as researchers have sought to define immune cells and molecules in ever greater detail. While cell subsets were once defined by microscopic appearance alone—lymphocytes are small and stain blue; neutrophils have lobed nuclei—today they are defined by the degree to which they express specific markers, such as the cell-surface protein CD4, which is present on the surface of cells infected by HIV. Hundreds of such cell-surface and intracellular markers, as well as scores of secreted proteins like perforin and the cytokines, are now known. While there are patterns to the expression of all these markers, the number of biologically meaningful subsets they define is large and growing.

Still greater complexity is to be found among antibodies and T-cell receptors (TCRs), of which the average person has not hundreds but tens of millions of different types or "clones" (Arnaout et al. 2011; Warren et al. 2011). Different clones bind different antigens, allowing the immune system to recognize, destroy, and remember innumerable potential threats like bacteria, viruses, and cancer cells. However, these protective abilities also pressure infectious agents and cancer cells to evolve away from immune control, resulting in an evolutionary "arms race" and dynamic equilibrium in which new clones are constantly replacing each other.

Amid the complexity of antibodies and TCRs, specificity is the key: when antibodies and T cells miss their marks, they can cause autoimmune diseases like Type I diabetes and lupus, transfusion reactions, and organ-transplant rejection. But when they identify their targets properly, the immune system's capacity to remember—so-called immunological memory—is what makes vaccination possible. Vaccines contain antigens that stimulate specific subsets of B and T cells to divide and expand in number. In this way, when the body is later re-exposed to these or similar antigens through infection or the development of cancer, it has sufficient antibodies and T cells to neutralize these threats before illness sets in.

Monitoring vaccine responses is an indispensible measure of efficacy. In practical terms, the ability to monitor the immune system is largely a matter of being able to count how many cells there are of a given subset, and to measure the concentration of secreted immune biomarkers, in the blood and in different tissues.

The way we count cells and measure concentrations today is the product of convergence among a number of twentieth century technologies, including microscopy, biochemistry, genetics, tissue culture, animal models, flow cytometry, and monoclonal antibody technology (itself a product of immunology research). The convergence of multi-channel flow cytometry and monoclonal antibody technology—the ability to make monoclonal antibodies against almost any molecule, including cell-surface markers—has been especially important because it has made it possible to count thousands of cells of various subsets at a time. In fact, in hospitals around the world, the most frequently performed immune monitoring tests—the white blood cell count and differential—are performed by flow cytometry, and diagnosis of leukemia and lymphoma is made using a combination of monoclonal antibody staining and flow cytometry (in conjunction with old-fashioned microscopy). Before these technologies, diagnoses related to the immune system depended mainly on physical signs such as fever, redness, swelling, pus, and phlegm.

Despite this progress, today our ability to monitor the immune system remains crude. Only a handful of different biomarkers can be detected simultaneously using most flow cytometers, and these machines remain expensive to buy and maintain. In addition, until recently there has been no way to monitor antibodies and TCRs in all their complexity. Combined with the inherent inertial conservatism of medicine, these factors have limited our clinical capability. Fortunately, a new convergence of technologies is now transforming immune monitoring technologies, with direct implications for vaccine development, in two ways.

First, *microfluidics and microfabrication* techniques are making it possible to monitor and perform experiments on single cells, hundreds at a time. This represents an important advantage over flow cytometry. For example, whereas a flow cytometer can count the number of B cells of a specific subset in a sample, microfluidics makes it possible to observe how each of many cells responds to a specific controlled stimulus, such as exposure to a potential vaccine or a sequence of such stimuli over time. Meanwhile, microarrays printed with vaccine antigens can be used to quantitate a person's antibody responses to each antigen. Both technologies require imaging and computing. Together, they offer high-resolution pictures of what different populations of cells are up to and what they can be expected to do in particular situations, with enough fine-graining to detect both new subpopulations and stochastic behaviors that might affect overall responses. Microarrays can routinely be printed with many thousands of different analytes, and microfluidic devices can be built that allow more-or-less independent control over hundreds or thousands of individual reaction chambers, with the limiting factor being the complexity of the control circuitry.

Second, so-called *high-throughput sequencing* is making it possible to count and describe the tens of millions of different B and T cell clones that may be present in a given biological sample. The antigen-binding specificity of each clone depends on the DNA sequence of its antibody or TCR genes. Unlike other genes, antibody and TCR genes are generated anew in each new B or T cell, which is what accounts for the extraordinary number of clones. Deep sequencing makes it possible to monitor or "profile" the immune system by providing a readout of which clones are present in a particular biological setting, such as during infection or exposure to a vaccine. Today it is routine to sequence hundreds of thousands or millions of clones in a single sequencing run. This represents a thousand-fold increase over the number of clones that could be sequenced in a single run in the years before next-generation sequencing was introduced in 2005. For years, low-throughput studies (with sample sizes measured in the dozens or hundreds of clones) had hinted that some infections and autoimmune conditions may have "immune signatures" defined by the presence of specific sets of clones (Arnaout 2005). While the existence of such signatures remains under debate, there is considerable excitement that deep sequencing can uncover them.

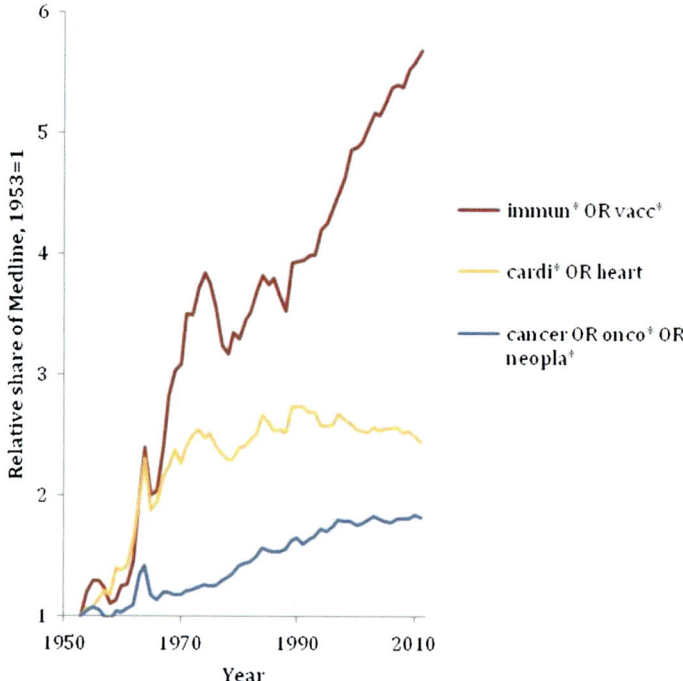

Fig. 5.5 Relative growth in medical research activity, 1953–2011. Shown is the relative growth in immunology and vaccine research compared with research in two other prominent research areas, heart disease and cancer, over the past 60 years. The relative "scientific market share" of immunology and vaccine research is on pace to have grown six-fold by the end of 2012, while the other areas shown will have only about doubled. Medline was searched for the indicated terms, and the number of studies in the indicated research area each year scaled by the total number of studies published in Medline that year and presented as a fold increase, with 1953=1 (Figure courtesy of R. Arnaout)

Outcomes of Convergence

Research interest and output in immunology and vaccines has been growing steadily and rapidly for two decades, even as other medical fields have plateaued (Fig. 5.5). This growth has coincided with the birth and expansion of the therapeutic monoclonal antibody industry. The U.S. Food and Drug Administration (FDA) approved the first therapeutic monoclonal antibody in 1986. By 2010, this group of drugs, which includes rituximab (Rituxan®) for cancer and infliximab (Remicade®) for autoimmune diseases, had become the best-selling class of biologic drugs, with U.S. sales of close to $20 billion on back-to-back annual growth rates of nearly 10 %, accounting for a third of all biologics sales (Aggarwal 2011). With $10 billion pledged by the Bill and Melinda Gates Foundation, vaccines, including cancer vaccines (Davis and Dayoub 2011), are generating similar excitement (Rappuoli et al. 2011). By allowing functional characterization of single immune cells, microfluidic and microfabrication technologies will contribute to vaccine development by offering precise readouts (Han et al. 2012) of the effects of potential vaccines.

Looking Ahead

In the decade since completion of the draft human genome, the combination of microarray-based genotyping and high-throughput sequencing have transformed knowledge about how genes contribute to disease. It made it possible to map genetic diversity on a global scale and offered the first large-scale data to test hypotheses about how genetic differences correlate with phenotypic differences. That the correlations are not yet clear is less important than that hypotheses are now enriched by data. These advances make it safe to predict a similar transformation in immune monitoring, despite foreseeable obstacles. For example, high-throughput sequencing of B and T cells is about to receive a significant boost as key obstacles are surmounted.

The key to high-throughput sequencing is parallelism. Life Technologies' 454, ABI Solid, and Ion Torrent platforms, and Illumina's Genome Analyzer all achieve their throughput by sequencing many millions of DNA molecules in parallel. Read length—the amount of sequence obtained from each molecule—is generally short, on the order of a few hundred bases. Short reads have been sufficient for genome resequencing and *de novo* genome assembly, the applications that have driven development of these technologies so far.

But antibody and TCR genes present special challenges. Like most genes, they are longer than the read length of most sequencing platforms, but unlike most genes, they are generated through a combinatorial mix-and-match mechanism involving dozens of gene segments, and differ from cell to cell. These factors make them hard to assemble from short reads. In addition, the number of clones in a biological sample may be too close to the number of reads for in-depth coverage. (As a yardstick, the amount of sequence in a million clones is equivalent to a quarter of the human genome.) Thus, the current challenge to sequencing antibody and TCR genes is that they are too long for short-read sequencers and too many for longer-read sequencers. As a result, to date, high-throughput immune profiling studies have had to focus on particular regions within the genes—for example the third antigen complementarity-determining region of the antibody heavy chain gene—and ignored the rest. However, this obstacle should disappear as early as 2013, as increases in read length and volume across the high-throughput platforms commoditize immune sequencing by making it possible to sequence whole genes in single reads for all the clones in a biological sample.

Attention will likely turn subsequently to association studies. Over a decade ago, the development of microarrays for detecting single-nucleotide polymorphisms (SNPs) across the genome led to genome-wide association studies for diabetes risk, heart disease, and numerous other conditions. Related technologies led to searches for, and analyses of, gene expression signatures. Whole-genome sequencing is producing ever more detailed views of different cancers and of infectious agents like *E. coli* (Rasko et al. 2011) and HIV (Tsibris et al. 2009). In the same way, immune monitoring technologies will soon be used to look for immune profiling signatures across numerous conditions and states of health. A likely focus will be chronic conditions, including chronic infections such as HIV, which has already been the topic of immune profiling studies (Scheid et al. 2011), but also autoimmune diseases, cancer, and non-disease states such as pregnancy, transplantation, and aging. In vaccine development, immune profiling will allow researchers to more easily identify and synthesize sets of antibodies and TCRs that are produced in response to antigen exposures (Cheung et al. 2012). The common goal of all these studies is predictive sequence-to-specificity mapping between antigens and the antibodies and TCRs that bind them. Such an understanding will make it possible to detect and design immune-based therapeutics for a wide range of conditions, including cancer immunotherapeutics (Fox et al. 2011)—or else reveal the fatal flaws of such a dream. At the same time, microfluidics and microarrays will provide windows into the cellular and soluble context of immune responses.

Even as association studies become technically feasible and affordable, obstacles regarding sampling and analysis will have to be overcome. One obstacle is determining to what extent, and in what situations, clones in different anatomic compartments reflect phenotypes of interest. For example, subsets of B and T cells circulate in the blood but are "home" to various tissues, including sites of inflammation, infection, and cancer. How well a blood sample will reflect these other sites is still unknown; it is a potentially important issue for monitoring or other applications that require frequent sampling. A second obstacle is to understand which subsets in a given compartment to analyze. There are also obstacles when it comes to computational biology. Association studies associate signals with phenotypes. However, it is not yet clear where in the sequence data the signals will be found. Antibodies and TCRs are three-dimensional structures whose avidity for antigen depends on multiple contact surfaces. The relationship between sequence and structure is complex. Different sequences can have convergent structures (Scheid et al. 2011), and the same sequence can encode multiple conformations (James et al. 2003). However, simpler properties like gene segment usage, CDR3 length and charge, and mutation frequency may also provide signals. The answers to these debates may emerge from computational analysis of large datasets.

Overall, the convergence of knowledge and technologies offers the hope of being able to monitor the immune system in all its complexity. The tools and techniques that the new technologies provide offer the chance to approach the immune system at its "native resolution"—at the level of individual cells and molecules, and on its immense scale. Inevitably, this convergence will contribute to deeper understanding across the spectrum of conditions in which the immune system plays a role, as well as to a new generation of vaccines.

5.4 Infrastructure Needs

Historically, the practice of medicine has been anchored in the knowledge base and traditions of taxonomic biology-based research. As a result, hypotheses are based upon phenomenological observations rather than fundamental laws. An interesting

5 Implications: Human Health and Physical Potential

analogy can be drawn with the Japanese postal service, a system for which there is no logical connection between addresses and locations. This organically developed system is effective but also requires the postal worker to memorize the mail route. That "local knowledge" is not translatable. For instance, if the postal worker is given a new route, the whole process of learning how to deliver the mail begins again. Biology (and medicine) are similar fields of intense specialization, where specific knowledge of a particular signaling pathway, for example, takes years to acquire and does not necessarily help when learning about a different area of biology or even signaling pathway! But disparate fields are converging, and the challenges of one discipline, (e.g., biology), can be offset by the specialization and tools of another (e.g., physics and engineering).

As an example, take single-cell analytics. Using new nanotechnology-based separation and analysis tools (Fan et al. 2008), it is now possible to perform quantitative measurements, for example, measuring for a particular biomolecule in its abundance either in copy number or in a cell-by-cell basis. This is becoming true for measurements of transcripts and proteins as well. However, a single cell, as compared to a pool of cells (or related piece of tissue), is a finite system. In a thermodynamic sense, this means that each cell exhibits individualized fluctuations; thus, a quantitative measurement of a specific parameter in one cell will yield a different answer when measured in an identical cell separated in time. Statistical variations of such measurements across many single cells define the fluctuation ranges, and these maps can provide the basis for predictions of behavior of large pools of cells.

Over the next 10 years, the major scientific infrastructure needed will be an effort to define these "laws of biology" within a convergence approach that nurtures engagement of the physics and physical sciences research communities. In an effort to do this, several universities and research institutions have been establishing interdepartmental institutes that draw on multidisciplinary science groups that are co-located within a single institute and work towards solutions jointly. Examples of contemporary institutes designed using this model include the Clark Center, which houses the Bio-X program at Stanford University, the Koch Institute for Integrated Cancer Research at MIT, the Wyss Institute at Harvard University, the Petit Institute for Bioengineering and Bioscience at Georgia Tech, the Molecular Engineering Institute at the University of Chicago, and the North Campus Research Complex at the University of Michigan (Sharp and Langer 2011). The successes of these models in improving productivity and innovation in research need to be better understood and then compared with "traditional" organizational research models. These new best practices can then be adopted across many more research institutions to enable the development of robust nationwide convergence-style research infrastructure.

5.5 R&D Strategies

Enabling and supporting convergence requires astute initial investment. To this end, governments should establish funding programs that support convergence of different research fields (engineering, physics, biology, medicine) in the attempt to benefit from communication and learning a "common language" and as a result building a "whole which is bigger than its contributing parts." When possible, these funding initiatives should leverage large multidisciplinary academic centers, like those described in the last section, that demonstrate high levels of innovation, creativity, and productivity as compared to smaller efforts led by single investigators. A companion objective should be to include scientists from social science research areas—economics, anthropology, psychology, and sociology, to name a few—to provide additional insight. One option to enable this integration would be a *National Converging Technologies Initiative*. This effort would help coalesce convergence experts and expose research opportunities and funding needs, much as the National Nanotechnology Initiative (NNI) and the High-Performance Computing and Communications (HPCC) effort have helped coordinate and spur activity in nanotechnology and computing, respectively.

There is no substitute for predictable and sustainable funding mechanisms to support the advance of convergence in medicine. However, that funding must be coupled with a rigorous peer-review process that includes reviewer groups with a broad array of expertise to handle the unique interdisciplinary complexities of convergence-related research. Realizing convergence-style reviewer groups may necessitate novel approaches such as providing opportunities for experienced program managers and reviewers from Federal agencies to take on short-term assignments in other research agencies. For example, think-tank environments could be established within each research agency, and agency detailees from across the Executive Branch could be invited to visit for several months to collaborate on the convergence-related research priorities of the host institution. This might be coordinated by the National Science and Technology Council.

A comprehensive, long-term R&D investment strategy for convergence, as with any paradigm-shifting scientific endeavor, involves bringing together stakeholders who did not work together jointly in the past (as is often the case for fields undergoing convergence), and it will take time to show results. A high expectation of a long-term funding commitment is required. As exemplified in Fig. 5.6, considerable and consistent Federal investment forms the basis for a system of feedback loops that enables industry to engage and reengage at various points.

Convergence is a new approach to science, collaboration, and cross-cutting interaction, and thus it is a lasting proposition that would be well served with consistent funding—funding that should be allocated within the "convergence account" to the most promising areas. Such funding must be flexible enough that it can be reallocated to high-impact opportunities as they emerge.

5.5.1 What Are the Overall Emerging Topics and Priorities for Converging Technologies Research and Education?

Integration of convergence into everyday life will require parallel engagement of society in defining acceptable objectives and setting agreed boundaries. These evolving standards will define how future technologies are developed and implemented in

5 Implications: Human Health and Physical Potential

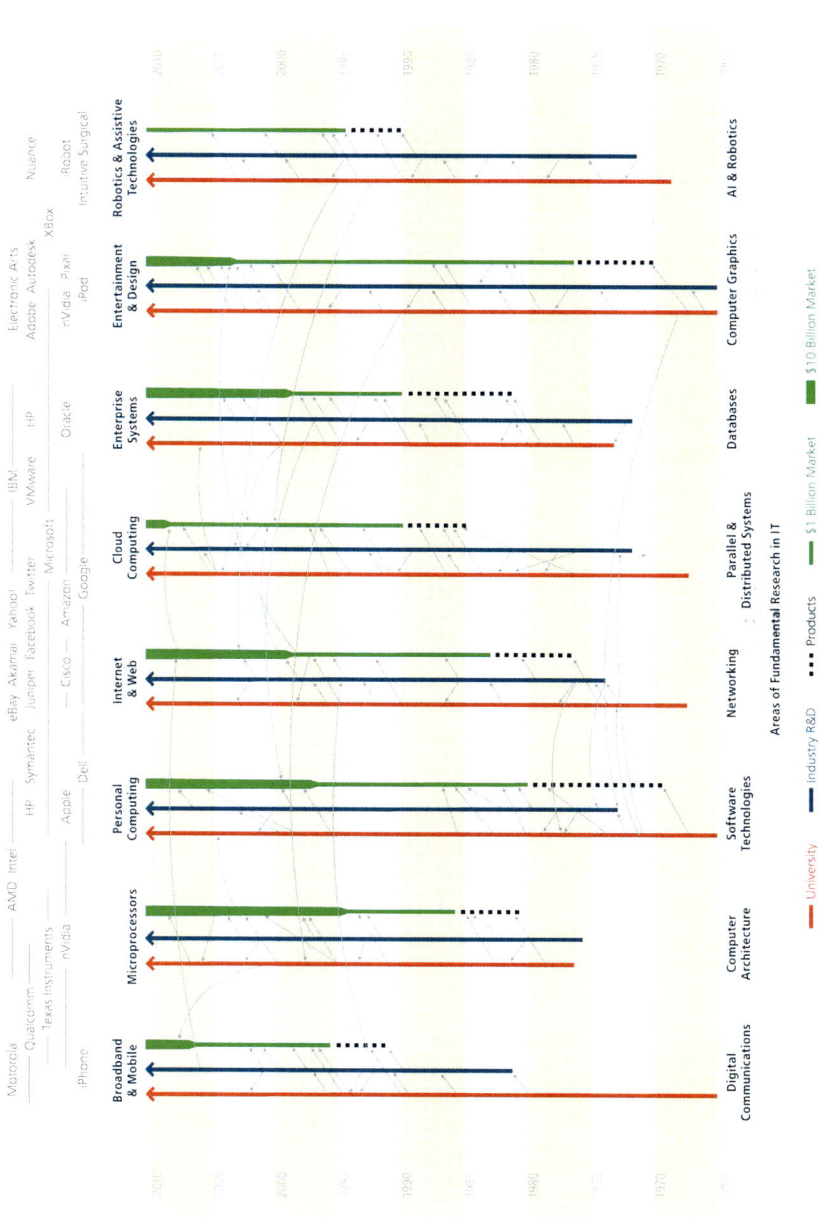

Fig. 5.6 Examples of academic/government-sponsored and industry-sponsored IT research and development efforts in the creation of commercial products and industries (NRC 2012, Fig. 1, p. 3; © 2012, The National Academies Press, used by permission)

Fig. 5.7 Public perspectives of science and technology spanning the human condition (images, Creative Commons)

everyday life and how future technology developers and scientists are being trained. Let us use an example of "nanotechnology," because the emergence of this term in public discourse provides some recent guidance. Whereas the term "nanotechnology" itself has little specific reference in popular culture, fear of the interface between human biology and physical machines is widely present in popular science fiction movies and television shows—from the *6 Million Dollar Man* and *RoboCop* to *Star Trek* and *The Terminator* (Fig. 5.7). To explore the public perceptions around the topic of nanotechnology, in 2008 the Center for Nanotechnology in Society at Arizona State University and its collaborators at North Carolina State University conducted the nation's first *National Citizens' Technology Forum* on the topic of nanotechnology and human enhancement. The study focused on citizens in six sites across the United States. Each of the groups consistently expressed concerns about the effectiveness of related regulations and doubt about equitable distribution and monitoring. The groups also placed greater importance on therapeutic rather than enhancement research and requested greater public information and education, especially in K–16 education, about these technologies (Hamlett et al. 2008; Guston 2010).

It is clear that the interaction between technology (human-made) and biology (nature-made, or God-made, depending upon one's perspective) can provoke responses that are strongly influenced by personal values. Research of Yale Law School's Cultural Cognition Project demonstrates that, "Cultural cognition … causes people to interpret new evidence in a biased way that reinforces their predispositions. As a result, groups with opposing values often become more polarized, not less, when exposed to scientifically sound information". Dan Kahan of Yale University

(Kahan 2010) argues that: "We need to learn more about how to present information in forms that are agreeable to culturally diverse groups, and how to structure debate so that it avoids cultural polarization. If we want democratic policy-making to be backed by the best available science, we need a theory of risk communication that takes full account of the effects of culture on our decision-making."

In addition to religious values, perspectives associated with the environment, economics, social justice, and other issues play key roles in the reactions that individuals and groups have to emerging and converging technologies. Working towards potential strategies, David Guston of ASU's Center for Nanotechnology in Society envisions a process of anticipatory governance, or an effort to manage the social understanding and reception to emerging knowledge-based technologies while such management is still possible. This includes three critical components: *foresight* (of plausible future scenarios), *integration* (of social science and humanities research with nanoscale science and engineering), and *engagement* (of publics in deliberations) (Guston 2010).

Various efforts have been undertaken to avoid a negative public perception of emerging technology in reference to nanotechnology. The Nanoscale Informal Science Education Network (NISE Net) conducted a series of public forums at five sites across the United States focused on nanotechnology and health and personal care. These public engagement programs enhanced attendees' understanding of nanotechnology, its potential impact, attendees' awareness of benefits and risks, and their confidence in expressing and supporting their viewpoints about nanotechnology (Flagg and Knight-Williams 2008). For instance, after hearing experts speak on the topic of nanotechnology improvements, they were asked whether they agreed or disagreed with the statement "new nanotechnology applications in medicine should be made available for use before we understand the possible risks." Participants generally disagreed with this statement when the application was sunscreens, but agreed with it when the application was cancer therapies (Kollmann and Reich 2011).

If there is an interest in addressing these concerns and in engaging and including societal understanding in emerging technologies, it could prove prescient to develop a *Center for Communication and Societal Engagement for Converging Technologies*. Such a center could offer expert assessment in anticipation of potential future applications and couple that information with public education efforts to encourage fact-based dialogue among the citizenry. This center could also drive opportunities for training in science communication and in the societal and ethical implications of technology into all graduate and undergraduate science curricula. This center would need to be an independent nonprofit organization working closely with funding agencies, national labs, universities, and industry to identify scientific frontiers emerging within public discourse.

Together these new elements of governance, outreach, and public engagement could develop a network capable of supporting educational enrichment in emerging frontiers among the general public and at all grade levels, especially 7–12.[2]

---

[2] See Chap. 8 of this report for a greater discussion regarding how converging technologies should shape science, technology, engineering, and mathematics (STEM) education in levels 7–16.

Widespread public engagement in policy considerations and in the related governance of converging science and technology development will be critical. Research of this sort is important in charting a course and establishing educational objectives associated with the development of converging technologies so that years of work and funding are not wasted by a mismatch between public support and research and development trajectories.

5.5.2 What Is the Impact of Converging Medical Technology on Society?

Among the most important societal implications for converging knowledge and technology in medicine will be the redesign of foundational decision-making and choice. The conventional practice of medicine today is reactive and seemingly in diametric opposition to the tenets of preventative medicine (NRC 2011). In fact, the current system is rather ineffective in fostering long-term wellness at the level of individuals or larger cohorts. This inadequacy is based on several fundamental issues. First, current systems are, by and large, based on the assumption that participants (doctors and patients) are rational actors who, when presented with adequate information, will make better decisions at an individual and group level. Unfortunately, "adequate information" about health and cost consequences of daily action and inaction is not yet truly available. Second, most current medical practitioners were trained in a "standard of care" context that is typically exemplified by reacting to illness rather than preventing onset. Third, "medicine" as an institution is risk-averse and opposed to dramatic change. In the short term, the system rewards consistency, (i.e., standard of care), while the long-term goals, especially those that present risk, liability, or potential sacrifice, are discouraged.

However, new models are emerging from research labs and also from ad hoc experiments in governance, especially those coming from virtual communities on the Web. Vast amounts of data about human behavior are making robust behavior modeling more useful. Leveraging this data and making it meaningful for decision-making will also be crucially important. Working with new insights about people—human behavior and cognitive capacities—combined with knowledge of organizing structures and processes will point to new ways of ordering society at a global level. Global modeling and modeling of complex systems will bring about another wave of decision-making tools for business and governments.

5.6 Conclusions and Priorities

As discussed throughout this chapter, convergence as a policy and a funding focus is central to driving biomedical progress. Converging technologies will impact economic competitiveness by enabling a better understanding of costly social health phenomena via better analysis and integration of big data. Convergence will also enhance human capacity with significant goals like developing cancer therapies that have no side effects. (Imagine approaching taking a chemotherapy drug like you would an aspirin.) Finally, while not discussed at length in this chapter, converging technologies have the potential to impact national security because many of the resulting advances will lead to the democratization of cheap and available tools that will enhance healthcare outcomes and save lives: As many social science studies have shown, there is a strong correlation between social justice and new medical and social technologies and between the dueling frameworks of access to healthcare and the potential for violence in a society (Farmer 2003).

Success in these areas, however, is dependent upon a clear and consistent course of action—a trajectory of research and development, public communication, education, and engagement—that enables scientists, engineers, and various stakeholders, including the public, to engage in development of the goals and strategies that will make the promise of convergence in health research a reality.

5.7 R&D Impact on Society

Convergence for health, the new kind of alchemy we discuss throughout this chapter, is not only promising for the reasons discussed above but is also our society's best option to avert challenges we face in the area of healthcare access and infrastructure.

The United States spends a larger portion of its gross domestic product (GDP) on healthcare than any other major industrialized country, and healthcare is one of the fastest-growing components of the Federal budget. Total expenditure on healthcare is now above $2.5 trillion, almost one-fifth (17.6 %) of the U.S. Gross Domestic Product (PBS News Hour 2012). These statistics continue to climb, with national health expenditures expected to increase an average of 5.7 % per year over the projection period of 2011–2021 (CMS 2011). This burden is increasingly felt at home, with the National Center for Health Statistics reporting that as of 2009, the per capita annual healthcare expenditures topped $8,086 (NCHS 2012, 370, Table 125).

The challenge to our society becomes clearer by looking at the contrast of high expenditures and low access. For instance, between 2000 and 2010, the percentage of adults with private insurance declined from 71 % in 2000 to 60 % in 2010, and persons 18–44 years of age who were uninsured increased from 22 to 27 % during the same period (NCHS 2012, 14, Figure 9).

In addition to high cost and poor access to care, our society also faces a huge demographic shift that will push our existing healthcare infrastructure to the brink. This is a two-fold issue. First, biomedical advances prolong American lives. Consider that Americans born at the end of the twentieth century can expect to live about 30 years longer than if they had been born in 1900 (Murphy and Topel 2006), and most retirees in their 60s and 70s are physically able to work (Maestas and Zissimopoulos 2009).

Second, this longer lifespan is translating into increasing costs for enhanced end-of-life care due to aging-associated complex diseases. Just as cancer recently replaced heart disease as the leading U.S. cause of death (thanks to major advances in heart-disease treatment), some expect brain diseases to displace cancer in upcoming years. Alzheimer's disease, for example, will affect the baby boomers at such an alarming rate that it is unclear how our current healthcare system will cope with it. According to the Alzheimer's Foundation (2010; Hebert et al. 2003), "the number of people aged 65 and older with Alzheimer's disease is estimated to reach 7.7 million in 2030—more than a 50 % increase from the 5.1 million aged 65 and older currently affected." Unfortunately, the average annual cost of a Medicare patient with Alzheimer's is three times that of a patient without. Medicare spent $91 billion on patients diagnosed with Alzheimer's disease in 2005. Future projections include $189 billion in 2015 and an increase to over $1 trillion by 2050 (Swartz 2010).

The best way for science to contribute to this alarming situation is to ease the demographic shift through the development of new health technologies to address disease burden and number of individuals affected, and to enable this wave of aging citizens to remain a productive part of the workforce for a longer time. Such an effort would allow the attending healthcare costs to be spread more evenly over an extended period and across the population base.

Converging knowledge and technology for society (CKTS), and the corresponding transformation of healthcare it can trigger, is a key component to buying our society the time we will need to innovate our way through the dramatic challenges outlined above.

5.8 Examples of Achievements and Convergence Paradigm Shifts

5.8.1 Wellness-Focused Contemporary Medicine

Contact: *Piotr Grodzinski, National Cancer Institute, National Institutes of Health*

A convergence of several disciplines to enhance capabilities of medicine has been occurring for several years. It has led to introduction of many technologies that have enabled breakthroughs in both research capabilities and practical clinical and medical practices. Several of these examples were discussed in the Vision (5.1) and Goals sections (5.3) of this chapter. To review a few important ones:

- Discovery and implementation of magnetic resonance imaging (MRI) and positron emission tomography (PET) have dramatically improved the capabilities of imaging techniques and have led to new modalities of detecting cancer and other diseases.
- Establishment of high-throughput, inexpensive sequencing techniques allowed for successful completion of the Human Genome Project; these techniques are used now to sequence several types of cancer under a large initiative, the Cancer Genome Atlas at the National Cancer and National Human Genome Research institutes.

- Sequencing efforts, which produce large amounts of data, in turn contribute to further expansion of bioinformatics and sophisticated data analysis techniques.

Such new technological and engineering advances will continue to enrich medicine and will continue to change ways future medicine is practiced. We should expect disease diagnosis to become more accurate, with several tests performed by patients at their homes and treatments becoming more effective.

Although implementation of new technologies into medicine improves people's lives significantly, it is not a decisive factor alone when it comes to society's health and wellness. As described in Sect. 5.7, R&D Impact on Society, of this chapter, the United States spends over 17 % of its GDP on medical care. This is partially due to the use of expensive and sophisticated medical technologies often used towards the end of a patient's life. For example, the United States ranks second, after Japan, on the availability of MRI units. The average MRI instrument costs over $1 million, and tests performed with this equipment are also expensive (Conference Board of Canada 2013). Also, despite the use of sophisticated technologies; life expectancy at birth in the United States ranks as only fortieth among all countries in the world (UN DESA 2011), most likely due to the uneven access to healthcare (see Sect. 5.7).

Lifestyle, eating habits, and access to medical care play very significant roles in determining overall wellness. In order to leverage all these factors in the future, models of medical care will need to shift from reactive—treating the disease after the patient already succumbed to it—to proactive and preventative—working hard on educating society, implementing healthy lifestyles, and expanding further approaches to screening and vaccination, when appropriate. We have already made significant strides in some of these areas; for example U.S. smoking rates are significantly down from the peak of 45 % of adults in the mid-1950s to about 20 % of adults today. Cigarette smoking is considered the leading cause of preventable death and is responsible for over 400,000 deaths per year (CDC 2012). Wellness also is a function of mental/neural health and mental happiness determined by a summation of factors that include societal relationships. This component is growing in importance with the expansion of societal interactions and with the population aging.

Healthy lifestyles are not only related to diet and personal habits, but also to the environment in which we live, work, and rest. There is an effort to improve construction practices to achieve high-performance and sustainable buildings with optimized energy efficiency, conservation of water use, enhanced indoor environmental quality, and reduced environmental impact of used construction materials (see http://www.wbdg.org/references/fhpsb.php). The Federal Government is aggressively implementing these guidelines in the management of the approximately 450,000 buildings it owns and in the construction of new ones. A correlation between workers' productivity and health with the design of a building with appropriate lighting, ventilation, and control of air contamination has been well documented (e.g., http://www.wbdg.org/design/promote_health.php). An implementation of novel construction methods to improve the in-house environment and conserve energy, in combination with distributed sensor technologies, will lead to the development of "smart homes" capable of sensing the local environment and then, in conjunction with knowledge of inhabitant behaviors, allowing for "smart" adjustment of indoor environmental conditions (Tonn, B, Personal communication, 2013).

Future efforts to improve health and maximize human potential will need to rely not only on convergence and implementation of new technologies into medical care, but foremost on taking a holistic approach to creating wellness environments in the places where we live and work.

5.8.2 Physical Sciences and Engineering Applied to Oncology

Refer to: *WTEC Assessment of Physical Science and Engineering Advances in Life Sciences and Oncology in Europe* (http://www.wtec.org/reports.htm)

R&D projects have begun in the United States and Europe to invite physical sciences and engineering researchers to work with cancer biologists to rethink biomedical approaches to understanding and curing cancer. These initiatives are a response to the fact that, globally, the extensive cancer biology programs, substantial funding, outstanding talent, and state-of-the art tools applied over four decades have been unable to meaningfully reduce cancer mortality rates. In the United States, a new NIH Office of Physical Sciences-Oncology began in 2008/2009 to award cooperative agreements to U.S. universities in a collaborative network of now-12 physical science-oncology centers that unite experts in the fields of biology, medicine, physics, chemistry, mathematics, modeling, informatics, engineering, and nanotechnology to address cancer. Interdisciplinary, transnational cancer research efforts also have begun in Sweden, France, Germany, and other EU countries, all with robust funding despite global economic uncertainties. U.S. and EU physical sciences oncology programs span the following broad categories of research (Jenmey et al. 2012):

- Information and complexity
- Extracellular and tissue microenvironments
- Cell and tissue mechanics
- Cell transport and metabolic waste removal processes
- Dynamics to understand and measure the rates and patterns of cell shape, change, migration, and division and their integration with biochemical and genetic information
- Application of new devices and diagnostic principles to exploit the physical properties of tissues for cancer diagnosis and treatment

The work of the NIH Physical Science–Oncology Centers (http://opso.cancer.gov/centers/) is helping to define a "soft-matter physics" or "physics of cancer," looking in part at tissue boundaries as comparable to fluid boundaries. It is hoped that such novel applications of physical science and engineering to oncology will bring breakthrough solutions to cancer mortality due to tumor growth and metastasis.

5 Implications: Human Health and Physical Potential   211

Table 5.1 Major achievements in the areas of human health and physical potential

CKTS domain	Example
Cancer detection and treatment with reduced side effects	Combining nanotechnology devices with recent advances of cancer biology to develop new platforms for delivery of localized therapies and new multiplexed and sensitive diagnostics
Health data analysis and delivery for real-time health monitoring towards wellness	Combining advanced electronics, sensor technologies, and physiology to develop self-based or embedded wearable body function monitoring and to design 'smart homes' for improved interaction of living environment and the individual
Regenerative medicine and advanced prosthetics	Using advanced fabrication technologies to create *in vitro* environments for cell and tissue growth to create artificial organs. Through combination of advanced electronics, design techniques, and ability to monitor brain signal activity, creating advanced prosthetics with seamless interconnectivity of the artificial limb and the human body
Harnessing the human immune system to develop next-generation vaccines	Using advanced technologies for functional characterization of single immune cells and monitoring of vaccine responses to measure new vaccine efficacy

5.8.3 List of Examples

Major achievements in addressing human health and physical potential are illustrated in Table 5.1.

5.9 International Perspectives

The following are summaries relevant to this chapter of discussions at the international regional WTEC NBIC2 workshops held in Leuven, Belgium, September 20–21, 2012; in Seoul, Korea, October 15–16, 2012; and in Beijing, China, October 18–19, 2012. Further details of those workshops are provided in Appendix A.

5.9.1 *United States–European Union NBIC2 Workshop (Leuven, Belgium)*

Panel members/discussants

Laura Ballerini (co-chair), University of Trieste (Italy)
Miloš Nesládek (co-chair), Hasselt University (Belgium) and Minatec
Jian Cao, Northwestern University (U.S.)
James Olds, George Mason University (U.S.)

Mark Lundstrom, Purdue University (U.S.)
Mira Kalish Marcus, Tel Aviv University (Israel)
Sylvie Rousset, CNRS (France)
Christos Tomakanis, EC (EU)
Robert Urban, Massachusetts Institute of Technology (U.S.)

The working group convened via the United States–European Union NBIC2 Workshop in Leuven, Belgium, resulted in three distinct focus areas that were proposed to represent tractable, high-impact opportunities for human health over a ten-year time frame:

1. Molecular and dynamic profiles used for personalized medical treatment
2. Quality-of-life enhancement: prosthetic devices and regenerative medicine
3. Personalized innovative education

These three opportunity areas are described in more detail below.

1. *Molecular and dynamic profiles used for personalized medical treatment* of cancer or neurodegenerative disease, as well as more acute conditions such as in context infection or response to trauma, represent important opportunities. Critical components to drive progress in these areas will be:

   - Data optimization in the sense of innovative and enhanced data harvesting, sharing, analysing, drug surveys, and management of personalized data from an ethical point of view
   - Improvement of technological tools for information at the molecular level (proteome, genome, bioinformatics, contactless tools, biomarkers groups for disease progression versus individuals, sensors, cellular imaging, and signalling in cell biology)
   - Tools for improving accuracy and efficacy towards predictive diagnosis
   - Cost-effective, personalized manufacturing of drugs
   - Distributed systems for delivery of personalized care

Boosting innovation in industrial systems is crucial for the European Commission: developing new knowledge, technologies, products, and applications that, bridging the divide between research and innovation needs, could turn today's societal challenges (e.g., health, well-being, ageing) into opportunities with high potential for competiveness, innovation, and growth is a pivotal strategy to achieve that. So, enhancing convergence of key enabling technologies (i.e., nanotechnology, nanoelectronics, advanced materials, biotechnology, photonics, etc.) should lead towards knowledge-, capital- and skill-intensive innovation cycles, driving the development of innovative industries. The joint NSF–EC workshop held in Leuven (September 20–21, 2012) pinpointed that developing and delivering more effective translatability of converging technologies "from bench to bedside" by achieving better personalized information for enhanced medical treatment could contribute to progress towards this goal. In this light, developing translational hubs that bring together converging technologies research from laboratory into applicable clinical trials can constitute a possible pathway to deliver personalized diagnostics and treatment. These can be done through different strategy lines:

- Mastering molecular information and dynamic progression of molecular profiles for personalized medical treatment using nano-enabled, nano-structured, and nanolayered structures, devices, processes, and systems is key to reducing the very high economic cost of treatments of cancer, neurodegenerative syndromes, and injuries, accounting for many billion Euros per year.
- The challenge is to bring together imaging, diagnostics, and therapeutics to provide integrated converging technologies-based "bench to bedside" solutions, translatable from lab into applications—to conceive, design, and develop innovative nanostructures, devices, and systems for the analysis and development of molecular information and dynamic progression of molecular profiles for personalized medical treatment of various diseases such as cancer, neurodegenerative disease, Parkinson's, and spinal lesion.
- Designing and developing, enhancing, and improving innovative nano-enabled tools for enhanced personalized data harvesting, sharing, analyzing, surveying and managing, and processing the information at the molecular level. Proteome, genome, bioinformatics, contactless tools, biomarkers groups for disease progression versus individuals, sensors, cellular imaging and signalling in cell biology should be used for developing tools with improved accuracy and efficacy for personalized predictive diagnosis and treatment by cost-effective and personalized manufacturing of drugs, supported by an affordable system for delivering of care. Ethical considerations should be included.

2. *Quality-of-life enhancement via prosthetic devices and regenerative medicine* primarily relates to engineering—"of" and "for" the human body—human bionic machine/organ interfaces as new frontiers in personalized regenerative medicine.
Critical components to drive progress in these areas will be:

- Nanotechnology, nanomaterials, nanoelectronics
- Artificial organs by design
- Artificial restoration of functions (sight, ear, motor functions…)
- Development of multidirectional interfacing, e.g., therapeutic interfaces
- Improved knowledge in intracellular signal transduction codes
- Addressing cognitive enhancement
- Restoring functions to restore productivity

The joint NSF-EC workshop also highlighted that convergence of key enabling technologies that can augment translatability of research "from bench to bedside" by delivering *innovative engineering systems "of" and "for" the human body, prosthetic organs "by design", and new human–bionic machine/organ interfaces*. New research concepts in cellular interactions with artificial man-designed nanomaterials are needed to provide innovative rules in the design of interfaces that can subsequently be applied to creating and integrating tissues *in vivo*, improving control of tissue development, and tuning tissue performance. These novel interfaces could also be based on nano-enabled and/or nanostructured surfaces providing innovative properties for new nano–bio functionalities. Tissue engineering aims at developing functional substitutes for damaged tissues and organs, where the increased understanding of the mechanical and physical environments that cells need to form functional tissues has contributed to promoting research into the converging technologies domain and into the nanoscaled features ultimately instructing tissue regrowth or repair.

The **core ideas** are to:

- *Conceive, design, and develop innovative engineering systems "of" and "for" the human body*[3]
- *Provide innovative functionalities* for delivering new human bionic machine/organ interfaces as new frontiers in personalized regenerative medicine
- *Apply and use* in a converging mode converging technologies, nanotechnology, nanomaterials, and nanoelectronics, making them work together to deliver "artificial organs by design" and new ways to provide affordable and viable artificial restoration of functions (e.g., sight, hearing, motor functions)
- Address the development of multidirectional interfacing (e.g., therapeutic interfaces)
- Design and deliver new improved knowledge in intracellular signal transduction codes and cognitive enhancement in order to recover functions to restore productivity

3. *Personalized innovative education* would represent a major paradigm shift in addressing formal and informal education with adaptive life-long learning systems. Although this topic is not limited to human health issues, it is certainly vital to being able to address the kinds of wide-ranging CKTS needs that are discussed in this chapter. Critical components to drive progress in these areas will be:

- Targeting aging needs for continuous education though informal education processes and support for sustainable employment
- Micro-education learning platforms based on personal needs: "I teach you the way you learn"
- Adaptability of workstations to the needs of individuals in terms of tailored content and the education system taking into account the man–machine relationship
- Haptic platforms for sensory and motor feedback for personalized improved learning
- Connecting informal with formal education through science centers on CKTS

---

[3] "Converging Nano-Bio-Info-Cognitive Science and Technology for Responsible Innovation and Society" Workshop, internal presentation, Panel 1 conclusions, IMEC, Leuven 20–21/09/2013.

5 Implications: Human Health and Physical Potential   215

- Reframing the formal educational system to promote convergence of disciplines by mutually enabling each other (improving disciplines' awareness of their limits and needs)
- Moving cognition to information distribution: synergies with public media
- Developing markets for targeted and personalized education

5.9.2 United States–Korea–Japan NBIC2 Workshop (Seoul, Korea)

Panel members/discussants

Takanori Ichiki (co-chair), University of Tokyo, (Japan)
Kyu Back Lee (co-chair), Korea University (Korea)
Robert G. Urban (co-chair), Massachusetts Institute of Technology (U.S.)
Kwon Wook Kang, Seoul National University (Korea)
Young Keun Kim, Korea University (Korea)
Ickchan Kwon, KIST Biomedical Research Institute (Korea)
Kuiwon Choi, KIST Biomedical Research Institute (Korea)

The working group convened via the United States–Korea–Japan NBIC2 Workshop in Seoul resulted in six distinct focus areas that were proposed to represent tractable, high-impact opportunities over a ten-year time frame:

1. Nanostructured materials for cancer vaccines
2. Theragnosis for personalized medicine via imaging
3. Human enhancement
4. Using bio-inspired nanomachines integrated with fabricated devices into systems
5. Optimizing stem cell function via nanotechnology-based structures
6. Developing specialized centers for health-related convergence

These opportunity areas are discussed in more detail below.

1. *Nanostructured materials for cancer vaccines* aim to improve efficiency of vaccine loading via nanotechnology via core–shell particles. Critical components to drive progress in these areas will be:

   - Iron-oxide cores with zinc-oxide coating resulting in cytoplasm-specific loading and also providing imaging modality to trace DC migration
   - Clarifications of toxicity profiles
   - Materials "safety platforms" that could be useful to expedite materials technology development
   - Multifunctional, biocompatible nanoplatforms
   - Personalization of antigens
   - Inclusion of toxicology and databases/predicative simulations
   - Leads to parenteral formulations

2. *Theragnosis for personalized medicine via imaging* will be needed to underwrite the use of expanding use of genetics diagnostics in clinical management. Critical components to drive progress in these areas will be:

   - Drug design that is based upon molecular imaging, not genotype, e.g., 90 % of drugs only benefit 40 % of patients
   - Investigation into why Her2 is overexpressed in about 25 % of breast cancer patients, but only 50 % of them benefit from Her2-specific therapy.
   - Enzyme-based diagnostics using imaging (e.g., caspase-based imaging) or folate receptor-based probes
   - Phenotype imaging and drug delivery that overlap
   - Change in regulatory requirements to facilitate a new development paradigm at the level of the individual
   - Individualized treatment, not just personalized
   - Imaging technologies within reasonable cost
   - "Safely done" imaging

3. *Human enhancement* using convergence to provide improved productivity and enjoyment. Enhancing human function is not new: glasses, teeth, and hearing aids have been available for decades, even centuries, but moving into new converging-technology-based advances, some example concepts driving progress will be:

   - "Human 3.0": healthy, fun, convenient
   - Blood substitutes via nanoparticle-based oxygen carriers, for example, can enable new capabilities such as improving human functioning in underwater activities
   - Enhancement of normal human power (recalling the "6-million-dollar man")
   - Early need to address societal concerns

4. *Using bio-inspired nanomachines integrated with fabricated devices into systems* integrated for detection, diagnosis, and treatment, including addressing the unsustainable cost burden of our aging society. Critical components to drive progress in these areas will be:

   - Noninvasive measurement of cell function via biomicroelectromechanical systems (bioMEMS)
   - Collaboration in mega-collaborative cluster interactions with rich diversity
   - Capability to track intra-vital recordings via transparent windows and interfaces

5. *Optimizing stem cell function via nanotechnology-based structures* to facilitate stem cell use in regenerative medicine and cell therapy. Critical components to drive progress in these areas will be:

   - Building nanoscience-based structures to drive the desired stem cell maturation and differentiation in cell culture
   - Removing the need for exogenous factors such as cytokines

5 Implications: Human Health and Physical Potential

- Use of nanoscience-based structures to optimize stem cells and to drive *in vivo* differentiation and paracrine effects
- Treatment synergies with RNAi, micro-RNA, hormones, and physical/biological/chemical signals
- Possible political benefits: may ease ethical constraints of using adult stem cells or induced pluripotent stem cells (iPS)

6. *Developing specialized centers* for health-related convergence and organizing "national need"-oriented centers to address the rapid demographic shift towards older populations. Critical components to drive progress in these areas will be:

- Global networks focusing on research diversity and clinical need
- High-level recruitment of experts
- Integration of medical and engineering students and trainees
- Exchange programs—brain exchange
- Partnerships with medical professionals brought in to work with scientists and engineers
- Space-sharing/collaborative teams
- Redesign of incentives at institutions to be successful
- Protocols for collaboration
- International interactions, e.g., between institutes
- "Shared-interest" person-to-person directories
- Industrial guidance with targeted research funding support

5.9.3 *United States–China–Australia–India NBIC2 Workshop (Beijing, China)*

Panel members/discussants

Xiaomin Luo (co-chair), BGI Healthcare (China)
Gordon Wallace (co-chair), University of Wollongong (Australia)
Robert G. Urban (co-chair), Massachusetts Institute of Technology (U.S.)
Ming Liu, Institute of Microelectronics, Chinese Academy of Sciences (China)
Wei Li (China)
Xiaomin Wang (China)

The working group convened via the United States–China–Australia NBIC2 Workshop in Beijing, resulted in four distinct focus areas that were proposed to represent tractable, high-impact opportunities over a ten-year time frame:

1. "From conception through childhood" medical supportive technologies
2. Improve nanotechnology-based formulations for clinical use
3. Enable translational medicine in the context of global collaboration
4. Address chronic neurodegenerative disease

These opportunity areas are discussed in more detail below.

1. *From "conception through childhood" medical supportive technologies* are focused on reducing pregnancy-related health issues, supporting birth, childhood, and motherhood via converging technology. Critical components to drive progress in these areas will be:

   - Improving *in vitro* fertilization (IVF)
   - Full term pregnancy monitoring
   - Early infancy and pre-term birth support
   - Facilitating the mother-child relationship via technology
   - Enhancing telecommuting options and flexibility
   - Preeclampsia as an example:

     o Finger postprandial glucose (PPG) waveform morphology is predictive
     o Fetal age effects PPG
     o Effects of gestational age and maternal age are predictive
     o Soluble endoglin at week 13 is a sensitive biomarker
     o All cell-free DNA in maternal DNA is also a sensitive biomarker

   - Convergence of nanosensors can be an effective new tool to improve reliability
   - New tools are also useful in measuring blood volume change related to stressful conditions

2. *Improve nanotechnology-based formulations for clinical use,* and use known physiological *in vivo* behaviors to inform specific nanomaterial design. Critical components to drive progress in these areas will be:

   - Working towards a set of known and shared standards
   - Establish fundamental correlations (between materials and biological needs)
   - Requires global data sharing and standardization and physical standard samples
   - Requires protocols for collaboration
   - Other components of the challenge:

     o High flow rate
     o Dilution is huge
     o Control degradation
     o EPR
     o Anaerobic and low pH
     o Critical micelle concentration

3. *Enable translational medicine in the context of global collaboration*, taking full advantage of large population-based science. Critical components to drive progress in these areas will be:

   - National biobanks (blood and other tissues)
   - Biomarkers-focused research efforts

- Large cohort studies
- Biggest of big data challenges
- Reduction to treatment in individual patient across *full network*
- Disease agnostic but driven by centers of excellence
- Will require NBIC-like driven leadership to facilitate (Genome Project, Particle Physics, Space in 2000)

4. *Address chronic neurodegenerative diseases* with Central Nervous System (CNS) embedded devices to monitor onset, deliver treatment, track efficacy, and eventually prevent disease altogether. Critical components to drive progress in these areas will be:

- Electronic devices and related software algorithms
- Long-term device biocompatibility
- Defining the optimal treatment combination
- Developing new surgical procedures
- Integration with physical therapies
- Measurements of progress: (1) healthcare cost per patient, (2) availability, (3) vitality
- Engagement with society and ethical teams early in development.

References

Aggarwal, S.: What's fueling the biotech engine–2010 to 2011. Nat. Biotechnol. **29**, 1083–1089 (2011)

Alzheimer's Foundation: 2010 Alzheimer's disease facts and figures. Alzheimers Dement. **6**, 14 (2010). http://www.alz.org/documents_custom/report_alzfactsfigures2010.pdf

Arnaout, R.A.: Specificity and overlap in gene segment-defined antibody repertoires. BMC Genomics **6**, 148 (2005)

Arnaout, R., Lee, W., Cahill, P., Honan, T., Sparrow, T., et al.: High-resolution description of antibody heavy-chain repertoires in humans. PLoS One **6**, e22365 (2011)

Booth, J.: A short history of blood pressure measurement. Proc R. Soc. Med. **70**, 793–799 (1977)

Caplan, L.: Posterior circulation ischemia: then, now, and tomorrow [and references therein]. Presented as the Thomas Willis Lecture at the American Heart Association 25th International Stroke Conference, New Orleans, LA, February 10, 2000. Available online: http://stroke.ahajournals.org/content/31/8/2011.short (2000)

CDC (Centers for Disease Control and Prevention): Fact sheet on smoking and tobacco use. http://www.cdc.gov/tobacco/data_statistics/fact_sheets/adult_data/cig_smoking/ (2012)

Chan, M., Campo, E., Estève, D., Fourniols, J.: Smart homes—current features and future perspectives. Maturitas **64**, 90 (2009)

Cheung, W.C., Beausoleil, S.A., Zhang, X., Sato, S., Schieferl, S.M., et al.: A proteomics approach for the identification and cloning of monoclonal antibodies from serum. Nat. Biotechnol. **30**, 447–452 (2012)

CMS (Centers for Medicare and Medicaid Services): National health expenditure fact sheet, Online at: http://www.cms.gov/Research-Statistics-Data-and-Systems/Statistics-Trends-andReports/NationalHealthExpendData/NHE-Fact-Sheet.html (2011)

Conference Board of Canada: Health spending: do countries get what they pay for when it comes to health care? Available online: http://www.conferenceboard.ca/hcp/hot-topics/healthspending.aspx (2013)

Davis, M.M., Dayoub, E.J.: A strategic approach to therapeutic cancer vaccines in the 21st century. JAMA **305**, 2343–2344 (2011)

Ding, D., Cooper, R.A., Pasquina, P.F., Fici-Pasquina, L.: Sensor technology for smart homes. Maturitas **69**, 131–136 (2011)

Elgert, P.A., Gill, G.W.: George N. Papanicolaou, MD, PhD, Cytopathology (biographical review). LabMedicine **40**, 245–246 (2009). Online at: http://labmed.ascpjournals.org/content/40/4/245.full

Fan, R., Vermesh, O., Srivastava, A., Yen, B.K., Qin, L., Ahmad, H., Kwong, G.A., Liu, C.C., Gould, J., Hood, L., Heath, J.R.: Integrated blood barcode chips. Nat. Biotechnol. **26**(12), 1373–1378 (2008)

Farmer, P.: Pathologies of Power: Health, Human Rights, and the New War on the Poor. University of California Press, Berkeley/Los Angeles (2003)

Farrell, D., Alper, J., Ptak, K., Panaro, N.J., Grodzinski, P., Barker, A.D.: Recent advances from the National Cancer Institute Alliance for Nanotechnology in Cancer. ACS Nano **4**(2), 589–594 (2010)

Farrell, D., Ptak, K., Panaro, N.J., Grodzinsk, P.: Nanotechnology-based cancer therapeutics—promise and challenge—lessons learned through the NCI alliance for nanotechnology in cancer. Pharm. Res. **28**, 273–278 (2011)

Ferrari, M.: Cancer nanotechnology: opportunities and challenges. Nat. Rev. Cancer **5**(3), 161–171 (2005)

Flagg, B.N., Knight-Williams, V.: Summative Evaluation of Nise Network's Public Forum: Nanotechnology in Healthcare. Multimedia Research, Bellport (2008). Available online: http://informalscience.org/reports/0000/0391/NISEForumSummativeEval.pdf

Fox, B.A., Schendel, D.J., Butterfield, L.H., Aamdal, S., Allison, J.P., et al.: Defining the critical hurdles in cancer immunotherapy. J. Transl. Med. **9**, 214 (2011)

Goetz, T.: Cancer and the new science of early detection. Wired. p. 82 (2008)

Grossman, C., Powers, B., McGinnis, J.M. (eds.): Digital Infrastructure for the Learning Health System: The Foundation for Continuous Improvement in Health and Healthcare. Institute of Medicine Workshop Series Summary. The National Academies Press, Washington, DC (2011)

Guston, D.: The anticipatory governance of emerging technologies. J. Korean Vacuum Soc. **19**(6), 432–441 (2010)

Hamburg, M.A., Collins, F.S.: The path to personalized medicine. N. Engl. J. Med. **363**, 301–304 (2010)

Hamlett, P., Cobb, M., Guston, D.: National Citizens' Technology Forum: Nanotechnologies and Human Enhancement. CNS-ASU Report #R08-0003. Center for Nanotechnology in Society, Arizona State University, Tucson (2008). Available online: http://cns.asu.edu/cns-library/type/?action=getfile&file=88

Han, Q., Bagheri, N., Bradshaw, E.M., Hafler, D.A., Lauffenburger, D.A., et al.: Polyfunctional responses by human T cells result from sequential release of cytokines. Proc. Natl. Acad. Sci. U. S. A. **109**, 1607–1612 (2012)

Heath, J.R., Davis, M.E.: Nanotechnology and cancer. Annu. Rev. Med. **59**, 405–419 (2008)

Hebert, L.E., Scherr, P.A., Bienias, J.L., Bennett, D.A., Evans, D.A.: Alzheimer's disease in the U.S. population: prevalence estimates using the 2000 census. Arch. Neurol. **60**, 1119–1122 (2003)

Hrkach, J., Von Hoff, D., Mukkaram Ali, M., Andrianova, E., Auer, J., et al.: Preclinical development and clinical translation of a PSMA-targeted docetaxel nanoparticle with a differentiated pharmacological profile. Sci. Transl. Med. **4**(128), 128ra39 (2012). http://dx.doi.org/10.1126/scitranslmed.3003651

James, L.C., Roversi, P., Tawfik, D.S.: Antibody multispecificity mediated by conformational diversity. Science **299**, 1362–1367 (2003)

Jenmey, P., Fletcher, D., Gerecht, S., Malllick, P., McCarty, O., Munn, L., Reinhart-King, C.: Executive Summary, WTEC Panel Report on Assessment of Physical Sciences and Engineering Advances in Life Sciences and Oncology (Aphelion) in Europe. World Technology Evaluation Center, Arlington (2012). Available online: http://www.wtec.org/aphelion/APHELION Report (Complete).pdf

Kadouche, R., Abdulrazak, B., Mokhtari, M., Giroux, S., Pigot, H.: Personalization and multi-user management in smart homes for disabled people. Intl. J. Smart Homes **3**(1), 39–47 (2009)

Kahan, D.: Fixing the communications failure. Nature **463**, 296–297 (2010)

Khademhosseini, A., Vacanti, J.P., Langer, R.: Progress in tissue engineering. Sci. Am. **300**(5), 64–71 (2009)

Kollmann, E.K., Reich, C.: NISE Network Forum: Nanomedicine in Healthcare Formative Evaluation. Available online: http://www.nisenet.org/catalog/evaluation/nise_network_forum_nanomedicine_healthcare_formative_evaluation (2011)

Maestas, N., Zissimopoulos, J.: How longer work lives ease the crunch of population aging. Working paper, December. RAND Corporation, Santa Monica. http://www.rand.org/pubs/working_papers/2010/RAND_WR728.pdf (2009). (Dec.)

Murphy, K., Topel, R.: The value of health and longevity. J. Polit. Econ. **114**(5), 871–904 (2006). http://www.journals.uchicago.edu/doi/full/10.1086/508033

Murphy, S.L., Xu, J., Kochanek, K.D.: Deaths: preliminary data for 2010. National Vital Statistics Reports **60**(4). Department of Health and Human Services, Centers for Disease Control and Prevention, National Center for Health Statistics, National Vital Statistics System. Available online: http://www.cdc.gov/nchs/data/nvsr/nvsr60/nvsr60_04.pdf (2012)

Murray, G., Holden, R.: Transplantation of kidneys, experimentally and in human cases. Am. J. Surg. **87**, 508–515 (1954)

NCHS (National Center for Health Statistics): U.S. Department of Health and Human Services: Health, United States, 2008. (Public-use file for 2008 deaths.) Available online: http://www.cdc.gov/nchs/hus/previous.htm (2008)

NCHS (National Center for Health Statistics): Chartbook section in Health, United States, 2011, with Special Feature on Socioeconomic Status and Health. U.S. Department of Health and Human Services, Centers for Disease Control and Prevention, National Center for Health Statistics, Hyattsville, MD. Available online: http://www.cdc.gov/nchs/hus.htm and http://www.cdc.gov/nchs/fastats/hexpense.htm (2012). Accessed Jan 2013

NRC (National Research Council): Continuing Innovation in Information Technology. The National Academies Press, Washington, DC (2012). http://www.nap.edu/catalog.php?record_id=13427

NRC (National Research Council) Committee on a Framework for Developing a New Taxonomy of Disease: Toward Precision Medicine: Building a Knowledge Network for Biomedical Research and a New Taxonomy of Disease. The National Academies Press, Washington, DC (2011)

O'Callaghan, T.: Introduction: the prevention agenda. Nature **471**, S2–S4 (2011). doi:10.1038/471S2a

PBS News Hour: Health costs: how the U.S. compares with other countries. Available online: http://www.pbs.org/newshour/rundown/2012/10/health-costs-how-the-us-compares-with-other-countries.html (2012)

Rappuoli, R., Black, S., Lambert, P.H.: Vaccine discovery and translation of new vaccine technology. Lancet **378**, 360–368 (2011)

Rasko, D.A., Webster, D.R., Sahl, J.W., Bashir, A., Boisen, N., et al.: Origins of the *E. coli* strain causing an outbreak of hemolytic-uremic syndrome in Germany. N. Engl. J. Med. **365**, 709–717 (2011)

Rosen, G.: A History of Public Health. Johns Hopkins University Press, Baltimore (1993)

Scheid, J.F., Mouquet, H., Ueberheide, B., Diskin, R., Klein, F., et al.: Sequence and structural convergence of broad and potent HIV antibodies that mimic CD4 binding. Science **333**, 1633–1637 (2011)

Sharp, P.A., Langer, R.: Promoting convergence in biomedical science. Science **333**(6042), 527 (2011). doi:10.1126/science.1205008

Sinha, R., Kim, G.J., Nie, S., Shin, D.M.: Nanotechnology in cancer therapeutics: bioconjugated nanoparticles for drug delivery. Mol. Cancer Ther. **5**, 1909–1917 (2006). doi:10.1158/1535-7163.MCT-06-0141

Swartz, K.: Projected costs of chronic diseases. Healthcare Cost Monitor (online). The Hastings Center, Garrison. http://healthcarecostmonitor.thehastingscenter.org/kimberlyswartz/projected-costs-of-chronic-diseases/#ixzz2AF5mXxOa (2010)

Tobert, J.A.: Lovastatin and beyond: the history of the HMG-CoA reductase inhibitors. Nat. Rev. Drug Discov. **2**(7), 517–526 (2003). http://dx.doi.org/10.1038/nrd1112

Tsibris, A.M., Korber, B., Arnaout, R., Russ, C., Lo, C.C., et al.: Quantitative deep sequencing reveals dynamic HIV-1 escape and large population shifts during CCR5 antagonist therapy in vivo. PLoS One **4**, e5683 (2009)

UN DESA (United Nations Department of Economic and Social Affairs Population Division): Table of Life Expectancy at Birth (years), UN World Population Prospects 2010. Available online: http://en.wikipedia.org/wiki/List_of_countries_by_life_expectancy (2011)

Wang, A.Z., Langer, R., Farokhzad, O.C.: Nanoparticle delivery of cancer drugs. Annu. Rev. Med. **63**, 185–198 (2012)

Warren, R.L., Freeman, J.D., Zeng, T., Choe, G., Munro, S., et al.: Exhaustive T-cell repertoire sequencing of human peripheral blood samples reveals signatures of antigen selection and a directly measured repertoire size of at least 1 million clonotypes. Genome Res. **21**, 790–797 (2011)

Yoo, D., Lee, J.-H., Shin, T.-H., Cheon, J.: Theranostic magnetic nanoparticles. Acc. Chem. Res. **44**, 863–874 (2011)

Chapter 6
Implications: Human Cognition and Communication and the Emergence of the Cognitive Society

James L. Olds, Philip Rubin, Donald MacGregor, Marc Madou,
Anne McLaughlin, Aude Oliva, Brian Scassellati, and H.-S. Philip Wong

6.1 Vision

6.1.1 Changes in the Vision over the Past Decade

Over the past 10 years, the vision of how convergent technologies can be utilized to positively affect society has undergone many changes (Adamson 2012). Among them are the deployment of ubiquitous noninvasive brain visualization technologies, the recognition of nonverbal communication (spatial cognition, alternative sense modalities, brain–brain, and brain–machine), and the emergence of neuromorphic engineering. At a societal level, these changes to the vision have manifested

---

Corresponding editors M.C. Roco (mroco@nsf.gov) and W.S. Bainbridge (wbainbri@nsf.gov).

J.L. Olds
Krasnow Institute for Advanced Studies, Fairfax, VA, USA

P. Rubin
Office of Science and Technology Policy, Washington, DC, USA

D. MacGregor
MacGregor-Bates, Eugene, OR, USA

M. Madou
University of California, Irvine, Irvine, CA, USA

A. McLaughlin
North Carolina State University, Raleigh, NC, USA

A. Oliva
Massachusetts Institute of Technology, Cambridge, MA, USA

B. Scassellati
Yale University, New Haven, CT, USA

H.-S.P. Wong
Stanford University, Stanford, CA, USA

themselves in the viral nature of social networking, the popularity of functional brain images in the mass media, and the ability to use nanoelectronics to emulate some aspects of how the brain functions. This change in vision has made it possible to massively enhance the readily assessable computational power of human brains: the current meme that one's smartphone "has more computing power than all of NASA did when it put a man on the moon in 1969" (Otellini 2012) reflects not only massive change in American society but a qualitatively different perspective on what convergent technologies mean for everyday life. This evolution in our perspective, then, is that convergent technologies are fully embedded in ordinary life, where they augment individual human connections, access to domain knowledge, and cognition in profound ways that their inventors (e.g., Google's Larry Page, and Facebook's Mark Zuckerberg) never conceived of at the time of their invention. In this chapter, we argue that this evolutionary trajectory is profound and use the term "The Cognitive Society" to represent a future global cognitive awareness that is emerging from the current progression.

6.1.2 Background: The Emergence of Cognitive Science as a Discipline

Contact: *W. S. Bainbridge, NSF*

The convergence of subdisciplines from the multiple fields that created cognitive science three or four decades ago had the effect of distancing this new discipline from others, notably clinical psychology and psychiatry, with which a rapprochement may now be in order. Founded in 1979, the Cognitive Science Society (http://cognitivesciencesociety.org/) lists the main constituent fields of cognitive science as artificial intelligence, linguistics, anthropology, psychology, neuroscience, philosophy, and education—but not such highly cognitive social sciences as sociology and political science. The inclusion of "psychology" refers primarily to cognitive psychology, which leaves ambiguous how deeply involved social psychology should be—and the Cognitive Science Society does not really include personality psychology and clinical psychology. The inclusion of neuroscience has not meant a strong connection to psychiatry, but it serves as a second bridge to applications of cognitive science in the general area of mental well-being.

By the mid-1980s, despite its great accomplishments and high prestige, the American Psychological Association came to be seen by many research psychologists as too heavily influenced by mental health practitioners, whose perspective may on occasion be unduly oriented toward what their clients or patients accept, rather than toward the results of rigorous research (Pinker 1997). In 1988, a group initially calling itself the American Psychological Society—now the international Association for Psychological Science—sought to establish the field on a more scientific basis, but without the convergent quality of the Cognitive Science Society. Now that cognitive science is well established as a field, and the leaders of many nations have come to recognize the need for reform of many aspects of their healthcare systems, cognitive science could make a major contribution in the area of mental and emotional well-being, whether or not this is conceptualized as a convergence with clinical psychology and psychiatry.

Applied mental health fields tend to follow a *disease model* of the problems they face, except to some extent in dealing with cases of mental retardation and autism where a *disability model* also comes into play. Conceptualizing a mental problem as a disease asserts that proper treatment could return a person to normal, and thus that a clear definition of normal exists. Critics have called the excessive imposition of the disease model *medicalization* and have suggested that despite the great benefits medicine often can offer, many kinds of "cases" could better be conceptualized as a poor fit between the innate mental characteristics of a person and the expectations of the surrounding society (Conrad and Schneider 1980). Unfortunately, most problems faced by psychiatrists can be diagnosed currently only by listing behavioral symptoms, which is superficial, imprecise, and makes it difficult either to select the correct treatment from the range of remedies currently available or to develop new and more appropriate treatments.[1] Here, cognitive science may often be in a better position than psychiatry to develop really rigorous modes of diagnosis to identify the kinds of cases where a medical model is inappropriate, whether through the use of sophisticated brain scan methods or genetic testing, both of which combine information technology with biotechnology. Principles from cognitive science could better decide when efforts to help a person should concentrate on new assistive technologies, educational programs to give the person new skills that can compensate for an incurable disability, and perhaps even make adjustments in society's expectations.

By listing anthropology among its constituent fields, the Cognitive Science Society chiefly included a number of theory-oriented anthropologists, but that wing of cognitive science can play a key role now by reversing a divergence that occurred decades ago, when the *culture and personality* school in anthropology, and a subfield within it called *ethnopsychiatry*, faded from prominence. Anthropologists had observed that societies differed significantly—but certainly not completely—in what they defined as normal (Benedict 1934; Ackerknecht 1943). By the 1960s, a considerable body of literature in ethnopsychiatry had developed, some of it incorporating rather sophisticated cognitive theories (Opler 1959; Kaplan 1961). However, the significance of this approach faded, because it did not offer practical treatments to assist maladapted people in adjusting to the lives in which they find themselves. A defining feature of this approach is the *variance model* of mental and emotional problems that some people currently suffering greatly could make a very satisfactory adjustment in a different society that happens to value the innate characteristics they possess. To provide the best modern response to the full range of problems, the three models—disease, disability, or variance—must converge (e.g., the spectrum disorder model, Wurzman and Giordano 2012; Kendler and Parnas 2008). A competent diagnostician must be ready to select the right model for the

---

[1] See Haslam (2002) and associated references for examples of articles that discuss psychiatric taxonomy and the ontological assumptions of the medical model.

particular case: disease, disability, or variance. Cognitive science can play the central role in developing rigorous diagnostic methods, and perhaps in designing new social environments, that would be especially helpful for people whose problems fit either the disability model or the variance model.

6.1.3 The Vision for the Next Decade

Below the level of human cognition, but subserving it, is the neurobiology of the brain. The neurobiology of the human brain plays out in the micro-domain of neurons and the nano-domain of synapses. While nanotechnologies are maturing along the trajectory laid out in the "NBIC" process (see Chap. 4), the vision for how they will change society has been altered by the new discipline of neurotechnology, which seeks to build on new knowledge of the brain's natural neural code to create brain–machine interfaces far more powerful than what had once been thought possible. In particular, nanotechnologies offer the vision of bringing the power of the smartphone and the social network into direct physical contact with brains, in sharp contrast to the current touchscreen interfaces that we now use. Such hybridization or blurring of the distinction between machines and biology, formerly the stuff of science fiction, will in the future create very significant and new governance challenges for humans, while at the same time offering up new potential for an enhanced human response to global challenges. It may also be possible to construct neuromorphic machines that complement and subsequently emulate the functions of the brain.

Furthermore, deeper knowledge of natural brain processes will usher in the possibility of reprogramming some of that inherent biology to open up new vistas. Sensory channels (such as gustation) may acquire new additional cognitive "meaning" through the use of artificial sensor–neural communication systems (Bach-y-Rita et al. 2003). Neural code translation devices may make possible direct brain-to-brain communication channels in addition to the ones that have been used for millennia, that is, language and culture.

The development of sophisticated brain–machine interfaces that make full use of neural coding will increase human capacity. New information channels with greater bandwidth will be a central characteristic of the technological driver for capacity growth. As capacity grows for individuals, it will also be reflected at the societal level. New modes of communication that take advantage of the above-described advances will change the very nature of group decision-making processes. Almost certainly they will add an unprecedented level of transparency. Such changes will affect education, markets, and policies. There will be both opportunities and risks associated with such a sea change.

In essence, the vision for the next 10 years moves beyond the notion of connectome (the wiring blueprint for the brain) to a new concept: the "cognome", a blueprint for higher cognition in both individual and socially interacting brains (Horn 2002). This new concept of cognome in turn will lead to a new science of "mind", which will bring together disciplinary threads that include most of the physical and social sciences. In short, the vision for the next 10 years is for a transdisciplinary Renaissance in human cognition and social communication.

6.2 Advances in the Last Decade and Current Status

6.2.1 *Advances in Converging Technologies*

The first decade of the twenty-first century saw a number of significant advances in convergent technologies that affected human cognition, communication, and quality of life. The common denominator in these advances was not so much qualitative shifts to new technologies but rather the notion of ubiquity: formerly cutting-edge technologies became smaller, faster, and embedded in the day-to-day life of Americans, particularly in the five specific domains of cognitive science of science, computing/cloud/social networking, functional brain imaging, social "enabling" of big data, and widespread general access to "maker" technologies such as 3D printing.

Cognitive Science of Science

Scientists in different fields, and researchers with different dispositions or backgrounds within each field, use different modes of cognition to do their work, and cognitive science has begun to identify alternative ways of thinking scientifically. For example, one tradition in the artificial intelligence branch of cognitive science conceptualizes engineering problems in terms of a predefined design space, with each dimension representing a different class of alternative choices, and then uses rigorous computer methods to search the design space for one or more optimal solutions (Simon 1996). This method assumes that the design space is already well defined, so it does not prepare a researcher well for making entirely fresh discoveries, yet many scientists may use a similar rubric even in pure research, whether or not they employ computers.

A standard view in the philosophy of science is that formal hypotheses must be stated within a rigorous theoretical structure and tested empirically, although debates have raged about whether true hypotheses can be unambiguously confirmed, and how cognitive habits may bias the process of evaluation (Popper 1959; Klayman and Ha 1987). But many scientists—in fields as diverse as paleontology and linguistics—spend most of their time collecting and categorizing specimens rather than searching a design space or testing hypotheses. Many of the difficulties that members of the general public have in understanding science may stem not so much from ignorance as from the fact that the modes of cognition that evolved to serve everyday needs are very different from those required in science. For example, much of the resistance to the theory of evolution by natural selection from random variation may reflect the human propensity to think in terms of narrative stories centered on protagonists who seek goals, face challenges, and gather resources to help them (Abbott 2003). Even among scientists, this may bias thinking about evolution, instinctively seeing it as goal-oriented, and possibly deterring convergence between evolutionary biologists and scientists in other fields where the human propensity to think in terms of conscious goals is insufficiently counteracted by their training.

Cognitive science needs to take a new and comprehensive look at the varieties of scientific cognition, and as each variety is better understood, cognitive science should develop an increasingly rigorous map to chart their variations and potential convergence. Positive results could include:

- Better designs for the tools used by the scientists
- Improved education of future scientists
- Insights that can help practicing scientists understand their own mental processes
- New principles for convergence between sciences based on much better understanding of how the various different kinds of scientists think

Computing/Cloud/Social Networking

What used to be called "high-performance" computing in the 1980s is now present in hand-held computers used by a majority of U.S. consumers (Lunden 2012). This ubiquity has been made possible by advances in semiconductor technologies as projected by Moore's Law and underlying innovations in nanotechnologies, particularly with regard to device density improvements of computer chips (Fig. 6.1).

With the "smartphone" revolution has come the advance of cloud computing and the deployment of mobile social networking applications that enhance human communication, knowledge retrieval, and citizen journalism. Regime change on a multinational scale via the "Arab Spring" appeared to be enhanced considerably by these convergent technologies.

Functional Brain Imaging

The current gold standard for noninvasive imaging of human brain activity remains the same: functional magnetic resonance imaging (fMRI) (Fig. 6.2). The use of high-field magnets, sophisticated multichannel receiver coils, and advances in analytical imaging methodologies have all enhanced fMRI's sensitivity, yet the essential challenge of mismatch to the neural code, both spatially and temporally, remains.[2] What has changed in the past 10 years is the use of fMRI in cognitive research. This technology has come to

---

[2] fMRI detects functional neuronal activity with a spatial resolution of millimeters and a temporal resolution of seconds. By contrast, neurons have spatial dimensions on the order of microns and fire action potentials that last milliseconds.

6 Implications: Human Cognition and Communication and the Emergence...

Fig. 6.1 The evolution of computation: An early 1980s image-acquisition board and an iPhone 4 (Photo by J. Olds)

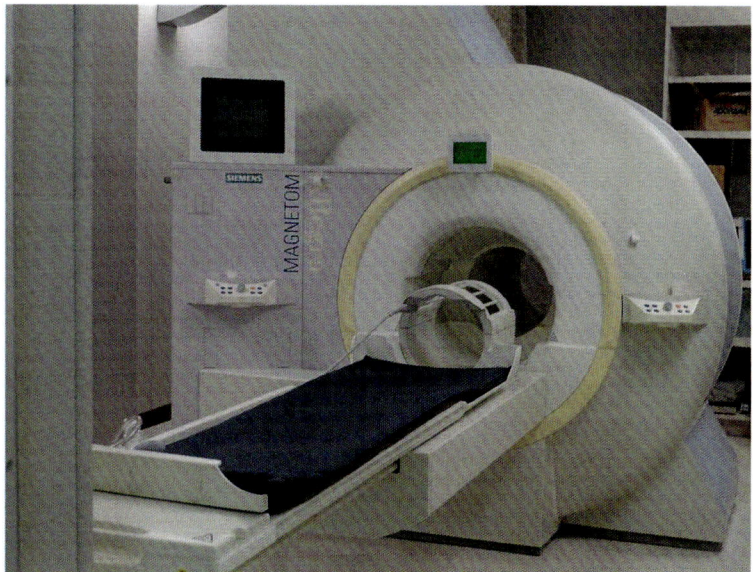

Fig. 6.2 Functional magnetic resonance imaging (fMRI) remains the most common method for imaging human brains as they think (Photo by J. Olds)

dominate cognitive and neurophysiological studies (Cabeza and Nyberg 1997, 2000; Rosen and Savloy 2012). The technology has also been widely adopted (potentially inappropriately) by the private sector in areas ranging from marketing to deception detection (e.g., see Farah and Wolpe 2004).

Social Enabling of Big Data/Social Networking

What has made Facebook a global phenomenon was not software innovation but rather its ability to engage users. As of 2012 Facebook had 500 million mobile users and 901 million monthly active users (SEC 2012). That engagement was compelling enough that users willingly enabled "Big Data" approaches to mining their personal information. A similar "value proposition," writ large, was also crucial to Google's success in the search domain—at least as important as its page rank algorithm. In May of 2011, Google had a U.S. audience of 155 million unique visitors (Machlis 2011).

What is the nature of that "value proposition"? At its most basic level, it is an exchange of information. Users exchange their own personal information (within a heterogeneous ecosystem of privacy protection schemes) for information about the world, including most importantly their own social networks.

The emergence of the above-described exchange has resulted in massively coupled social networks. These networks played central roles in the Arab Spring, the "Occupy" movement, and recovery efforts after the earthquake in Haiti. The field of Computational Social Sciences is a new discipline that has arisen, at least partly, because this "enabled" Big Data has become both available via public APIs (application programming interfaces) and amenable to computational approaches.

Widespread General Access to "Maker" Technologies

The "maker" phenomenon, a grassroots movement that represents a technology-based extension of the do-it-yourself culture, has been accelerated by the advent of inexpensive 3D printing, genomics, and microcontroller chips. While chip fabrication facilities (fabs) are still beyond the reach of the maker culture, a vast array of other technologies is now within the reach of the motivated lay public: a low-end 3D printer costs about $1,500 (Peck 2012). In a sense, science and engineering skills are being democratized in the same way that software engineering was several decades ago. The societal implications of this shift are not yet clear but promise to be profound.

6.2.2 *Relationship of Advances to the Tools for Converging Technologies*

Each of the above-described advances acts to enhance the impact of converging technologies on the American polity. Thus, while the societal value of these advances is still open to debate, they each act as an "amplifier". Taken together, the common denominator of these advances is to accelerate societal change.

6.3 Goals for the Next Decade

The overarching goal is to enrich individual human lives while at the same time sustaining the biosphere that humans share with all other living things.

6.3.1 Goal 1: Assistive Robotic Technologies in a New Context: Cognitive and Social Support

Robotics has the potential to impact our daily lives substantially in the next decade, although perhaps not in the ways we envisioned 10 years ago (Tapus et al. 2007). We typically have viewed the promise of robots as automated manual laborers, a vision that matches both their early capabilities (industrial automation) and fictional visions. The difficulties faced in developing these kinds of systems are well known; for example, perception is deceptively challenging, manipulation lacks flexible and compliant actuators and control algorithms, and planning requires both fine detail and extensive computational power. However, robots have the potential to offer other forms of support—cognitive, social, and behavioral support. This change in vision offers a substantial intersection for converging technology that can impact quality of life for many individuals. An important area of convergence in the next 20 years will be of robotic systems that (a) offer support for individuals to navigate the cognitively challenging society that the information revolution has produced; (b) enhance social support for individuals to allow for a more connected and more natural personal life experience; and (c) help to coach, train, and support healthy behavior and educational goals.

Embodied robotic systems that provide social and cognitive support to individuals have the potential to address a diverse range of populations and some of the most critical societal issues facing the world today. For example, consider the following three target areas:

1. *Aging populations*: It is estimated that in 2050 there will be three times more people over the age of 85 than there are today (UN/POPIN 2012). A significant portion of the aging population is expected to need physical and cognitive assistance. Yet, space and staff shortages at nursing homes and other care facilities are already an issue today. As the elderly population continues to grow, a great deal of attention and research will be dedicated to assistive systems aimed at promoting "aging in place," i.e., living independently in one's own home for as long as possible. Assistive robotic systems for the elderly, therefore, require technologies capable of being commanded through natural communication (e.g., speech, gestures), of fetching items, and of assisting with daily activities (e.g., dressing, feeding, moving independently).
2. *Early childhood education*: Early caretaker interaction with prekindergarten children helps to promote healthy life habits, raises the quality of the future work force, enhances the productivity of schools, and helps to reduce crime, teenage pregnancy, and welfare dependency (Campbell et al. 2002). While many studies

support the societal benefits of early childhood education, there is surprisingly little technological innovation enabling new and distributed forms of such training for young children, especially outside of traditional classroom settings or for children who need aid in areas not covered by typical classroom instruction. Furthermore, increases in class size, decreases in the availability of trained pre-K teachers, and the increasingly hectic pace of life for many parents have resulted in opportunities for technology to supplement existing formal and informal educational programs. Robotics technology has the potential to augment the skills of parents and educators rather than supplanting them. Technologies for tracking skill development, assessing skill competence, and individualizing behavior and habit shaping and instruction could have substantial benefits for this young population.

3. *Individuals with disabilities.* Individuals with cognitive disabilities, developmental disabilities, and social disorders constitute another growing population that can benefit from socially assistive robotics through special education, therapy, and training (Scassellati et al. 2012). Current research suggests that an average of 1 in every 110 children in the United States will be diagnosed with autism spectrum disorder (ASD) (DHHS/CDC 2012). Early intervention is critical for a positive long-term outcome, and many individuals with ASD need high levels of support throughout their lives (Volkmar et al. 2004). Robotics offers the potential for long-term supportive interactions that motivate individuals to maintain appropriate therapeutic activities, support human–human interactions through novel technological interfaces, and enrich human experience throughout an individual's lifespan.

One of the primary challenges in providing these kinds of technological support for social and cognitive tasks is the development of systems that adapt to the individual needs of their users. Just as there is no single educational approach or training technique that is guaranteed to succeed with all students, a single constant design for a cognitive support technology will fail if it cannot adjust to the unique needs, personality, and capabilities of each individual user. The converging technology solution to this requirement relies upon leveraging the collective information of many users, identifying commonalities among groups of users, and applying likely approaches from one user to "jump start" the interaction options for a new user, all while maintaining the privacy and identity of each of the users.

Over the longer-term, one might have a vision for assistive software/machine interfaces that assist individuals' ability to understand the world around them. Such interfaces might customize in real-time approaches to communicate scientific information. An application might be to communicate information about medicine and treatments. Such interfaces might also distill and synthesize Internet communications about an issue of importance (e.g. climate change) using words and concepts appropriate for the user. Analogous tools might help users better perceive the world around them (e.g., perception of the feelings of family members, friends, coworkers; of potential hazards—while driving, recreating, eating; of sounds and electromagnetic stimulus beyond normal human senses).

6.3.2 Goal 2: Cognome: A Theory-Based Rule Set for Understanding Human Cognition

By any account, humanity's rate of progress is breathtaking: in 1969 we celebrated mankind's first walk on the moon. Fifty years later, we could have billions of people "reaching into the cloud" to access a compendium of human knowledge via noninvasive mind–machine interfaces.

Today, we are standing at the threshold of producing paradigm-shifting discoveries by taking what we learn from basic sciences to address practical problems so as to drastically improve quality of life and augment personal capacity. However, an immense obstacle remains: we are surrounded by an exponential growth of processing units connected with the outside world (via web-enabled devices), yet we do not know how our own "smart device", the human mind, works. We are rapidly approaching an era in which the benefits of living in a highly technologized society will be put at risk unless we are able to understand how we, as a single individual or a group, process and retain information, make decisions, and perform actions. Getting to that point will require a much deeper understanding of the rule set that subserves decision-making. We term that rule set "cognome".

Cognitive technologies have started to inundate our everyday life: the Internet and wireless access put the world's information in the hand of the individual. The availability of mobile web-enabled devices—noninvasive artificial interfaces between a user's mind and the external world (e.g., smartphones)—is enabling everyone to accomplish tasks that used to take days or minutes in a fraction of the time. As web-enabled technology compresses time, there is zero lag time between an event's occurrence in the world, that information flowing to the world, and the world's response. At such a time scale, errors can also be corrected with no lag, which will revolutionize personal medicine, the global economy, and education, and force us to reconsider issues of national and international security.

But as technology and access to information have grown exponentially, our understanding of the human brain and the human mind has not. The wider the gap grows, the harder it will be to close, and the harder it will become to create the next technology wave. Hence, closing the gap is necessary in order to realize the potential utility of convergent technologies.

There are, however, ways forward. For instance, noninvasive intelligent mind–machine (and body–machine) interfaces will become increasingly beneficial when we know the code used by the brain to communicate with the external world—to improve both physical and mental capabilities. But it will be necessary to first significantly expand our understanding of how the brain interfaces with the world.

In the direction of devising assistive devices that can enhance human physical capabilities, a breakthrough is underway in a surgery-free method for treating blindness by Dr. Sheila Nirenberg (Nirenberg and Pandarinath 2012; Nirenberg et al. 2012). By cracking the retina's code—that is, the code the retina normally uses to communicate with the brain—Dr. Nirenberg designed a noninvasive neuroprosthetic that allows the restoration of quasi-normal vision for completely blind retinas, by producing normally coded signals of faces, landscapes, people walking, etc., that the brain can understand. Another breakthrough is underway in integrating human perceptual science with material engineering. Dr. Edward Adelson (Johnson and Adelson 2009) has developed a new tactile sensing technology able to sense the shape of the surface it touches with extremely high spatial resolution and with compliance similar to that of a human fingertip. Such a technology has direct applications for medical, robotic, and industrial domains where the mechanical properties of material or tissues touched needs to be recovered with fine detail and in a minimally invasive manner (e.g., robotic applications with sensitive gripping surfaces, surgery, brain–machine tactile interfaces, and wearable computing). With similar implications, recent advances in computational data sciences (Torralba et al. 2008), can allow smart devices to "see" better than we are currently able to, remembering massive amounts of information that we cannot store in our brain. Augmenting our memory capacity can potentially be achieved by improving the perceptual and cognitive capabilities of our smart devices so that they can "perceive" the world without our assistance and in sensory modalities that we do not naturally possess.

Considering the evolution of assistive devices that can learn from the human brain, our ability to build very-large-scale nanoelectronic systems will let us emulate functions of the brain. Further advances in nanomaterials and nanodevices enable such brain simulations to reach capabilities at the functional level. At that point, the notion of reverse-engineering the brain to find out how it works may become reality.

As perceptual, cognitive, and social sensors and technologies become increasingly ubiquitous, they will enable us to transcend current human limitations and improve life from birth until old age.

6.3.3 Goal 3: "My Own Genomics": Convergent Technologies to Enhance Wellness, Quality of Life

When considering human cognition, communication, and quality of life in converging technologies it is critical to include individual empowerment through tools and systems designed to allow such empowerment. Understanding the self at all levels, from the nanoscale level up through one's behavior within a society, requires access to data and the ability to interpret and comprehend those data. With the advent of low-cost technologies, citizen scientists with access to low-cost education can build tools to collect massive datasets and analyze them at the individual level. But better tools need to be created (Fig. 6.3) as well as an educational infrastructure supporting the minds that will build and utilize such tools, as described below.

There are numerous current examples of individual data easily available to the layperson and inexpensive or free tools for interpreting these data. For example, the 23andme Company (http://23andme.com) can provide a complete genotype given just a cheek swab. This can reveal health and ancestry information. Further, the data provided are not only of the individual's DNA results, but those results in the

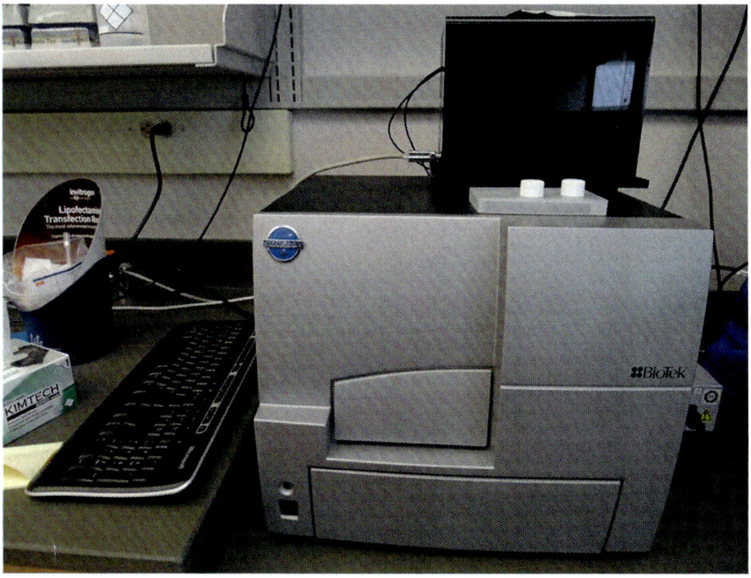

Fig. 6.3 Current gene sequencer; next version the size of a thumb drive? Current genomic technology is undergoing a rapid transition. Microfluidics and other nanotechnologies may radically reduce the footprint and cost for human genomics (Photo by J. Olds)

context of others; for example, a user might learn that s/he possesses a genetic variant linked to hemochromatosis. In addition, the context would also be provided, namely, that it is the most common genetic disease in the United States (Siddique and Kowdley 2012). Finally, that user would gain access to educational materials about genetics in general.

As an example of what may lie ahead in the near future, the SpikerBox technology has opened electrophysiology to citizen scientists. Created by Timothy Marzullo and Gregory Gage through Backyard Brains (2012; http://backyardbrains.com/), the device allows one to record single-neuron activity from an insect, using a smartphone as a data-visualization device. In the future, such do-it-yourself technology may be applicable to human brains using noninvasive sensors.

Another example is the personalized health record. Via smartphone apps and other devices, it is possible to record real-time health information, from heart-rate to skin cancer identification, from blood sugar readings. Although summary information of these and other vital signs is helpful for bringing to the attention of a doctor, this information will be most helpful in context, for example, in helping people to understand the relationships between differing foods and activities and their blood sugar levels and in exploring patterns of reactions to foods, hormones, and environmental stimuli. There are concerns about such data, from privacy and security issues to misinterpretation and medical errors (El Boghdady 2012). In this case, the ability to collect data has outstripped the ability of most users to analyze it. There are specialized programs for a few variables, but these data need to be examined in context with each other, in spatial context, in temporal context, and in social context.

The barriers to a lay understanding of combining genetics, current and historical vital statistics, and other individual variables are the same barriers as for scientists: not only are the data sets large, but one must have advanced knowledge in the many areas from genetics to behavior in order to comprehend and correctly interpret the data. Tools for analysis that make data more comprehensible to humans (Halford et al. 2005) should be created. An initial step is to enable analysis across converging domains. Interpretation will require extensive knowledge and logical skills by scientists capable of asking the right questions and comprehending the answers.

Training such scientists is currently an expensive process, for both the students and the educators. However, online courses may be one avenue toward reducing education costs for students and the costs of time on the part of instructors. Faculty members at prestigious universities have created online courses that allow them to have in a class hundreds of thousands of students from across the globe (Friedman 2012; see also Chaps. 4 and 8). Dissemination of knowledge on such a scale has literally never before been possible. It remains to be seen what forms of knowledge are best disseminated this way (i.e., can critical thinking skills and research methods be taught online as well as machine learning or algorithm design?). Further, as the United States generally lags in math and science education (WEF 2011), interventions will likely have to occur before the college level.

In conclusion, deep understanding and empowerment of the individual and society may come from assisting human cognition and communication through technological tools as well as educational opportunities across disciplines.

6.3.4 Goal 4: It's Time to Lead Again: CKTS Can Provide a Means and a Trajectory for National Renewal

With globalization has come a need for our nation to focus on maintaining and advancing its competitive edge in the world order. When we look to our assets for accomplishing this objective, convergent knowledge, technology, and society (CKTS) offers itself as a potential trajectory we can follow that provides both a means for sustaining economic growth and a resource for national renewal. The full power of the potential of convergence can only be manifested by addressing the wants and needs of stakeholders from all walks of life, perhaps best characterized as the broad population of the nation. In short, CKTS provides an opportunity to increase global leadership by improving quality of life on a large and demonstrable scale. How it does this and the pathways it takes will determine its success as a platform for providing heretofore unattained life benefits that transform our nation through richer lives that are characterized by both longevity and prosperity. Of course, for this to happen, society must view supporting such renewal as a critical investment.

When we look at human life span over the past century, we see a continuing increase in longevity. If we define quality of life in terms of numbers of years (quantity) of potential life, our quality has been increasing. Over these same years, however, we have come to gauge quality of life less by its quantity and more by other dimensions that relate to life's meaning and its enjoyment (e.g., MacGregor 2003). Living longer needs to be accompanied by living well and happily, extending not only the number of years we live but also the number of productive and socially engaged years we live (e.g., Veenhoven 1996). In addition, these longer life spans need to be experienced in terms of states of mind and emotion that are enriched by positive affect that promotes human flourishing (Fredrickson and Losada 2005). It is in this realm of human needs that CKTS offers real potential to improve quality of life through technologies and their implementation that accomplish objectives along the lines of the following:

- Extend the range of *productivity* of individuals to include what we now consider the retirement years, thereby providing not only income streams that supplement (or supplant) our current Social Security system, but also provide *meaningful* engagement with our society
- Increase levels of safety and security so that that normal decline of physical abilities is either lessened or ended in key areas, with a longer life being translated into longer independent living, perhaps through robotics, social networking, and computerized health maintenance
- Improve methods of wealth development and attainment that rely on large-scale computerized systems to improve personalized financial planning and that reduce sources of uncertainty in wealth accumulation and distribution
- Develop enhanced modeling of social milieus that provide both continued opportunities for social experiences and a meaningful contribution to people's emotional lives

6.4 Infrastructure Needs

To meet the promise of convergent technologies in the area of cognition, communication, and the quality of life, significant investments must be made over the next decade in the enabling infrastructure.

First, we must prime the pipeline of new scientists and engineers with the transdisciplinary training that is needed to understand these new convergent technologies and then exploit them. NSF's Integrative Graduate Education and Research Traineeship (IGERT) program is an excellent example of past successes. These types of investments must continue.

Second, the tools for understanding the human cognome must be developed. This includes noninvasive brain imaging better matched to the neural code, but it also encompasses new methodologies for interacting with ensembles of neurons—such as via optogenetics, transcranial direct current stimulation, and transcranial magnetic stimulation—and the ability to build large-scale nanoelectronic systems with sufficient power efficiency to enable large-scale simulations of the brain. These developments must be guided by theoretical advances in our understanding of the fundamental psychological and physiological principles that underlie behavior.

Third, the governance structures to support massive data-sharing between research groups must be developed. Not only will they enhance the economic efficiency of research and development, they also will make possible a great acceleration in the pace of deployment of convergent technologies.

6.5 R&D Strategies

Over the course of the next decade, significant sustained (or renewable) investments are needed in the following areas:

- *Research and development funding gaps that currently exist need to be filled.* There is a critical intersection of technology development that falls in between current funding coverage by the National Institutes of Health (NIH) and the National Science Foundation (NSF), that is, technology development for cognitive, behavioral, and social support. This includes research aimed at cognitive support for aging populations as well as support for individuals with disabilities and support for day-to-day activities of typical adults. While basic technology development (without evaluation) can often be funded via NSF, and clinical evaluation of an extant technology can be funded via NIH, it is often difficult to develop technology aimed specifically at a clinical application, because it cannot be both developed and evaluated under the same funding programs.
- *Researchers in academia need to be properly incentivized.* Truly interdisciplinary work can be difficult to conduct and maintain in an academic setting, for both cultural and funding reasons. While early career funding programs can be found in nearly every Federal funding agency, early career researchers are not well positioned to conduct interdisciplinary or integrative work. Tenure processes as well as expectations from departmental and professional organizations tend to favor disciplinary success, in part because interdisciplinary work can be difficult to evaluate. Investment in mid-career researchers (rather than in specific research projects) would boost exploratory and "high-risk, high-reward" investigations in converging technology and interdisciplinary areas. In short, we need more mid-career research awards.
- *Stakeholders* (*such as the large aging population*) *need to be properly involved* in the ethical, legal, and social issues, and given a voice to respond to proposed lines of convergent research.
- *Metrics of success need to be standardized across disciplines.* Incentives and rewards vary enough across disciplines that it discourages interdisciplinary work. As an example, in the humanities, a book is valued; in the social sciences, a journal article is valued; in computer science, a conference paper is valued: If these three types of academics work together on a project, who decides how it get published?

- *Participatory governance will be critical for an open-ended system.* No one can foresee the regulations that would be required by emerging technologies, so there must be another way than simply regulations to guide decisions.
- *Funding should lead a transdisciplinary project to develop a new "science of the mind"* aimed at understanding higher cognition and behavior (of humans and animals) on the basis of the biological activity of brains.
- *An international effort begun in 2007 aimed at evolving a science of the mind should be continued.* The complexity and scale of the problem requires the same "big science" approaches as those used in high-energy physics and polar exploration.
- *The White House neuroscience initiative should continue to encourage coordination and cooperation* across the agencies of government and the creation of public–private partnerships to accelerate progress in neuroscience, cognitive science, and related areas.

6.6 Conclusions and Priorities

The human brain has, throughout human history, interacted with other human brains and the biosphere with vast effect. The emergent phenomena of culture, history, agriculture, technology, and anthropogenic planetary changes are all the results of human brains interacting as agents. Now, convergent technologies are making it possible for human brains to interact with each other in entirely new ways that may well shape the future of human cognition, communication, and the quality of life (Carr 2008). Such a future would be characterized by a greater cognitive awareness as emerging from the processes described above. We call that future *The Cognitive Society*. The priority must be to shape this future to be positive, for individuals and also for society at large.

The four goals listed in Sect. 6.3 provide a roadmap for creating this positive future (Fig. 6.4): (1) Augmented and embodied cognition will enhance and support human interaction and understanding of society and planet. (2) The human cognome will enable a fuller understanding of our own human potential and limitations. (3) "My Own Genomics" will enhance human wellness as well as the aspects of our biology that play a role in our individual phenotype. Finally, (4) "It's Time to Lead Again" creates an agenda for national renewal that leverages convergent technologies towards a better national future.

Above all, the overarching vision is that a deeper understanding of human cognition will play a central role in how human beings live and how they interact, both in the biosphere and then later, perhaps, during sustained human space exploration, on spacecraft, and in other solar system locales. The idea that our interacting cognitive brains represent an "information field" that can be explored using the same scientific principles that we have used to explore other fields may have a deep significance for how humans collaborate to solve complex problems in the future. As such, the study of cognition and its related convergent technologies represents a grand opportunity for humanity.

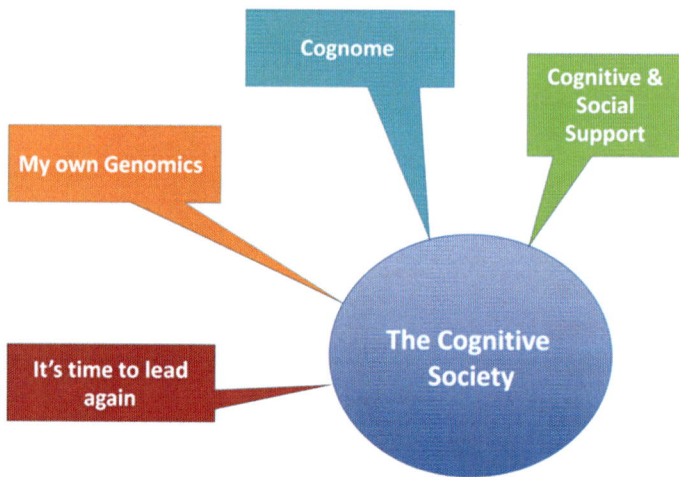

Fig. 6.4 Four goals, cognitive and social support, the cognome, "My Own Genomics", and "It's time to lead again" all play a role in creating the Cognitive Society (Figure courtesy of J. Olds)

6.7 R&D Impact on Society

Human society faces massive cognitive, communicative, and physical challenges over the next decades. Fundamentally, human decision processes, whether at the level of the individual or the collective, now affect the sustainability of the entire biosphere. The quality of human decision-making is both a function of natural human cognitive capabilities and the emergent effects of human-to-human interactions. The welfare of humans, in turn, feeds back onto human cognitive capabilities and hence to human decision-making. Thus, the converging technologies that affect human cognition, communication, and quality of life have very significant broad societal implications for the future.

As the new generation of convergent technologies becomes fully embedded within the science of cognition, it is inevitable that our own view of what it means to be a human being will change fundamentally (e.g., see Giordano 2012). This change in view will be a central characteristic of *The Cognitive Society*. At an obvious level, the boundary between brain and machine will become increasingly blurred as machines are increasingly made out of materials that mimic those of the brain. At a deeper level, as humans become increasingly more deeply connected to one another, the notion of "self" may evolve in as yet unpredictable ways that could have profound effects on society. The ethical, social, and legal implications for such trajectories in human development will be extraordinarily important for serious consideration—by scientists, policymakers, and the larger global human polity.

6.8 Examples of Achievements and Convergence Paradigm Shifts

6.8.1 Possibilities for a Cognitive Society Initiative

Contact person: James Olds, George Mason University

The Problem and Background

In the first two decades of the twenty-first century, the general recognition of the brain as the main engine of human behavior and thought has converged with advances in cerebral sensor–activator technology. This convergence has resulted in a now-ubiquitous technological ecosystem whereby the relationship between functional brain activity and human behavior can be routinely studied, not in some animal model, but in conscious human subjects. This new ecosystem of tools has qualitatively changed how we view cognition as a society.

In future decades, as this convergence accelerates innovation cycles, it can be expected that a much deeper knowledge of the human cognition will be elucidated. This deeper understanding will be enabled by the same technological advances (e.g., Moore's Law) that are driving Big Data and the various "-omics" fields. At the same time, brain sensor–activator technologies will improve to the point where they are better matched to the actual spatial–temporal dynamics of the human brain neural code. These sensor–activator technologies will become increasingly integrated into biological brain tissue as brain–machine interfaces become both smaller and more sophisticated. Jointly, we term these "cognitively enabled technologies." At some point soon, human societies, including our own, will begin to consider deploying this new knowledge and these new convergent technologies beyond the simple goal of creating new brain knowledge. This emergence will be catalyzed by the complexity and span of human challenges, ranging from public health to climate change.

The sustained societal conversation about the ethical, legal, and social issues related to future use of cognitively enabled technologies and knowledge will constitute the early stages of what we term The Cognitive Society. A cognitive society is a human society that engages with deep knowledge about human cognition and brain sensor–activator technologies in order to more fully realize its potential. The Cognitive Society Initiative ("SCI-C") is a program designed to assure that the arrival of a cognitive society here in the United States is both accelerated and a positive development.

What Is Proposed

It is proposed that SCI-C begin the process of facilitating and understanding how twenty-first century America will be transformed as it becomes a truly cognitive society. SCI-C will engage with scientists, ethicists, and philosophers across many disciplines—it will be transdisciplinary. SCI-C will also reach out to members of the lay public and their elected representatives so that societal consensus will emerge from SCI-C activities. The SCI-C will have three major scopes:

1. Access and forecast trends for the embedding of cognitive knowledge and technologies into American society
2. Fund transdisciplinary research designed to enhance this embedding
3. Consider the ethical, legal, and social issues

In (1), SCI-C will bring together scientists, technologists, and futurists to perform a cognitive forecast for American decision-makers. These activities will facilitate situational awareness about the processes by which cognitively enabled technologies are becoming embedded in current-day America while at the same time establishing a variety of forecasting methodologies to elucidate likely futures. The forecasting function will be dually focused on both challenges and opportunities.

In (2), SCI-C supporting agencies will come together to fund projects that can catalyze the enhancement of American potential through advances in either cognitive knowledge (e.g., hierarchal linked frameworks for brain architectures) or brain sensor–activator technologies. Such research funding will require cross-cutting agency support and will be designed to fill the "gaps" between the classical funder portfolios.

In (3), SCI-C will convene members of the lay public, ethicists, politicians, philosophers, scientists, technologists, and decision-makers at all levels to engage with the ethical, legal, and social issues regarding the migration of cognitively enhanced technologies into the "wild."

Rationale

The rationale for SCI-C stems from the challenges presented by the already ongoing rapid introduction of convergent technologies related to cognition into American society. These current technologies include functional brain imaging, transcranial brain stimulation, and brain–machine interfaces. Their successors may include human optogenetics, nano-level computing, and advanced robotics. Because individual human experience and all human social interactions are driven by cognitive processes, the introduction of these technologies could pose a systemic risk to society. A primary aim of SCI-C is to avoid unintended consequences, or tripping hazards. As convergent technologies accelerate the innovation cycle, this risk will only grow. Rather than respond reactively, SCI-C is a proactive approach to addressing an American future as a cognitive society.

At the same time, SCI-C represents a national opportunity to both shape and accelerate the implantation of these cognitive technologies in such a way as to optimize their potential for positive impact on American society. Whether it be by enhancing social communication capabilities or ameliorating the effects of traumatic brain injury through novel brain–machine interfaces, SCI-C can play a crucial role in catalyzing America's worldwide leadership as a cognitive society.

6 Implications: Human Cognition and Communication and the Emergence... 243

Deliverables

There are several key objectives for SCI-C:

- *Marked improvement in brain health with a resultant public health and economic benefit for the United States.* Brain diseases represent an enormous drag on national well-being. This drag will only increase as the Baby Boomer generation enters retirement and becomes vulnerable to chronic neurodegenerative disorders such as Alzheimer's or Parkinson's disease. Central to SCI-C initiatives are the development of new therapies, technologies, and interventions to reduce this public health and economic load on American society.
- *Deployment of cognitively enabled technologies to improve the well-being of seniors so that productive lifespan increases significantly.* As America ages, life can retain meaning and productivity for seniors with the embedding of assistive technologies. Many of these technologies (e.g., assistive robots) represent the convergence of cognitively enabled machines or software with sensor–activators. SCI-C investment in these technologies will create concrete opportunities for the enhancement of geriatric care so that a greater proportion of the American life-span will be spent enjoying good health.
- *Improved outcomes for K–20 education (and beyond) as a result of embedding cognitively enabled technologies in our schools and colleges.* Learning across the human lifespan constitutes one of the most obvious cognitive activities, and yet currently there has been little movement for educators to take advantage of what cognitively enabled technologies may have to offer. SCI-C research investments will seek to catalyze the use of such technologies in lifespan learning to enhance the competitive edge of American citizens as they compete in a globalized world.
- *Avoidance of strategic surprise over the next decade in the area of cognitive enabled technologies.* Formal cognitive forecasts that engage with ethical, legal, and social implications will be performed by SCI-C supported researchers. The lay public and decision-makers at all levels of government will also play key roles in this process. Success will be measured by the discovery and societal avoidance of tripping hazards related to the transition of the United States to a cognitive society.

6.8.2 Possibilities for a Cognitive Technology Initiative

Contact person: *R. Stanley Williams, Hewlett-Packard Laboratories*

Advancing the nascent discipline of cognitive science into engineering and development practices from which economically and societally important technology can be launched will require a concerted transdisciplinary effort. A successful model for achieving critical mass in an emerging field and applying a wide range of expertise to create commercial outlets has been established by the U.S. National Nanotechnology Initiative (NNI).

Cognition is now recognized as a critical component of future applications ranging from healthcare to information technology. Just one example is that information

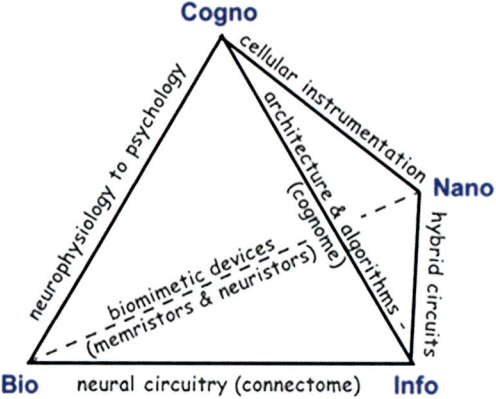

Fig. 6.5 The cognohedron (Adapted from R. W. Leland, Sandia National Laboratory, with permission) illustrates the nano-bio-info-cogno (NBIC) convergence to address the challenge of understanding how cognition arises in biological systems and how that understanding can be applied to advances in human health, economic competitiveness, and national security

analytics is moving from mainly arithmetical operations on highly structured databases to finding meaning in vast quantities of unstructured data. This requires a new computation paradigm in which cognitive systems and algorithms recognize context and intent without having to be programmed by experts in arcane software languages, but rather learn in an uncertain and changing environment. The one successful example we know that is capable of such analysis is the human brain. However, at this point, cognitive science, which includes quantitative psychology, is a highly specialized research field that is mainly focused on fundamental understanding of an incredibly complex system. To bring the insights developed from cognition to information technology (IT) and other fields will require encouragement in the form of research funding that is specifically targeted at transdisciplinary teams (see Fig. 6.5). Given the realities of the funding of today's disciplinary silos and agencies, the only way to break down the barriers to collaboration is to fund an effort that is specifically tasked to dramatically advance cognitive technology through focused research programs.

What would such an effort look like? Taking the successful example of the NNI, we understand that there are two important components for accelerating a field and encouraging applications.

The first critical component of a focused cognitive technology initiative is targeted research funding for transdisciplinary groups—which may include participants from academia, national laboratories, and commercial entities—to address fundamental discovery, national needs, and economic development. The primary goal here is to address the "Pascal quadrant" of research with specific goals—in the IT example, building systems based on new hardware and software that address real-world problems far more efficiently than present computing platforms by emulating models of brain function. At this stage, the two major U.S. institutions with portfolios related to cognitive technology are NIH and NSF, but the Department of Energy (DOE), the Department of Defense (DOD), and the intelligence communities are major stakeholders in the outcomes of such efforts, so coordination in funding efforts across these agencies is critical to success. However, new methods of encouraging collaborative efforts should be explored. An example could be to use expert crowd sourcing to identify, prioritize, and fund some research programs. A sponsoring agency would host a website on which researchers describe their expertise and recent results. Other researchers could then propose projects that would advance the state of the art in a dramatic way if particular experiments could be performed and/or theories be developed. The participants in such a forum could vote on which ideas they believe are most important, and motivated research groups from appropriate institutions could sign up to obtain starter grants to deliver tangible results based on predetermined statements of work. Those groups that deliver high-quality results, as determined by the research community, would then receive longer-term grants to complete a project.

The second critical component of a focused cognitive technology research initiative is establishment of a core research and user facility that can act as a center of excellence in transdisciplinary research, builder of prototype systems, first user of the technology that is created, and supplier of the technology and facilities to individual investigators until commercial markets are established (similar to a DOE Nanoscale Science Research Center). Such a center (which can be geographically distributed) should be hosted by an organization that has a strong foundation in a broad range of relevant disciplines, for example, expert capabilities in cognitive science; computer architecture; algorithms and modeling; facilities for designing, building, and testing instrumentation and hardware; and ability to host researchers from all the required disciplines to interact with each other to advance the field. A center can provide a focus for the grand challenges faced by the cognitive science community, such as the question of multiscaling—how to measure and model critical aspects of cognition from the ion channel and synapse to memory and learning. The tools for probing brain function are now developed in academic research labs on an individual basis—everything from reporter nanoparticles to multiwire electrical probes to miniature wearable microscopes to functional MRI. What is needed is serious engineering development of some of these tools to standardize and integrate them into systems so that results from multiple scales can be correlated to specific behavior.

Just as important are theoretical models of what happens at each scale in the brain, how these models can be connected together to provide a unified picture of function across scale, and most importantly, how the models compare with the multiscale experimental results. For example, one of the more intriguing models of memory involves sparse coding: What lower-level models of neuron function can provide additional insight to the mechanism of long-term memory storage, and what experimental measurements are required to confirm or refute the model? The relevant comparison is to the multiscale modeling painstakingly developed by the nanotechnology community over the past dozen years to bridge the length scales from atoms to nanoparticles, to composite materials, to aircraft fuselages. Development of multiscaling tools and theories is greatly advanced by providing smaller-scale research groups the opportunity to interact with user communities and centers; however, the primary focus of the centers should be the support of, rather than replacement of, the individual research groups.

What are the desired outcomes? The long-range goals for any national-scale initiative must be measured in terms of dramatic improvements in the lives of the people who support the initiative through their taxes. Three significant areas will be impacted by cognitive convergence:

- *Health.* At present, there are a large number of brain-based diseases that afflict people, and they are becoming more prevalent as the population ages. Greatly improved understanding of human cognition and preventing the biological factors that degrade it will mean that increased quantity of life can also translate into increased quality of life and productivity.
- *Security.* As our world both shrinks and becomes more complex, the number and severity of threats to our security are increasing dramatically. We need cognitive tools that can exponentially improve our ability to sense our cyber and physical environments, and enable us to respond to threats before they become incidents.
- *Economic opportunity.* We now live in a global technological society, where the most important economic assets of a nation are the inventiveness and productivity of the people. The countries that successfully understand and harness cognition to amplify their human assets will be far more competitive than those that do not. In a world of exponential technological advances, not being the leader risks becoming irrelevant.

6.8.3 List of Examples

Major achievements in addressing issues of cognition, communication, and human quality of life are illustrated in Table 6.1.

6.9 International Perspectives

The following are summaries relevant to this chapter of discussions at the international regional WTEC NBIC2 workshops held in Leuven, Belgium, September 20–21, 2012; in Seoul, Korea, October 15–16, 2012; and in Beijing, China, October 18–19, 2012. Further details of those workshops are provided in Appendix A.

6.9.1 United States–European Union NBIC2 Workshop (Leuven, Belgium)

Facilitator and Rapporteur: *Laura Ballerini*

Discussants:

Milos Nesladek, Academy of Sciences of the Czech Republic (EU)
Mira Kalish, Tel Aviv University (EU)
Sylvie Rousset, CNRS (EU)
Jian Cao, Northwestern University (U.S.)
James Olds, George Mason University (U.S.)
Mark Lundstrom, Purdue University (U.S.)
Christos Tomakanis, EC (EU)

6 Implications: Human Cognition and Communication and the Emergence... 247

Table 6.1 Major achievements in the areas of cognition, communication, and human well-being

CKTS Domain	Example
Communications	*Smartphone revolution*: The release of Apple's iPhone in 2007 brought about a qualitative change in the use of mobile computing across the globe
Quality of Life	*Personal genomics*: The massive reduction in the marginal cost of DNA base sequencing has brought whole genome analysis within the reach of ordinary citizens
Cognition	*Transcranial magnetic stimulation*: The ability to noninvasively activate (and inhibit) specific human brain areas has opened up new avenues for neural therapies
Quality of Life	"*Maker tools*": New additive manufacturing technologies are profoundly affecting innovation and the ability of empowered individuals to bring new intellectual property to market
Communications	*Twitter*: The searchable and real-time nature of Twitter has altered the definition of who is a journalist and greatly reduced the importance of the 24-h news cycle. Real-time twitter feeds played a major role in the Arab Spring and have the potential to have major political influence in the future
Communications	*Facebook*: With over 500 million users, Facebook represents one of the most ubiquitous examples of how the convergence of computing and social networking has changed the world
Cognition	*Brain imaging technologies*: The embedding of brain imaging technologies into medicine has brought new diagnostic tools and therapies to the fore

This group of scientists assessed the role of convergences in nano-bio-info-cognitive science on human development. Consensus emerged around three main topics:

1. Use of molecular and personalized information to enhance treatment of disease and improve human cognition
2. Quality-of-life enhancement, including prosthetic devices and regenerative medicine
3. Personalized innovative education

In the area of molecular and personalized information to enhance treatment and cognition, the core idea was that the technology for developing dynamic and personalized molecular profiles for medical treatments will emerge over the next 10 years. Diseases that might be better treated include cancer, neurodegenerative illness, and spinal lesion. The convergent technology tools that will make this possible include:

- Data optimization (in the sense of innovative and enhanced data harvesting, sharing, analyzing)
- Improvement of technological tools for information management at the molecular level (genome, proteome, bioinformatics, contactless tools, biomarkers, sensors, cellular imaging)
- Tools for improving accuracy and efficacy towards predictive diagnosis
- Cost-effectiveness (including personalized manufacturing of drugs)
- A distributed system for delivery of care

Thinking about enhancement of quality of life, the central idea was engineering "of" and "for" the human body, including human bionic machine/organ interfaces as new frontiers in personalized regenerative medicine. The convergent technologies that are emerging to make this possible include:

- Advances in nanotechnology (including nanomaterials and nanoelectronics)
- Artificial restoration of functions
- Development of multidirectional therapeutic interfaces
- Improved knowledge of intracellular signal transduction codes
- Addressing cognitive enhancement
- Restoring function to restore human productivity

Turning to personalized innovative education, the group saw that the next decade will bring a paradigm shift in addressing formal and informal education with adaptive life-long learning systems. The convergent technology tools that will make this possible include:

- Targeting aging needs for continuous education through informal pedagogical processes
- Micro-education learning platforms based on personal needs ("I teach you in your own personal learning style")
- Adaptability of workstations to the needs of each individual in terms of a human–machine "relationship"
- Haptic platforms for text-based sensory motor feedback
- Connecting informal with formal education via science centers on convergent technologies
- Reframing formal education systems to promote convergence of disciplines by mutual enabling—improving disciplines' awareness of their limits and needs
- Enabling "smart" information distribution channels using synergies with public media
- Developing a market for targeted personal education

Cross-cutting across each of these topics was the notion that governance is critical for enabling the technological convergence necessary for societal improvement in each of the above three topics.

The joint NSF–EC workshop held in Leuven also highlighted that convergence can provide new job opportunities for a competitive and sustainable economy by promoting responsible innovation, which is an increasingly important challenge for diverse policies in the EU, United States, and beyond. Promoting *new skills for "scientific social responsibility"* could contribute to progress towards this goal. Research, innovation, and education are a living ecosystem, where education contributes to bridging the gap between research and innovation. So, there is a need to enable the next generation of scientists to meet the needs of industry by developing the appropriate skills and competencies. In this context, society at large is showing a progressive need to better understand the pros and cons of converging technologies so as to make informed responsible choices about their new products and processes, while young people are becoming increasingly inclined to engage in studies on converging technologies. This requires adapting both education and training by developing inter- and cross-disciplinary education to develop an inter- and cross-disciplinary "*scientific social responsibility*" scheme for future scientists and technologists, by integrating generics (e.g., nanosciences, nanomoleculars, nanoengineering, nanophotonics) with specifics (e.g., nanoelectronics, nanomaterials, nanobiotechnology, nanomedicine, ethicalities, safety, standards, regulation, entrepreneurship). This can be done by conceiving, designing, and developing new ways to promote paradigm shifts by addressing formal and informal education with adaptive life-long learning systems; disseminating and developing versatile learning/teaching materials for multipliers (e.g., teachers, trainers, science communicators, explainers); including innovative learning methods and scientific experimentation; and designing new formats for staff exchanges between labs, academia, and industry.

6.9.2 United States–Korea–Japan NBIC2 Workshop (Seoul, Korea)

Panel members/discussants

Myung Joon Kim, Electronics and Telecommunications Research Institute (ETRI, Korea)
Mitsuo Kawato, ATR Brain Information Communication Research Laboratory (Japan)
James Olds, George Mason University (U.S.)
H.-S. Philip Wong, Stanford University (U.S.)
Byoung-Tak Zhang, Seoul National University (Korea)
Young Jik Lee, ETRI (Korea)
Jon-Won Lee, Hanyang University (Korea)
Tsuyoshi Hasegawa, National Institute for Materials Science (NIMS, Japan)
Takanori Ichiki, Tokyo University (Japan)
Takeshi Kawano, Toyohashi University of Technology (Japan)
Shinya Nakamoto, Japan Science and Technology Agency (JST, Japan)
Takahiro Fujita, NIMS (Japan)
Wan Seok Kim, ETRI (Korea)
Hong Soon Nam, ETRI (Korea)
Young-Jae Lim, ETRI (Korea)

The group reached consensus around five core ideas:

- The concept of a "Trust Society" (e.g., computer-enhanced privacy)
- The concept of the Aging Society and the need to provide a meaningful and healthy aging experience
- The notion of bricks-and-mortar "convergence centers" bringing multidiscipline investigators together under one roof

- The coming ubiquity of brain–machine interfaces to cure brain diseases
- The utility of national buy-in to support NBIC technologies through policies and grant support

The team saw both risks and the potential for enhanced well-being from the advent of convergent technologies. Specifically, the team noted the potential for the systematic erosion of privacy as a result of Big Data and the commoditization of computing. With those risks in mind, the team advocated for the development of a Trust Society built upon novel CKTS tools that act to enhance privacy, such as novel easy-to-use encryption schemes.

There was also the general agreement that all of the participating nations (Korea, Japan, and the United States) face an increasingly aging population as a result of demographic shifts. This realization led to consensus around the notion of an Aging Society—where CKTS is deployed from NBIC technologies to not only enhance health but also a sense of life having real meaning during the last decades of the human lifespan.

The team learned about Seoul National University's Advanced Institute of Convergent Technology (AICT) that brings together NBIC researchers under a single, non-virtual "roof." The institute, formally supported by both national and provincial governments, serves as an excellent model for promoting the types of paradigmatic changes so critical to cognitive well-being and communication. There was strong consensus that such examples should be facilitated elsewhere.

It was also felt that brain–machine interfaces were on a trajectory to become ubiquitous (and therefore embedded) in society in such a way as to completely change the nature of both cognition and communication between individual humans and machine artefacts. The team felt that, if pursued in a careful manner that is sensitive to the relevant legal, ethical, and social issues, such interfaces would be very positive.

Finally, there was a realization among team members that a formal recognition of the positive role for CKTS in promoting a healthy society would be a good thing. The Korean members of the team brought up the idea of formal legislative support, and such a notion was endorsed by the group.

6.9.3 United States–China–Australia–India NBIC2 Workshop (Beijing, China)

Panel members/discussants

Tingshao Zhu, Academy of Sciences (China)
Jonathan Manton, University of Melbourne (Australia)
James Olds, George Mason University (U.S.)
Tanya Monro, University of Adelaide (Australia)
Xioming Wang (China)
H.-S. Phillip Wong, Stanford University (U.S.)
Tianzi Jiang, Academy of Sciences (China)

This group came together over a shared vision for how NBIC advances might improve human cognition and communication in the next decade. That vision included:

- An increased focus on educating the public about mental health issues
- Leveraging communications and wearable devices to improve decision-making and patient management
- Increasing recognition of the *Aging Society*, supported by technologies and services to enable people to remain in their communities
- Novel approaches to diagnosis and therapy of psychiatric and neurological diseases
- Transdisciplinary focus
- Open access to learning

Scientists and research agency directors from China, Australia, and the United States discussed NBIC in the context of cognition and communications. Consensus was reached over the following achievable goals for the next decade:

- Embedding of transdisciplinary education into research and vice versa, leading to improved health outcomes
- Establishing the physical underpinning of cognition, i.e., understanding how cognition emerges from the chemical, physical, and biological activities of brains
- Wearable devices for personalized medical services and monitoring
- Reduced cost and increased efficiency of access to medical services
- Reduced incidence, medicalization, and impact of autistic spectrum disorders and other cognitive challenges in children

The group felt that the following R&D strategies would be critical towards achieving the above goals:

- Investing in programs that address "bigger" scientific problems
- A significant increase in prioritization of government support
- Educating government and political decision-makers about NBIC scientific priorities
- Application-oriented R&D targeted towards real societal problems
- Filling interagency gaps
- Creating a new "brand" for NBIC science in the area of cognition and communication
- Improved communication to the general public about NBIC science.

References

Abbott, H.P.: Unnarratable knowledge: the difficulty of understanding evolution by natural selection. In: Herman, D. (ed.) Narrative Theory and the Cognitive Sciences, pp. 143–162. Center for the Study of Language and Information, Stanford (2003)

Ackerknecht, E.H.: Psychopathology, primitive medicine and primitive culture. Bull. Hist. Med. **14**, 30–67 (1943)

Adamson, G.: Socially beneficial technology: can it be achieved in practice? IEEE Technol. Soc. Mag. (Summer): 20–27 (2012)

Bach-y-Rita, P., Tyler, M.E., Kaczmarek, K.A.: Seeing with the brain. Int. J. Hum. Comput. Interact. **15**, 285–295 (2003)

Benedict, R.: Anthropology and the abnormal. J. Gen. Psychol. **10**, 59–80 (1934)

Cabeza, R., Nyberg, L.: Imaging cognition: an empirical review of PET studies with normal subjects. J. Cogn. Neurosci. **9**(1), 1–26 (1997)

Cabeza, R., Nyberg, L.: Imaging cognition II: an empirical review of 275 PET and fMRI studies. J. Cogn. Neurosci. **12**(1), 1–47 (2000)

Campbell, F.A., Ramey, C.T., Pungello, E., Sparling, J., Miller-Johnson, S.: Early childhood education: young adult outcomes from the Abecedarian Project. Appl. Dev. Sci. **6**(1), 42–57 (2002)

Carr, N.: Is Google making us stupid? Atlantic Magazine (July/August). Available online:http://www.theatlantic.com/magazine/archive/2008/07/is-google-making-us-stupid/306868/ (2008)

Conrad, P., Schneider, J.W.: Deviance and medicalization. Mosby, St. Louis (1980)

DHHS/CDC (Department of Health and Human Services, Centers for Disease Control and Prevention, National Center of Birth Defects and Developmental Disabilities). How many children have autism? (based on 2006 statistics). Available online: http://www.cdc.gov/ncbddd/features/counting-autism.html (2012). Accessed 7 Sept 2012

El Boghdady, D.: Health-care apps for smartphones pit FDA against tech industry. *The Washington Post* Business section. June 22 (2012)

Farah, M.J., Wolpe, P.R.: Monitoring and manipulating brain function: new neuroscience technologies and their ethical implications. Hast. Cent. Rep. **34**(3), 35–45 (2004). Published online 6 March 2012: http://onlinelibrary.wiley.com/doi/10.2307/3528418/full

Fredrickson, B.L., Losada, M.: Positive affect and the complex dynamics of human flourishing. Am. Psychol. **60**, 678–686 (2005)

Friedman, T.L.: Come the revolution. *The New York Times* Op-Ed, 15 May 2012

Giordano, J. (ed.): Neurotechnology: premises, potential and problems. CRC Press, Boca Raton (2012)

Halford, G.S., Baker, R., McCredden, J.E., Bain, J.D.: How many variables can humans process? Psychol. Sci. **16**(1), 70–76 (2005)

Haslam, N.: Kinds of kinds: a conceptual taxonomy of psychiatric categories. Philos. Psychiatr. Psychol. **9**(3), 203–217 (2002). http://dx.doi.org/10.1353/ppp.2003.0043

Horn, R.E.: Beginning to conceptualize the human cognome project. A paper prepared for the National Science Foundation Conference on Converging Technologies. Available online: http://www.stanford.edu/~rhorn/a/topic/cognom/artclCncptlzHumnCognome.pdf (2002)

Johnson, M.K., Adelson, E.H.: Retrographic sensing for the measurement of surface texture and shape. In *Proceedings, IEEE Conference on Computer Vision and Pattern Recognition* 1070–1077 (2009)

Kaplan, B. (ed.): Studying Personality Cross-Culturally. Harper and Row, New York (1961)

Kendler, K.S., Parnas, J. (eds.): Philosophical Issues in Psychiatry: Explanation, Phenomenology, and Nosology. Johns Hopkins University Press, Baltimore (2008)

Klayman, J., Ha, Y.-W.: Confirmation, disconfirmation, and information in hypothesis testing. Psychol. Rev. **94**(2), 211–228 (1987)

Lunden, I.: Nielsen: Smartphones used by 50.4% of U.S. consumers, Android 48/5% of them. AOL Tech online: http://techcrunch.com/2012/05/07/nielsen-smartphones-used-by-50-4-of-u-s-consumers-android-48-5-of-them/ (2012)

Machlis, S.: Google vs. Facebook by the numbers. Computerworld (July 7). Online: http://www.computerworld.com/s/article/9218177/Google_vs._Facebook_by_the_numbers (2011)

MacGregor, D.G.: Psychology, meaning and the challenges of longevity. Futures **35**, 575–588 (2003)

Marzullo, T., Gage, G.: The SpikerBox: a low cost, open-source bioamplifier for increasing public participation in neuroscience inquiry. PLoS ONE **7**(3), e30837 (2012). http://dx.doi.org/10.1371/journal.pone.0030837

Nirenberg, S., Pandarinath, C.: Retinal prosthetic strategy with the capacity to restore normal vision. PNAS Published online before print, August 13. http://dx.doi.org/10.1073/pnas.1207035109 (2012)

Nirenberg, S., Pandarinath, C., Ohiohenuan, L.: Retina prosthesis. WO Patent WO/2012/030625 (filed under the international Patent Cooperation Treaty) (2012)

Opler, M.K. (ed.): Culture and Mental Health. Macmillan, New York (1959)

Otellini, P.: As quoted by M.J. Miller. Intel enters smartphone chip race for real. PC Magazine, Forward Thinking, online: http://forwardthinking.pcmag.com/ces/292745-intel-enters-smartphone-chip-race-for-real (2012)

Peck, M.E.: 3-D printing revolution stymied by high prices. Txchnologist (August 27). Online: http://txchnologist.com/post/30829846467/3-d-printing-revolution-stymied-by-high-prices (2012)

Pinker, S.: How the Mind Works. Norton, New York (1997)

Popper, K.R.: The Logic of Scientific Discovery. Basic Books, New York (1959)

Rosen, B.R., Savloy, R.I.: fMRI at 20: has it changed the world? Neuroimage **62**(2), 1316–1324 (2012)

Scassellati, B., Admoni, H., Matarić, M.: Robots for use in autism research. Annu. Rev. Biomed. Eng. **14**, 275–294 (2012). http://dx.doi.org/10.1146/annurev-bioeng-071811-150036

SEC (U.S. Securities and Exchange Commission): Amendment Number 4 to Form S-1 Registration Statement: Facebook, April 2012: http://www.sec.gov/Archives/edgar/data/1326801/000119312512175673/d287954ds1a.htm (2012)

Siddique, A., Kowdley, K.V.: Review article: the iron overload syndromes. Aliment. Pharmacol. Ther. **35**(8), 876–893 (2012). http://dx.doi.org/10.1111/j.1365-2036.2012.05051.x

Simon, H.A.: The sciences of the artificial. MIT Press, Cambridge (1996)

Tapus, A., Matarić, M., Scassellati, B.: The grand challenges in socially assistive robotics. IEEE Robot. Autom. Mag. **4**(1), 35–42 (2007)

Torralba, A., Fergus, R., Freeman, W.T.: 80 million tiny images: a large dataset for non-parametric object and scene recognition. IEEE Trans. Pattern Anal. Mach. Intell. **30**, 1958–1971 (2008)

UN/POPIN (United Nations Population Information Network, United Nations Population Division, Department of Economic and Social Affairs). United Nations Statistical Yearbook. http://www.un.org/popin/, http://unstats.un.org/unsd/syb/ (2012). Accessed 7 Sept 2012

Veenhoven, R.: Happy life-expectancy: a comprehensive measure of quality-of-life in nations. Soc. Indic. Res. **39**, 1–58 (1996)

Volkmar, F.R., Lord, C., Bailey, A., Schultz, R.T., Klin, A.: Autism and pervasive developmental disorders. J. Child Psychol. Psychiatry **45**(1), 1–36 (2004)

WEF (World Economic Forum): Global competitiveness index. Available online: http://reports.weforum.org/global-competitiveness-2011-2012/ (2011)

Wurzman, R., Giordano, J.: Differential susceptibility to plasticity: a 'missing link' between gene-culture co-evolution and neuropsychiatric spectrum disorders? BMC Med. **10**, 37 (2012). Available online: http://www.biomedcentral.com/1741-7015/10/37

Chapter 7
Implications: Societal Collective Outcomes, Including Manufacturing

Jian Cao, Michael A. Meador, Marietta L. Baba, Placid Mathew Ferreira,
Marc Madou, Walt Scacchi, James C. Spohrer, Clayton Teague,
Philip Westmoreland, and Xiang Zhang

7.1 Vision

The fundamental tools and approaches in converging knowledge and technology described in earlier chapters will have a profound impact on societal collective outcomes. This chapter is focused on the impact of these outcomes on manufacturing, innovation, and long-term societal development.

7.1.1 Changes in the Vision over the Past Decade

"Passive" convergence of knowledge and technologies has had an important effect on manufacturing development. Manufacturing enterprises have evolved for the most part as centralized and concentrated urban complexes, primarily driven by the need to minimize manufacturing costs. Examples include the automobile industry of southeastern Michigan and the electronics manufacturing complex in Shenzhen,

---

The authors would like to express their appreciation for the insights and helpful discussions with the late Prof. Richard E. DeVor of University of Illinois at Urbana-Champaign, Prof. Ehmann, Bruce Carruthers and Leslie McCall of Northwestern University, and Profs. Joseph Beaman, Jr., and David Bourell of University of Texas, Austin.

Corresponding editors M.C. Roco (mroco@nsf.gov) and W.S. Bainbridge (wbainbri@nsf.gov).

J. Cao
Northwestern University, Evanston, IL, USA

M.A. Meador
National Aeronautics and Space Administration, Washington, DC, USA

M.L. Baba
Michigan State University, East Lansing, MI, USA

China, that hosts companies such as Foxconn with its estimated 450,000 employees (Focus Taiwan 2010). These manufacturing complexes have created a variety of societal problems, including those related to the environment, the conspicuous consumption of resources, land utilization, urban transportation, and continually changing patterns of employment needs and opportunities that precipitate other sets of social problems.

Converging knowledge and technology has supported a transition from mass production to mass customization in the "post-Fordism" of the last two decades, as evidenced by the shift from large-batch mass production in centralized factory settings to small-batch customized production of high-quality goods in more widely distributed locales (Kotha 1995; Vallas 1999; Zysman 2004). This transition has been driven by a number of factors, including mass customization needs (e.g., medical devices tailored to particular patients), cyberinfrastructure that changes the way people communicate, emergence of point-of-use (POU) technologies, and emergence of miniaturization technologies, to name a few. The evidence is strong that the mass customization model is well in place (Federal Reserve Bank 1998); for example, already between the 1970s and 1990s, the number of national TV channels rose from 5 to 185, the number of running shoe styles increased from 5 to 285, and the number of contact lens types increased from 1 to 36. These developments might be suggesting a swing to a more *distributed manufacturing model that might coexist with the centralized model* (DeVor et al. 2012). The United States has played a dominant role in this development due to its entrepreneurial spirit and research enterprise. The synergy that can emerge from the confluence of these forces has the potential to radically alter the future course of the world of manufacturing based on converging technologies.

Long-term societal development outlined in the first NBIC report (Roco and Bainbridge 2003) had a focus on science and technology feasibility. The adoption of advanced science and technologies raises a number of social issues that need to be explored, which include labor skill requirements and training. In the $70 million

---

P.M. Ferreira
University of Illinois at Urbana–Champaign, Champaign, IL USA

M. Madou • W. Scacchi
University of California, Irvine, CA, USA

J.C. Spohrer
IBM Corporation, Armonk, NY, USA

C. Teague
Independent Nanotechnology Consultant, Gaithersburg, MD, USA

P. Westmoreland
North Carolina State University, Raleigh, NC, USA

X. Zhang
University of California, Berkeley, CA, USA

National Additive Manufacturing Innovation Institute established in August 2012, workforce training constitutes one critical cornerstone (NNMI 2012). It involves nine universities and colleges, five community colleges, and professional society and nonprofit institutes specialized in standards and certificates.

7.1.2 The Vision for the Next Decade

Convergence of knowledge and technology is envisioned to feed the gradual emergence of highly integrated and flexible manufacturing processes, enhanced by the extensive communication capabilities offered by the Internet and by their integration with nanotechnology, biotechnology, and cognitive technologies; ultimately, these changes may give rise to a fundamental shift in the way manufacturing is performed. The emergence of machines and systems with a much higher degree of autonomy, coupled with *in situ* functional metrology as well as remote diagnostics and maintenance capabilities, etc., allows one to draw parallels to the development of the personal computer (PC). Like the PC, these new emerging systems could be deployed in a highly dispersed manner and evolve into general-purpose technologies that enable new forms of production and creativity through the adoption of a more distributed approach to the creation of the future generations of manufacturing systems and enterprises. Such a "Distributed Manufacturing" model could become a disruptive and transforming paradigm that could forever change the landscape of the world of manufacturing that has been dominated by the centralized model. The interconnectivity and dispersion of such systems would allow new business models such as "manufacturing at the mall," and "point-of-need-manufacturing" to coexist with the centralized model.

The distributed scenario with multiple technology sources would *shift manufacturing capabilities from the hands of the few into the hands of the many*, with a dramatic impact on the standard of living of a large portion of the planet's population. This paradigm shift is likely to produce highly nonlinear behaviors with regard to both the engineering and socio-economics of the challenge. Without taking a broad perspective when managing the process, our ability to guide the process and identify early signals of potentially dangerous patterns and effects will be greatly inhibited.

Figure 7.1 illustrates the envisioned transformation in knowledge generation spurred by a distributed manufacturing and technology development landscape. In the past and at present, the advancement of knowledge has started from the basic understanding of fundamental physical sciences, typically initiated by people with doctoral degrees and many years of research experience. For example, the laser was first researched and characterized by physicists, among them, the 1981 Physics Nobel Laureate Arthur Schawlow. Once the physical science is better understood, technologies are developed through innovations by *many* researchers and engineers, such as those who dramatically increased the types, power, and sizes of lasers while lowering their costs—further stimulating new inventions, and adoption of laser applications by millions of users. The laser has found applications in laser cutting,

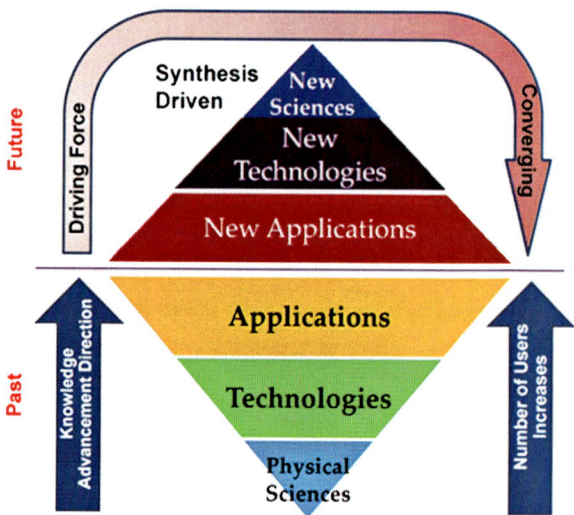

Fig. 7.1 Schematic of current and future knowledge development and adoption paths. The *right up-arrow* shows an increase in number of users as applications increase (Courtesy of Jian Cao)

laser welding, scanning, and printing, to name a few. The bottom half of Fig. 7.1 summarizes this currently prevailing scenario. The left arrow shows the path of knowledge advancement from physical sciences to technologies to applications, while the size of each block symbolizes the number of users in each category, i.e., fewer people engaged in physical sciences knowledge development than in application areas.

The distributed nature of manufacturing and use of converging technology platforms will alter the path of how knowledge is created, as illustrated in the upper half of the diamond in Fig. 7.1. For example, school-age students can now create "apps" (applications) for smart phones. New additive manufacturing processes will increasingly allow users to locally create one-of-a-kind parts, even from multiple materials, to meet specific needs across a wide range of applications, from aerospace to fashion to tissue engineering; data can be shared in the cloud. The ease of the user–machine interface will allow the general public to adopt technologies and to invent new technologies, which, in turn, may lead to new scientific discoveries and new sciences, in particular, synthesis-driven sciences. As indicated in the top half of Fig. 7.1, the driving force for knowledge creation may start from grass-roots applications then broaden to technologies and to converging sciences. This is in stark contrast from how it has been done in the current and past centuries. Newly discovered and developed converging sciences will support distributed manufacturing for economic, human-potential, and societal developments.

Finally, *new methodologies have to be developed to ease the knowledge burden in distributing convergence knowledge*. Critical challenges to be addressed are how to translate information into knowledge and education, how to accelerate learning and make it fun to learn, and how to restructure universities to provide quality education to all human beings who are interested in learning.

Long-term development of society will be affected by convergence that will become proactive and more systematic. Applications are envisioned to affect all sectors of human activity in the next 10 years and beyond. For example, understanding of how the brain functions can lead to new ways of defining manufacturing processes and systems tailored to people's choices, and hence spin off new manufacturing processes or create new enterprise business models.

7.2 Advances in the Last Decade and Current Status

The role of convergence in the last decade was generally was in response to coincidental collaborations. Main technological advances in the last 10 years are flexible manufacturing processes and systems, particularly in terms of *additive manufacturing* (Sect. 7.2.1) and *small-scale multifunctional manufacturing* (Sect. 7.2.2); *robots* for assisting in human motor control (Sect. 7.2.3); and *universal access to quality education* (Sect. 7.2.4). The weakness of social science advancement in the last 10 years has limited the role that social sciences could be playing in policy and physical sciences for convergence and human benefits. The discussion on aspects related to social science will be more focused in Sect. 7.3, where the goals for the next decade are discussed.

7.2.1 Additive Manufacturing

Additive manufacturing (AM) creates parts layer-by-layer directly from the digital computer-aided design (CAD) files in conjunction with a layered material deposition technology (Prinz et al. 1997; Beaman et al. 2004; Bourell et al. 2009). AM has evolved steadily over the last 30 years beyond its initial application for rapid prototyping; particularly in the last 10 years, it has grown to be competitive as a valid process for rapid manufacturing of final functional products. For example, Fig. 7.2 is a demonstration part of a hinge bracket used on the Airbus A320 in its original form (*rear image*) and the optimized design produced by AM (*front image*). Using the HyperWorks topology optimization tool, OptiStruct, engineers were able to

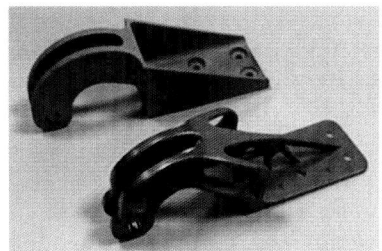

Fig. 7.2 Airbus hinge bracket in its original form (*rear*) and optimized form achieved by additive processing (*front*) (Courtesy of Jian Cao)

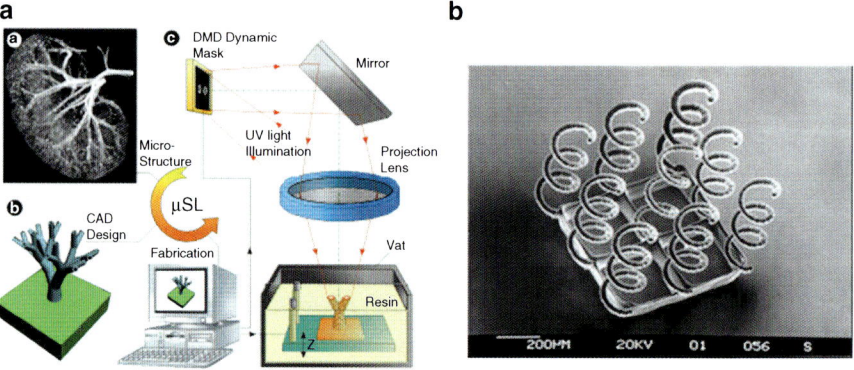

Fig. 7.3 (**a**) Fabrication of 3D microvasculature for tissue engineering and fractal antenna process using a dynamic mask projection system; (**b**) 3D micro-spring made by photolithography to demonstrate the ability to fabricate 3D structures at the micro-scale (Courtesy of Prof. Xiang Zhang of University of California, Berkeley)

merge AM capability in design to achieve a significant weight reduction of 64 % while retaining the same characteristics in terms of stiffness and bolt loading.

Additive manufacturing can also be performed with an increasing variety of materials, including polymers, ceramics, and metal alloys such as aluminum, titanium, and copper alloys. Innovations in high-throughput parallel dynamic mask projection enables AM at the microscale and even at the nanoscale to overcome time-to-manufacturing limitations of conventional serial AM processes, as shown in Fig. 7.3.

One major impact of additive manufacturing is that it links designers directly to manufacturers, putting manufacturing capabilities in the hands of even school-age individuals. Further research and development can further strengthen AM by providing seamless design tools, increasing processing speed, improving the surface finish, increasing accuracy, providing more material choices and means to recycle materials, strengthening the capabilities of predicting product performance, and reducing cost.

7.2.2 Small-Scale Multifunctional Manufacturing

The downsizing of manufacturing equipment and systems has the potential to totally change the application protocols of various sectors of the economy. For example, a desktop manufacturing (DM) system can literally reside in a room adjoining a hospital operating room, ready to manufacture a diagnostic probe or implant that is tailored to the precise size and needs of the person on the operating table. Such technology can not only significantly reduce healthcare costs but improve the quality of healthcare delivery as well. An example of a similar technology already in use is ceramic dental reconstruction, as practiced at Sirona Dental Systems (http://www.cereconline.com/). In a single visit, a patient's tooth can be scanned

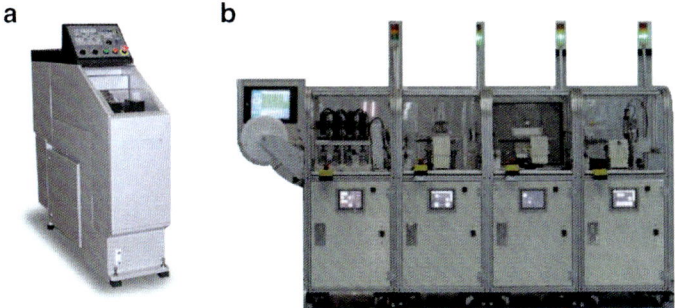

Fig. 7.4 Small-scale multifunctional machines developed in recent years: (**a**) Slim Lathe by Takamaz (300 (W) × 1,225 (L) mm); (**b**) On-demand MEMS desktop factory (each unit or "cell" is 500 (W) × 800 (D) × 1,200 (H) mm), at the National Institute of Advanced Industrial Science and Technology (AIST), Japan, 2008 (Courtesy of Kornel Ehmann)

with a 3D CAD system, and a tabletop computer numerical control (CNC) machine tool in the next room can then manufacture a ceramic crown that is immediately put in place in the patient's mouth. Point-of-use-manufacturing, and therefore DM, also has considerable applications for activities in remote locations, including military deployments or in space, (e.g., the International Space Station).

Ehmann et al. (2005) conducted a WTEC study on micromanufacturing that found that in 2004–2005, the trend toward miniaturization of machines was already evident in both Asia and Europe, with commercialization of desktop machine tools, assembly systems, and measurement systems well underway. Examples today of commercial developments of machines that fit the desktop manufacturing paradigm are numerous and are epitomized by significantly downscaled versions of their macro-level counterparts. One example is the "SlimLathe" by Takamaz (Fig. 7.4a). Another example is the Takashima Sangyo Company's DM "plant," where 120 desktop-sized machines operate in a mere 300 m^2 space (Endo 2005). Concurrently, a number of Japanese and European companies are starting to offer specialized products ranging from assembly, joining, and metrology, to NEMS (nanoelectromechanical systems) and MEMS (microelectromechanical systems) processing (e.g., Fig. 7.4b) and other equipment, along with supporting component technology products such as sensors, actuators, and controllers that support the desktop manufacturing paradigm (AIST 2006).

In the United States, a notable example is the microfactory developed at the University of Illinois at Urbana-Champaign (UIUC) in collaboration with Northwestern University, which has integrated a number of desktop-sized cutting, forming, and metrology devices. Figure 7.5 gives a photograph and schematic of the microfactory (Honegger et al. 2006a, b). Additional desktop or small-scale flexible manufacturing facilities have been developed at other universities, including but not limited to, the University of Michigan (Prof. Jun Ni); Georgia Institute of Technology (Prof. Shreyes Melkote); Carnegie Melon University (Prof. Burak Ozdoganlar); and Massachusetts Institute of Technology (Prof. Martin Culpepper).

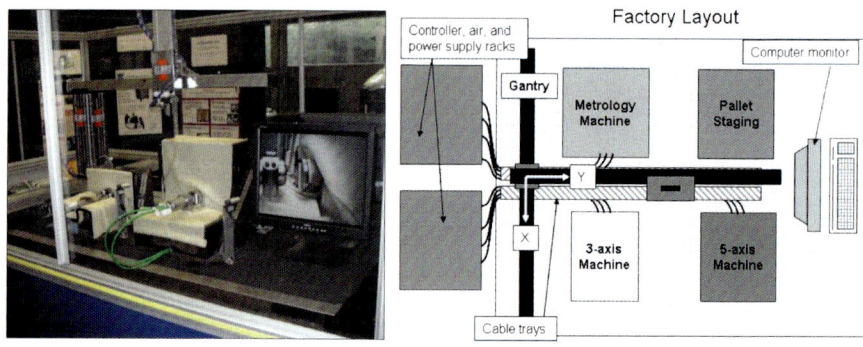

Fig. 7.5 Microfactory developed at University of Illinois (Prof. Shiv Gopal Kapoor) in collaboration with Northwestern University (Courtesy of Profs. Kornel Ehmann and Jian Cao)

Small-scale, distributed chemical production is emerging as a powerful development within industry. Such distributed production, localized production, and reduced capital investment models run counter to the maxim of "economies of scale," but they offer economic and safety benefits. As Meredith Kratzer of UIUC has noted, "Novel microchemical systems will reduce the need for hazardous reactants, generating them [only] on demand."

The above-mentioned flexible and compact systems possess the compactness, flexibility, modularity, and *in situ* metrology required for distributed manufacturing. The technology can be further enhanced by adding reconfigurablity, multifunctional and multimaterial processes, multi-domain scale, high reliability and low maintainability, self-diagnosis capability, Internet connectability, and remote monitoring and/or control, to name a few.

7.2.3 Robots for Assisting in Human Motor Control

Development of actuators, sensors, lightweight materials, and power systems for limb prosthetics is making great strides toward achieving the goal of producing a prosthetic that is so human-like that a patient or an observer could not tell the difference from the wearer's own limb. The march toward highly effective brain–prosthetic interfaces for brain–prosthetic communication of signals and the better understanding of brain processing of information is occurring at an amazing pace. The limitations lie within our inadequate understanding of the brain. Research has demonstrated the preliminary success of restoring various functions to persons with impairments such as spinal cord injury, brain injury, and stroke. Kevin Lynch at Northwestern University has been using functional electrical stimulation (FES), as shown in Fig. 7.6, to restore reaching motions to persons with high spinal cord injuries who have little or no voluntary control over their upper extremities. Obtaining signals from the brain with electroencephalography (EEG), magnetoencephalography (MEG), and functional magnetic resonance imaging (fMRI) has progressed at surprising rates, with research on interpreting these signals also progressing well.

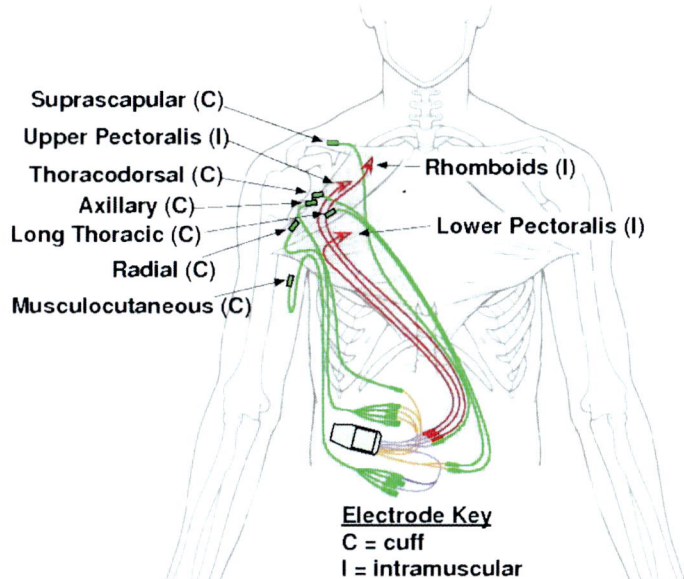

Fig. 7.6 Use of functional electrical stimulation to restore limb movement (Courtesy of K. M. Lynch, Northwestern University)

Co-robots such as the ASIMO robot by Honda (Fig. 7.7) demonstrate great progress in achieving human-like performance for assistive robots. This robot has 34 degrees of mechanical freedom in movements in its head, arms, hands, hips, legs, and feet (see demonstration video at http://asimo.honda.com/). "Intelligence" is demonstrated by ASIMO in its ability to see, hear, distinguish sounds, recognize faces and gestures, and chart a route to a specified location.

Even this level of co-robots promises huge advances in assistance for "direct support of and in a symbiotic relationship with human partners," as envisioned by the National Robotics Initiative supported by multiple agencies of the Federal Government, including the National Science Foundation (NSF), NASA (the National Aeronautics and Space Administration), the National Institutes of Health (NIH), and the Department of Agriculture (USDA). For example, in space, profitable areas like geostationary satellite maintenance and resource extraction (e.g., mining minerals on asteroids or mining helium-3 on the moon) demand smart robotics. Integrating high-impulse and high-thrust rockets, robotics can be applied to industries beyond low Earth orbit.

Such co-robots as those described above will serve as critical elements in distributed manufacturing or point-of-need manufacturing in terms of extending human intelligence to difficult frontiers. However, much research is still needed in interpreting brain signals, from a multiplicity of sensors placed on either the brain or the skull, for prosthetics, bio-inspired sensing, and robot technologies.

Fig. 7.7 Honda's co-robot ASIMO "offers tea" and gestures (© 2007 Honda Motor Co., Ltd., courtesy American Honda Motor Co., Inc.)

7.2.4 Universal Access to Quality Education

Free online Internet access to thousands of high-quality lectures in science, technology, engineering, medicine, history, arts, philosophy, etc., is now being provided by a growing number of first-rate institutions around the world. Chapters 4 and 8 provide a number of examples in the United States. Such examples of universal access to quality educational materials are challenging conventional approaches to student learning systems. Delivery of high-quality lectures to students is being combined with such efforts as those by the Knewton Company to obtain real-time feedback from students' learning activities in order to customize subsequent educational activities and materials. Could a computer tutor be sufficiently "intelligent" that a student could not tell the difference in comparison to a human tutor (Fig. 7.8)? Can a student selectively pick courses that will adequately prepare him/her for a unique career, for example, in the IT-assisted manufacturing framework? Answering these questions will require a new model of our education system.

It is worth noting that computer games and virtual worlds will also be part of the future educational landscape. For example, efforts at University of California–Irvine and Intel Research have produced and demonstrated how existing commercially available entertainment-oriented computer games and their associated "software development kits" can be modified into ones that are suitable for training adult

Fig. 7.8 Man or computer? (©*Smithsonian Magazine* July 2012, used by permission)

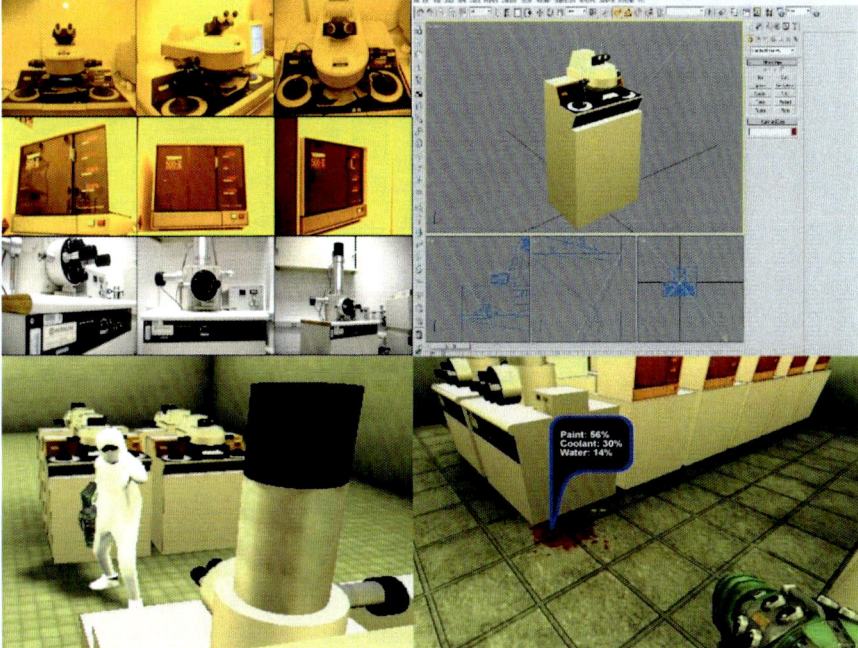

Fig. 7.9 Computer game for diagnosing problems in semiconductor fabrication operations (Scacchi 2010; courtesy of Walt Scacchi)

technicians in the operation and diagnosis of semiconductor and nanotechnology fabrication equipment and operations, as seen in Fig. 7.9. In this figure, the images in the upper left are photographs of different devices involved in fabrication operations (e.g., an electron microscope); the upper-right image is a 3D CAD model of one such device; at the lower left is a view into a virtual factory with different virtual devices visible, along with an operations technician; and at lower right is a diagnostic view of a liquid material that is leaking out of the bottom of a device. Similar techniques have also served to support the development and deployment of science learning games that address National Science Education Standards for targeted student grade or skill levels (Scacchi 2010).

7.3 Goals for the Next Decade

The attendees of the U.S. NBIC2 workshop in Breakout Session 7 identified three major challenges in convergence of knowledge and technology for manufacturing and societal benefits:

1. High productivity and automation reduces employment opportunities in a centralized manufacturing model; however, process flexibility, automation, and the Internet together have created an opportunity for making things that individuals want. How can this be done autonomously so that members of the general public can execute their own designs to meet their individual needs?
2. Weakness in social science development in recent years limits the role that social science could be playing in policymaking and physical science R&D for convergence and human benefits.
3. Advanced knowledge has been created in both traditional ways and in new ways, as indicated in Fig. 7.1. How do we translate new information into knowledge and education? How can we learn faster, and how do we make it fun to learn so much new information? How should universities adapt to this need?

To address those challenges, this section summarizes the goals for the next decade in distributed manufacturing to empower individuals and communities (Sect. 7.3.1), "manufacturing process DNA" (Sect. 7.3.2), integrated social and physical sciences (Sect. 7.3.3), and the individualized education model (Sect. 7.3.4).

7.3.1 Distributed Manufacturing to Empower Individuals and Communities

Innovation for advanced manufacturing, jobs, and competitiveness will be provided by personalized manufacturing, diy (do-it-yourself) manufacturing, and availability of manufacturing information as a service, which are all very good examples of the trend of integrating information, communications, materials, and economics in an evolving new paradigm for manufacturing. With automated processes for creating artifacts and assembling them, the skills and actions needed to create products are heavily discounted relative to the knowledge and information needed to create them. Downstream activities such as *design* and *manufacturing* can serve as important means of integrating diverse knowledge bases and converging knowledge and technology (CKT), specifically in the following areas:

- *Scalable models for distributed manufacturing*: The emerging idea here is the integration of small distributed facilities to obtain scaled-up manufacturing. Crowdsourcing is one new idea related to this trend. Technologies needed for this include tools built upon coherent design, manufacturing, and service platforms for easy transitions among sketch, manufacturability, and serviceability. The model needs to be able to service both paths of knowledge generation illustrated in Fig. 7.1.

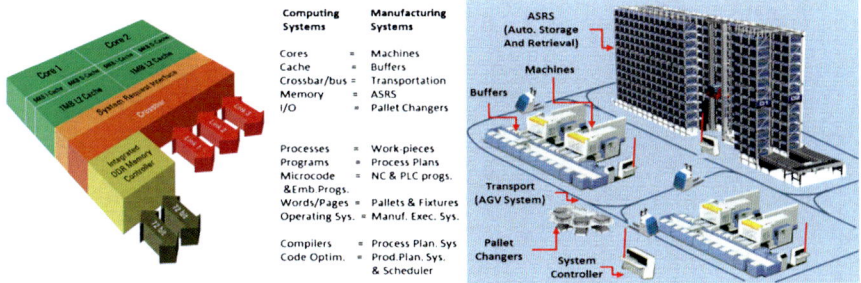

Fig. 7.10 Analog of distributed manufacturing systems with PCs (Courtesy of P. Ferreira, University of Illinois Urbana-Champaign)

- *Info-inspired manufacturing systems*: Distributed manufacturing can follow an architecture similar to that used in computer systems, as illustrated in Fig. 7.10, using the concepts of cores, buses, peripherals, operating systems, networks, etc., in PCs (left column in Fig. 7.10) to design/architect the advanced manufacturing systems (right column in Fig. 7.10). What the new architecture enables is the interchangeability, integration, and hybridization of various manufacturing processes using different materials and operating at various length scales.
- *Translational science-driven micro-/nanomanufacturing*: Technology innovations face prime challenges to continuously scale-down the devices in mass production and to heterogeneously integrate multiple functionalities for applications such as electronics, data communications, personalized medicine, and energy. Continuous harvesting of the capabilities of micro- and nanotechnology has to be developed. We especially need to pay attention to the new scientific-breakthrough-driven manufacturing innovations. Translational manufacturing examples in lithography technology—such as micro-factory, nano-imprinting, and plasmonic lithography based on scientific discoveries of metamaterial superlenses—are of great significance to many other technological sectors (Zhou et al. 2011; Fang et al. 2005; Liu et al. 2007; Valentine et al. 2008). Multidisciplinary research efforts bring together physics, mechanics, metrology, chemistry, material science, and electrical engineering to provide synergetic routes towards development of emerging manufacturing innovations. Key developments include achieving high-performance designs, improving reliability, improving management of data flow, and integrating additional ultra-high-accuracy metrology with automatic positioning control systems (Srituravanich et al. 2008). As key drivers, translational science-driven micromanufacturing and nanomanufacturing innovations will provide a foundation for economic growth.
- *Predictive science and engineering*: One of the backbone elements in realizing the grand vision for distributed manufacturing is the ability to predict with confidence. For example, take additive manufacturing: bottlenecks for AM to replace existing processes in fabricating, particularly high-value-added metallic parts, are: (1) material powders design, and (2) ability to predict part performance such as fatigue behavior based on powder composition and process conditions, because

most AM processes involve complex melting and solidification or solid-state phase transformation. Only with credible microstructure prediction and residual stress prediction can a processing strategy be autonomously developed to compensate for potential shape distortion to achieve desired performances. Thus, new methods are needed in integrating physics-based models with various manufacturing processes and with new materials.
- *Human–machine interactions*: This technology will have a broad impact on society in terms of improving human performance, reducing risk and exposure of humans to dangerous situations, and improving the quality of life of people with physical challenges. For example, future space exploration mission scenarios include a greater reliance on human-directed robots to explore a planet surface, aid in the construction of permanent bases or habitats on a planet surface, or function as surrogates for humans in situations where it may not be possible to adequately protect astronauts from exposure to hazardous environmental conditions such as temperature extremes or radiation. Technologies developed for space exploration applications could also be utilized by manufacturing operations, firefighters, police, and in homeland security and defense applications.

7.3.2 Manufacturing Process DNA

Manufacturing, in this study, is defined as processes and systems that alter the structure, shape, surface, and function of materials. It includes a combination of mechanical, optical, and chemical processing in a broad domain, including but not limited to subtractive processes, additive processes, net-shape processes, pharmaceutical processes, refining processes, and self-assembly processes. This broad definition of manufacturing plays a particularly important role in the framework of establishing converging technologies and distributed manufacturing because scientists and engineers from different disciplines need to be able to communicate with each other using a common language. Here, we call this common language "manufacturing process DNA".

The establishment of manufacturing process DNA will enable the co-design of products and processes; will integrate, intertwine, and synchronize multiple processes; and will require collaboration among multiple disciplines.

7.3.3 Integrated Social and Physical Sciences

Realizing artifacts and services of benefit to society necessarily requires the consideration of multiple factors: economics, technology, sustainability, the environment, societal effects, etc. As an example, nanomanufacturing not only seeks to create artifacts that exploit physical nanoscale phenomena, but it also motivates the study of these phenomena as the basis of manufacturing technology.

7.3.4 *Individualized Education Model*

The distributed manufacturing framework may also demand different job-performance skill set requirements from individuals. To address this need, the following recommendations were made by the attendees of the U.S. NBIC2 conference:

- Create participatory games and tools that enable new ways of learning science, engineering, technology, and manufacturing.
- Educate *all* people so as to enable even "disabled" and "disadvantaged" individuals to have the capabilities to be part of society's growth and to participate in the distributed economic model.
- Reinvent higher education, i.e., to improve the weakest link in the higher education system, to attract bright new minds to all universities, not just the top-ranked ones.
- Establish a culture of renewed appreciation of "making things." From a technological/societal point of view, it is important to be able to translate CKT inventions and R&D into valuable products that are accessible to as many people as possible. Many advanced countries are losing manufacturing skills because of outsourcing and over-emphasis on design, marketing, and financial "manipulations." Making things is not very popular anymore. In the case of MEMS, the United States is not the main beneficiary of the tremendous investments it has made in the field. To reverse that trend will require that we reengineer the pathway from basic research to product realization, and reconnect with community colleges, the public, and the general workforce. We also need to find a way to influence political decisions more effectively.

7.4 Infrastructure Needs

Cyberinfrastructure has become an essential part of process-based manufacturing, and its role will grow dramatically. One of the more vivid projections is that of zero-risk, zero-emissions process manufacturing. Safe, clean operations will rely on high levels of process automation, abundant small sensor devices, and data analytics to detect key variances during steady-state operations and during transient operations like startup and shutdown. "Risk" refers here to safety, but economics will also be aided by real-time supply-chain management feeding relevant and timely data into process operations.

Similarly, *IT-supported citizen science* will enable ordinary people to engage more actively in economic and/or social development via cyberspace. The impact of realizing the scientific and technological goals described in Sect. 7.3, i.e., distributed manufacturing, process DNA, synthesis of social science and technology, and individualized education, will not be possible without a well-developed cyberinfrastructure.

The levels of advanced scientific equipment in most labs of U.S. institutions are now behind those of international counterparts in Korea, China, Japan, Switzerland, Germany, etc., as experienced by educators and indicated in the National Academy of Engineering's *Rising above the Gathering Storm Revisited* report (NRC 2010). We need to reinvest now in the best available equipment here in the United States.

7.5 R&D Strategies

7.5.1 Investment Strategies

It was perhaps expected that private industry would invest in R&D and better equipment at R&D institutes once they started retrenching their own internal efforts away from R&D (see Bell Labs, IBM, etc.). That has not happened. As the current financial crisis has made clear, that will also not happen in the near future. In the meantime, competitors Samsung, LG, and many others are doing what those U.S. R&D houses used to do. The more applied U.S. national laboratories should perhaps be reconverted into nimble engines of economic survival. The more fundamental research arms of NSF, NASA, and the Defense Advanced Research Projects Agency (DARPA) should go back to investigating the most daring and far-reaching science goals instead of trying to address short-term industrial needs. We must expect vigorous international competition for these achievements, which worldwide are funded primarily by government entities rather than by industry. The advantages of such investment for the United States—and other countries—will be to develop the needed science, technology, and people with knowledge and skill to meet pressing emerging challenges and to transport society to improved collective technological constructs and broadly shared decent standards of living.

7.5.2 Implementation Strategies

Over the last decade, the vision and reality of converging technologies have gained momentum and acceptance among scientists and engineers; however, we have lost ground in terms of broad public support, and we are even less able to implement and exploit our national intellectual property (IP) position in converging knowledge and technologies than we were 20 years ago. At a time when sciences are converging with the promise of undreamed-of new technological opportunities, our nation is ill-prepared in terms of necessary resources, the support of politicians, and adequate workforce training. One cannot implement a long-term vision in an economic and political climate of shorter and shorter time horizons. Those societies that still have the resources and "educated" politicians with the patience and foresight to stick with a longer-term plan will reap the benefits of converging technologies.

We need to make the process of generating cutting-edge knowledge much more inclusive, and while deepening scientific interdisciplinary R&D, also place much more emphasis on the realization of physical products that can be seen by the public as both exciting and beneficial, for example, personalized medical devices or assistants. The goal of *universal access to quality education* needs to have two directions, pull and push. The technology and framework discussed in Sect. 7.2.4 illustrates the push direction, i.e., how to *push* knowledge to people who have the desire to learn. The challenge is in the *pull* direction, how to obtain the attention of people without strong backgrounds in science, technology, engineering, and mathematics (STEM)—including the majority of politicians—to educate and excite them about these fields. The STEM community has to be more proactive to ensure that goals are measured and achieved, that none of the fundamental research institutes become political footballs, and to fill in the current blank to create a long-term political vision regarding the value of adequate and sustained U.S. investment in these areas.

7.6 Conclusions and Priorities

The U.S. NBIC2 workshop attendees evaluated the current manufacturing landscape, defined what "advanced manufacturing" means to society, recognized the emerging pattern of knowledge creation and absorption, and explored the possibilities for radical change in the landscape of manufacturing over the next decade or two via the emerging paradigm of distributed manufacturing. The research and education priorities they identified are noted briefly below.

Research priorities: Advancing manufacturing through converging technologies will be essential to progress in economic and quality-of-life indicators. Research needs were summarized by participants as (1) distributed manufacturing enabled by process flexibility, modularity, in-process metrology, predictive sciences and technologies, human–machine interaction, etc. (refer to Sect. 7.3.1); (2) manufacturing process DNA (refer to Sect. 7.3.2); (3) integration of the social and physical sciences (refer to Sect. 7.3.3); and (4) individualized education (refer to Sect. 7.3.4).

Education priorities: Financial pressures have pushed educational institutions into making shortcuts that are leading to students obtaining degrees in less time with weaker content. Understanding of advanced topics like structural colors (caused by the interaction of light with nanoscale structures in materials) requires not only a mastery of physics but also insights into advanced manufacturing. Therefore, education will require a better balance between theory and practice and between computation and experimentation. To meet the challenges of the modern era—i.e., how to gain knowledge accurately and fast enough to ease the knowledge burden, and how to disseminate STEM knowledge to the general public, including politicians—new methodologies of individualized education will be needed.

7.7 R&D Impact on Society

A science-based understanding of "making things" in a distributed fashion will enable an economic and societal transformation of productivity and creativity. The past decade of research and development on cyberinfrastructure, an essential part of all the converging technologies, has transformed public and private communications, business transactions, the conduct of science, and manufacturing, largely for the better. Convergence of technologies creates innovative ideas and products that benefit all people.

At the same time, a consequence of these trends has been to redirect the wealth generated by industrial workers, altering the source of middle class income to be based more on service-sector employment. Industrial employment as percentages of U.S. and of international overall employment appears to have fallen due to a natural progression similar to the monotonic decrease in agricultural employment. Eventually, it will likely fall to some level consistent with high manufacturing productivity. A side effect is that industry-generated wealth then flows mainly to upper management and owners rather than into worker wages and salaries, altering the traditional flow paths of wealth back into the general population through consumer spending by the industrial employees. In order NOT to further the gap between rich and poor (or educated and uneducated), we will have to become much more inclusive and work on building a middle class that can participate in the benefits of CKT.

The impact on society of envisioned converging knowledge and technologies, and particularly distributed manufacturing, will be of the same order as the PC revolution. Three profound elements of the grand vision are to:

1. Shift manufacturing capabilities from the hands of the few into the hands of the many to spark innovations
2. Alter the paths of how knowledge and technologies are created
3. Ease the knowledge burden in distributing knowledge quickly and accurately

7.8 Examples of Achievements and Convergence Paradigm Shifts

Above we have discussed what the converging technologies are for manufacturing, and their societal impacts. Below, we will illustrate the process and impact through the case study of Xerox, including hybridization of disciplines, lessons from big data, and healthcare information systems for the future of social science research in industry. Then we will present an example of technology integration for wearable computers and sensors.

7.8.1 Social Science Research and Technology-Based Industry

Contact person: *Marietta L. Baba, Michigan State University*

Hybridization of Disciplines: The Initiation

Over the past 30 years, experts in anthropology, linguistics, psychology, design, and other disciplines linked to the social sciences have worked together with members of the computer science and industrial engineering disciplines to form new interdisciplinary fields of research and practice referred to as *design ethnography* and *participatory design* that have reinvented the processes by which new products, services, systems, and spaces are conceptualized, designed, and implemented by industry (Blomberg et al. 1996; Suchman et al. 1999; Squires and Byrne 2002; Cefkin 2009). Design ethnography has its own annual conference (EPIC, or Ethnographic Praxis in Industry, in its eighth year in 2012), with global rotation of venues and refereed proceedings published through the American Anthropological Association. Sponsors include Intel, Google, Microsoft, and several other large technology firms. The emergence of this hybrid profession was one consequence of criticism regarding mainstream academic anthropology, especially charges that the discipline was too much wedded to studying former colonies (Marcus and Fisher 1986). These charges gave legitimacy to students who wanted to pursue research in venues other than typical remote locations. In the process of repatriation, anthropologists spread out over many different field sites and came into contact with a number of other disciplines and professions, and they hybridized their knowledge and practices as they worked with each other. Participatory design is a related field that is more specialized; its development will be discussed below.

The industrial roots of design ethnography and participatory design in the United States may be traced to the earlier efforts of Frederick Taylor and the Human Relations School to improve interactions between people and equipment in the production process. These early streams of investigation on the "human factor" in production eventually gave rise to the subfield of *"human factors"* research, a multidisciplinary offshoot of psychology that identifies aspects of human psychology and its context that must be taken into account in the design and development of new products. "Human factors" is both a field of study and an area of research and development in corporations that produce goods and services based on advanced technology.

Xerox PARC: A Case Study

In the 1970s, one company that was committed to pushing the envelope of knowledge on human factors (broadly defined) surrounding advanced computing was Xerox's Palo Alto Research Center (PARC). PARC was interested especially in human–computer interaction and the development of artificial intelligence to support this interface; it also funded graduate student interns to work in this and other areas related to computing. One of the first interns was Lucy Suchman, who came to PARC in 1979, at first to study office work practices, but later she became intrigued with the idea of machine intelligence in a computing context.

In her doctoral research, Suchman conducted a Garfinkle-inspired ethnomethodological study of computer-supported work: she videotaped pairs of engineers attempting to make copies of documents using an expert help system and then compared the users' conversations and actions during this process with the machine's automated instructions (Suchman 1987). Contrasting the two points of view side-by-side (i.e., the users' and the machine's), Suchman portrayed communication breakdowns between them, as humans moved fluidly among several different levels of conversation (e.g., simple requests for action, "meta" inquiries about the appropriateness of a procedure, and embedded requests for clarification of procedures), while the machine was severely limited to producing responses that its designers had programmed into it in anticipation of stereotypical responses that users "should" make. While these observations might not seem revolutionary now, they were a lightning bolt at Xerox PARC and led the corporation to change the design of its copiers to make them simpler to use. This research also gave Suchman a reputation for bold and fresh insight and enabled her to expand the role of social science at Xerox PARC.

Suchman attracted other social scientists and computer scientists to her group. A breakthrough came in 1989 when the Doblin Group of Chicago asked Xerox PARC to partner on a project for the office furniture manufacturer Steelcase (Reese 2002). Steelcase managers wanted to understand how the workplace of the future would evolve and what kinds of work environments and designs it should be thinking about over the long term. PARC agreed to co-fund the project that came to be known as the Workplace Project. The project was situated in an airport, which was believed to have properties reflecting the workplace of the future (e.g., high fluidity of people and information and workflow extending into multiple kinds of space via electronic means). Suchman served as lead on this project for several years, and through it she assembled a talented team of social scientists and designers who would revolutionize the design industry. One of the individuals involved was Rick Robinson, then at Doblin (an innovations consultant firm), who subsequently co-founded E-Lab in Chicago, an entrepreneurial firm explicitly dedicated to the concept of equally balancing product design projects with ethnographic research and design talent. The notion that all new product and service concepts should emerge from a contextually rich understanding of the client's natural world, developed through ethnographic field research at client sites, captured on videotape and analyzed using social science theory and method, was first conceptualized by Suchman's group at the Workplace Project, but it was Rick Robinson who took the concept to market. This was the beginning of design ethnography.

Suchman's group at Xerox PARC also established the Work Practice and Technology area in 1989, which mobilized arguments centered upon the value of ethnographically informed design of prototype technology in research and development (i.e., participatory design). This group engaged in productive collaborations with members of other disciplines over more than a decade to address many important questions that relate to computer-supported cooperative work (CSCW), including engagement with an international network of computer scientists and systems designers committed to systems development in a more participatory form involving workers and other users.

This latter model of participatory design was drawn from and strengthened by interactions with colleagues in Denmark, Norway, and Sweden; academic computer scientists collaborating with Scandinavian trade unions developed union-sponsored demonstration systems informed by values of quality of working life and workplace democracy. At Xerox PARC, these values were not the ones emphasized; rather, the potentially superior design outcomes were stressed, to produce information systems better suited to working practices.

One project that stands out was conducted by Xerox in the late 1990s at the headquarters of a state department of highways (Suchman et al. 1999). For 2 years, Suchman's group engaged in a collaborative effort with engineers charged with designing a bridge scheduled for completion by 2002. The prototyping effort was focused on collecting the engineers' documents, a heterogeneous assortment of paper documents, and understanding whether digital media might provide new and useful ways to access the information. As it was assembled on site, the prototype stood as a kind of developing description of how it was that engineers were interested in accessing their documents, and a provisional proposal for a way of working. The prototypes were working artifacts, not given in their specifications, but in the unfolding activity of cooperative design in use.

This approach illustrates one of the ways in which ethnographic practices became a resource for participatory design—two distinct fields that are complementary but not necessarily fully compatible. Despite the potential for incongruities, more than two decades of collaboration between ethnographers and participatory designers have enabled the two fields to develop several creative approaches to working their ways of knowing together, and this collaboration created a widening interest in ethnography in the field of design that has had important implications for industry and anthropology. A number of major technology corporations today have incorporated anthropologists and ethnographers into their R&D staffs (e.g., IBM and Intel) to support design operations, and smaller boutique firms that provide research and design for larger companies also employ anthropologists and ethnographers. This represents a significant development in industrial practice that does not appear to be a passing phase, but the permanent incorporation of a new set of knowledge and skills that heretofore were not available to industry.

Science and Technology Studies: A Critical Turn

Suchman's group at Xerox PARC was identified with an interdisciplinary tradition in the social sciences sometimes known as Science and Technology Studies or Sociology of Knowledge Studies, which became well established in anthropology during the 1980s and 1990s. This subfield included social scientists engaged in the study of artificial intelligence, human–computer interaction, and other related subjects sufficient to form an interest group in the American Anthropological Association.

The anthropologists and linguists who were involved conducted empirical field studies in which they attempted to document what was going on in science, engineering, and technology-intensive venues (e.g., laboratories); however, some of the researchers appear to have been rather naïve in failing to realize that their subjects would read what they wrote and interpret it through their own lenses (Hess 2001).

As it turned out, the social science "discovery" of what were considered "social" influences within the laboratory was sometimes taken as a discrediting maneuver by the some of the scientists (not those at Xerox) and led to a special kind of "science war" between social scientists and natural scientists, one of the results of which has been a sharpening of the philosophical differences between these fields. Some social scientists lost their access to scientific sites as a result (e.g., see Rabinow 2012).

Science and Technology Studies continued into its second and third generations, but it has developed a very different form of practice because in many cases it is no longer welcome in scientific laboratories. The way this field works now is by attending scientific conferences, attending schools, going on virtual chat rooms, reading the technical literature, interviewing outsiders and laypersons about their perspectives on products, becoming part of social and activist organizations, and providing services to help the social issues community, such as by writing or lecturing on the social, historical, or policy aspects of the community (Hess 2001). In some cases, practitioners have developed a sustained engagement that has lasted for 5–10 years or more, gaining deep knowledge of the field and becoming expert commentators. These methodological practices have been developed at the same time that interpretive and critical theories were reaching their zenith within American anthropology and other academic fields.

Transcending the Distance: Lessons from Big Data and Healthcare Information Systems for the Future of Social Science Research in Industry

There are many reasons for hope that social scientists, both in industry and academia, will continue their collaboration with members of other disciplines such as computer science and engineering, following the transition that has been described above. It may not be that the cooperation occurs in the area of Science and Technology Studies, but there are many other areas of opportunity created by converging technology. At present, due to trends noted in the Xerox PARC case study, there are numerous social scientists in industry, both as a result of growing demand for understanding global consumers and the downsizing of academia. Industry practitioners such as those involved in EPIC (see Sect. 'Hybridization of Disciplines: The Initiation') are reaching out to academic social scientists to build intellectual bridges (Cefkin 2009).

More importantly, there are emerging challenges and opportunities for the social sciences in which other disciplines, and industry, must play a vital role. The most clearly articulated challenge is that of data analytics (or "big data") and the role of computational approaches to the aggregation, analysis, and interpretation of such data. There is an increasing volume and detail of digital information captured by organizations with the rise of multimedia, social media, and the "internet of things" that is predicted to be involved in innovation and economic growth. A major challenge for social scientists will be the development of means to access, aggregate, analyze, and interpret the significance of such data for social and economic problems. These are not challenges that the social sciences can tackle on their own. They will need to partner with other disciplines in science and engineering, and academics will need to cooperate with industry and government, where much of this data is being collected (Manyika et al. 2011).

Social scientists are already working on large digital data sets in the areas of environmental science, and there are possibilities in cognitive science as well (i.e., analysis of fMRI data). An area with potentially more interest for industry may be in electronic health records, which have been mandated by the HITECH Act of 2009 (Health Information Technology for Economic and Clinical Health Act, Title IV of the American Recovery and Reinvestment Act). Currently, thousands of healthcare organizations around the United States are engaged in the process of adopting the "meaningful use" of electronic health records, as required by the HITECH Act. Electronic health records can improve the quality of healthcare and reduce its costs; McKinsey Global Institute has estimated that the creative and effective use of data analytics in healthcare could create an additional $300 billion in value per annum, two-thirds of which would be in reducing healthcare expenditures in the United States (Manyika et al. 2011). The aggregation of health records in electronic form could provide avenues for research collaboration among social scientists and health industry practitioners, as researchers pose questions and search for patterns that help achieve the goals of quality care and cost reduction. Yet another somewhat ironic opportunity exists in the disappointing adoption rate of electronic health records (EHR) in the United States; hundreds of thousands of healthcare practices have not yet adopted the new technology, despite economic incentives and the threat of penalties (Fiegl 2012). The reasons for this delay are not well understood; however, they could be related to difficulties being experienced by adopters of EHR systems, for example, being plagued by embedded software errors and usually adopted by healthcare staff through "workarounds" that are inefficient and themselves the source of other errors (Koppel et al. 2005, 2008). Case studies of successful health information technology reveal that large healthcare organizations have developed their own information technology after lengthy learning processes or as a result of internal growth and development with a large base of consumers (see for example Scholl et al. 2011).

The vendor community for EHR generally has not been one of the leaders in participatory design. Much of the software used as a basis for EHR was developed for the accounting function, and this could explain some of the basis for its errors (Koppel and Kreda n.d.) Also, contracts for external vendors contain "hold harmless" clauses such that vendors are not liable for software errors (Koppel and Kreda 2009). These issues present important problems for society at large that cannot be resolved by any single discipline.

Studies of organizations that have been successful in developing EHR over long periods of time (e.g., Kaiser Permanente and the Veterans Administration) could disclose the principles and processes through which such systems have become institutionalized (including cognitive models) and possibly suggest alternative approaches to the implementation of EHR. The software industry, healthcare practices, and the social science community need a better understanding of the ways in which complex information systems can become successfully embedded in services systems that provide medical care to people, regardless of scale. This challenge will not be addressed without collaboration across disciplines, including the social sciences.

Lessons from the healthcare context for the social sciences in industry could be of value for other areas of research and development. Healthcare is a highly fragmented industry with many diverse actors whose interests are not necessarily aligned, and it is increasingly under public scrutiny, so there are opportunities for social science to become engaged. The same features characterize global supply chains that incorporate numerous and increasingly diversified actors, including transnational corporations, developing world entrepreneurs, migrant workers, affluent and conscientious consumers, and NGOs acting as self-styled certifiers of labor rights or other types of standards (Partridge 2011). Global supply chains require the development of standards to ensure quality and/or alignment of other values, yet standards are embedded within social contexts and relations of power, and thus their development and enactment may be fraught with conflict that requires understanding and amelioration (Busch 2011). The intertwining of technical and social factors in the construction and sustainment of global supply chains for manufacturing industries is one of the areas in which social science could collaborate with the natural sciences and engineering in the future.

7.8.2 Example of CKT: Wearable Computers and Sensors

Contact person: *Dongyi Chen, Mobile Computing Research Center, University of Electronic Science and Technology, China*

Wearable computers and sensors are becoming a greater possibility with the field of nanotechnology expanding and with the integration of body sensor networks, fiber assemblies, and interactive textile devices. Wearable electronics and photonic systems can be manufactured in regular facilities using low-cost, environmentally friendly processes. Once assembled, these wearable electronics and photonic systems can be programmed to operate as an extension of the humans wearing them, to enhance their capabilities, as well as to provide a unique mechanism of monitoring body conditions.

This field of wearable technology extended to implantable, or "body computing" devices can be classified into three categories: ultra-miniature electronic devices and sensors, flexible and planar printed electronics, and e-textiles. The *ultra-miniature electronic devices and sensors* are the implants that monitor physical body functions. The goal of these devices would be to improve intelligence (if implanted in the brain), power (if implanted in muscles), or basic functions (if implanted in cells or organs). A body sensor network would work together with the micro implants to create "intrabody networks" that could be used to measure pressure, flow, temperature; to stimulate nerves or tissues; and even to activate drug therapies.

The two categories flexible and planar printed electronics, and e-textiles, expand the areas where research and development will make body computing possible. "Printed electronics" can be flexible, stretchable materials with special single-cell-based "nano-ink" that can function as monitors, printed organic memories, electrochemical capacitors, or even batteries. "E-textiles," also referred to as smart textiles, refers to

7 Implications: Societal Collective Outcomes, Including Manufacturing

"intelligent" fibrous materials that can be developed to incorporate sensing as well as actuating, control, and transmitting of wireless data. There are still many opportunities to discover, develop, and expand this field of wearable computers/sensors and photonic devices. Dr. Xiao Ming Tao from The Hong Kong Polytechnic University divides these opportunities into four major research areas:

- *Intelligent fibers and fibrous assembly structures*: This entails creating fibers and fibrous assemblies from smart materials with single or multifunctional intelligence.
- *Interactive textile devices*: This entails engineering of fibrous products that can be used for sensing, communicating, memorizing, actuating, and energy harvesting.
- *Product integration and fabrication technology*: This entails investigating and developing ways to incorporate existing fabrication technologies with intelligent fibers and textile products.
- *Human-machine-clothing interaction and wearable technology*: This entails developing efficient sensing functions in order for the wearable computers to detect various human functions.

Developing these areas of research will expand the application of wearable computers/sensors to the extent that they will become essential to quality of life.

7.9   International Perspectives

The following are summaries relevant to this chapter of discussions at the international regional WTEC NBIC2 workshops held in Leuven, Belgium, September 20–21, 2012; in Seoul, Korea, October 15–16, 2012; and in Beijing, China, October 18–19, 2012. Further details of these workshops, including workshop agendas and experts who attended, are provided in Appendix A.

7.9.1   United States–European Union NBIC2 Workshop (Leuven, Belgium)

The discussions of NBIC2's impact on manufacturing, innovation, and societal outcomes spread over several breakout sessions, for example, Working Group #1 on research and Working Group #2 on innovation. The summary below incorporates findings from the several breakout sessions.

NBIC2 was identified by Working Groups 1 and 2 to have direct impacts on water, energy and environment, and food and agriculture. The challenges are obviously so great that no single discipline can solve them alone. For example, in the area of food and agriculture for the EU, there is a balance between social acceptance of technology-based optimum distributions and culture-based existing distributions—a tension between sustainable development and individual preferences. Therefore, convergence technologies can contribute more in using technologies to enhance productivity, such as refrigeration, sensors for water delivery systems, bio-systems to induce lateral root growth in desert plants or reduction of water usage, and in assisting governance by conducting comprehensive life cycle analysis.

While the above group discussions focused on the impact of NBIC2 on physical and environmental systems, another group examined the impact of NBIC2 on human capacity in terms of (1) using machines effectively to increase human capabilities, even perhaps, passing some human activities over to machines, including neuromorphic engineering; (2) understanding how the machines we create and use will modify ourselves, and understanding how learning occurs; (3) enhancing the way we do a variety of tasks (including imagination) and doing tasks we cannot currently do; (4) enhancing the ability of humans and machines to communicate with each other beyond language; and (5) changing the education models at all levels to advance personalized learning. NBIC2 technologies can effectively tie the tremendous work in understanding how the brain works with our engineering systems, for example, manufacturing processes and systems, and cyberspace security.

7.9.2 United States–Korea–Japan NBIC2 Workshop (Seoul, Korea)

Panel members/discussants

Co-Chairs:

Hak Min Kim, Korea Institute of Machinery & Materials (KIMM, Korea)
Masafumi Ata, National Institute of Advanced Industrial Science and Technology (AIST, Japan)
Jian Cao, Northwestern University (U.S.)

Others:

Eung-sug Lee, Korea Institute of Machinery & Materials (KIMM, Korea)
Sung-Hoon Ahn, Seoul National University (Korea)
Dae Maun Kim, Korea Institute for Advanced Study (KIAS, Korea)
Takashi Kohyama, Hokkaido University (Japan)
Kazunobu Tanaka, Japan Science and Technology Agency (JST, Japan)
Seiichiro Kawamura, JST (Japan)
Kuiwon Choi, Korea Institute of Science and Technology (KIST, Korea)
Yong Joo Kim, Korea Electrotechnology Research Institute (KERI, Korea)
Jo-Won Lee, Hanyang University (Korea)

The Korean government has invested heavily in convergence technologies, defined as the convergence of nanotechnology, biotechnology, information technology (IT), environmental technology, space technology, and culture technology. The Twenty-First Century Frontier R&D program was established about 10 years ago to fund R&D centers such as the Center for Nanoscale Manufacturing and Equipment

(CNMT). Each center received $10 million per year for 10 years. The program was replaced by the Global Frontier R&D program, which has a funding level of $10-30 million per year per center. Three such centers were established in 2010 (the Center for Bionics, Center for Biomaterials, and Center for Theragnosis) and four centers in 2011 (the Center for Multiscale Energy Systems, Center for Advanced Soft Electronics, Multi-dimensional Smart IT Convergence Systems, and Design and Synthesis of Biosystems). The new initiative "Nano-convergence 2020" was established for creating new industries and markets through the commercialization of nano-convergence technologies. The project duration will be 9 years, from 2012 to 2020 with a total budget of $440 million, of which $370 million in government funding is matched with $70 million in private funding.

Panelists from Korea and Japan took the view that *in the last ten years* there has been particularly rapid and widespread integration among many individual manufacturing technologies, nanoscience, and biology, resulting in various hybrid processes for applications such as structural color and bio-inspired lotus leaves as self-cleaning mechanisms for glass, textiles, etc. With the aid of computer and information technologies, design and analysis of manufacturing have taken on more prominent and positive roles in increasing productivity. Conventional manufacturing processes have also been extended to biomanufacturing areas and small-scale multifunctional manufacturing. The semiconductor industry continues to push Moore's Law. Robots have been increasingly advanced in terms of accuracy and flexibility. Success stories include automation in manufacturing environment, the da Vinci surgical system, and others.

Similar to the observations made at the U.S. NBIC2 workshop, panelists at this workshop also believe that mass customization of manufacturing and additive processing using functional materials will be promising *in the next 5–10 years*. Distributed IT-assisted manufacturing can empower individuals and communities. Specifically, research needs include scalable models for distributed manufacturing, info-inspired manufacturing systems, translational science-driven micro-and nanoscale manufacturing, predictive science and engineering from the nanoscale to the macroscale, human–machine interactions, and a unified description language for manufacturing processes. Robotics can assist human health and functionality. There will be more development in a system-level biomimetic-ecomimetic approach for societal and technology development, for example, convergence of a self-organization approach in developing new manufacturing processes.

To facilitate these changes, robust standards for data/material formats and faster IT network speeds were identified as infrastructure needs. Particularly, urban-appropriate technology with emphasis on practice and implementation was identified as an emerging topic with high priority. The impact of converging technology will be rapidly experienced by society in general, including the bottom majority of the income pyramid.

The R&D investment and implementation strategies suggested by this group include establishing pull-driven (needs-driven) R&D strategies; creating global open access to designer, engineer, and manufacturer; inspiring politicians and their staff about the NBIC concept and potentials through increasing communication channels and effectiveness; having a better plan for disseminating NBIC information to the public; and establishing an intergovernmental platform for addressing Earth-scale issues in developing NBIC technologies, e.g., energy, pollution, water, etc.

7.9.3 United States–China–Australia–India NBIC2 Workshop (Beijing, China)

Panel members/discussants

Dongyi Chen (Co-Chair), University of Electronic Science and Technology (China)
Gordon Wallace (Co-Chair), University of Wollongong (Australia)
Jian Cao (Co-Chair), Northwestern University (U.S.)

Others:

Bin Hu, Lanzhou University (China)
Calum Drummond, University of Melbourne (Australia)

The Chinese government has invested heavily in nanotechnology and biotechnology. One example is the Suzhou Science and Technology town jointly established by the national Ministry of Science & Technology and the Jiangsu provincial and Suzhou municipal governments (SND 2010). The investment has been about US$800 million. In the last 10 years, the integration among different disciplines has been increasingly seen in practice, for example, in integration between materials, design, and manufacturing. One example is 3D printing on soft materials such as textiles (Jost et al. 2011) or printing of single cell and growth factors. Manufacturing enterprises have evolved for the most part as concentrated urban phenomena.

A 2012 cartoon titled, "How babies will be born in the future" (http://www.makeuseof.com/tech-fun/how-babies-will-be-born-in-future/), and the key character Sun WuKong in the classical Chinese novel *Journey to the West* by ChengEn Wu in 1550, inspired lively discussions on the vision for the next 10 years. In summary, four areas were identified:

1. Distributed IT-assisted manufacturing to empower individuals and communities, used for point-of-need manufacturing, for example, printing conduits in surgery for nerve, muscle or organ repair in an operation room, or personalized medicine with doctor-prescribed formulation
2. Enhancing human sensing beyond current capabilities, e.g., body sensor networks, such as for biological signal sensing and environmental sensing, on jewelry, on the body, on hair mousse, or on watches
3. Integrating brain research with manufacturing research, e.g., electronic circuits/devices for understanding brain function, and then using the knowledge to enhance the speed and the process of determining successful outcomes for robots or for extension of human performance

4. Integrating environmental, health, safety, ethical, legal, and societal issues with technology development, e.g., regulation in machinery fabrication in the new paradigm where printers print printers; the new role of pharmaceutical companies in an era of personalized medicine

Cyberinfrastructure, collocated facilities, and personnel were identified as the *infrastructure needs* for the future development. In addition, new standards for NBIC needs, and manufacturing equipment in universities, community colleges, and high schools were also identified as needs.

The suggested R&D strategies included establishing mission-driven challenges to lead convergence; engaging scientists and engineers in the importance of manufacturing if real outputs from the convergence of NBIC are to be realized; having an open architecture to encourage the engagement and the contributions from scientists and engineers, e.g., application software and hardware modules; inspiring resource providers about the NBIC concept and potentials through increasing channels and effectiveness in communication; and establishing ongoing education and retraining of manufacturing workforce. Among those, the emerging priorities are identified as new manufacturing processes and systems for integrating different scales and the convergence of cognitive science and IT to create innovative manufacturing processes rather than the current ad hoc approach.

In terms of social impact, other than previously identified positive effects, one cautious point identified is the risk of having technology (for example, advanced 3D printing technology) land in the wrong hands, and measures to mitigate this.

References

Beaman, J., Atwood, C., Bergman, T., Bourell, D., Hollister, S., Rosen, D.: WTEC Report on Additive/Subtractive Manufacturing. Study of European Research, Baltimore (2004)

Blomberg, J., Suchman, L., Trigg, R.: Reflections on a work-oriented design project. Hum. Comput. Interact. **11**, 237–265 (1996)

Bourell, D.L., Leu, M.C., Rosen, D.W. (eds.): Roadmap for Additive Manufacturing; Identifying the Future of Freeform Processing. University of Texas at Austin, Laboratory for Freeform Fabrication, Austin (2009)

Busch, L.: Standards: Recipes for Reality. MIT Press, Cambridge, MA (2011)

Cefkin, M. (ed.): Ethnography and the Corporate Encounter: Reflections on Research in and of Corporations. Berghahn, New York (2009)

DeVor, R.E., Kapoor, S.G., Cao, J., Ehmann, K.F.: Transforming the landscape of manufacturing: distributed manufacturing based on desktop manufacturing (DM)2. J. Manuf. Sci. Eng. **134**, 041004 (2012). http://dx.doi.org/10.1115/1.4006095

Ehmann, K.F., Bourell, D., Culpepper, M.L., Hodgson, T.J., Kurfess, T.R., Madou, M., Rajurkar, K., DeVor, R.E.: WTEC Panel Report on International Assessment of Research and Development in Micromanufacturing. WTEC, Baltimore (2005). Available online: http://www.wtec.org/micromfg/report/Micro-report.pdf

Endo, C.: Desk top factory at Takashima. Presentation by Chiaki Endo, Managing Director, Takashima Sangyo Co. Ltd, 21 Oct 2005, Suwa, Japan (2005)

Fang, N., Lee, H., Sun, C., Zhang, X.: Sub-diffraction-limited optical imaging with a silver superlens. Science **308**(5721), 534–537 (2005)

Federal Reserve Bank of Dallas: The Right Stuff: America's Move to Mass Customization. Annual report (1998)

Fiegl, C.: Stage 2 meaningful use rules sharply criticized by physicians. American Medical News online: http://www.ama-assn.org/amednews/2012/05/14/gvl10514.htm (2012)

Foxconn (Focus Taiwan News): Foxconn plans to increase China workforce to 1.3 million. Available online: http://focustaiwan.tw/ShowNews/WebNews_Detail.aspx?ID=201008190012&Type=aECO (2010, August 19)

Hess, D.: Ethnography and the development of science and technology studies. In: Atkinson, P., Coffey, A., Delamont, S., Lofland, J., Lofland, L. (eds.) Handbook of Ethnography, pp. 234–242. Sage Publications, London (2001)

Honegger, A., Langstaff, G., Phillip, A., VanRavenswaay, T., Kapoor, S.G., DeVor, R.E.: Development of an automated microfactory: Part 1 – Microfactory architecture and sub-systems development. Trans. N. Am. Manuf. Res. Inst. SME **XXXIV**, 333–340 (2006a)

Honegger, A., Langstaff, G., Phillip, A., VanRavenswaay, T., Kapoor, S.G., DeVor, R.E.: Development of an automated microfactory: Part 2 – Experimentation and analysis. Trans. N. Am. Manuf. Res. Inst. SME **XXXIV**, 341–348 (2006b)

Jost, K., Perez, C.R., McDonough, J.K., Gogotsi, Y.: Carbon coated textiles for flexible energy storage. Energy Environ. Sci. **4**, 5060–5067 (2011)

Koppel, R., Kreda, D.: Health care information technology vendors' "hold harmless" clause: implications for patients and clinicians. JAMA **301**(12), 1276–1278 (2009)

Koppel, R., Kreda, D.: Health IT usability and suitability for clinical needs: challenges of design, workflow, and contractual relations. Working Paper (n.d.)

Koppel, R., Metlay, J.P., Cohen, A., Abaluck, G., Localio, A.R., Kimmel, S.E., Strom, R.B.: Role of computerized physician order entry systems in facilitating medication errors. JAMA **293**(10), 1197–1203 (2005)

Koppel, R., Wetterneck, T., Telles, J.L., Karsh, B.-T.: Workarounds to barcode medication administration systems: their occurrences, causes and threats to patient safety. J. Med. Inf. Assoc. **15**, 408–423 (2008)

Kotha, S.: Mass customization: implementing the emerging paradigm for competitive advantage. Strateg. Manage. J. **16**, 21–42 (1995)

Liu, Z., Durant, S., Lee, H., Pikus, Y., Fang, N., Xiong, Y., Sun, C., Zhang, X.: Far-field optical superlens. Nano Lett. **7**(2), 403–408 (2007)

Manyika, J., Chui, M., Brown, B., Bughin, J., Dobbs, R., Boxburgh, C., Byers, A.H.: Big Data: The Next Frontier for Innovations, Competition and Productivity. McKinsey Global Institute Report. Available online: http://www.mckinsey.com/Insights/MGI/Research/%20Technology_and_Innovation/Big_data_The_next_frontier_for_innovation (2011)

Marcus, G., Fisher, M.: Anthropology as Cultural Critique: An Experimental Moment in the Human Sciences. University of Chicago Press, Chicago (1986)

NNMI (National Network for Manufacturing Innovation): National Additive Manufacturing Innovation Institute (description of the pilot institute). http://www.manufacturing.gov/nnmi_pilot_institute.html (2012)

NRC (National Research Council): Rising Above the Gathering Storm, Revisited: Rapidly Approaching Category 5. The National Academies Press, Washington, DC (2010)

Partridge, D.J.: Activist capitalism and supply-chain citizenship: producing ethical regimes and ready-to-wear clothes. Curr. Anthropol. **52**(Suppl. 3), S97–S111 (2011)

Prinz, F.B., Atwood, A.L., Aubin, R.F., Beaman, J.J., Brown, R.L., Fussell, P.S., Kramer, B.M., Lightman, A.J., Narayanan, K., Sachs, E., Weiss, L.E., Wozny, M.J.: JTEC/WTEC Panel Report on Rapid Prototyping in Europe and Japan Volume I Analytical Chapters. Loyola College in Maryland, Baltimore (1997)

Rabinow, P.: The Accompaniment: Assembling the Contemporary. University of Chicago Press, Chicago (2012)

Reese, W.: Behavioral scientists enter design: seven critical histories. In: Squires, S., Byrne, B. (eds.) Creating Breakthrough Ideas: The Collaboration of Anthropologists and Designers in the Product Development Industry, pp. 17–43. Bergin and Garvey, Westport (2002)

Roco, M.C., Bainbridge, W.S. (eds.): Converging Technologies for Improving Human Performance: Nanotechnology, Biotechnology, Information Technology, and Cognitive Science. Kluwer Academic Publishers (now Springer), Dordrecht (2003)

Scacchi, W.: Game-based virtual worlds as decentralized virtual activity systems. In: Bainbridge, W.S. (ed.) Online Worlds: Convergence of the Real and the Virtual, pp. 225–236. Springer, New York (2010)

Scholl, J., Syed-Abdul, S., Ahmed, L.A.: A case study of an EMR system at a large hospital in India: challenges and strategies for successful adoption. J. Biomed. Inform. **44**(6), 958–976 (2011)

SND (Suzhou National New and High-Tech Industrial Development Zone): http://www.cnssz.com/cnssz/xqyw/xwnews.asp?newsid=413. Accessed November 2012 (2010)

Squires, S., Byrne, B. (eds.): Creating Breakthrough Ideas: The Collaboration of Anthropologists and Designers in the Product Development Industry. Bergin and Garvey, Westport (2002)

Srituravanich, W., Pan, L., Wang, Y., Sun, C., Bogy, D.B., Zhang, X.: Flying plasmonic lens in the near field for high-speed nanolithography. Nat. Nanotechnol. **3**, 733–737 (2008)

Suchman, L.: Plans and Situated Actions: The Problem of Human-Machine Communication. Cambridge University Press, Cambridge (1987)

Suchman, L., Blomberg, J., Trigg, R.: Reconstructing technologies as social practice. Am. Behav. Sci. **43**(3), 392–408 (1999)

Valentine, J., Zhang, S., Zentgraf, T., Ulin-Avila, E., Genov, D.A., Bartal, G., Zhang, X.: Three-dimensional optical metamaterial with a negative refractive index. Nature **455**(7211), 376–379 (2008)

Vallas, S.P.: Rethinking post-Fordism: the meaning of workplace flexibility. Sociol. Theory **17**(1), 68–101 (1999)

Zhou, R., Cao, J., Ehmann, K., Xu, C.: An investigation on deformation-based micro surface texturing. ASME J. Manuf. Sci. Eng. **133**(6), 061017 (2011). http://dx.doi.org/10.1115/1.4005459

Zysman, J.: Creating Value in a Digital Era: How Do Wealthy Nations Stay Wealthy? Berkeley Roundtable on the International Economy, Berkeley (2004). BRIE Working Paper No. 165

Chapter 8
Implications: People and Physical Infrastructure

James Murday, Larry Bell, James Heath, Chin Hua Kong, Robert Chang, Stephen Fonash, and Marietta Baba

This chapter addresses two different issues: the complementary needs for an appropriately educated populace and for specific kinds of physical facilities to facilitate development of converging knowledge and technologies. Based on this dual focus, each of the major headings in this chapter has two sub-headings, one for education and one for physical infrastructure.

An "educated populace" must be further delineated to address the different educational needs as one grows older: (a) K–12, (b) community college/technical college (CC/TC) programs leading to associate degrees, (c) university/college programs leading to bachelor's and master's degrees, (d) research university programs leading to Ph.D. degrees, (e) informal education, and (f) continuing education. Since everyone should be in a position to make informed decisions on the benefits/risks of converging technologies, the K–12 and informal education efforts—efforts that potentially reach everyone—must confer that knowledge. Workers on the

---

Corresponding editors M.C. Roco (mroco@nsf.gov) and W.S. Bainbridge (wbainbri@nsf.gov).

J. Murday
University of Southern California, Los Angeles, CA, USA

L. Bell
Museum of Science, Boston, MA, USA

J. Heath
California Institute of Technology, Pasadena, CA, USA

C.H. Kong
Indiana University, Bloomington, IN, USA

R. Chang
Northwestern University, Evanston, IL, USA

S. Fonash
Pennsylvania State University, University Park, PA, USA

M. Baba
Michigan State University, East Lansing, MI, USA

Table 8.1 University departments and their involvement in converging technologies

Converging tech.	Report chap.	Illustrative science departments				Illustrative engineering departments			
		Phys	Chem	Bio	Psych	ChE	EE	CSE	MSE
Nanotech		x	x	x		x	x	x	x
Biotech		x	x	x		x	x	x	x
Infotech		x	x	x	x	x	x	x	x
Adv Prosthetics	5	x	x	x	x	x	x	x	x
Cancer therapy	5	x	x	x	x	x	x	x	x
Cognition	6	x	x	x	x	x	x	x	x
Social networking	6				x		x	x	
Assistive robotics	7	x			x		x	x	x

Key: *Phys* Physics, *Chem* Chemistry, *Bio* Biological Sciences, *Psych* Psychology, *ChE* Chemical Engineering, *EE* Electrical Engineering, *CSE* Computer Science and Engineering, *MSE* Materials Science and Engineering

factory floor or in repair shops require specialized training beyond the traditional K–12 class work; CC/TC must address that need, as well as being a stepping-stone into bachelor's degree programs. At the bachelor's/master's degree level, a range of people such as engineers, business persons, politicians, and government regulators will need sufficient knowledge about converging technologies to expedite the transition of research discovery into innovative, socially benevolent products in the market. To ensure new ideas emanate from research discovery, it will be necessary to have Ph.D.s with sufficient breadth and flexibility to bridge the traditional disciplines. Finally, the growing pace in generating new knowledge and technical innovation will compel continuing education at all levels of the workforce.

The general theme of the NBIC2 workshop was converging technologies, i.e., convergence of knowledge from traditional academic science disciplines in order to engineer innovative technologies for the benefit of humanity. Table 8.1 itemizes some illustrative traditional academic disciplines (departments)—science focuses on the fundamental knowledge of natural things, whereas engineering focuses on the application of that fundamental knowledge. The rows show some selected technology goals that depend on the convergence of knowledge being developed in the disciplines (indicating, when pertinent, chapters in this book). The challenge for an educational system is to devise appropriate mechanisms such that the converging technology aspirations can be effectively and efficiently realized, in spite of any barriers imposed by the disciplinary silos (NRC 2012c; The Third Revolution 2011; Blackwell et al. 2009).

It should be noted that convergence is happening between the traditional disciplines. For instance, the subfields of "physical chemistry" in chemistry and "chemical physics" in physics highlight the fuzzy boundary between much of chemistry and physics. Much of modern biology—computational and molecular biology—draws heavily on mathematics, physics, chemistry, and computer science/engineering. It should also be noted that a converging technology can morph over time into an academic discipline of its own. Materials Science and Engineering began as a convergence of disciplinary topics in metallurgy, ceramics, chemistry (especially polymers), and physics. Starting in the 1960s, with Defense Advanced Research Projects Agency (DARPA) and the National Science Foundation (NSF) funding stimulus, it morphed into its own discipline, largely subsuming the older metallurgy and ceramics departments.[1] Computer Science and Engineering has a similar story; the progression from earlier mechanical inventions and mathematical theories towards the modern computational concepts and machines draws on contributions from physics, chemistry, mathematics, psychology, and several engineering disciplines.[2] The first Department of Computer Sciences in the United States was established in 1962. There are presently experiments at various universities to see if the continuing growth in nanoscale science and engineering (nanotechnology) might warrant a similar transition (Murday 2009).

8.1 Vision

8.1.1 Changes in the Vision Over the Past Decade

Education

Science, technology, engineering, and mathematics (STEM) education is now viewed as a key to the economic future of the United States (PCAST 2012a, b; Wladawsky-Berger 2012). While unemployment in the United States remained at ~8 % overall in 2012, jobs for skilled workers are going unfilled due to lack of qualified applicants (Engler 2012). That being said, we must also be attentive to changes in workforce needs so that an excess of any given skill set is not generated (Vastag 2012; ACS 2012; NIH 2012). The scope of the challenges mentioned in this chapter, and the potential cost to address them, is compelling partnerships among Federal/state/local governments, industry, and foundations (PCAST 2012a; NRC 2012c).

As one approach to remedy this situation, companies have recognized they must be more proactively involved to encourage STEM interests, influence the education system curricula, and project future employment needs (Schiavelli 2011). In 2010, Change the Equation was launched (http://changetheequation.org), a nonprofit that matches Federal and corporate funding to programs that promote science, technology, engineering, and math. State and foundation contributions to STEM programs are also growing (Holt et al. n.d.; http://stemgrants.com/).

The U.S. predominance in science and engineering (S&E) research is being challenged by dramatic improvements in other nations' programs (NRC 2012c). It is recognized that this will require attention to (a) the education of a native STEM workforce, (b) retention of those highly trained non-native students graduating from

---

[1] http://en.wikipedia.org/wiki/Materials_science

[2] http://en.wikipedia.org/wiki/History_of_computer_science

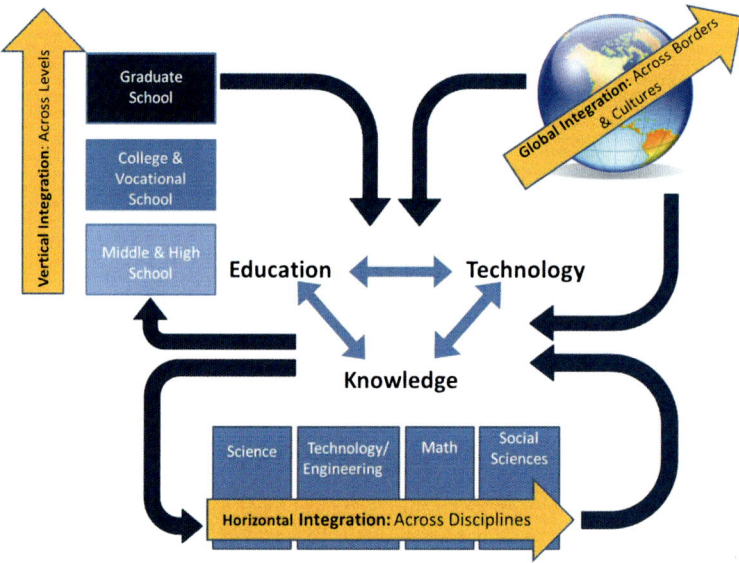

Fig. 8.1 Schematic highlighting the axes of integration required for maximal success in the convergence of technology and knowledge to further education (Courtesy of Robert Chang)

our universities, and (c) improved mechanisms to be aware of, and collaborate with, efforts in other nations.

Today the world is facing unprecedented global challenges that will take more than one country or region to solve. With population growth comes the need for more of the world's diminishing resources, which include energy, food, water, minerals, and many more. In addition, with large-scale transportation and growth in construction and industry, the environment and human health are under stress. Technology is helping, but it alone will not be able to solve these big problems. It will take the collective participation of all citizens to deliver the solutions. Everyone needs to have a big picture of the world and what is needed to build a sustainable future. This is where a cross-disciplinary approach is needed. An integrated education platform that includes STEM subjects and the social sciences is needed to teach and train the connectedness of STEM and society (see Fig. 8.1 and Case Study 8.8.1 that elaborate on this challenge and the approaches taken by several countries to address it). Both formal and informal education are needed in this regard.

Over the past decade, frontier scientific movements have occurred at the intersections of traditional disciplines, and that trend will certainly continue over the next decade. As an example from the National Nanotechnology Initiative (NNI), the National Cancer Institute-funded Cancer Centers for Nanotechnology Excellence (CCNEs) and the Physical Sciences Oncology Cancer Centers (PSOCs) have brought physical scientists and engineers into medical schools at an unprecedented level. Other programs also have played significant roles, for instance, the human genome project has brought together engineering, computer science, mathematics, biology, and social science. In another example, the convergence in the sciences and technologies over the last several decades has brought the social sciences to the brink of a transition from a phenomenological basis into disciplines based on chemical/physical laws underlying brain function, i.e., cognition.

Both educational deficiencies at the K–12 levels and the skyrocketing cost of university/college degrees (*push*), coupled with the revolutionary advances in digital information (*pull*), compel the introduction of new approaches to education (NSF 2011, 2007; Oblinger 2012; Molebash 1998; see also Chaps. 2 and 4 and Case Study 8.8.3). If the cost of digital capabilities can be lowered sufficiently, one can even entertain the possibility for personalized education (Evans 2012). It will be necessary for education, cognition, information technology, and subject matter experts to work together toward the most effective implementation of information technology in education. Further, one might argue that the communications skills traditionally embodied in reading and writing must now be supplemented by the development of digital communication skills, especially in post-secondary education (Adobe Education 2011).

K–12. There has been concern over the state of the U.S. K–12 education system for decades (NRC 2007a; NSB 2007; The Opportunity Equation 2009), including its lack of any instruction in engineering (NRC 2009). The importance of improvement in U.S. STEM education has been highlighted by an additional three recent reports (NRC 2007b; NSB 2010; PCAST 2010). A recent study, *Successful K–12 STEM Education: Identifying Effective Approaches*, focuses on means to achieve successful K–12 STEM outcomes (NRC 2011).

In the last decade the National Governor's Association has initiated a vision of common core standards for K–12, which seeks to identify up-to-date standards that can be adopted by all (or most) of the states. Given a mobile society, this vision is long overdue (Schmidt and McKnight 2012); students moving from one location to another should find essentially the same expectations. Common core K–12 standards for Mathematics and English Language Arts have been established (http://www.corestandards.org/). The present state science standards have been given a collective grade of "C" (Lerner et al. 2012). Common core standards for science, named Next Generation Science Standards (NGSS), are under development and are to be integrated, be progressive, and incorporate engineering and technology (http://nextgenscience.org/next-generation-science-standards/).

Community and Technical Colleges. The importance of community colleges/technical colleges as an approach toward a skilled workforce has been recognized (NRC 2012b). Manufacturing jobs requiring no more than a secondary education are declining at a rapid pace. Workers on the factory floor or in repair shops require specialized training beyond the traditional K–12 class work; CC/TC must address that need, as well as provide a stepping-stone into bachelor's degree programs. The CC/TC level of education is considered the "sweet spot" for reducing the skills gap in manufacturing; increased investment in this sector is recommended, following the best practices of leading innovators (NSTC 2012). In rapidly evolving, converging science/technology areas, community colleges/technical colleges will need faculty training, technical support, and access to advanced laboratory facilities to fully perform their critical role in today's education needs.

Colleges and Universities. There is growing recognition that the traditional academic departments at universities and colleges are: (a) losing too many students, (b) restrictive in their subject matter, (c) discouraging interdisciplinary collaboration, (d) not providing adequate knowledge about the skill sets required for innovation, and (e) not making effective use of the evolving information technologies (NRC 2002; Board on Science Education 2012; Young 2012). There are other indicators that the role of digital technology in education has also become more paramount (Center for Digital Education, http://www.centerdigitaled.com/; Oblinger 2012; Sacramento Bee 2012). Surveys of employers reveal that what they prize most in future managers are excellence in written and spoken communication, critical and creative thinking, an ability to collaborate across distances and cultural differences, breadth of knowledge and experience that takes students out of localism and provincialism, basic technical skills, quantitative literacy, and an ability to be flexible and take risks in changing environments (Davidson n.d.). In addition to breaking down barriers in the sciences/engineering disciplines, it's time to transform the focus, mission, and rhetoric of liberal arts by infusing a focus on cross-disciplinary critical thinking with real-world technological approaches to problem solving.

Research Universities. While acknowledged as among the best in the world, the U.S. research universities are facing serious challenges: (a) Federal/state funding for research has been unstable and is unlikely to grow significantly in the short term; (b) business and industry has not yet fully partnered to effectively translate, disseminate, and transfer into markets the new knowledge and ideas that emerge from research; and (c) cost efficiency must be improved (NRC 2012c).

Recent trends have led to increasingly blurred boundaries between academia and industry, including the demise of basic research in industrial laboratories and the subsequent reliance on universities for the new ideas that might transition into innovative technologies. NSF and other agencies have long encouraged academic–industrial collaborations, but the basic nature of those collaborations has changed over the past decade. For example, 15–20 years ago a typical university–industrial project might have been a collaboration between scientists from a large industrial lab (e.g., GE, IBM, or DuPont) and a university group to work on a project that, by its nature, would have been considered precompetitive. Thus, intellectual property issues and conflicts of interest were easily identified and resolved. Today, a typical collaboration is likely to be between a small startup company and the lab of the faculty member who started that company, perhaps fueled by Small Business Innovation and Technology Transfer Research (SBIR/STTR) resources, or perhaps fueled by the fact that the faculty member sees the company as an extension of his or her lab. This is a huge difference. The IBMs of the world didn't rely upon their university collaborators for success in the marketplace, but small startup companies often need the leverage provided through an academic partnership for both short-term survival and long-term success.

These trends are likely to accelerate over the next decade. How a given university positions itself within this changing world may end up making the difference in the types of science it supports, the faculty members it can recruit and retain, and the economic impact that the university has on its surrounding community.

Informal Education. Ten years ago, informal education was not explicitly in the vision for converging nano-, bio-, info-, cognitive (NBIC) and other technologies, nor were converging technologies explicitly in the vision for informal education. The subject of nanotechnology alone was just beginning to be on the radar screen for informal education, with several television and radio media projects including nanotechnology in their programming and a very small number of science museums either planning exhibits related to nanotechnology and/or doing live presentations on the topic (Flagg 2005).

In 2009, the National Research Council (NRC) published its report *Learning Science in Informal Environments: People, Places, and Pursuits* (Bell et al. 2009). The report identifies six strands of science learning that encompass a broad, interrelated network of knowledge and capabilities that learners can develop in these environments.

- *Strand 1*: Experience excitement, interest, and motivation to learn about phenomena in the natural and physical world
- *Strand 2*: Come to generate, understand, remember, and use concepts, explanations, arguments, models, and facts related to science
- *Strand 3*: Manipulate, test, explore, predict, question, observe, and make sense of the natural and physical world
- *Strand 4*: Reflect on science as a way of knowing; on processes, concepts, and institutions of science; and on their own processes of learning about phenomena
- *Strand 5*: Participate in scientific activities and learning practices with others, using scientific language and tools
- *Strand 6*: Think about themselves as science learners and develop an identity as someone who knows about, uses, and sometimes contributes to science

The strands are distinct from, but necessarily overlap with, the science-specific knowledge, skills, attitudes, and dispositions that can be developed in schools. The strands are of special value in informal learning environments. The NRC report has helped the informal science education community to understand better the kinds of educational goals and impacts it should work toward in science learning. While much of this knowledge existed in the field prior to the report, it had not been pulled together into a coherent vision.

The Center for the Advancement of Informal Science Education (CAISE; http://caise.insci.org/) published its first inquiry group report early in 2009, *Many Experts, Many Audiences: Public Engagement with Science*, which focuses on the emergence of public engagement with science (PES) within the informal science education (ISE) field (McCallie et al. 2009). While the field uses the term "public engagement" quite broadly, the CAISE report focuses specifically on a paradigm for communication between scientists and the public that has developed in the science communication and public policy domains.

Over the past decade, many in the science communication and public policy arenas have argued that engaging the public in two-way dialogues about science-related public policy issues, in a way that allows scientists to learn from the public as well as the public to learn from the scientists, is a good strategy for strengthening communication and mutual support (McCallie et al. 2009; Leshner 2003; Yankelovich 2003). Today, public engagement with science in informal educational settings and as a formal component of policymaking is more prominent in Europe than in the United States. The approach gained momentum there as public opposition to genetically modified organisms and alarm at the BSE (mad cow disease) epidemic swelled in the late 1990s. Pointing to these public controversies about risk, scholars of "post-normal science" argue that the traditional decision-making model that relies on communicating the findings of scientific and technical research to receptive policymakers is inadequate in cases where "facts are uncertain, values in dispute, stakes high, and decisions urgent" (Turnpenny et al. 2011). These scholars assert that lively public engagement with science and technology across society is a strategy for managing long-term risks and realizing the benefits of new discoveries and technologies.

This "social technology," science and technology decision-making through public engagement focuses on quality as it stresses substance, process, and deliberation, as well as analysis (Jasanoff 2003). This approach introduces a "new organizing principle"—quality, rather than truth, as the value of scientific knowledge applied to problems in society (Luks 1999). Scholars in the United States have taken up and extended methodologies for post-normal science in the context of emerging technologies. David Guston at Arizona State University's Center for Nanotechnology in Society discusses the need for *anticipatory governance* (Barben et al. 2008, 992–993):

> … [It] comprises the ability of a variety of lay and expert stakeholders, both individually and through an array of feedback mechanisms, to collectively imagine, critique, and thereby shape the issues presented by emerging technologies before they become reified in particular ways. Anticipatory governance evokes a distributed capacity for learning and interaction stimulated into present action by reflection on imagined present and future sociotechnical outcomes.

Public engagement with science as it has emerged in the field of ISE encompasses a range of activities and techniques that can help to address this need. Training science museum educators for dialogue and discussion rather than for making presentations is a current joint project of the Nanoscale Informal Science Education Network (NISE Net) and the Center for Nanotechnology in Society (CNS). This work is presenting a new vision for informal science education.

Newspapers and magazines retain high prestige among scientists, educators, and the general public, but now is a difficult time to use traditional print media to increase public awareness of converging technologies. Science journalism positions are being cut from newspaper and magazine staffs across the nation. The growing influence of social networking sites such as YouTube, Second Life, MySpace, and Facebook provide new opportunities to replace and/or supplement traditional media.

Continuing Education. Longer careers enabled by extended life spans and unfilled STEM-related jobs in a period of high unemployment are directing attention to continuing education as a retraining mechanism, including for military veterans

(Wetzel 2009). In 2012 at the University of Southern California, 4,800 graduate students were enrolled in accredited online master's degree programs spanning nine USC schools. Total annual revenues for online USC professional, graduate, and continuing education programs were expected to reach $114 million in 2012 (Balassone 2012).

Physical Infrastructure

With the proliferation of digital sensing and the subsequent growth in digital data has come the opportunity to use the revolutionary advances in information technologies in new and more effective ways (Atkins et al. 2003; NSF Cyberinfrastructure Council 2007), including the concept of networked science/citizen science (Wikipedia: *Reinventing Discovery: The New Era of Networked Science* (Nielsen 2011a); Cook 2011; CAISE 2012).

The success of the various user facilities hosted by NSF,[3] the Department of Energy (DOE),[4] and the National Institute of Standards and Technology (NIST),[5] along with the new national focus on advanced manufacturing, has led to efforts toward advanced manufacturing user facilities.[6] NSF is also introducing the National Ecological Observatory Network, the Ocean Observatory Initiative, and the Earth Cube Initiative.

8.1.2 The Vision for the Next Decade

A vision for converging technology education in the coming decade is building out the nation's human and technical infrastructure to better enable the rapid, effective creation and manufacturing of innovative, converging-technology-enabled products and processes that can address our shared social needs. To realize this vision it will be necessary:

1. For our education system to provide

    - Personalized education, customized to an individual's learning style, enabled by effective use of nanoscale information technologies, and informed by new cognition insights
    - The convergence of information technologies, coupled with deeper understanding of cognition, leading to highly effective, widespread, web-based education aids

---

[3] Various centers and networks, such as the Nanoscale Science and Engineering Centers (NSEC), the National Nanotechnology Infrastructure Network (NNIN), and the Network for Computational Nanotechnology (NCN).

[4] NanoCenters at Argonne, Brookhaven, Berkeley, Sandia, and Oak Ridge National Labs.

[5] Center for Nanoscale Science and Technology (CNST) at NIST Gaithersburg.

[6] For example, manufacturers can find user facilities and assistance at the websites http://www.nist.gov/mep/ and http://www.ornl.gov/user-facilities/mdf/.

- A distributed converging knowledge and technology network with a multidomain database, education modules, and facilities, including regional centers capable of leading efforts to enable/foster converging technology education
- New science, engineering, and technology knowledge incorporated more quickly into the standards/curriculum for K–20
- A working sustainable partnership structure within and between universities (including facilities, equipment, and expertise) and community/technical colleges to enable effective and efficient delivery of a broad education in converging technology areas
- A skilled workforce for converging-technology-enabled jobs with specific attention to the roles for community colleges and technical colleges
- Foundational courses in the college freshman year for non-science/engineering students that introduce the broader concepts of science and technology, including the expected impacts of converging knowledge and technologies
- Joint curricula developed by education colleges and science departments working in tandem and tied to a system of outcome metrics
- Restructuring of colleges away from traditional disciplines towards transdisciplinary approaches to the solution of societal needs, allowing focus on apprenticeship/mentoring in interactive workspaces while retaining rigorous student skill sets
- Awareness and understanding of societal implications of research in areas of converging technologies, and enhancement of science communication skills, both as key components of formal training for science and engineering graduate and undergraduate students
- Industrial participation in determining the graduate/undergraduate education experience to better prepare students for workforce needs
- Shift of graduate education programs from traditional departments into interdisciplinary programs, administered centrally with cross-departmental chairing of committees; removal of institutional barriers to interdisciplinary research
- Students given exposure to business and information technology (IT) best practices, in environments that encourage entrepreneurship
- Science museums and other venues of informal education interested in and capable of engaging public audiences in the areas and developments of current research, including converging technologies
- Convergence between physical/engineering sciences, the social/political sciences, and the science of science communication, leading to more effective public outreach and engagement efforts in informal learning environments

2. For our physical facilities to provide

- An infrastructure that enables rapid access to converging technology fabrication facilities ("advanced machine shops") implemented at research universities, with distributed special user facilities for the most expensive items
- Expanded geographically distributed user facilities such as the National Nanotechnology Infrastructure Network (NNIN) and Network for Computational Nanoscience (NCN) but focused more broadly on converging technologies

- Advanced manufacturing facilities with the diverse capabilities necessary to accelerate transition of converging technologies into products
- Utilization of distance technology to give teachers and, more importantly, students direct access to, and control of, expensive university-based instrumentation within K–14 classrooms

3. For the social sciences to be more fully integrated with physical and information sciences, and to be based more on fundamental laws rather than phenomenological observation.

8.2 Advances in the Last Decade and Current Status

The social sciences today are roughly where the biology and geology disciplines were 50 years ago, on the cusp of becoming sciences based on fundamental chemical/physical laws rather than phenomenological observations. The social sciences are now largely based on statistical inferences based on human behaviors but are beginning to migrate toward statistical inferences based on understanding cognition/emotion processes in the brain. The increasing availability of large digital data sets and computational analysis (i.e., data analytics) is helping to accelerate this progress, as it already has in areas such as climate and land-use in geography and some sub-fields of economics.

8.2.1 Education

Numerous examples illustrate the introduction in the last decade of new educational materials addressing the remarkable advances in individual converging technologies, especially those represented by the nano-, bio-, and info- topics. There are fewer examples that address the boundaries between any two converging knowledge/technology areas (see Fig. 8.2).

The cost of public elementary, secondary, and college education in the United States is high—estimated at ~$900 billion for the 2004–2005 school year (DOD 2005). Figure 8.3 illustrates the relative funding provided by state, local, and Federal sources for K–12 in 2009.

The U.S. Department of Education has about $70 billion annually to support education (not including loan programs). But recent documents show only ~$3 billion across the entire Federal government is devoted to STEM education (OSTP 2012a, b). Foundations provide additional support for education of about $2–3 billion, with examples in Table 8.2; some of that funding addresses STEM education. Industry also provides some monies for STEM education. A challenge is to make the most effective use of the limited STEM development funds, which are distributed across diverse sources. Open Education Resources (OER) and other digital education aides are viewed as a key mechanism to bring down the cost of education (Wiley et al. 2012).

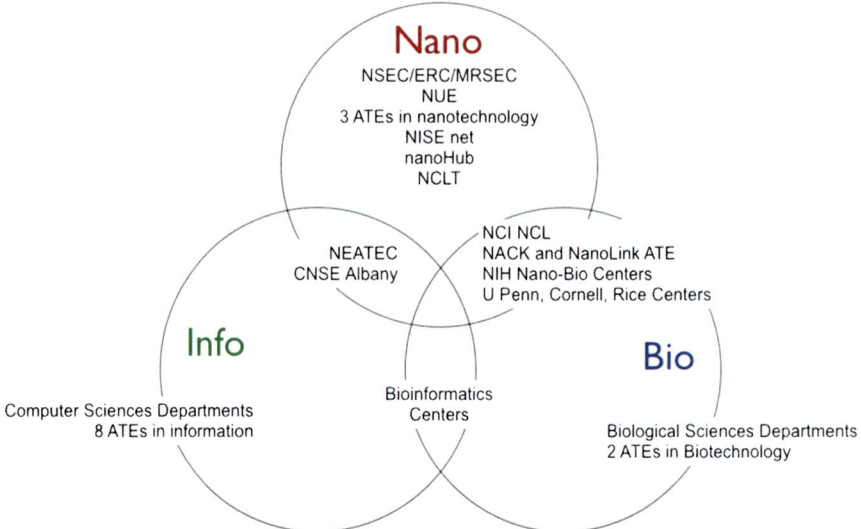

Fig. 8.2 Venn diagram illustrating education infrastructure components added in the last decade that address nano-, bio-, and info- needs, and those that provide capabilities at the intersections (Courtesy of James Murday)

Key: *NSEC* Nanoscale Science and Engineering Center, *ERC* Engineering Research Center, *MRSEC* Materials Research Science & Engineering Research Center, *NUE* Nanotechnology Undergraduate Education, *NCLT* National Center for Teaching and Learning, *NEATEC* Northeast Advanced Technological Education Center, *CNSE* College of Nanoscale Science and Engineering, *NCI* National Cancer Inst., *NCL* Nanotechnology Characterization Lab, *NACK* Nanotechnology Applications and Career Knowledge network, *ATE* Advanced Technology Education centers

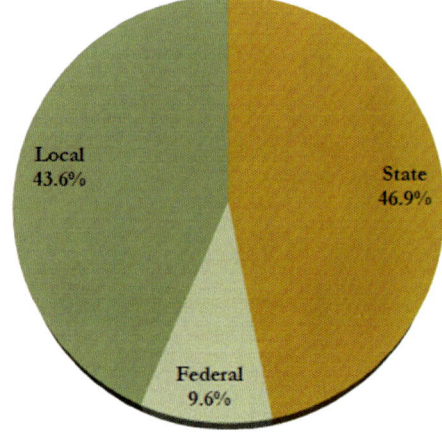

Fig. 8.3 Revenue sources for public-supported K–12 education in 2009, $590 billion total (Gai and Dadayan 2012; http://www.rockinst.org/government_finance/; ©Nelson A. Rockefeller Institute of Government, used by permission)

8 Implications: People and Physical Infrastructure

Table 8.2 Examples of foundation grants for education circa 2010

Foundation	Total awarded ($M)	No. of grants
Bill & Melinda Gates	357	314
Walton Family	223	289
W.K. Kellogg	137	101
Silicon Valley Community	80	630
Andrew W. Mellon	72	174
Michael and Susan Dell	63	191
Carnegie Corp. of NY	58	76
Ford	49	127
William and Flora Hewlett	49	116
Susan Thompson Buffett	47	103
Starr	45	84

Source: The Foundation Center (2012)

K–12

In 2010 the National Academy of Sciences released a report *A Framework for K–12 Science Education: Practices, Crosscutting Concepts, and Core Ideas* that identifies the key scientific ideas and practices all students should learn by the end of high school (NRC 2012a). This document provides the basis for the Next Generation Science Standards (NGSS)[7] that is currently under development by Achieve, Inc. The NGSSs are distinct from prior science standards in that they integrate three dimensions (Science and Engineering Practices, Disciplinary Core Ideas, and Crosscutting Concepts) within each standard and have intentional connections across standards.

In the last decade, NSF funded about a dozen projects that focused on nanotechnology education for K–12 and public audiences. These included K–12 formal education projects like *NanoTeach: Professional Development in Nanoscale Science* (DRL award #0822128), *Nanosense: The Basic Sense Behind NanoScience* (DRL award #0426319), *Probing the Nanoworld* (DRL award #0426401), and *NCTL: A Center to Develop Nanoscale Science and Engineering Educators with Leadership Capabilities* (DRL award #0426328). The projects in this portfolio of awards created innovative educational resources that were used and yielded evidence of learning in their target audiences of more than 10,000 students and teachers, but there is limited evidence of impact beyond those who directly accessed the resources produced by these projects (Nielsen 2011b).

With the growing cost of printed textbooks, there is now experimentation with electronic media to supplement or replace traditional textbooks.[8, 9] There are

---

[7] http://www.nextgenscience.org/next-generation-science-standards/
[8] http://www.ck12.org/flexbook/
[9] http://theeducationcafe.wordpress.com/2010/02/03/online-textbooks-for-middle-school-and-high-school/

programs that are fully adaptive and data-driven, such as Pearson's SuccessMaker for K–8 reading and math.[10] In January 2012, Pearson, McGraw-Hill Education, and Houghton Mifflin Harcourt announced partnerships with Apple to produce exclusive content through the new iBooks 2 platform.

The digital format has an advantage for the topic of converging technologies in that changes to paper textbooks requires more than a decade, but e-books can be updated at lower cost and can be made available far more quickly. To assist teachers, there is software that focuses on unifying public cloud, private cloud, and local device applications into a single, secure solution.[11]

New information-technology-based instruction is being introduced beyond e-books. There are experiments with a "flipped classroom" whereby information technology is used to present content (lectures) at home and the classroom is used to do problem-solving activities with the teacher as personal coach rather than lecturer.[12] The National STEM Video Game Challenge is a multi-year competition whose goal is to motivate interest in STEM learning among America's youth by tapping into students' natural passion for playing and making video games.[13] The World Wide Workshop has developed a game design curriculum being used in five states.[14] There are fully accredited online private schools.[15] (See also Chap. 4.)

Community and Technical Colleges

NSF has funded a number of Advanced Technology Education Centers (http://atecenters.org/) addressing nano-, bio-, info- and advanced manufacturing technologies (see Table 8.3). The Nanotechnology Applications and Career Knowledge network (NACK Network; see the case study in Sect. 8.8.2, which illustrates how convergence at the nanoscale is being addressed in education) and the Midwest Regional Center for Nanotechnology Education (Nano-Link), emphasize the importance of teaching the cross-disciplinary aspects of nanoscience and nanotechnology. For example, the NACK courses available for free downloading at http://www.nano4me.org stress the converging knowledge/technology reality of today's world and are designed for students coming from very diverse science and technology disciplines (including biology, biotechnology, chemical technology, manufacturing/industrial technology, physics, chemistry, electronics, etc.). The approach is to add a nanotechnology skill set to students' educational backgrounds, no matter what

---

[10] http://www.forbes.com/sites/jamesmarshallcrotty/2012/08/21/the-tech-driven-classroom-is-here-but-grades-are-mixed/

[11] For example, see https://customer.stone-ware.com/site/solutions/education/index.html, and https://getclever.com/about/

[12] htpps://sites.google.com/a/cloud.stillwater.k12.mn.us/flipped-classroom/about

[13] http://www.stemchallenge.org/about/Default.aspx

[14] http://www.worldwideworkshop.org/

[15] Examples include K12 International Academy, http://www.k12.com/int, and CalPac http://calpacschools.rtrk.com/?scid=2506134&kw=12030127&pub_cr_id=8435962594

Table 8.3 NSF-funded advanced technology centers addressing NBIC topics

Program name	Institution	Location
Nano		
Maricopa Advanced Technology Education Center	Maricopa Community Colleges	Phoenix, AZ
Northeast Advanced Technological Education Center	Hudson Valley Community College	Troy, NY
Midwest Regional Center for Nanotechnology Education (Nano-Link)	Dakota County Technical College	Rosemount, MN
Nanotechnology Applications and Career Knowledge (NACK)	Pennsylvania State University	University Park, PA
Bio		
Bio-Link National Center	City College of San Francisco	San Francisco, CA
Northeast Biomanufacturing Center and Collaborative	Montgomery County Community College	Blue Bell, PA
Info		
Boston Area Advanced Technological Education Connections	University of Massachusetts	Boston, MA
Cyber Security Education Consortium	University of Tulsa	Tulsa, OK
Center for Systems Security and Information Assurance	Moraine Valley Community College	Palos Hill, IL
Convergence Technology Center	Collin College	Frisco, TX
Creating the Next Generation of Cybersecurity Professionals	Prince George's Community College	Largo, MD
Information and Communications Technologies Center	Springfield Technical Community College	Springfield, MA
Mid-Pacific ICT Center	City College of San Francisco	San Francisco, CA
Advanced Manufacturing		
Florida Advanced Technological Education Center	Hillsborough Community College	Tampa, FL
National Center for Manufacturing Education	Sinclair Community College	Dayton, OH
National Center for Rapid Technologies	Saddleback College	Mission Viejo, CA
Regional Center for Next-Generation Manufacturing	CT Community Colleges' College of Technology	Hartford, CT
Technology and Innovation in Manufacturing and Engineering	Community College of Baltimore County	Baltimore, MD

field they come from. The students acquire, for their educational toolboxes, the ability to understand nanoscale concepts and to use them in other knowledge areas, as well as acquire invaluable "hands-on" ability to work at the nanoscale.

Colleges and Universities

The traditional lecture format is outdated and unscalable; there is need for standardized online delivery platforms. A growing number of experiments in the utilization of digital information technologies in college education (Brooks 2012; Sitzmann et al. 2006; Katsouleas 2012) includes web-based scaled-up courses offered by edX (Harvard, Massachusetts Institute of Technology, University of California–Berkeley, and University of Texas; https://www.edx.org/), Coursera (http://coursera.org), nanoHUB (http://nanohub.org/resources/), and the Kahn academy (http://khanacademy.org). To make the web-based approaches most effective, the nanoscale will be essential for information technology miniaturization and affordability, while cognitive studies will provide insight into the best way to implement the new technologies. Case study 8.8.3 provides some interesting ideas regarding the potential implications of e-learning for universities.

In 2004 the University of Albany, State University of New York (SUNY), created a College of Nanoscale Science and Engineering, an aggressive experiment in university nanoscale education with BS, MS, and Ph.D. degree programs. By leveraging its resources in partnership with business and government, CNSE supports accelerated high-technology education and commercialization and seeks to create jobs and economic growth for nanotechnology-related industries. (See the CNSE Case Study in Sect. 8.8.4.)

Engendered in part by converging technologies, there is growing experimentation in the engineering curricula to address the different demands on engineering graduates in the modern marketplace. For instance in 2010, select departments in the MIT School of Engineering launched a degree option that responds to the evolving desires of undergraduate students and to emerging changes in the engineering professions. Students satisfy department-based core requirements and declare an additional concentration, which can be broad and interdisciplinary in nature (energy, transportation, or the environment), or focused on areas that can be applied to multiple fields such as robotics and controls, computational engineering, or engineering management (http://engineering.mit.edu/education/engineeringdegree.php). Another example is the Franklin W. Olin College of Engineering in Needham, MA (http://www.olin.edu/), initiated in 2002, that has no academic departments and a commitment at all levels to active learning and interdisciplinary courses built around hands-on projects.

Research Universities

As a partnership paradigm, in 2005 the Semiconductor Industries Association (SIA) entered into a partnership with NSF and NIST in the Nanoelectronics Research Initiative (NRI; http://www.src.org/program/nri/). Industry monies funded four university centers: Southwest Academy for Nanoelectronics led by University of Texas–Austin, Institute for Nanoelectronics Discovery and Exploration led by University at Albany, Midwest Institute for Nanoelectronics Discovery led by Notre Dame, and Western Institute for Nanoelectronics led by University of California–Los Angeles. The Semiconductor Research Corporation (SRC) has also teamed up with NSF to fund research projects at existing NSF nanoscience centers and networks at universities across the country. In addition, there are six center research programs (STARnet, which supercedes the older Focused Center Research Program) jointly funded by SRC, industry, and DARPA. (See the SRC case study, Sect. 10.8.3 in Chap. 10.) Another partnership paradigm, reflecting medicine/health, is presented in Case Study 8.8.5.

NSF has implemented a variety of programs to encourage the development of new approaches to converging technology (e.g., solicitations 12-549 Computationally & Data Enabled Science & Engineering; 12-509 Science, Technology and Society; and 12-515 Advanced Health Services through System Modeling as recent examples), and other programs to foster innovation (e.g., solicitations 12-586, Innovation Corps–Regional Node Program, and 12-012, Creative Research Awards for Transformative Interdisciplinary Ventures [CREATIV]).

Informal Education

In 2005 with NSF funding, the Museum of Science in Boston, in partnership with the Exploratorium in San Francisco and the Science Museum of Minnesota in St. Paul, assembled a group of museums and nanoscale research centers to establish NISE Net (http://www.nisenet.org/). The focus of NISE Net's work has been to build the capacity of science museums and research centers to raise public awareness, understanding, and engagement with nanoscale science, engineering, and technology. (See the NISE Net Case Study, Sect. 8.8.6.) However, over the last 10 years there have been few advances in informal education and public engagement explicitly addressing converging technologies and their potential impacts, except for the work that has been done in connection specifically with nanotechnology.

In the last decade NSF funded several projects that focused on nanotechnology education for public audiences, including media projects like Earth and Sky's *Nanoscale Science and Engineering Radio Shows* (DRL award #0426417), Oregon Public Television's *Nanotechnology: The Convergence of Science and Society* (DRL award # 0452371), and Twin Cities Public Television's *Dragonfly TV GPS: Investigating the Nanoworld* (DRL award #0741749); exhibit projects like Cornell's *Too Small to See* (DRL award #0426378), Sciencenter's *It's a Nano World* (ECCS award #9876771), and the Materials Research Society's *Strange Matter* (DRL award #0000586); and NISE Net (DRL awards #0532536 and #0940143).

The Centers for Nanotechnology in Society (CNS) at Arizona State University (http://cns.asu.edu/) and at the University of California at Santa Barbara (http://cns.ucsb.edu/) have provided research on public perceptions of nanotechnology and ideas about the implementation of anticipatory governance. These organizations and the NISE Net have explored, both independently and collaboratively, how formal and informal education might include content related to the societal implications of emerging technologies. In 2009, a newly formed Society for the Study of Nanoscience and Emerging Technologies (http://www.thesnet.net/Welcome.html), dedicated to open intellectual exchange that is aimed at the advancement of knowledge and understanding of nanotechnologies in society, held its first professional conference in Seattle and has alternated between European and U.S. conference sites annually.

The Center for Nanotechnology in Society at Arizona State University and its collaborators at North Carolina State University conducted the nation's first National Citizens' Technology Forum on the topic of nanotechnology and human enhancement in 2008. Groups of citizens at six sites across the United States expressed concerns about the effectiveness of regulations, equitable distribution, and monitoring; expressed greater preference for therapeutic over enhancement research; and identified the need to provide public information, including more public deliberative activities and K–12 education, about nanotechnologies (Hamlett et al. 2008).

The same group also conducted a national survey about nanotechnology and human enhancement. The data showed that the public differentiates between different applications of converging technologies, giving strong support for some and opposing others (Hays et al. 2013).

Collaborations between informal science educators and researchers in science communication and the social and political sciences have led to growing capacities for public engagement on topics of societal and ethical implications of emerging technologies (Bell 2008; Bell et al. 2009; Reich et al. 2006, 2007; McCallie et al. 2009; Flagg and Knight-Williams 2008). Collaborations between CNS social and political scientists and NISE Net educators have led to the development of a series of professional development workshops aimed at helping informal educators incorporate dialogue about the societal and ethical implications of nanotechnologies into their ongoing program activities.

Another collaboration between CNS social and political scientists, NISE Net educators, and others concerned with science, technology, and public policy has led to the establishment of the ECAST Network (Expert and Citizen Assessment of Science and Technology; http://ecastnetwork.org/). While ECAST's mission is fundamentally a governance mission "to support better-informed governmental and societal decisions on complex issues involving science and technology," one of its methodologies involves broad public engagement aimed at helping the public learn about new technologies and their societal and ethical implications.

While the developments described here are not specifically about converging technologies, they describe newly developed capacities, structures, and convergence within academic and informal educational institutions that can be valuable contributors to informal education and public engagement about converging technologies in the decade ahead.

Research in science communication is another discipline that could support better public outreach and engagement. On May 21–22, 2012, the National Academies of Science held a colloquium on the Science of Science Communication.[16] The meeting surveyed the state of the art of empirical social science research in science communication and focused on research in psychology, decision science, mass communication, risk communication, health communication, political science,

---

[16] http://www.nasonline.org/programs/sackler-colloquia/completed_colloquia/science-communication.html

sociology, and related fields on the communication dynamics surrounding issues in science, engineering, technology, and medicine. There were five distinct goals:

- To improve understanding of relations between the scientific community and the public
- To assess the scientific basis for effective communication about science
- To strengthen ties among and between communication scientists
- To promote greater integration of the disciplines and approaches pertaining to effective communication
- To foster an institutional commitment to evidence-based communication science

Speakers at the colloquium said that research suggests that scientists are highly trusted by the public) and that credibility is bestowed by the audience and is based on perceived common interest and perceived relative expertise. Despite a science goal of objectivity, personal ideologies also affect scientists' decisions and attitudes, but these may not be immediately visible to the public. Furthermore, psychological research suggests that we make decisions using one system of the mind and explain them using another. This suggests that our explanations may not be effective in influencing the decisions of others, especially if personal ideologies differ.

One intriguing and confounding idea in science communication comes from the work of the Cultural Cognition Project at the Yale Law School. One of its researchers in cultural cognition, Dan Kahan, says that it "causes people to interpret new evidence in a biased way that reinforces their predispositions. As a result, groups with opposing values often become more polarized, not less, when exposed to scientifically sound information" (Kahan 2010, 296). This leads him to conclude that, "we need to learn more about how to present information in forms that are agreeable to culturally diverse groups, and how to structure debate so that it avoids cultural polarization" (Kahan 2010, 297).

Public engagement efforts about converging technologies could benefit from designs that are grounded in this evidence-based communication science. They would also benefit from science communication training for science communicators and scientists. Collaborations between informal science educators and nanoscale research centers and professional scientist organizations has led to a number of activities that foster science communication training, especially for early career scientists and graduate and undergraduate students in science and engineering. The Materials Research Society in collaboration with NISE Net has offered a growing number of workshops at its two conferences each year. These include sessions such as *Mastering Public Presentations*, *Technical Poster Design*, and *Making the Most of Broadcast Media*.

The Boston Museum of Science developed materials for dissemination of two professional development activities on science communication for early career scientists and posted them in the NISE Net catalog in September 2011. The *Research Experience for Undergraduates* [REU] *Science Communication Workshop* provides REU students with training in how to communicate their research to the other students in their programs with skills also useful for communicating with the public. The *Sharing Science Graduate Student Workshop and Practicum* provides similar training for graduate students but is tied to participating in public presentations during NanoDays activities. In June and July 2012, the Museum of Science conducted a Dissemination Workshop for Education and Outreach Faculty and REU Directors at NNIN and other nano-research centers. Eight institutions attended and learned how to implement these science communication workshops for their own students.

Beginning in 2007, the Pacific Science Center developed Portal to the Public "…to assist informal science education institutions as they seek to bring scientists and public audiences together in face-to-face public interactions that promote appreciation and understanding of current scientific research and its application." The Portal to the Public *Implementation Manual* includes the *Catalog of Professional Development Elements*, a practical guide to creating and facilitating professional development experiences for scientists" (http://pacificsciencecenter.org/Portal-to-the-Public/).

Continuing Education

Continuing education has always been important, but the rapid growth in fundamental knowledge and the resultant new technologies, coupled with their multidisciplinary aspects (convergence) combine to exacerbate this need. There has been a notable increase over the past decade in the availability of web-based remote learning. A variety of courses, certificate programs, and even graduate degree programs can be accessed either completely online or in a blended combination of online and classroom instruction. Examples of providers include the University of Washington (http://pce.uw.edu/online-learning/), the University of Illinois (http://oce.illinois.edu/), the University of Massachusetts–Lowell (http://continuinged.uml.edu/online/), the University of Southern California (http://continuingeducation.usc.edu/), and the e-Learning Center (http://e-learningcenter.com/). The online approach provides two critical advantages for people in the workforce: (1) material is available in time frames that fit their personal schedules, and (2) at least theoretically one can gain access to the best instruction, independent of geographical location.

Professional science and engineering societies also play a role in continuing education. For example, the American Chemical Society has introduced Sci-MIND (http://proed.acs.org/products-services/sci-mind/), an innovative and interactive training product designed to challenge chemists to continuously invest in their professional scientific development.

8.2.2 Physical Infrastructure

Over the last decade there has been significant growth in physical infrastructure in each of the individual NBIC disciplines, with some of that growth addressing convergence between at least two disciplines (see Fig. 8.4 and Table 8.4).

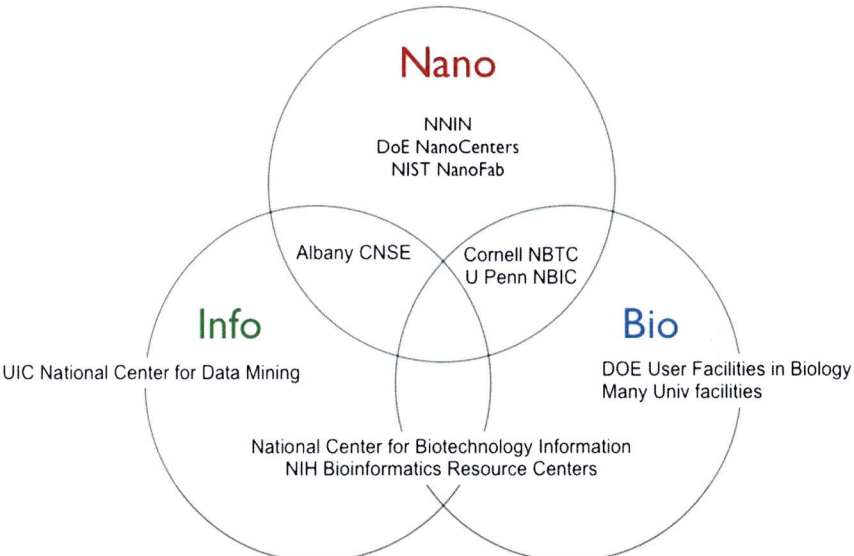

Fig. 8.4 Venn diagram illustrating facility infrastructure added in the last decade that addresses nano-, bio-, and info- needs, along with those that provide capabilities at the intersections (Courtesy of James Murday)
Key: *NNIN* National Nanotechnology Infrastructure Network, *CNSE* Center for Nanoscale Science and Engineering, *NBTC* Nanobiotechnology Center, *NBIC* Nano/Bio Interface Center, *UIC* University of Illinois Chicago

The NISE Net community of researchers and informal science educators is dedicated to fostering public awareness, engagement, and understanding of nanoscale science, engineering, and technology. With that goal, it has created and implemented exhibits and programs in a number of science museums (see the NISE Net case study, Sect. 8.8.6).

Innovative research depends on the infrastructure necessary to fabricate and characterize new devices and systems. In the past, university machine shops provided the wherewithal to do those tasks locally, with a minimum of time, effort, and expense needed to access the shop. The converging technologies frequently require highly sophisticated machines, too expensive to house within each university's "machine shop." While the existing user facilities illustrated above do provide the needed capability, for most researchers they require considerable travel time and delays waiting for facility permission.

The Network for Computational Nanotechnology is an "Infrastructure and Research Network" that strives to broaden researchers' access to sophisticated modeling and simulation tools and collaborations in the advancement of nanoscience and nanotechnology. It engages in research in the areas of (nano) electronics, mechanics, biology, photonics, and materials, primarily through leveraged funding and deployment of web-based research software tools on nanoHUB (http://nanohub.org/). (See the nanoHUB case study, Sect. 8.8.7).

Table 8.4 Sampling of center-scale projects with integrated capability for two or more NBIC fields

Affiliated agency	Program type	Member institution	Program title
Nano–Bio			
NSF	NSEC	Univ. Pennsylvania	Nano/Bio Interface Center
	NSEC	Rice Univ.	Center for Biological and Environmental Nanotechnology
	NSEC	Northwestern	Integrated Nanopatterning and Detection Technologies
AFOSR	MURI	Northwestern	Bioprogrammable 1-2-and 3-Dimensional Materials
	MURI	Harvard Univ.	Bio-Inspired Optics
NIH	NCI Alliance	Natl. Cancer Inst.	Nanotechnology Characterization Laboratory
	CCNE	Cal Tech	Nanosystems Biology Cancer Center
	CCNE	Dartmouth Col.	Dartmouth Center for Nanotechnology Excellence
	CCNE	UTexas Health Sci.	Texas Center for Cancer Nanomedicine
	CCNE	Harvard/MIT	Center of Cancer Nanotechnology Excellence
	CCNE	Northwestern	Nanomaterials for Cancer Diagnostics and Therapeutics
	CCNE	Stanford Univ.	Center for Cancer Nanotechnology Excellence and Translation
	CCNE	Johns Hopkins	Center for Nanotechnology Excellence at Johns Hopkins
	CCNE	Univ. N. Carolina	Carolina Center of Cancer Nanotechnology Excellence
	CCNE	Northeastern	Center for Translational Cancer Nanomedicine
	CNPP	Northeastern	Combinatorial-designed Nano-platforms to Overcome Tumor Resistance
	CNPP	Univ. Nebraska	High-capacity Nanocarriers for Cancer Therapeutics
	CNPP	Univ. Utah	Magnetoresistive Sensor Platform for Parallel Cancer Marker Detection
	CNPP	Cedar Sinai Med	Nanobioconjugate Based on Polymalic Acid for Brain Tumor Treatment
	CNPP	UNC Chapel Hill	Nanoscale Organic Frameworks for Imaging and Therapy of Pancreatic Cancer
	CNPP	Univ. New Mexico	Peptide-directed Protocells and Virus-like Particles: new Nanoparticle Platforms for Targeted Cellular Delivery of Multicomponent Cargo
	CNPP	Rice Univ.	Preclinical Platform for Theranostic Nanoparticles in Pancreatic Cancer

CNPP	Univ. Cincinnati	RNA Nanotechnology in Cancer Therapy
CNPP	Univ. S. Calif.	Targeting SKY Kinase in B-Lineage ALL with CD-19 Specific C-61 Nanoparticle
CNPP	Emory Univ.	Theranostic Nanoparticles for Targeted Treatment of Pancreatic Cancer
CNPP	Emory Univ.	Toxicity & Efficacy of Gold Nanoparticle Photothermal Therapy in Cancer
CNPP	Northwestern	Tumor Targeted Nanobins for the Treatment of Metastatic Breast and Ovarian Cancer
NHLBI-PEN	Burnham Institute	Nanotherapy for Vulnerable Plaque
NHLBI-PEN	Georgia Tech.	Nanotechnology: Detection & Analysis of Plaque Formation
NHLBI-PEN	Mass General Hosp.	Translational Program of Excellence in Nanotechnology
NHLBI-PEN	Washington Univ.	Integrated Nanosystems for Diagnosis and Therapy
NDC	Baylor Col. Med.	Center for Protein Folding Machinery
NDC	NYU Med	Nanomedicine Center for Mechanobiology Directing the Immune Response
NDC	Georgia Tech	Nanomedicine Center for Nucleoprotein Machines
NDC	UC Berkeley	NDC for the Optical Control of Biological Function
NDC	UCLA	Center of Cell Control
NDC	UC San Francisco	Engineering Cellular Control: Synthetic Signaling and Motility Systems
NDC	Univ. Cincinnati	Phi29 DNA-Packaging Motor for Nanomedicine
NDC	Univ. Illinois-Urbana-Champagne	Biomimetic Nanoconductors
Nano–Info		
NIST		
NRI	UTexas Austin	Southwest Academy of Nanoelectronics
NRI	UCLA	Western Institute of Nanoelectronics
NRI	Notre Dame	Midwest Institute for Nanoelectronics Discovery
NRI	SUNY/Albany	Institute for Nanoelectronics Discovery and Invention
NSF		
NSEC	Columbia Univ.	Center for Electron Transport in Molecular Nanostructures
NSEC	Harvard Univ.	Science for Nanoscale Systems and their Device Applications
NNIN	Stanford Univ.	Stanford Nanofabrication Facility
nanoHUB	Purdue Univ.	Network for Computational Nanotechnology

(continued)

Table 8.4 (continued)

Affiliated agency	Program type	Member institution	Program title
DARPA	STARnet	UCLA	Function Accelerated Nanomaterial Engineering Center
	STARnet	Univ. Minn.	Center for Spintronic Materials, Interfaces and novel Architectures
	STARnet	UIUC	Systems on Nanoscale Information Fabrics Center
AFOSR	MURI	UC Santa Barbara	Investigation of 3D Hybrid Integration of CMOS/Nanoelectronic Circuits
	MURI	Stanford Univ.	Integrated Hybrid Nanophotonic Circuits
	MURI	Stanford Univ.	Robust and Complex On-Chip Nanophotonics
	MURI	Columbia Univ.	New Materials Approaches for Future Graphene-Based Devices
	MURI	U. Wisc. Madison	Adaptive Intelligent Photonic/Electronic Systems Based on Silicon Nanomembranes
	MURI	UTexas Austin	Three Dimensionally Interconnected Silicon Nanomembranes for Optical Phased Array (OPA) and Optical True Time Delay
ONR	MURI	UC Berkeley	Functionalized Nanoscale Graphene: A Platform for Integrated Nanodevices
	MURI	Univ. MD	Tailoring Electronic Properties of Graphene at the Nanoscale
	MURI	MIT	Graphene Approaches to Terahertz Electronics
Bio–Info			
NIH			PubMed Central
			PubChem
			National Center for Biotechnology Information
–	University centers	Univ. Washington	Department of Biomedical Informatics and Medical Education
–		Columbia Univ.	Department of Biomedical Informatics
–		Univ. Michigan	Department of Computational Medicine and Bioinformatics
		Univ. Pittsburgh	Department of Biomedical Informatics
ARO	MURI	UTexas Arlington	Revolutionizing High-Dimensional Microbial Data Integration
	MURI	Albany Medical Col	A Brain-Based Communication and Orientation System

ONR	MURI	Conductive DNA Systems and Molecular Devices
	MURI	Roll-to-Roll High-Speed Printing of Multifunctional Distributed Sensor Networks for Enhancing Brain–Machine Interface
Cogno–Info		
DOE	PNNL	Cognitive Informatics
—	UTexas Health Sci UCenter	National Center for Cognitive Informatics and Decision Making
	Northwestern	
	Univ. Minnesota	

Key (in order of use): *NSEC* Nanoscale Science and Engineering Center, *AFOSR* Air Force Office of Scientific Research, *MURI* Multidisciplinary University Research Initiative, *CCNE* Centers of Cancer Nanotechnology Excellence, *CNPP* Cancer Nano Platform Partnership, *NHBLI-PEN* National Heart, Lung & Blood Institute Programs of Excellence in Nanotechnology, *NDC* Nanomedicine Development Centers, *NRI* Nanoelectronics Research Initiative, *STARnet* Semiconductor Technology Advanced Research Network, *NNIN* National Nanotechnology Infrastructure Network, *ONR* Office of Naval Research, *ARO* Army Research Office, *PNNL* Pacific Northwest National Laboratory

It is not only the physical/chemical/biological science facilities that are being affected by converging technologies; the earth sciences are in an analogous revolution, with a number of new facilities under development. In 2012 the National Center for Atmospheric Research (NCAR) opened on a new data center in Cheyenne, Wyoming. Scientists will use the supercomputing center to accelerate research into climate change, examining how it might impact agriculture, water resources, energy use, sea levels, and extreme weather events, including hurricanes. In 2013 the National Ecological Observatory Network (NEON) will open; it is a continental-scale research instrument consisting of geographically distributed infrastructure, networked via cybertechnology into an integrated research platform for regional- to continental-scale ecological research. Also in 2013, the Ocean Observatory Initiative will open; it is a program to provide sustained ocean measurements to study climate variability, ocean circulation and ecosystem dynamics, air–sea exchange, seafloor processes, and plate-scale geodynamics. To manage the data emanating from these new facilities, NSF is developing the Earth Cube concept (http://www.nsf.gov/geo/earthcube/), seeking transformative concepts and approaches to create integrated, convergent data management infrastructures across the geosciences.

8.3 Goals for the Next Decade

8.3.1 Education

It will be important to create a coordinated partnership of Federal/state/local governments, industry, and foundations addressing the challenges of STEM education in general and converging technologies specifically. This partnership must address an effective, affordable program to foster student interest in STEM careers, new approaches to STEM education, and accelerated transition of research discovery into technology innovation.

Converging technologies can certainly benefit from diverse perspectives. The United States is implementing efforts to entice underrepresented groups into STEM fields (e.g., see NRC 2012c and NSF programs such as BRIGE[17]). Converging technologies will benefit and may also assist in the recruitment, because they are expected to play major roles in ameliorating societal problems, a feature of interest to underrepresented groups (Sjoberg and Schreiner 2010; see also http://www.ils.uio.no/english/rose/).

Utilizing the vast capabilities of the Internet, a portal-like converging-technology website for educational research should be created to broadcast the latest innovative research news, to provide access to converging technology resources, to expand

---

[17] Broadening Participation Research Initiation Grants in Engineering, http://nsf.gov/pubs/2007/nsf07589/nsf07589.htm

8 Implications: People and Physical Infrastructure

public interest in converging trends, to guide the introduction of converging technology into education, to showcase works in converging technology, and to introduce availability of converging technology-related programs (e.g., workshops and training events).

K–12

More timely, cost-effective approaches to including progress in science, engineering, and convergent technologies should be integrated into K–12 education. Teachers must be provided the materials and professional development needed to implement effective, hands-on activities on this topic. Experience with K–12 nanotechnology educational efforts (e.g., see Murday 2011) underscores the challenge to widely disseminate the outputs of innovative educational research on topics like converging technologies throughout local U.S. school systems nationwide.

There must be mechanisms for professional development of existing teachers. But it is also essential to improve the training of new STEM teachers. University/college education departments (education-process-oriented) and science/engineering departments (content-oriented) must work in tandem to develop joint curricula, with appropriate metrics to monitor and ensure their effectiveness.

A consortium effort should be established, enabled by Federal incentives and guidance, to provide teachers with training and materials they can use to enrich curriculum units with hands-on activities, media, and other resources to illustrate the potential future outcomes of converging technologies. The consortium should include such stakeholders as the National Science Teachers Association (NSTA), universities, science museums, and other community organizations.

Affordable, interactive, personalized digital education aides should be developed and disseminated as a means to improve the rate and extent of learning.

Community and Technical Colleges

Enabling and fostering a partnership approach to converging knowledge and technology (CKT) education programs at community and technical colleges across the United States is vital to meeting the growing converging technology workforce needs. The approach of forming partnerships in CKT areas between research universities, small colleges and universities, and community/technical colleges appears to be a very viable methodology for the future. It should address the issues of community/technical college access to the latest developments and directions, availability of a state-of–the art equipment base, and availability of staff expertise. The infrastructure and human capital efficiencies gained through this approach are substantial. This education partnership approach will help ensure the availability of a broad converging technology education (synthesis, fabrication, characterization, and applications) at 2-year community and technical colleges in every region of the United States.

Converging technology concepts also must be integrated into classical science and technology classrooms so that the students become aware of the many career opportunities available to them in these exciting fields. In order to do this, we need to continue to make significant investments in the professional development of current and future educators and administrators, and create activities and lessons that enable them to easily bring converging technology awareness into their classrooms.

Web-accessible high-cost equipment resources must become an integral element of all partnership programs across the nation. Interconnecting converging technology programs with local CC/TC feeder K–12 schools will enable students to explore converging knowledge/technology areas via hands-on distance technology. Web access is also envisaged to be a cost-effective and efficient method to bring twenty-first century tools as well as scientists and engineers into classrooms all across the country. The ideal outcome would be to spark interest in a larger number of students to explore education and careers (at all levels of study) in STEM fields in general and CKT areas in particular.

Universities and Colleges

There is a general consensus that degree programs providing a fundamental and thorough education in a core scientific discipline are highly effective at training students for cross-disciplinary research. That being said, a rigorous course of education does not need to be confined to an existing academic discipline. As any converging technology gains momentum and impact (as did materials and computing in earlier generations), there should be experiments in developing transdisciplinary curricula focused on the educational needs for that technology.

Widespread use of information-technology-based courses, underpinned by cognitive understanding, will enable the best instruction at lower cost to all students, independent of the matriculation school.

Research Universities

Within universities, effective organizational structures should be implemented that permit effective cross-disciplinary research to support graduate student/mentor relationships based purely upon scientific interests, regardless of the departmental affiliation of either person. Graduate training should be administered centrally with cross-departmental chairing. Many universities have significant barriers to these types of structures (Blackwell et al. 2009), typically because of financial boundaries, that can prevent graduate students from crossing departments or schools. For example, a department may recruit a graduate student and put that individual on a 1-year fellowship during the coursework period. This is viewed as an investment. If that graduate student moves to a different department, the investment is viewed as lost. This view is shortsighted. Universities want to position themselves to more

8 Implications: People and Physical Infrastructure 315

effectively compete for research resources, and they also want to serve their mission of training students who can successfully enter an increasingly competitive workforce. Any barrier to cross-disciplinary research is going to ultimately prove to be a disservice to these key mission areas. Departmental and divisional structures provide useful frameworks for recruiting faculty, assigning teaching loads, and recruiting graduate students. However, they should not provide restrictive barriers to science and graduate education.

The social sciences, by virtue of continued progress in converging sciences and technologies, should be integrated more fully with the physical/engineering sciences and become disciplines founded in the fundamental chemical/physical laws governing brain function.

Informal Education

Effective public education about converging technologies should be widespread, informed by evidence-based communication science, include societal perspectives, and also include professional science communicators, informal educators, scientists, and engineers. The knowledge and skills needed to achieve this goal are dispersed across different fields, and a form of convergence is needed for success. In this case the fields are not only those directly engaged in physical sciences and engineering research, but also social and political science, educational psychology, and both formal and informal education. At present, converging NBIC technologies are not on the radar screens of either informal educators or the public. So there is a challenge, as there was for nanotechnology, to capture people's interest and attention.

Work on nanotechnology over the last decade has paved the way for public outreach and engagement about converging technologies. In addition, it will be important to bring social and political science researchers doing research on the societal implications of converging technologies together with (a) social scientists and psychologists doing research on science communication, and (b) practitioners in informal science education, science communication, and educational outreach.

Continuing Education

By the end of the next decade, there should be viable, web-based converging technology programs in place. With extended human lifetimes, it is projected that people will be more mobile in their careers, changing directions several times. That and the rapid changes in disciplinary knowledge and technological capabilities will make effective approaches in continuing education more important in the next decade. There should be a growing availability of web-based instruction, an education approach that is "friendly" to employed individuals in that there is flexibility as to when course work is accomplished and no travel time and lower costs required to "get to school."

8.3.2 Physical Infrastructure

Manufacturing demonstration facilities need to be created that provide the capability for facility users to experiment with different approaches to the affordable, large-scale manufacturing of new convergent technologies.[18] A range of such facilities will be needed to reflect the different approaches to manufacturing,[19] e.g., additive manufacturing or bottom-up assembly rather than stamping or machining. There should be a geographical distribution of these facilities to minimize travel time. It will also be important to develop appropriate intellectual property agreements and access arrangements that are conducive to small and medium enterprises (SMEs).

It will not be affordable for every university to have every machine, so there needs to be a mechanism established to determine (a) the best distribution of those machines, and (b) appropriate routes for virtual access to them. Funds will be necessary to implement the findings.

8.4 Infrastructure Needs

Facilities that support modern biological, physical, and engineering research are expensive to build, support, and maintain at the state-of-the-art level. They are also critical resources if research university faculty members are going to compete effectively for grants. A microchip fabrication facility may be utilized as much by chemistry and bioengineering faculties as it is by electrical engineering faculty; it now plays the role that machine shops played a generation ago. Similarly, microscopy facilities and molecular characterization facilities can also serve as centers that bring disparate disciplines together. In addition, these facilities can provide valuable resources for companies that are spun out of university research programs. Most of these facilities can operate only partially on user fees; they require additional commitments from other sources.

There is a huge need for standardized online education delivery platforms, yet to date no provider makes an interactive and fully integrated delivery–testing–management learning platform using tested best practices. While it is clear that digital education will grow to be mainstream, it is not clear what the best solutions will be. Support will be needed for infrastructure that experiments with a variety of approaches and provides means to expand the most successful approaches to full-scale use. A government competition to provide such a platform could be effective.

---

[18] As an example see http://engineering.mit.edu/research/labs_centers_programs/novartis.php

[19] On August 15, 2012, the White House announced the launch of a new public–private institute for manufacturing innovation in Youngstown, Ohio. the National Additive Manufacturing Innovation Institute (NAMII), including manufacturing firms, universities, community colleges, and nonprofit organizations from the Ohio-Pennsylvania-West Virginia "Tech Belt." In his State of the Union address on 12 February 2013, the President announced the launch of three more of these manufacturing hubs, aiming for a network of 15 of such hubs.

8.5 R&D Strategies

8.5.1 Education

There must be improved mechanisms for the transition of NSF education discovery projects into full-scale usage. There are many stakeholder organizations involved, including professional education societies, parent–teacher associations, the U.S. Department of Education, state/local education bureaucracies, the Business-Higher Education Forum, Educate to Innovate, and various foundations. Attention must be paid to coordinating these organizations and to making their efforts collectively more effective.

The recommendations made in the 2012 PCAST report, *Engage to Excel: Producing one Million Additional College Graduates with Degrees in Science, Technology, Engineering, and Mathematics*, should be implemented. The report has five overarching recommendations to transform undergraduate STEM education during the transition from high school to college and during the first 2 years of undergraduate STEM education:

1. Catalyze widespread adoption of empirically validated teaching practices
2. Advocate and provide support for replacing standard laboratory courses with discovery-based research courses
3. Launch a national experiment in postsecondary mathematics education to address the math preparation gap
4. Encourage partnerships among stakeholders to diversify pathways to STEM careers
5. Create a Presidential Council on STEM Education with leadership from the academic and business communities to provide strategic leadership for transformative and sustainable change in STEM undergraduate education (Blumenstyk 2012)

A network of regional hubsites should be established—the "Converging Technologies Education Hub Network"—as a sustainable national infrastructure for accelerating converging technology education. Just as the National Nanotechnology Infrastructure Network has been critical for shared access to leading-edge equipment in the early days of the NNI, shared converging technology resources, expertise, and training are imperative to preparing a skilled workforce and well-educated innovation leaders of the future. Each hubsite would connect and serve a number of research universities, 2-year and 4-year colleges, public school districts, and government and industrial laboratories. Each hubsite would focus on activities most pertinent to its locale. The hubsites would host teams of visiting professors, school teachers, and researchers from around the country to carry out (a) integrated content development for K–16 from the R&D stage through publication and dissemination; (b) related professional development, assessment and evaluation, learning research, and networking for teachers, faculty, and other stakeholders; (c) work on the science of science communication; and (d) effective implementation of digital education technologies using new insights stemming from the cognition sciences. These hubsites should focus on bigger issues, with a goal to nurture research, energize teaching, and build partnerships. For example, the Arizona State University Origins Project is a transdisciplinary initiative offering new possibilities for exploring the most fundamental of questions.

Even a decade into the NNI, nanotechnology is only marginally noticeable by the general public; it is likely other converging technologies will suffer the same fate. A prime goal for the next decade's infrastructure is to unfold the "secret magic" behind the converging technology, transmitting the knowledge to educators from middle school and beyond, and providing media for access to this knowledge. This could be done by boosting innovative technology in the rankings of popular search engines such as Google, Microsoft Bing, and Yahoo! and by facilitating dynamic information exchanges via social networks such as Twitter, Second Life, and Facebook.

Investment should be focused by NSF, the Department of Education (an "ARPA-ED" would be a logical entity), and foundations to enable effective, affordable e-teaching aides and better use of information technologies such as mobile devices and web applications. There are a number of extant efforts, but no overarching leadership. The 2012 U.S. Department of Education Race to the Top district competition has "Personalized Learning Environment" as priority one. DARPA has an Education Dominance program[20] focused on several key approaches:

- Replicating expert tutor behavior using knowledge engineering techniques
- Modeling intrinsic motivation and memory to optimize learning and consolidation
- Building student/tutor models based on abstractions of a wide range of student behaviors with live tutors
- Incorporating remediation strategies to enable the digital tutor to provide targeted reinforcement

The Innosight Institute has developed "A Guide to Personalizing Learning"[21] with guidance toward the blending of information technology and traditional learning modes (Evans 2012). Gooru (http://www.goorulearning.org/) and Silverback Learning (http://silverbacklearning.com/) have announced a partnership to enable educators to pinpoint the exact educational areas where students individually excel or struggle, to identify student needs, and to immediately suggest the most appropriate and high-quality learning materials available.

K–12

The National Governor's Association has been proactive in establishing common core standards for K–12 curricula (http://nga.org/cms/center/edu). That effort should expand to develop programs and partnerships (involving state and

---

[20] http://www.darpa.mil/Our_Work/DSO/Programs/Education_Dominance.aspx

[21] http://www.innosightinstitute.org/innosight/wp-content/uploads/2012/09/A-guide-to-personalizing-learning.pdf

local jurisdictions, Federal efforts, industry, and foundations) that will smoothly transition the continually evolving converging technology information seamlessly into the K–12 education system. The STEMx effort to transform STEM education and workforce development in the states, by the states, is an example (http://www.stemx.us/).

U.S. programs such as the Department of Education's Race to the Top (http://www2.ed.gov/programs/racetothetop/) should focus on the innovative capabilities inherent in the development of digital education, but with careful attention to its appropriate role(s) in learning.

Educators play an important role in introducing the reality of converging technology to younger generations and can encourage their interests to pursue higher education in converging technology fields. Government should support middle school educators' professional development opportunities by creating converging technology training programs and workshops, and encouraging the participation of the educators. Funding for innovative tools to assist converging technology teaching at middle schools should be increased. There is a huge gap between middle school skills and the skills needed in higher education research, such as the common analysis and visualization techniques used in graduate research. These analysis and visualization skills could be easily learned and used by middle school students with the support of new innovative visualization tools such as the Science of Science tool (http://sci2.cns.iu.edu) and Gephi (http://gephi.org).

A central website should be created to provide a registry for converging technology education materials; the Gooru/Silverback Learning Solutions partnership provides a step in this direction. The NSTA might serve as the evaluator for quality control to ensure website materials are of high quality, are in a format readily utilized by K–12 teachers, are carefully indexed to the common core standards (or the various state learning standards), and can be readily accessed from the NSTA website. Teachers must be made aware of the registry, find it useful, and find it easy to use.

There is a proposed STEM teacher core program to establish an elite corps of master teachers to boost U.S. students' achievement in science, technology, engineering, and mathematics.[22] It should explicitly incorporate converging technology, and it would benefit from a joint curricula developed by interaction between education and science/engineering departments.

Education partnerships such as the Triangle Coalition (http://trianglecoalition.org/), Business/Education Partnership Forum (http://biz4ed.org/), and Education Partnerships (http://educationpartnerships.org/) need to be fostered and better coordinated.

To attract underrepresented groups into STEM careers, it has been suggested that linking curricula directly with societal impact can be a motivator. Converging technologies will be at the heart of many, if not most, of the technological contributions toward the solution of societal challenges. This feature can be exploited to help remedy the underrepresentation. In addition, students, including minority

---

[22] http://www.whitehouse.gov/blog/2012/07/18/white-house-office-hours-stem-master-teacher-corps

students, should be made aware of the financial benefits that may be realized from a STEM education (Langdon et al. 2011; Melguizo and Wolniak 2012).

Community and Technical Colleges

A consortium of regional CC/TC and associated industries should be instituted to frame converging technology curricula around industry skill requirements and link these to the proposed network of regional hubsites. In addition, national core skill education standards based upon national industry requirements need to be developed to ensure that students graduating from converging technology programs are familiar with key core concepts and can easily move from region to region of technology, as well as from region to region of the country, armed with practical skill sets.

Colleges and Universities

As noted in the section on advances in the past decade, advances in information technology are leading to many new and promising approaches to effective education. There should be an effort to partner Federal agencies toward developing plans and/or processes for sharing the most effective and efficient means to educate undergraduate and graduate students regarding converging technologies. There will not be a standard model. There is a need for flexibility and adaptability to respond to the variety of educational contexts in the United States. The NSF Integrative Graduate Education and Research Traineeship (IGERT) program, explicitly tasked to address converging technology opportunities, is one approach to watch (see the IGERT case study, Sect. 8.8.8).

The several massive open online course (MOOC) experiments now in progress should be closely watched by NSF and the U.S. Department of Education, with funding for new experimentation and then attention to large-scale utilization of the more successful efforts.

Research Universities

A center should be established to address the societal and ethical implications of converging technologies. The center should conduct expert assessment in anticipation of potential future applications. It could be focused on human health and physical potential (although other converging themes are important as well). The public consistently expresses interest in new medical advances, so this may be effective subject matter for gaining the attention of educators and the public. Such a center could play an important role in the governance of converging technologies, but it could also provide information and perspectives valuable both in formal K–12 education and in informal public outreach and engagement.

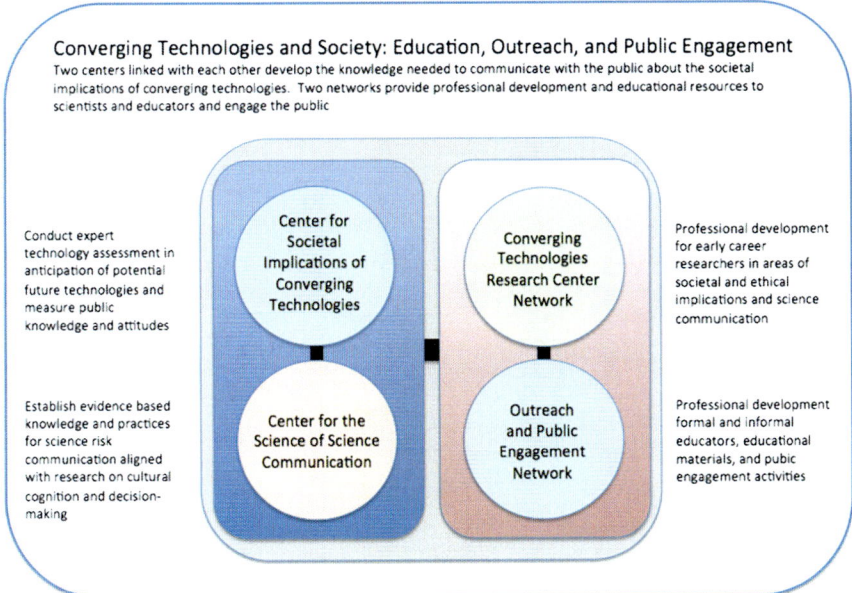

Fig. 8.5 Converging technologies and society: education, outreach, and public engagement (©Boston Museum of Science; courtesy of Larry Bell)

A center should also be established to address the science of science communication. This center should develop evidence-based knowledge and practices for science communication aligned with research on cultural cognition and science-informed decision-making. Such a center will be able to provide valuable scientific knowledge and also knowledge that can be useful to science communicators, educators, and scientists as they interact with lay audiences in their communities.

The treatment of tax-free bond-funded facilities at universities should be refocused to enable greater and stronger interactions between universities and industry (PCAST 2012a).

Informal Education

Industries can play a part in educating the public by increasing the appearance of the "converging technology" phrase in their advertisements. Leading converging technology IT companies are encouraged to include converging technology portals on their company websites.

An outreach and public engagement network, building on the experience of the NISE Net, should be created to establish a network of informal educators, science communicators, and university-based educational outreach coordinators (see Fig. 8.5).

They should be tasked with creating and disseminating professional development activities and materials to support educational enrichment experiences at all grade levels in both formal and informal education. This will provide students and the public with knowledge and experience with citizens' civic roles in connection with science and technology development in a democracy, as well as supporting general interest in science and science-related careers, all centered on converging technologies. This network would also create mechanisms to support widespread public engagement in policy considerations related to converging science and technology development, working with the center for societal and ethical implications and the center for science communication, to advance public participation in governance of converging knowledge and technologies.

Training in science communication and in the societal and ethical implications of technology must be incorporated into graduate and undergraduate science curricula and into sessions and workshops at meetings of professional societies. The initial focus should be on university centers that are conducting research in areas of converging NBIC technologies. The network should connect the educational outreach coordinators or other appropriate individuals who can implement student professional development activities that are built upon the work of the two centers and the outreach and public engagement network. The network would also disseminate information about established university-based courses in science communication for scientists as well as workshop activities designed for professional meetings.

Continuing Education

Partnering with the Business-Higher Education Forum and similar organizations will help to define the needs for continuing education focused on converging technologies.

Over the next 5 years, there will be up to a million military personnel returning to civilian life and getting into the U.S. civilian workforce. Juxtapose this with the 600,000 jobs currently open in advanced manufacturing and the anticipated need for ten million skilled workers by 2020. Programs, such as the Society of Manufacturing Engineers and U.S. Army example,[23] are needed to provide certifications to validate existing military personnel skills for civilian manufacturing jobs.

Portability and modularity of the credentialing process in advanced manufacturing is critical to allow coordinated action of organizations that feed the talent pipeline (PCAST 2012a).

8.5.2 Infrastructure

Two eminent programs that might be utilized to establish converging technologies manufacturing facilities are NIST's National Network for Manufacturing Innovation

---

[23] http://www.sme.org/MEMagazine/Article.aspx?id=66817&terms=Army%20veteran%20training

(NNMI) and the Clean Energy Manufacturing Initiative (CEMI) of DOE's Energy Efficiency and Renewable Energy program.

8.6 Conclusions and Priorities

8.6.1 Education

New information-technology-based instruction must be fully exploited as means to (a) make high-quality education available via access to proven, effective digital education tools; (b) provide personalized, interactive education aides for faster, more comprehensive learning enabled by using all the learning modes (audio, text, video, tactile, motion); (c) provide competitions to excite STEM interest in youth; (d) level the access to quality education, especially for underrepresented minorities; (e) reduce the cost of introducing the new knowledge emanating from converging technologies; and (f) reduce the cost per person at all levels of education.

U.S. demographics clearly show that underrepresented categories of students in STEM must be addressed, but with attention to where the job openings will likely be (e.g., see Vastag 2012; ACS 2012; NIH 2012). Since a key driver for education is employment, it will be important for industry to participate in defining appropriate areas of concentration for K–14/undergraduate/postgraduate STEM education.

The United States must establish an integrated, progressive STEM education in K–20. The imminent K–12 Next Generation Science standards are only a first step that must be followed by curricula development, assessments, teaching aides, and teacher training. Further, those standards are clearly deficient in the area of information/computing technology. That deficiency reflects in part the problems of incorporating new information into an education system already struggling with triage—what topics must be deleted to provide room for newer topics.

Education must address the ways in which converging technologies are responding to urgent societal challenges such as the need for alternative energy sources and adaptation/mitigation for climate change (these are only two examples; many more could be added). Embedding converging technology education in important public policy debates could facilitate its introduction to a wide range of educational levels and disciplinary foci.

Along with the present emphasis on creation of new jobs, there should be emphasis on entrepreneurship as part of the education process.

A portal-like website for educational research related to converging technologies should be created to broadcast the latest news about innovative science and technology research and to provide access to converging technology resources, expand public interest, showcase work in converging technology, and advertize availability of converging technology-related programs and events. Social networking also should be exploited in this endeavor.

The behavioral sciences, as represented by cognitive science and psychology, are evolving toward the status of sciences based upon fundamental or nomothetic laws. This evolutionary trend will influence other areas of social science, especially those that interact strongly with information and computational sciences. Converging technologies will provide an opportunity to accelerate this evolution of the social sciences toward the advances that have been made by the hard sciences.

8.6.2 Facilities

Without affordable, reliable approaches to manufacturing, many converging technologies will never enter the marketplace. Building a range of user facilities that provide opportunities to explore manufacturing approaches is important to accelerating innovation.

Ready access to a wide range of converging technology characterization and fabrication capabilities should be implemented at each major research university, with user centers providing open access to those instruments too expensive (to build and operate) for wide dissemination.

Cyber infrastructure will require continual investment, both to keep it current and to extend the capacities to more people and institutions.

8.6.3 Visionary Priorities

- Using new insights into cognition, utilize new, inexpensive, nano-enabled digital education assets toward more effective, interactive approaches to personalized education with:
  - Sensing of the environment and the student to ascertain learning readiness
  - Interactive modalities—3D video/text display, oral (including interactive dialog), motion (haptic and large motor), virtual reality—all tailored to personal learning modes and motivations
  - Awareness of the learner's present state of knowledge and comprehension
  - Constant assessment of comprehension to ensure an efficient rate of progress
  - Downloadable new materials from a central storage site into a local memory for real-time accessibility
  - Web-basing to enable access to worldwide information
  - Affordability
- Create a distributed international "converging knowledge and technology network" with a multidomain database, education modules, and facilities
- Expand the concepts of geographically distributed user facilities, such as NNIN and NCN, to create converging technologies clusters, and introduce test bed/manufacturing user facilities to accelerate the translation of the research discoveries into innovative technologies.

8.7 R&D Impact on Society

Without the necessary infrastructure, especially the educated manpower, the promises and expectations in the rest of this report are doomed to slow progress if not to outright failure. Success in the education endeavors will provide the skilled workforce at all levels of the innovation chain, including the factory floor (CC/TC education), business professionals/public officials, including regulators (BA/MA education), implementers (BS/MS education), and research discovery (Ph.D. education). Access to facilities will provide the tools to accelerate transition of research discovery into innovative technology, thereby improving people's quality of life.

Converging technologies have the potential to produce revolutionary innovations that could have profound impacts on individuals and society. Yet we must more deliberately address the benefits and issues attendant to such changes, whether individual or collective. The National Academies report *Technically Speaking* observes that,

> Americans are poorly equipped to recognize, let along ponder or address, the challenges technology poses or the problems it could solve …. Although our use of technology is increasing apace, there is no sign of a corresponding improvement in our ability to deal with issues related to technology. (Pearson and Young 2002, 1–2)

Technically Speaking identifies five benefits to society of technological literacy, which are supported by the kinds of K–12 enrichment and public outreach and engagement activities proposed here and which address the three overarching themes of this report:

1. Improve economic competitiveness

   - Increase the economic benefit from a workforce interested in and prepared for work that emerges from research in converging technologies

2. Develop human capacity

   - Improve decision-making as consumers who are able to make more critical assessments of new technologies
   - Increase citizen participation in a democratic society in ways that are well informed
   - Narrow the technological divide by making knowledge about the opportunities provided by emerging technologies known to everyone

3. Increase society's sense of security about change

   - Enhance people's sense of social well-being as they are empowered by acquiring the tools to make sense of the world even as it changes around them

These benefits would apply for science and innovation broadly, but also specifically for converging technologies, through implementation of the educational and public outreach and engagement activities outlined above.

Converging technologies, of which NBIC is only one important example, will pose benefit/risk challenges for society. A better-informed public debate and decisions based on more quantitative appreciation of the balance of benefit/risk in new technologies will be important to attain the best while minimizing the worst.

The convergence of NBIC technologies holds the promise of transformative, personalized education and entertainment based on information systems interacting with humans. The science of communication, honed by the challenges of CKT and a growing hard science basis for social sciences, will provide insights that accelerate the learning process. This will be even more important in the future as the pace of new knowledge accelerates.

8.8 Examples of Achievements and Convergence Paradigm Shifts

8.8.1 Vision for Changing Education Through Integration of Knowledge and Technology: A Vertical, Horizontal, and Global Approach

Contact person: *R. P. H. Chang, Northwestern University*[24]

Introduction

Throughout human history it has been shown over and over that information is "king." Better information or data has helped to improve all aspects of life, including economic forecasts, business management, health, and security. Starting with the invention of the transistor, integrated circuit, laser, and nanotechnology-based discoveries, the volume of information gathered, stored, and processed has been increasing exponentially. In particular, the cost of information has also come down dramatically, with 87 % of the world's population now having cell phones (ITU 2011) that also can provide entertainment and business transactions.

The ability to process tens of gigabits of information on a personal computer or tablet has dramatically changed the way of life for most citizens. In particular, information-based knowledge can be quickly gathered and processed for learning and training. Interactive digital information in the form of video, 3D simulation, and virtual reality has greatly enabled everyone to learn through visualization and real-time experience. As our knowledge of processes in the world grows, it pushes at the boundaries of what can be taught from textbooks and classic educational paradigms, demanding that we apply that knowledge and use our technology to create better educational tools. Examples include cockpit simulation for flight training and the study of nanoscience through interactive digital games and large-scale simulation of nanoparticle interaction dynamics.

---

[24] This work was supported by NSF (DMR award #0843962). The author thanks Jennifer Shanahan and Kathleen Cosgrove for their help in preparing this case study.

A concrete example of how technology and human knowledge are strongly coupled, reinforcing one another, can be found in understanding how the human brain learns and perceives. Consider the case of surgical residents learning to perform complex surgeries for the first time. Many classic learning paradigms have followed the practice of "see one, do one." Basic neuroscience research (knowledge) in some ways bears this out as an effective mechanism. *Watching* someone perform a task activates parts of the brain that will later *do* the task—a literal mental rehearsal (Rizzolatti et al. 2001). Thus, when asked to physically do the task after seeing it, the initial learning pathways have been forged; the parts of the brain that will have to cooperate to enable this task have at least introduced themselves. When learning a surgery, however, there are limited resources to allow for a safe first practice, and limited opportunities to "see one." Though residents watch videos, study diagrams, and attend lectures, it is likely that these 2D representations do not fully engage their learning circuits. Newer virtual surgery (technology-based) 3D training systems allow interaction and even provide haptic feedback, and allow for better learning and better-prepared surgeons who can learn more and be more precise in their practice—creating new knowledge (Hart and Karthigasu 2007; Wong and Matsumoto 2008).

These findings, that full simulations and virtual learning environments enhance learning, highlight an important aspect of the integration of technology, neuroscience, and education: that is, to be most effective and best utilize natural learning circuits, the learning must engage the student not just with sounds or images, but with responsive, reciprocal interaction. As our technology improves to allow for better human subject testing in more natural environments, we will begin to understand and target our educational technologies to ensure that we are tapping in to these full learning circuits, which will prepare the next generation to make their technological advances. This in turn can help make good brains great!

Importance of Convergence for Economic Development; The Infrastructure of Translating Education into a Technology Workforce

It is interesting to observe how countries have developed their economic growth over the past 50 years. While not all leading economists agree that education, science, and technology are the engines behind all economic growth, a few successful examples are given below to demonstrate how the use of the concept and implementation of the principle of convergence of education, knowledge, and technology have brought economic success.

Example 1: *Japan*. Japan is a country without many natural resources and whose economic success depends heavily on its skills to turn imported resources into technology-driven products. Soon after the Second World War, with the help of the United States, it started to rebuild itself by hard work and a strong emphasis on STEM education. Over the decades, it relied on perfecting foreign technology and innovation of its products for export. In order to accelerate this development and with limited financial resources, the Japanese government took the lead in the 1970s to apply the principle of convergence and set up integrated hubsites throughout Japan to boost economic development. Each of these hubsites had at least one nearby national research lab, industrial lab, and university. This was an effective, efficient way to streamline high-tech product development from concept to manufacture. Professors and their students perform research in the nearby labs, and this also serves as a model of workforce development. These initial hubsite models proved to be very successful and, as a consequence, Japan further established science cities, one of which is Tsukuba, a leading location for advanced research and technology development in Japan today.[25]

Example 2: *Taiwan*. Taiwan is an island much smaller than Japan. In many ways it is very similar to Japan in its economic development strategy. Over the past 50 years through a focused industrial development plan led by its government, Taiwan has become a leader in high-tech electronic products. With initial government subsidies, Taiwan built Hsinchu as its first convergence city with a science and technology park,[26] along with a powerful Industrial Technology Research Institute (ITRI; http://www.itri.org.tw/) surrounded by several leading national universities. Again, the main route to success has been high-level and in-depth education at universities coupled with training in the local industries for the strong workforce that is needed. Many of the researchers from ITRI, and university professors and their students, have launched start-up companies. Taking the knowledge gained from their economic success, large manufacturing plants have been launched in China by Taiwanese business leaders.

Example 3: *China*. In the early 1990s China took the above convergence models to another level of sophistication and grandeur. A partnership between the Singapore and Chinese governments was established in 1994 to launch a China–Singapore Suzhou Industrial Park (CS-SIP) with area of 288 km^2.[27] In addition to government-sponsored research labs and industrial labs, there are many satellite campuses of top universities from other parts of China. In May of 2006, the SIP was the first location for the joint China–UK university known as the Xi'an Jiaotong–Liverpool University (http://www.xjtlu.edu.cn/en/), which offers degree courses in Architecture, Electronics, Computer Science, Communications, and Management. Today, similar models have been used throughout China to establish regional and sometimes international high-tech convergence hubsites.

Example 4: *Mexico*. The Governor of Nuevo Leon, Mexico, in 2003 led the way to reposition the city of Monterrey as the "Monterrey International City of Knowledge" with goals to establish a model for Mexico and increase its per-capita GDP from

---

[25] For example, see *Tsukuba, Ibaraki*. http://en.wikipedia.org/wiki/Tsukuba,_Ibaraki and University of Tsukuba. *Prospectus*, http://www.tsukuba.ac.jp/english/about/index.html

[26] *Hsinchu Science and Industrial Park*: http://en.wikipedia.org/wiki/Hsinchu_Science_and_Industrial_Park

[27] See *China-Singapore Suzhou Industrial Park*: http://www.sipac.gov.cn/english/ and *Suzhou Industrial Park*: http://en.wikipedia.org/wiki/Suzhou_Industrial_Park

$10,000 to $35,000 by 2030. To do so, the Research & Innovation Technology Park was established with a spread of 172 acres and a phase-one investment of $145 million (Engardio 2009). Many U.S. and Mexican companies and national labs have already established their presence in the park. In addition, park leaders have made sure that they are in close collaboration with local universities to help provide the future workforce that is needed for the growth of local industry. They are also collaborating with the University of Texas in research and education, and using Northwestern University's Material World Modules Program for STEM course development and training for precollege students. This is an excellent model of technology, knowledge, and education convergence.

Example 5: *The United States.* The College of Nanoscale Science and Engineering (CNSE) at SUNY-Albany is perfect example of how to have a focused approach in the integration of education and training with research and development in the area of nanoscience and technology all under a single roof. With a total projected investment of $14 billion and 800,000 square feet of premier research space including Class 1 cleanrooms and fabrication facilities, this is a very large focused investment. The CNSE has over 300 global corporate partners and nearly 3,000 R&D jobs on-site, making it one of a kind in the world for nano education, technology, and knowledge generation.[28] This is also an example of double convergence, with nanoscale science and engineering already being a focused integration of the STEM field (see also the CNSE case study, Sect. 8.8.4).

Example 6: *Germany.* All of the earlier examples of development have started from ground-zero and with large investments. This is an example where existing successful components of education, knowledge, and technology were successfully integrated based on the principle of convergence. Karlsruhe Institute of Technology (KIT) was established in 2009 as a result of a merger between the 187 year-old University of Karlsruhe and the 181 year-old *Forschungszentrum Karlsruhe* (a federal research lab).[29] With such a merger, KIT instantly became a powerhouse of integrated education and technology development in Germany and Europe. It allows the continued training of a STEM workforce along with the advancement of discovery, innovation, and entrepreneurship. This is a unique example for others to emulate.

Vision for Change

As alluded to earlier, the world needs leaders with a global perspective and knowledge of how to implement an integrated approach to produce solutions to the

---

[28] See *College of Nanoscale Science & Engineering.* http://www.cnse.albany.edu/Home.aspx and Center for Economic Growth. *College of Nanoscale Science & Engineering*: http://www.ceg.org/economic_development/sites/the-college-of-nanoscale-science-and-engineering/

[29] See *Karlsruhe Institute of Technology*: http://en.wikipedia.org/wiki/Karlsruhe_Institute_of_Technology

impending global issues in energy, environment, health, security, and resource management. The fundamental building blocks are education, knowledge, and technology. These building blocks need to be mixed in a synergistic way to establish a basis for global sustainability.

In considering the results of these examples, successful implementation of convergence has already started to a certain degree with the consideration of both vertical and horizontal, and even bilateral integration between countries, as seen in the example of China. Mexico shows us that a vertical, ground-up effort in the education system to engage pre-college students in an integrated STEM curriculum has a positive impact on students' preparedness for college and their ability to innovate in industry. In the United States we have been working hard at the university level to break down departmental barriers to provide a horizontally integrated approach to education and research. NSF has been taking the lead in establishing cross-disciplinary centers for research and education.

Moving forward by combining different methods of vertical, horizontal, and global integration will allow us to conserve resources and maximize the development of a stronger, science-literate, global citizenship. I would argue that the potential of our knowledge and technology is not fully reached until citizens can learn how to use it and improve upon it. This requires a new push to revolutionize education, increasing its accessibility and using our technology know-how to improve learning and teaching. The technological advances in cyber infrastructure and pervasive access of digital media create an opportunity to implement near-free and massive formal and informal education to all citizens. This creates a science-literate global citizenry who can best use a variety of networks to share practices in solving global issues, starting from such simple practices of recycling and conservation of materials use, or disseminating information about transmittable disease. Similarly, collaboration will not be as dependent on physical proximity, creating opportunities for countries to have virtual collaboration centers, shared technology and resources, and access to specialized knowledge. By starting at the bottom with innovative, freely accessible education, we can find ways to integrate education, technology, and knowledge and fully reach our potential as a global community to tackle global challenges. Thus the vision for the future requires a changed education practice that utilizes the integration of knowledge and technology in multiple directions: vertical, horizontal, and global.

8.8.2 Pennsylvania State University Center for Nanotechnology Education and Utilization—NACK Network (http://www.nano4me.org)

Contact person: *Stephen J. Fonash, Pennsylvania State University*

With the support of the National Science Foundation's Advanced Technology Education (ATE) program, Penn State has developed a nationwide partnership of research universities and community colleges with the goal of bringing meaningful core-skills nanotechnology workforce education to technical and community colleges

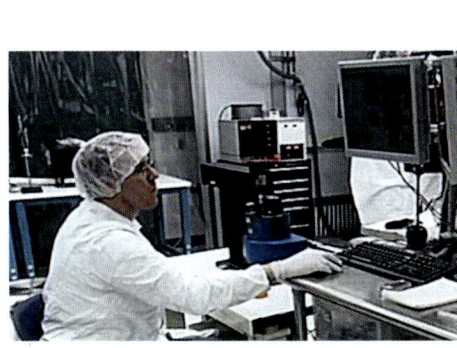

Fig. 8.6 NACK's laboratory practice is part of the Nanofabrication Manufacturing Technology Capstone Semester (Courtesy of S. Fonash)

across the United States, This partnership, the NSF National Nanotechnology Applications and Career Knowledge (NACK) Network, is focused on (1) resource sharing among community colleges and research universities for nanotechnology workforce development, (2) the availability of course materials, for web or in-class use, covering a core set of industry-recommended nanotechnology skills, (3) student awareness of converging knowledge and technologies, and (4) broad student preparation for careers in the wide spectrum of industries utilizing micro- or nanotechnology (Fig. 8.6).

In addressing the widespread need for a workforce possessing strong nano- and microtechnology 2-year degrees (Fonash 2009), NACK has created and offers continually updated, free-of-charge core-skills course lecture and lab materials, web-accessible equipment capability, and faculty development workshop curricula. Since the inception of the nationwide effort in 2008, NACK research university–community college partnership hubs have been set up and are functioning in Puerto Rico, New York, Indiana, Minnesota, Texas, and Arizona. Others are underway. These are in addition to the Pennsylvania hub comprised of 30 Pennsylvania schools and funded by the State of Pennsylvania since 1998.

The Penn State nanotechnology workforce development programs began as a Pennsylvania focused activity with the founding of Pennsylvania Nanofabrication Manufacturing Technology (NMT) Partnership funded by the State of Pennsylvania in 1998. The Pennsylvania NMT program encompasses 29 academic institutions in Pennsylvania, which offer a total of 53 2-year and 4-year nanotechnology degrees (Hallacher et al. 2002). In 2003 the additional component of a National Science Foundation (NSF) ATE regional center for nanotechnology workforce education was added. In 2008 this NSF ATE activity evolved into the NACK Network nationwide workforce development partnership. By creating education pathways from high school to skilled manufacturing careers across the country, the NACK Network is working to train the U.S. nanotechnology manufacturing workforce.

The NACK Network has introduced a number of paradigm shifts designed to give the United States a well-trained nano- and microtechnology workforce. These shifts address four key issues faced by many community and technical colleges as they consider developing courses for converging knowledge/technology areas. Specifically, in the case of the converging knowledge/technology area of nanotechnology, NACK has addressed the following:

- *Economic pressures.* To alleviate the economic burden of each institution creating and sustaining four semesters of new courses, NACK has designed a suite of standard courses to give students from a variety of science and technical fields a meaningful immersion in the converging knowledge/technology area of nanotechnology. This suite, which may be offered at a regional research university center or at a community college, taken online, or used in any combination of modes, gives students from biology, engineering, technology, chemistry, physics, and other programs a broad, meaningful experience in nanotechnology. There is a skill set requirement rather than a course set requirement for entry into these courses. The skill set requirements can be met by traditional biology, chemistry, engineering technology, math, materials science, and/or physics courses available at most 2-year institutions. Institutions thus do not need to develop four semesters of new courses. Students emerge from this suite with an exit skill set developed by the NACK Industry Advisory Council.
- *Student enrollment pressures.* The nanoscience/nanotechnology course suite approach eliminates the pressure to maintain a baseline student enrollment in a high-tech program. Students move from traditional programs into the nanotechnology immersion suite which, as noted, may be taken at a community college, at a university, online, or in any combination. The critical mass of students needed to economically maintain a nanotechnology education experience must be attained only for the course suite.
- *Pressures on faculty, staff, and facilities resources.* The NACK approach to the faculty, staff, and facilities issues faced by 2-year degree-granting institutions is one based on resource sharing. It entails several components: (1) sharing facilities and (2) sharing courses. Sharing facilities means 2-year-degree students using the facilities at a research university to obtain hands-on nanotechnology exposure, or it means community colleges themselves setting up a teaching cleanroom facility to be shared among institutions in a given area. In the NACK approach, sharing courses has the following possible implementations: a research university assuming responsibility for teaching the capstone semester for students attending from community colleges, community college students using web-accessible courses provided by NACK, and community college faculty using units from NACK's web-accessible courses.
- *Geographic isolation of some 2-year degree institutions.* The NACK approach to overcoming the drawbacks of geographic isolation in teaching students interested in nanotechnology is twofold: offering the nanotechnology suite of courses online for downloading, and providing web-based access to equipment. The NACK philosophy is that it is best to operate a tool (e.g., field-emission scanning electron microscope, scanning probe microscope, etc.) using a computer right

next to it, but second-best is to operate the tool with a computer via the web—even if the tool and computer are separated by thousands of miles.

8.8.3 Brave New University: How Convergence Might Transform Academia

Contact persons: *Michael E. Gorman, University of Virginia; Erwin P. Gianchandani, Computing Research Association; James C. Spohrer, Director, IBM University Programs*

The path from the transformation of institutions (online academe) to the transformation of individuals (connected and collective intelligence) has been politically charged. When the rector of the Board of Visitors at the University of Virginia recently decided to force the resignation of the University's president, one of her reasons was that the university was not on the leading edge of the recent online education movement. The politics surrounding the resignation, then restoration, of the president illustrate the high-stakes politics involved in the adaptation of universities to socio-technological change.

Online and distance education represent important challenges to the 1,000-year-old university model, but their potentially disruptive impact pales in comparison to other promising developments in convergent technologies, which could transform human bodies, brains, and social systems at a pace faster than anything experienced in human history.

Education Anywhere

Mr. Jefferson's goal in founding the University of Virginia was to create "an academical village" in which students and faculty could be part of a learning community, living right next to one another on Jefferson's historic Lawn, an architectural marvel that itself would be a constant source of inspiration. For Jefferson, having students and faculty together in the right surroundings was essential to intellectual growth.

Nearly 200 years later, technologies permit students to take courses—taught by world-class experts—from their laptops anywhere in the world with good Internet access. Students can even participate in discussion groups and joint projects online. Can this online experience replicate the kind of special bond created between students and a university—an experience that is part of a university's "brand" and ensures long-term loyalty and connection with alumni? Which aspects of a village are captured online, and which are not captured? What new affordances going beyond "an academical village" does online offer?

Inspired by "World of Warcraft," a gaming environment in which teams of 40 or more participants from all around the world work together to solve problems like defeating tough resilient monsters, the first author developed a simulation of the NNI for a recent course. Through this, students engaged in role-playing representing

Congress, funding agencies, selected Federal agencies, laboratories (both industrial and academic), nongovernmental organizations (NGOs), and a newspaper (Gorman et al. 2013). Readings and class discussions prepared the students to develop their own goals for the NNI and even a branching tree diagram of the technologies they had to develop to reach these goals. The laboratories competed or cooperated to develop and own technologies, writing proposals to funding agencies, navigating regulatory agencies and NGOs, etc.

Simple software was implemented to scaffold this experience, keeping track of the technology tree, budgets, intellectual property, patent owners, all proposals to the funding agencies, etc. The scaffolding, when combined with software to manage online groups, made it easier to offer this simulation over the Internet.

An alternative would be to capture this environment on an "island" on Second Life, in which students would have avatars and virtual buildings for their labs, agencies, etc. The first author recently had the opportunity to serve on an NSF review panel on Second Life and thought it worked nearly as well as being together physically. (NSF specially designed its island for this sort of work and put considerable effort into making it practicable—partly because this collaboration mode is much less expensive than having all panelists come to NSF.)

Convergent Technologies

Distance learning and Second Life are simply the beginning; there are technologies emerging with even greater transformative potential. Consider the ubiquity of smart phones and tablet PCs—indeed, the increasing dependence of today's youth upon these devices—as capable of enabling an entirely new learning modality. What sorts of "apps for learning" could we engineer? How might they improve students' perception and understanding? How might they catalyze changes in students' behaviors, perhaps by piquing interest and enhancing motivation at the right time in the learning process? These devices still require looking down at a screen, often while walking (or driving or biking)—but perhaps not for much longer: Google is pioneering modalities that would provide this information hands-free, and implants and neural interfaces are on the horizon. Indeed, these and other learning platforms envisioned just a decade ago are becoming realities more rapidly than many anticipated (Spohrer 2002).

Crowdsourcing—the process of engaging a distributed group of agents (humans and/or computers) to complete complex tasks—similarly offers the potential to engage active learners in solving actual problems. Researchers at the University of Washington recently developed a collaborative, web-based video game called Foldit that allows players to manipulate virtual molecular structures following real chemical rules. The more elegant the structure created, the higher the score attained. In the fall of 2011, players identified the structure of an enzyme critical for reproduction of the AIDS virus within weeks of the game's launch, a challenge that had stumped scientists for decades. Similarly, the search for Genghis Khan's tomb demonstrates the power of crowdsourcing to leverage advanced visualization technologies to pinpoint, in this case, interesting archeological sites in Mongolia. To be part of a university education, these crowdsourcing experiences would have to incorporate reflecting on the learning experience, and learning about the relevant fields, e.g., biochemistry and archeology. Here Dewey's vision of learning by doing could potentially become learning by solving. How many unsolved problems are a small number of inferences away from a properly prepared learner's mind?

And how about acquiring and transferring hands-on tacit knowledge (Gorman 2002)? Haptic interfaces could one day allow students to do a laboratory experiment together over a distance, moving the appropriate instruments and getting better at skills like creating a scanning tunnel microscope (STM) tip or dissecting a (virtual) frog or participating in an archeological dig.

Convergent technologies promise a day when prospective students might have neural interfaces and other enhancements that would allow them to experience a virtual environment as if it were real (Roco and Bainbridge 2003; Robinett 2003).

Similar technologies could be used to enhance "NNIsim" by turning it into an augmented reality, with virtual offices and equipment, and the ability to visit laboratories and foundations. Representatives from government or NGOs or actual labs could provide input on how to design avatars representing roles not actually played by simulation members—like the President. One could use crowdsourcing to turn NNIsim into a tool for public engagement. Imagine multiple NNIsims mixing scientists, policymakers, public stakeholders, and university students, each simulation following its own course to a unique set of technologies and goals, with details of the arguments preserved.

University-quality education could be available round the clock, right when a user needs it to get background and current information. Instead of classrooms and courses, learning communities might evolve, centered on specific problems or issues.

Steve Jobs designed the workspace at Pixar to maximize the possibility of chance encounters between employees at random times during the day, knowing that these encounters would lead to new ideas and collaborations (Isaacson 2011). Would a virtual community of this sort be as good as a real one? Virtual communities like Second Life can be designed to encourage random encounters. Empirical work needs to be done on the richness of these encounters, especially as the technology improves (Gorman and Spohrer 2010).

Perhaps most importantly, convergent technologies create the opportunity to capture more data on how students—and by extension, human beings more generally—learn. Recent and ongoing advances in recognition, tracking, and recording allow us to capture and pool data at low cost and make it available to an analytical pipeline. We can observe when a student pauses a video, what he/she reviews in his/her smart goggles, etc. We can notice when many students arrive at the same wrong answer, so that we can refine our teaching methodology accordingly. We can predict how well students will do before they even set foot in the classroom. And we can achieve the ultimate quest of personalized education.

So imagine a virtual University of Virginia, with a replica of some of Mr. Jefferson's historical buildings in an advanced equivalent of a Second Life island—perhaps goggles, retinal implants, or even a silicone interface linked with the brain and coupled with goggles to transmit the right stimuli to trigger the optic nerve and appropriate parts of the brain. Collectively these could be used to enter the Lawn from any location. Now students could, from afar, "stroll" the Lawn and discuss ideas as if they were there. Goggles or retinal implants could also be used while walking the actual Lawn to provide additional information on the social, intellectual, and architectural environment (Spohrer 1997). The University of Virginia runs a Semester-at-Sea Program to take students around the world. Convergent technologies create the possibility of sending a small number of students and linking a much larger number online through an immersive environment that allows them to share parts or all of the experience, interact with locals, and participate fully in lessons learned.

What Is Lost and What Is Gained?

In this way, convergent technologies could potentially save universities money on infrastructure: students would not have to fly or drive to a physical campus, stay in a residence, eat in dining halls, have health clinics, etc.; rather, their cost would be whatever technology is required to sign onto the courses. And in a world where goggles or retinal implants were readily available, much of the costs would be borne by the students—as is happening at universities where students are required to buy or bring their own computers. Like the students, the faculty could reside anywhere the appropriate technology was available—no longer incurring costs to relocate. Faculty could even be available asynchronously to share a learning experience with a student or a group of students.

But the cost savings from creating a virtual university of this sort may be less than one might expect, and may also create a divide between those students who can afford the best enhancements and those who cannot. Moreover, the primary focus should be on the quality of the university experience. Undergraduates are going through significant developmental changes within and outside of the classroom, learning a great deal from their peers and forming lifetime bonds (including marriages). They participate together in sports and clubs and service organizations. A virtual university cannot emulate all of this.

Laboratories and field research are not just places where students learn hands-on tacit knowledge: they are also places where they learn the collective tacit knowledge involved in becoming part of a disciplinary culture (Collins 2010), particularly when undergraduates and graduate students are engaged with faculty in making discoveries and creating new technologies. Would this kind of enculturation be as well developed in a virtual university?

Could virtual universities have unique "branded" experiences and co-evolve improvements with students? Or will the university as an institution disappear, with students simply selecting from a wide range of courses and experiences they can meld together into a degree? Universities are not just locations for knowledge transfer: they are places where knowledge is created, applied, and integrated across multiple fields. Students apprentice themselves to faculty to become members of the scientific community, build reputations, and sometimes even launch new entrepreneurial firms based on those newly found understandings of the world. Universities are also places of reflection on our society, our technologies, and ourselves. One advantage of convergent technologies is that they force us to examine the role of universities as institutions in society from many perspectives beyond knowledge transfer (teaching). By removing constraints and imposing new ones, convergent technologies allow deeper explorations of what are the essential characteristics of both institutions and individuals.

Equal Educational Opportunity for All?

At the same time, as tuition costs rise, it becomes harder and harder for those of modest means to complete college, which undermines the American ideal of a meritocracy in which anyone who works "hard enough" has an opportunity to succeed. The availability of education anywhere through online tools could theoretically reduce this gap, but only if the infrastructure were put in place and the convergent technologies were inexpensive or subsidized. Otherwise, students might have to pay more to link into a virtual university than to attend the real one.

What about equal access to convergent technologies like cochlear, retinal, and neural implants? Students who are blind, deaf, or have other sensory, motor, or cognitive issues that make learning more challenging will benefit from these and other convergent technologies. As these capability augmentations improve, it is possible that someone initially born blind, deaf, or with cognitive disorders may actually surpass the average capability levels of people born without those disabilities. It is also likely that these technologies will be adapted to augment the capabilities of learners not diagnosed with any special needs.

Ideally, such augmentations should be available for all students and faculty, but what if they cannot afford them? How would the presence or absence of enhancements affect the admissions process? Universities alone cannot be expected to guarantee universal access. To benefit from knowledge today you sometimes have to have the wealth to buy an iPhone or live somewhere with wireless Internet access. Amartya Sen (2000) describes development as freedom because those in poverty worldwide have few options compared with those of even modest means.

Today's capability divides (e.g., wealth, place of birth, etc.) exist in part due to the lack of governance systems that would ensure everyone has access to the applications of knowledge. Spohrer (2002) explores the meaning of learning in an age of rapid technological change, and concludes that new rules and regulations must co-evolve with new technologies brought about through convergence. Advanced civilizations must be as good at developing new governance systems (rules) as they are at developing technological systems (capability augmentations).

Universities are great places to discuss the ethics of equal access to these technologies, the governance systems required to create such access, and also the values issues that arise (for example, not all those born deaf want cochlear implants for themselves or for their children). Universities are also going to be deeply involved in creating the breakthroughs that lead to these technologies and have an interdisciplinary community that could conduct research on, and evolve solutions to, emergent problems.

Resiliency in Our Systems

Another challenge is resiliency. Anyone who uses distance education now knows that even well-known technologies have glitches. Two of the first author's classes were canceled one semester because of a failure of the University of Virginia's campus-wide online system for scaffolding student collaboration—it simply crashed. There were other failures caused by interactions among mutually dependent IT systems and human actors, to the point where as much time was spent troubleshooting the systems with support people as was spent teaching.

Imagine super-performing students whose tightly coupled convergent technologies crash, and who may no longer be able to operate without them. Then add the possibility of hacking into these systems—of hacking into neural-IT interfaces and stealing information, substituting memories, or just disrupting thought and action with viruses.

Resilient systems are critical—and part of this resilience depends on the ways in which our cognition, our minds, are distributed across a range of technologies and shared with other human beings (Gorman 1997). Can this kind of resilience be maintained or even enhanced by a more virtual university? If modern university campuses are turning from academic villages into mini-cities (including increasing numbers of student linked online), then how can innovativeness, equity (competitive parity), sustainability, and resilience of these nested, networked systems be enhanced (Spohrer and Giuiusa 2012)? As the amount of knowledge on which societies depend for a high quality of life grows, so, too, does the knowledge burden to maintain sustainability and resilience, thereby stressing the responsible individuals and institutions that must either carry more knowledge or communicate and interact with larger populations of entities across which the knowledge is distributed (Jones 2005).

Transcending the "Online" Debate

In the recent push toward online education, what's striking is what has been largely missing from the conversation. The online education debate up until this point has largely been one-dimensional, focused on the ability to put classes online. Certainly there are related issues: the significance of this movement upon higher education and the broader public, as well as the viability in terms of profit. But online education is about much more than simply transcending the boundaries associated with traditional universities—brick-and-mortar classrooms, large, lecture-based classes, increasingly rigorous enrollment requirements, etc. It is also about the wealth of convergent technologies that have transformative potential for augmenting and enhancing human performance and capabilities. Indeed, if we pause and think about the educational opportunities enabled by the mix of mobile devices, virtual reality goggles, neural interfaces, crowdsourcing platforms, haptic tools, and the like, we can start to see just how revolutionary this transformation can be. Not only institutions but also individuals will be transformed—their capabilities and their opportunities. In many ways, these are the questions that should be asked of university presidents and by which we should judge their tenures—not simply whether they are partnering with today's big-name online education providers, but how are they partnering with their local communities to enhance quality of life for everyone (Trani and Holsworth 2010). Convergent technologies will create additional opportunities for university–business partnerships, with universities providing breakthrough R&D and education not only in how to create new technologies (Lécuyer 2006) but also in how to study and manage their impacts. Universities will have to evolve policies for the conflict-of-interest and intellectual property disputes that can arise from these collaborations (Cole 2009).

Properly managed, the most likely impact of convergent technologies on universities is the potential to deliver a quality education off-campus to a much more diverse student population globally—but those students will have to be living somewhere, supported by local infrastructure and local institutions. Wise universities will begin to experiment with these new capabilities, just as they are now doing with online education, comparing this kind of education with the campus experience. Those who learn at a distance could be required to spend some time on-site, just as those on a campus are encouraged to go abroad. All of these experiments must be done with careful attention to social and ethical issues.

The university as an institution is over a thousand years old. Will it still exist in another thousand years? If we survive our own tendency to go to war with each other, human beings at that point may have diverged into what amounts to multiple species, based on the capabilities and choices available to people. We may be mining asteroids, settling other planets, and having at our disposal a "utility fog" that allows self-assembly of desired items (Spohrer and Engelbart 2004). Each generation will live longer, and the capability gap between generations may grow over time, depending on multiple nanotechnology, biotechnology, information technology, and cognitive science (NBIC) (and beyond) enhancements that interact in complex ways. There will be an increased need for institutions that study and teach this process of transformation, and encourage deep reflection on what sort of future we are evolving, and why. Universities should not simply react to these changes; they will be discovering and inventing many of the technologies that will produce transformations (including educational ones) and are therefore in a unique position to study and manage their impacts, reflecting on what kind of brave new worlds we want to create.

8.8.4 University at Albany, College of Nanoscale Science and Engineering (http://cnse.albany.edu/)

Contact person: Alain C. Diebold, University at Albany, State University of New York

The College of Nanoscale Science and Engineering of the University at Albany, State University of New York (SUNY) is the first college in the world dedicated to education, research, development, and deployment in the emerging disciplines of nanoscience, nanoengineering, nanobioscience, and "nanoeconomics." At CNSE, academia, industry, and government have joined forces to advance atomic-scale knowledge, educate the next-generation workforce, and spearhead economic development. The result is an academic and corporate complex that's home to world-class intellectual capital, unmatched physical resources, and limitless opportunities.

CNSE has reshaped the traditional "silo"-type college departmental structure into four cross-disciplinary constellations of scholarly excellence in nanoscience, nanoengineering, nanobioscience, and nanoeconomics. Through this game-changing paradigm, students engage in unique hands-on education, research, and training in the design, fabrication, and integration of nanoscale devices, structures, and systems to enable a wide range of emerging nanotechnologies. Students are supported by internships, fellowships, and scholarships provided by CNSE and its array of global corporate partners. CNSE complements its groundbreaking bachelor's, master's and Ph.D. programs in nanoscale science and engineering with educational outreach to elementary, middle, and high schools; partnerships with community colleges and academic institutions around the world; and certificate-level technical training. This unprecedented effort is designed to educate the next generation of nanotechnology-savvy professionals and build the foundations of a skilled nanotechnology workforce at every level.

Buoyed by its unparalleled combination of intellectual know-how and leading-edge technological infrastructure, CNSE's Albany NanoTech Complex is the site of some of the world's most advanced nanoscale research, development, and commercialization activities. Here, academic and corporate scientists engage in innovative research in a variety of fields, including clean energy and advanced sensor and environmental technologies; advanced CMOS and post-CMOS nanoelectronics; 3D integrated circuits and advanced chip packaging; ultra-high-resolution optical, electron, and EUV lithography; and nanobioscience and nanomedicine.

CNSE has built the world's first "nano mall," known as CNSE's Albany NanoTech Complex (Fig. 8.7). With more than $14 billion in high-tech investments, it serves as the hub of the world's most advanced university-driven research enterprise, offering students a one-of-a kind academic experience and providing over 300 corporate partners with access to an unmatched ecosystem for leading-edge R&D and commercialization of nanoelectronics and nanotechnology innovations. CNSE's footprint spans upstate New York, including its Albany NanoTech Complex, an 800,000-sq-ft megaplex with the only fully integrated, 300 mm wafer, computer

8 Implications: People and Physical Infrastructure

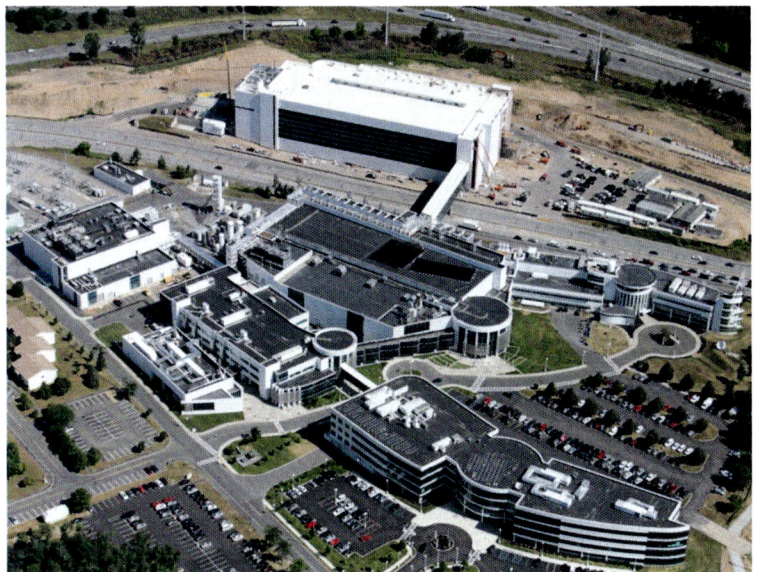

Fig. 8.7 Photo of the CNSE Albany NanoTech Complex (Courtesy of S. Janack)

chip pilot prototyping and demonstration line, with 85,000 sq ft of Class-1-capable cleanrooms. More than 2,700 scientists, researchers, engineers, students, and faculty members work here, including representatives from such companies as IBM, Intel, GlobalFoundries, SEMATECH, Samsung, Taiwan Semiconductor Manufacturing Company (TSMC), Toshiba, Applied Materials, Tokyo Electron, ASML, and Novellus Systems.

In September 2011, New York Governor Andrew M. Cuomo announced a $4.8 billion investment to develop a new era of computer chip technology in New York, highlighted by creation of the world's first Global 450 Consortium (G450C), headquartered at and managed by CNSE, through which Intel, IBM, Samsung, GlobalFoundries, and TSMC will work collaboratively to lead the industry transition from 300 mm wafer to 450 mm wafer production. A concurrent expansion will add nearly 500,000 sq ft of next-generation infrastructure, an additional 50,000 sq ft of Class-1-capable cleanrooms, and more than 1,000 scientists, researchers, and engineers from CNSE and global corporations.

In addition, CNSE's Solar Energy Development Center in Halfmoon, New York, is an 18,000-sq-ft facility that features a state-of-the-art, 100 kW prototyping and demonstration line for next-generation copper indium gallium selenide (CIGS) thin-film solar cells. CNSE's Smart System Technology and Commercialization Center of Excellence (STC) in Rochester offers the largest array of world-class solutions in the industry related to microelectromechanical systems (MEMS), including a 140,000-sq-ft facility with over 50,000 sq ft of certified cleanroom space with 150 mm wafer production, complemented by a dedicated 8,000-sq-ft MEMS and optoelectronic packaging facility. CNSE also co-founded and manages operations at the Computer Chip Commercialization Center at SUNYIT in Utica and is a co-founder of the Nanotechnology Innovation and Commercialization Excelerator in Syracuse.

And, in partnership with SEMATECH, CNSE leads and serves as headquarters for the $400 million U.S. Photovoltaic Manufacturing Consortium (PVMC), an industry-led collaboration created through the U.S. Department of Energy's (DOE) SunShot initiative to accelerate next-generation photovoltaic (PV) technologies, with a goal of reducing the total installed cost of solar energy systems by 75 % over the next decade.

8.8.5 The David H. Koch Institute for Integrative Cancer Research at MIT (http://ki.mit.edu/)

Contact person: *Amanda J. Arnold, MIT*

As noted throughout this study, convergence will be key to advances in many crucial areas. This case study describes how Massachusetts Institute of Technology (MIT) is literally internalizing the convergence approach and rebuilding itself to tackle the innately interactive challenges of cancer.

In 2007, MIT launched a partnership between the faculty of the former MIT Center for Cancer Research (CCR) and an equivalent number of distinguished engineers drawn from across MIT's science and engineering departments. By 2010, the David H. Koch Institute for Integrative Cancer Research was born in Building 76 (Fig. 8.8).

Infrastructure: The Importance of Physical Facilities

Building 76 or The Koch Institute (KI) was built specifically to foster the cross-disciplinary approach of convergence. The building features roughly 180,000 sq ft of state-of-the-art laboratory and work space. The floor plans are specially designed to foster interaction and collaboration among biologists and engineers—both in terms of dedicated lab space and in the common areas, where informal talks lead to new collaborations and spontaneous information-sharing.

Importantly, KI features 13 core facilities[30]—now being rededicated as the Swanson Biotechnology Center—to provide centralized technical services to faculty and students. From routine (though essential) support services to advanced technical and consulting services, these cores facilitate, support, and enhance KI research. Many of the facilities also offer training programs that enable Koch Institute staff, students, and postdoctoral fellows to acquire the additional technical and intellectual expertise needed to advance both their work and their careers.

---

[30] The full list of 14 KI core facilities is available at http://ki.mit.edu/sbc

8 Implications: People and Physical Infrastructure

Fig. 8.8 MIT's Koch Institute (photograph from http://ki.mit.edu/approach/ki)

Examples of these core units include:

- *Applied Therapeutics & Whole Animal Imaging*: This provides assistance with design, approval, and execution of relevant preclinical trials, in addition to instrumentation for *in vivo* and whole animal imaging.
- *Bioinformatics & Computing*: This provides Koch Institute researchers with assistance with experimental design and subsequent analysis of next-generation sequencing (Illumina platform) and microarray experiments, genome annotation projects, and other sequence and phylogenetic analysis applications, in addition to critical data backup as well as desktop hardware installation and maintenance.
- *Biopolymers & Proteomics*: This provides integrated synthetic and analytical capabilities for biological materials, including DNA, proteins, and nanoparticles. Services include sequencing, mass-spectrometry-based proteomics approaches for identification, characterization, or quantitation of proteins from simple to complex mixtures, and high-pressure liquid chromatographic analysis and purification.
- *High-Throughput Screening*: This supports development of a wide range of experimental assays into high-throughput screening strategies and includes liquid-handling robotic systems, high-content imaging capabilities, and management of RNAi and small molecule libraries.
- *Histology*: This assists investigators in producing quality histological slides from frozen, paraffin-embedded, and resin-embedded tissues, thus enabling

investigators to better evaluate the pathologic consequences of various mutations or treatments.
- *Nanotechnology Materials*: This is one of the newest cores; it supports new materials discovery and optimization, enabling development of drug- and gene-delivery vehicles, imaging, nano- and microparticles, and devices.

The costs of these core services are partially defrayed by funds from a number of sources, most notably a Cancer Center Support Grant to KI from the National Cancer Institute (NCI) at the National Institutes of Health (NIH). To the extent capacity will allow, a number of the cores are available to the broader MIT community and other collaborators.

Workforce Preparation and Training

In addition to the unique combination of scientists, clinicians, and engineers utilizing shared core facilities, the faculty and trainees cross-collaborate via research focus areas that drive the convergence of life, physical, and engineering sciences at KI. The main research focus areas at KI are:

- Nano-based formulation
- Detection and monitoring
- Metastasis
- Analysis of pathways of sensitivity and resistance
- Cancer immunology

The educational philosophy of the convergence model at KI requires that a deep disciplinary background remains vital but that a robust cross-disciplinary education is essential. Trainees and faculty at KI learn a kind of "convergence culture" to help them communicate across disciplinary lines and to become fully "multilingual" among and between disciplines.

The Funding Model

Federal funding is a key component of KI sustainability. Specifically, KI efforts are propelled in many ways by innovative funding models offered by the National Cancer Institute (NCI), part of the National Institutes of Health (NIH).

KI benefits from its designation as 1 of 12 Physical Sciences in Oncology Centers (PS-OC) awarded by NCI. The MIT PS-OC is collaboration among MIT, Harvard University, University of California–San Francisco, Harvard Medical School, Boston University, Hubrecht Institute, and Brigham and Women's Hospital. The overarching goal of this team is to use both theoretical and experimental approaches inspired by physics and engineering to attack important problems in cancer biology by developing novel technology and analytical and computational methods to track the dynamics of cancer at the single-cell level.[31]

KI is also a key component of the MIT-Harvard Center for Cancer Nanotechnology Excellence (CCNE). This is a collaborative effort among MIT, Harvard University, Harvard Medical School, Massachusetts General Hospital, and Brigham and Women's Hospital. It is one of eight Centers of Cancer Nanotechnology Excellence awarded by NCI and focuses on developing a diversified portfolio of nanoscale devices for targeted delivery of cancer therapies, diagnostics, noninvasive imaging, and molecular sensing. In addition to general oncology applications, the consortium focuses on prostate, brain, lung, ovarian, and colon cancer.[32]

A third and key component of Federal funding that is driving work at KI is MIT's participation in the NCI Integrative Cancer Biology Program (ICBP). MIT was first awarded this grant for the period September 2004–February 2010 to develop and effectively apply systems biology approaches to fundamental problems in cancer biology and therapy. Building on the cross-disciplinary success of the ICBP at MIT, KI received NCI funding to become a Center for Cancer Systems Biology (CCSB), a part of NCI's ICBP, in March 2010. The purpose of the NCI CCSB initiative is, "to stimulate the development and application of the integrative systems approaches and mathematical/computational modeling to cancer research … specifically in the areas of (a) cancer biology, (b) experimental therapeutics, (c) early interventions, and (d) cancer susceptibility."[33]

However, Federal funding does not make up the whole picture. Creative exploration at the leading edge of cancer research has often led to important, transformative new discoveries that themselves can lead to major improvements in patient care. Yet early-stage ideas often do not always qualify for funding from traditional government sources. To support ideas considered too risky for Federal research, the Koch Institute Frontier Research Program is a funding model that supports boldly conceived, highly innovative, and highly collaborative research proposals from faculty. Project areas funded in Koch's Frontier Program include:

- Inhaled nanoparticle formulations to deliver small interfering RNA (siRNA) molecules into the lungs of cancer patients
- Surgical tools that facilitate the real-time detection of residual cancer cells on surgical margins
- Methods to reactivate the tumor-specific immune cells in melanoma patients

There is no doubt that industry partnerships are also an important aspect of the KI funding model. For instance, TRANSCEND is a partnership between KI and

---

[31] For more information about the MIT PS-OC, visit http://ki.mit.edu/approach/partnerships/psoc

[32] Additional information about the NCI Alliance for Nanotechnology in Cancer, MIT-Harvard CCNE is available online at http://nano.cancer.gov/action/programs/mit/

[33] For more information about KI and NCI's Integrative Cancer Biology Program see http://ki.mit.edu/approach/partnerships

Ortho-McNeil-Janssen Pharmaceuticals, Inc. The initial 5-year collaborative agreement fosters oncology research and technology development in the areas of cancer diagnostics, cancer biology premalignancies, genetic models of disease, and profiles of the tumor microenvironment. A Joint Scientific Steering Committee composed of MIT faculty members and Ortho-McNeil Janssen employees jointly reviews and selects proposals from MIT researchers for funding. In addition, there is a provision for visiting scientists from Ortho-McNeil Janssen to participate in projects within the investigators' laboratories at the Koch Institute.[34]

University Partners

At a national level, more institutions are developing courses and programs that prepare students, postdoctoral researchers, and fellows for convergence-driven research. These efforts should be supported with concentrated Federal funding. Emerging university efforts working at the frontier of converging technologies include (Sharp and Langer 2011):

- Stanford University: The Clark Center (2003) houses the Bio-X program
- Harvard University: The Wyss Institute (2009)
- Georgia Tech: The Petit Institute for Bioengineering and Bioscience (1995)
- University of Chicago: Molecular Engineering Institute (2011)
- University of Michigan: North Campus Research Complex (2009)

8.8.6 Informal Education: NISE Net and NanoDays (http://www.nisenet.org/nanodays)

Contact person: *Larry Bell, Museum of Science, Boston*

The Nanoscale Informal Science Education Network (NISE Net) was launched in 2005 with National Science Foundation support to foster public awareness, knowledge, and engagement with nanoscale science, engineering, and technology through establishment of a network, a national infrastructure that links science museums and other informal education organizations with nanoscale science and engineering research organizations. As with converging technologies today, nanotechnology at that time was little known by the public or by educators and exhibit developers in informal educational institutions.

Informal educational activities about nanoscale science and engineering were virtually nonexistent 10 years ago, and even fundamental ideas about the behavior of matter at the nanoscale were mostly absent from informal science education. For the most part, informal science educators felt that content of this sort was too

---

[34] More information about partnerships at KI is available at http://ki.mit.edu/approach/partnerships

difficult for the public to understand, not really exhibitable, and of little interest. As a result, prior to 2005, nanotechnology was covered in only a small number of ISE institutions. Inverness Research Associates reported that at that time there was little expertise, experience, or incentive to provide nanoscale science education for the public (St. John et al. 2009). Neither science museums nor science research institutions had all the requisite capacities to carry out high-quality nanoscale science education, and they found little incentive to develop such capacity, as it was unclear that their audiences had a driving interest in the topic.

To address these challenges, NISE Net sought the convergence of expertise from the informal educational community and the nanoscale research center community, including not only scientists and engineers but also educational outreach specialists and experts from the social and political sciences and the arts. Subsets of this group worked together to create a catalog of accessible and engaging programs, activities, exhibits, forums, media, tools, and guides to support making informal education about topics related to nanotechnology openly available to all. By July 2012, the catalog included 236 entries, all downloadable from http://www.nisenet.org. Accompanying the downloadable activities are 118 evaluation and research reports that provide knowledge about what was learned in developing these educational materials and other activities of the NISE Net.

With the initial entries to this online resource in place by 2008, NISE Net launched a nationwide NanoDays festival of nanoscale informal education at 100 sites across the United States, with the distribution of a kit of educational material featuring several of the most accessible activities from the broader catalog and all of the supporting material an institution would need to host a NanoDays event. The NanoDays kit was designed to make it as easy as possible for a science museum or a research center to have an experience with nanoscale informal science education that is successful for the public participants, the researcher participants, and the informal educators. From 2009 to 2011, 200 physical kits were distributed each year, and in 2012 the total was increased to 225.

Inverness Research has found that national NISE Net efforts, in particular NanoDays, have been catalytic in engaging new informal science education institutions and scientists to enter into nanoscale science education (St. John et al. 2009):

> I think the NanoDays kit communicates that nano is do-able and that it's a shared initiative and the fact that you're doing the same thing that 200 other people around the country are all doing is kind of empowering. (Alexander et al. 2012)

Between 2008 and 2012, NanoDays Kits were distributed to 375 organizations, including nearly 200 science and children's museums, which represent 53 % of all such organizations listed in the database of the Association of Science-Technology Centers. Another 122 went to university research centers/departments, many of which collaborated with science museums and other informal educational organizations. Kits went to all 50 states, Washington, DC, Puerto Rico, and the Virgin Islands (Fig. 8.9).

NanoDays has not only introduced science museums to hands-on activities related to nanotechnology but also has created collaborations between science

Fig. 8.9 NanoDays map showing the distribution of NanoDays kits to 375 sites in all 50 states, Washington, DC, Puerto Rico, and the Virgin Islands from 2008 to 2012 (http://nisenet.org/nanodays; ©NISE Net, used by permission)

Fig. 8.10 NanoDays teams from just a few of the 225 sites in 2012, comprised of informal educators and research center students (http://nisenet.org/nanodays; ©NISE Net, used by permission)

museums and nanoscale research centers as informal educators, scientists, and university students worked together to provide nano educational experiences for the public (Fig. 8.10). The public has benefited from the knowledge and expertise shared by researchers and graduate students, and the students have gained valuable experience in communicating with the public about their areas of research.

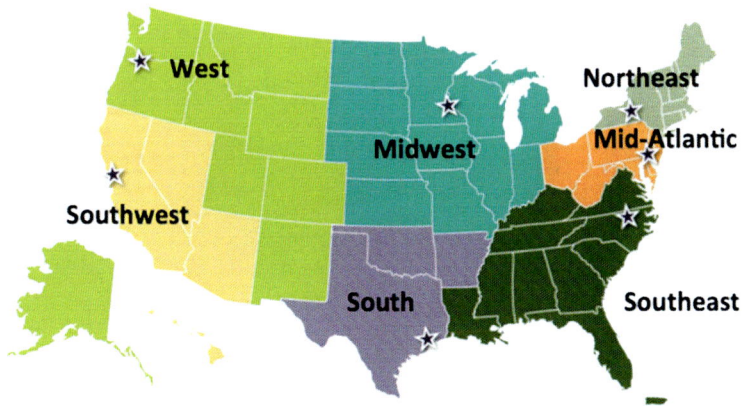

Fig. 8.11 NISE Net Regional Hub Structure (http://nisenet.org/nanodays; ©NISE Net, used by permission)

Like converging technologies—NBIC and beyond—today, nanotechnology was a new and unfamiliar topic to most science museums a decade ago, and creating a sense of a community with a common "cause" was key to getting widespread participation. The NISE Net created a hub structure with seven regional hubs to support the introduction of nanoscale science and engineering content to informal educational organizations nationwide and the deeper infusion of nanoscale content into ongoing programs, exhibits, and other educational activities (Fig. 8.11). This hub structure is a resource that could be used to disseminate other content, including content specifically focused on converging technologies.

Whereas the work of the NISE Net focused on nanotechnology, much of the underlying science is appropriate for a broad range of converging technologies, particularly those that involve work at the nanoscale. Many of the applications anticipated in educational programs about nanotechnology involve nano in convergence with other technologies such as biotechnology or information technology. Finally the NISE Network was made possible by the convergence of the fields of informal education, nanoscale science and engineering, and the social and political sciences. The convergence of these various players, development of the catalog of educational materials, development of the network itself, research and evaluation, and a wide range of activities carried out by the network has been supported by NSF by approximately $4 million/year in two- to five-year awards, beginning in October 2005 and expected to continue through September 2015.

The public has benefited by having educational activities and resources available at sites across the country where people can have a one-on-one, hands-on experience with engaging activities related to nanotechnology. Such work supports both the interest of students in a new field like nanotechnology or converging technologies more broadly, and provides the pubic with a trusted local resource for questions that may arise from current and future applications.

Fig. 8.12 nanoHUB.org users' locations in the United States (© nanoHUB.org, used by permission)

8.8.7 Network for Computational Nanotechnology (nanoHUB) (http://www.ncn.purdue.edu/home/; http://nanohub.org/)

Contact person: *George B. Adams, Network for Computational Nanotechnology*

The Network for Computational Nanotechnology supports the National Nanotechnology Initiative by designing, constructing, deploying, and operating the nanoHUB.org national cyber-infrastructure for nanotechnology theory, modeling, and simulation. NCN was established in September 2002 and is funded by the National Science Foundation to support the NNI.

nanoHUB.org is a science gateway where users can run any of over 250 nanotechnology simulation programs using their web browsers with just the click of a button. In the 12 months ending June 2012, nanoHUB users ran 570,000 such simulations (see map of U.S. users, Fig. 8.12). They also learned about nanotechnology from 3,300 educational resources, including state-of-the-art seminars and complete courses authored by over 1,000 members of the nanotechnology research and education community.

Purdue University's open-source HUBzero® software platform powers the nanoHUB website, and the open-source Rappture Toolkit is used to build the browser-compatible, interactive, graphical user interfaces for software programs published on nanoHUB.

nanoHUB has already had a strong impact on U.S. nanoscale science and engineering (NSE) research. The simulations and resources of nanoHUB have directly supported 885 research papers in the nanotechnology literature to date. To assess the quality of these papers, the number of citations that each has accumulated was counted and the h-index for the 885-paper "nanoHUB collection" computed. Figure 8.13 presents the results, which show an impressive h-index rating.

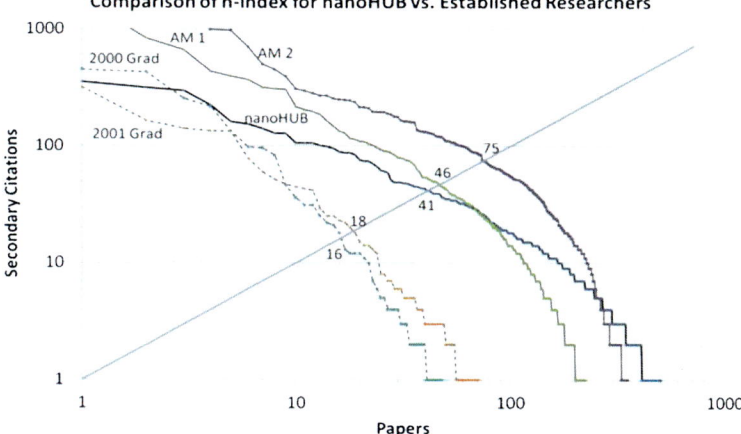

Fig. 8.13 nanoHUB.org h-index compared to two typical researchers earning their Ph.D. in 2000 (2000 Grad) and 2001 (2001 Grad) and two national academy of engineering members (AM 1 and AM 2). The abscissa coordinate is the number of papers that have citations greater than that number. There are 41 papers of the 885 that cite nanoHUB that are in turn cited in the literature at least 41 times. If 10-year-old nanoHUB.org were considered a co-author on these 885 papers, it would have an h-index of 41 (© nanoHUB.org, used by permission; https://nanohub.org/about/contact)

nanoHUB has also had a strong impact on U.S. NSE education. Faculty at 189 institutions have used nanoHUB in 761 classes totaling 14,521 students to date, including all top 50 U.S. engineering schools and 88 % of the top 33 physics and chemistry schools. It should be noted that nanoHUB is reaching students at all academic levels, and it has assumed a strong role in the science education of minority and nontraditional students. For the 449 Minority Institutions listed by the U.S. Department of Education, including 90 Historically Black Colleges and Universities (HBCU), and 215 High Hispanic Enrollment institutions, nanoHUB has been used by 18 %, 34 %, and 23 %, respectively. Bruce Barker, President of Chippewa Valley Technical College, Eau Claire, Wisconsin, has pointed to nanoHUB's value to his students, "We have a high percentage of nontraditional students, many of whom are older and starting new careers, or who are coming from disadvantaged families; nanoHUB provides them with a toehold to a wider academic world." (from telephone interview January 2010 with George B. Adams, Purdue).

nanoHUB is enabling a revolutionary convergence of U.S. NSE research and education. It moves newly published tools into the engineering classroom within months, even weeks, rather than years. Typically, research innovations take about 4 years to enter engineering course content through textbooks. Completely new concepts often take longer. However, simulation programs, primarily authored by researchers, that are published on nanoHUB, and that then reach a classroom do so with a median time of 174 days (5.7 months) (Fig. 8.14).

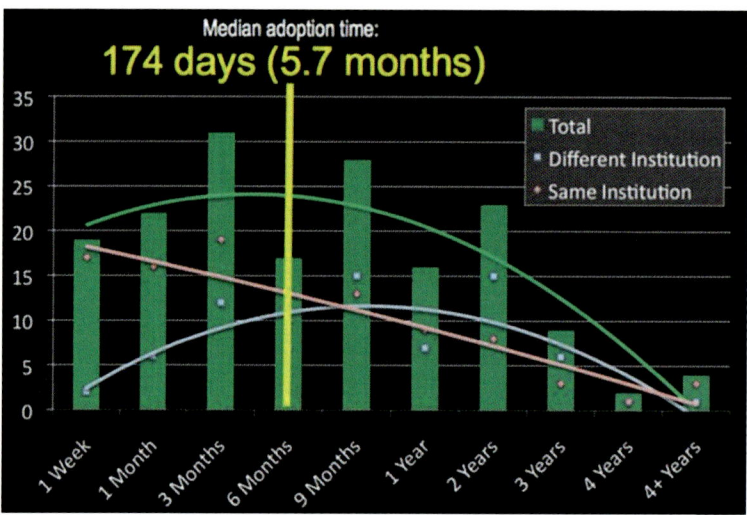

Fig. 8.14 Histogram of time from software program publication to classroom adoption via nanoHUB.org (© nanoHUB.org, used by permission; https://nanohub.org/about/contact)

NCN is leading an effort to rethink electronic devices from the nanoscale perspective. With support from Intel Foundation, NCN has created "Electronics from the Bottom Up," courseware that may reshape the teaching of nanoelectronic technology and will train a new generation of engineers to lead the Twenty-first Century semiconductor industry. Using these new concepts, NCN is building an electronic device simulation platform that powers several tools on nanoHUB and runs efficiently on the largest computers on the national grid. Dr. Dmitri E. Nikonov, Components Research, Technology, and Manufacturing Group, Intel Corporation, who is charged with using simulation to evaluate beyond-CMOS electronic devices (the next generation of transistors), asserts that, "nanoHUB tools are indispensible to the mission of my department."

NCN built the HUBzero and Rappture Toolkit not as special-purpose software but as infrastructure. Infrastructure is a general, fundamental service platform analogous to community water distribution systems that enable drinking, showering, and irrigation conveniently throughout the community; roadway pavement that provides heavy duty, all-weather pathways for a myriad of vehicles from bicycles to semis; and the electric power distribution system that delivers accurate timekeeping information to clocks and the energy to move elevators from floor to floor. The nanotechnology research and learning communities are using nanoHUB to publish, and thus share, their ideas and the fruit of their labor in ways not before possible. This sharing has brought the members of the community closer together than ever before. Infrastructure is empowering, reduces costs, and ultimately supports applications beyond the imaginings of its creators.

New communities are using HUBzero to carry out their missions, including cancer care engineering, healthcare delivery, pharmaceutical product development and manufacturing, environmental systems analysis, volcano research and risk mitigation, earthquake and tsunami research and risk mitigation, and energy from biomass. These and other communities in the years to come will leverage hub capabilities to help users carry out their missions more rapidly and effectively.

8.8.8 Integrative Graduate Education and Research Tr3aineeship (IGERT) (http://igert.org)

Contact: Mihail C. Roco and Melur K. Ramasubramanian, NSF

NSF's Integrative Graduate Education and Research Traineeship aims to better address grand challenges in science and engineering that are inherently interdisciplinary and complex. It involves not just science and engineering, but also policy, government, and geopolitics. The IGERT program provides support to students to work with three or more advisors on an interdisciplinary topic. The program has been developed to advance the challenges of educating U.S. Ph.D. scientists and engineers with interdisciplinary backgrounds, deep knowledge in chosen disciplines, more holistic research themes, and technical, professional, and personal skills. The program has been intended to establish new models for graduate education and training in a fertile environment for collaborative research that transcends traditional disciplinary boundaries. It is also intended to facilitate diversity in student participation and preparation, and to contribute to a world-class, broadly inclusive, and globally engaged science and engineering workforce.

IGERT creates opportunities to better tackle grand challenges. For this purpose, it includes systematic training for future scientists and engineers to take on these challenges and strategic support for research, thereby building an innovation ecosystem. Education needs to couple to policy, business, and law and must be student-focused, as well as enhance student interest in engineering, science, and technology entrepreneurship (http://summit-grand-challenges.pratt.duke.edu).

The IGERT program aims at developing career skills desired by both academic and nonacademic employers, catalyzing sustainable institutional change in graduate education for the training of future scientific research workforce, and broadening participation. It provides a framework wherein institutions, through principal investigators (PIs), can propose programs with enough flexibility to accommodate students' desires to design education plans to match their career goals.

The awards are made to institutions ($3–3.2 million for 5 years) with senior PIs who distribute the funds to Ph.D. students typically for 2–3 years. Since 1997, 278 awards were made to 122 different lead institutions in 43 states, Washington, DC, and Puerto Rico. There are about 25 trainees per award, typically supported for 2 years each. IGERT has supported 6,500 Ph.D. students. Examples of multidisciplinary themes supported are nanotechnology, smart sensors and integrated devices, biosphere-atmosphere research, molecularly designed materials, assistive technology, sequential decision-making, urban ecology, astrobiology, and alternate energy sources.

8.8.9 The Graduate School of Convergence Science and Technology (GSCST) at Seoul National University, Suwon, Korea (http://gscst.snu.ac.kr/eng/)

Contact: *Dr. Y. Eugene Pak, Director Convergence Research Division, Seoul National University*

In the midst of rapid growth in a knowledge-based economy, demand is rising for field-oriented experts with interdisciplinary integration of knowledge in not only basic sciences such as physics, chemistry, and mathematics but also cultural, social, life, and medical sciences. Creative experts are sought in the fields of newly emerged convergence technology such as information technology (IT), biotechnology (BT), and nanotechnology (NT).

The Graduate School of Convergence Science and Technology at Seoul National University was founded in 2009 with the following missions: (1) act as a knowledge-producing base for global standards, (2) lead the development of new technologies for future industry, (3) cultivate creative experts with international competitive power, (4) develop new technologies for industries through the promotion of academic-industrial cooperation, (5) train field-oriented experts, (6) nourish experts with both interdisciplinary integration of knowledge and practical professionalism, and (7) foster creativity as well as field experts in the new convergence technologies such as IT, BT and NT, thereby promoting the creation and development of new industries. GSCST will act as a role model for nourishing creative experts in convergence studies. In order to encourage interaction and convergence among different disciplines, traditionally separate academic departments are loosely divided into programs. GSCST currently consists of four programs—Nano Science and Technology, Digital Contents and Information Studies, Intelligent Systems, and Radiation Biomedical Sciences—and one department: Molecular Medicine and Biopharmaceutical Sciences (see Fig. 8.15).

In addition, all students are required to take an introductory course in Convergence Science and Technology. This course first deals with the definition and classification of convergence science and technology and then introduces the fundamentals of nano-convergence technology, digital contents convergence technology, and intelligent convergence systems technology. Students are assigned a term paper or a term project that is carried out in teams consisting of students with diverse backgrounds. The main point of this course is to learn to collaborate with people with different specialties in solving complex problems as a team effort. In its fourth year running, currently GSCST has about 217 graduate students in master's and Ph.D. programs. So far, 47 master's students have graduated, most of whom found jobs in high-tech industries such as Samsung, LG, KT, Hyundai, and some even in overseas companies. About a third continued on in a Ph.D. program.

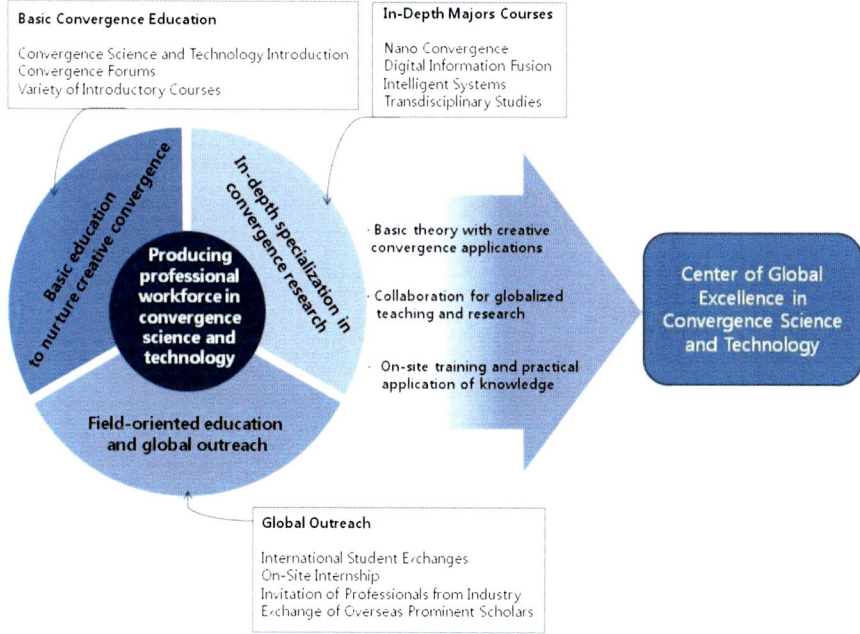

Fig. 8.15 Vision of the graduate school of convergence science and technology (Courtesy of Y. E. Pak)

Program in Nano Science and Technology. Nanoscale science, engineering, and technology (NSET) encompasses a wide range of traditional academic disciplines such as physics, chemistry, biology, electrical engineering, materials science, and mechanical engineering. This wide scope of the emerging NSET presents both promises and challenges: NSET, via convergence, has the potential of providing novel solutions and technologies unavailable from the traditional scientific disciplines individually, thereby greatly improving the ways humans obtain and process information, diagnose and treat diseases, and address energy and environmental issues. The challenge is that we must have the intellectual foresight to identify potential convergence areas having the greatest academic and social impacts. We also must turn those opportunities of convergence into reality.

The mission of our NSET program is to educate next-generation engineers and scientists who possess a solid understanding of the principles of nanoscience and are capable of making real-world impact. The department offers interdisciplinary courses and research opportunities at master's and Ph.D. levels in the following three loosely defined areas: Nanophysics and Nanodevices, Nanomaterials and Nanochemistry, and Nanobiological Science.

Program in Digital Contents and Information Studies. This addresses the radical changes in the information-oriented society that have greatly affected and altered our behavior, our life, and the whole social structure. In order to prepare for such changes, we must take "information-convergent" approaches by converging what was once regarded as independent fields of study, including information science, computer science, communication studies, business administration, and so on. Such convergent approaches require education and research institutes that can train analysts, directors, designers, and managers who not only have a high understanding of human society and culture, but who also are familiar with information technology.

In many countries, an Information School or iSchool already exists, providing a program for "information-convergence". These schools originate from library science and computer science; however, there are a number of factors in these programs that do not fit well with Korea's reality. In order to meet Korea's needs, Seoul National University launched its Information Convergence program as a collaborative effort between the Computer Science Department that provides education and research programs relating to information technology studies, and the Information Technology & Contents Technology (ITCT) Department that has capability for designing and analyzing the cultural contents with in-depth understanding of information technology. The goal is to understand people's needs and usage of information, analyze the role of information within our society, and to construct the information infrastructure for cultural understanding. Throughout this, we intend to create a better environment where people can create and share knowledge more effectively. Furthermore, we plan to bridge the gap between academia and industry by taking reality-based approaches that will maximize the students' potentials and research minds. We believe that such efforts will help train students to be creative and skilled enough to be able to work in the IT, knowledge-based, or cultural contents industries.

Program in Intelligent Systems. This combines mechanical engineering, computer engineering, human engineering, electrical and electronics engineering, intelligent systems engineering, business, and industrial design. One good example of intelligent systems is self-driving vehicles, which encompasses a synergistic combination of future smart cars with various information technology infrastructures such as cloud computing. The program seeks to establish intelligent systems that link humans, engineering, and markets together to foster experts who can apply the knowledge to reality, and to cultivate experts with professional knowledge who are capable of creating an innovative thought process.

Program in Radiation Biomedical Sciences. This utilizes a wide range of cutting-edge interdisciplinary academic study encompassing radiology, biophysics, radiopharmacology, radiochemistry, nanomolecular imaging, and imaging science, and it exploits the new biomedical science fields by converging a variety of traditional academic majors such as medicine, physics, pharmaceutics, chemistry, nuclear engineering, electrical engineering, materials science, and mechanical engineering. Recently, a part of the program (i.e., interdisciplinary program in Radiation Applied Life Science) became the world's first accredited medical physics graduate course outside of North America. This program has been developed into world-class education and research par excellence.

The curriculum of Radiation Biomedical Sciences focuses on the training of creative interdisciplinary specialists based on the expert knowledge of radiation science and biomedical science. We provide the education and practice for these experts to become core human resources who would manage medical institutions and national laboratories. We are creating a talent pool who will eventually nurture the domestic industry of radiation and medical appliances. The graduate program in Radiation Biomedical Science has the educational purpose to foster world-class human resources who have the interdisciplinary expertise, capability for combining that knowledge, and creative research skills. They are expected to lead the future interdisciplinary studies-oriented convergence science and technology field of Radiation Biomedical Convergence.

The *Molecular Medicine and Biopharmaceutical Sciences Program.* This program belongs to the World Class University (WCU) project, is a newly established division of the Graduate School at Seoul National University. It aims to become the world's top-ranked program in molecular medicine and biopharmaceutical science through collaborative research with the greatest scholars in the world. It is a graduate program to train professionals for translational research, combining fundamental knowledge in medical life science with applied knowledge such as pathophysiology, preclinical and clinical trials, and clinical medicine.

Up to now, research in the biomedical field has had limitations with respect to immediate clinical application because previous studies concentrated only on basic research rather than considering clinical applications. However, the importance of translational research has recently gained interest as a means of forming a bridge between research results and clinical medicine and related industries. Basic research scientists propose new methods or materials to clinicians to treat patients. Clinicians deliver clinical information or results of illnesses to basic scientists so that they can induce more meaningful results from their basic studies. Therefore, the necessity of translational research is gaining strength as a means to implement new technology in clinical medicine and pharmacology, and to apply the results of basic life science to clinical fields and development of new drugs. Translational research can provide a better understanding of the cause and mechanisms of a certain disease through studies of clinical medicine at the molecular level.

One approach of clinical medicine is to treat illnesses through pharmaceuticals. In our current aging society, our dependency on medicines is deepening every day. In order to develop a new medicine, we first need to understand the etiology of the disease, discover target molecules, and validate the efficacy of the produced medicine. In order to develop a medicine with therapeutic efficacy, we need to find the active compound that acts on the target molecules. To discover such compounds, many have to be screened and their efficacy, toxicity, and bioavailability analyzed to find the leading compound. This process is a main area of pharmacology. If medicine and pharmacology are combined, discovery and validation of targets and deduction of active compounds are done in the same place, making the environment for developing new medicines optimal. Also, when pharmacology and medicine are combined, the clinical information of the efficacy of active compounds can be exchanged swiftly, definitely making it possible to efficiently procure leading compounds.

8.8.10 Building a Next-Generation Convergence Research Hub at the Advanced Institutes of Convergence Technology (AICT), Seoul National University (SNU), Suwon, Korea (http://aict.snu.ac.kr/eng/)

Contact: Dr. Y. Eugene Pak, Director Convergence Research Division, Seoul National University

Established in 2008 with investment from Gyeonggi Province, the AICT is located in the Gwanggyo Techno Valley (GTV) near Korea's capital Seoul. More than 50 % of Korea's population and much of its industry and intellectual infrastructure are located in the province. This recently established "mini cluster," which combines features of a research institute, business incubator, and learning institution, has chosen to specialize in the convergence of key technologies deemed crucial to the future of the competitiveness of the Korean economy. To this end, the Seoul National University has established the Advanced Institutes of Convergence Technology (AICT) and the Graduate School of Convergence Science and Technology (GSCST) on the GTV campus. AICT's mission is to combine cutting-edge research on convergence technology with the education of a new generation of innovators trained in interdisciplinary science and engineering—all with the purpose of creating a next-generation high-tech hub based on convergence technology applications. In addition, the Korea Advanced Nano Fab Center (KANC), the Gyeonggi Bio Center, and the Gyeonggi Small & Medium Business Center (GSBC) have also been located in the valley to provide a comprehensive infrastructure to conduct convergence technology research and business development. GTV is close to large multinational companies such as Samsung, Hyundai, SK, and KT, as well as many high-tech SMEs. With this strategic location, GTV can act as a "corridor" or "link pin" connecting other technology clusters situated near Seoul and in other parts of the country.

Locating the research institutes away from the main campus of SNU has the advantage of starting a new organizational structure with minimal or no boundaries between disciplines or departments. Such a configuration encourages fresh thinking outside the traditional boundaries and allows new ideas to develop from intermixing of traditional disciplines, which is further enhanced by experts working in many jointly appointed teaching and research positions.

The four institutes organized around 17 research centers cover convergence technology areas such as nano, bio, IT, and transdisciplinary studies (see Fig. 8.16). Through the process of providing seed funding by the AICT and assistance with moving from early planning to the pilot stage, the Bio Convergence Institute was able to win a 10-year $100 million government grant to develop key technologies for fast and low-cost drug development (BioCon). This project brings together many researchers from diverse disciplines such as molecular biology, bioinformatics, micro and nano engineering, as well as robotics.

8 Implications: People and Physical Infrastructure

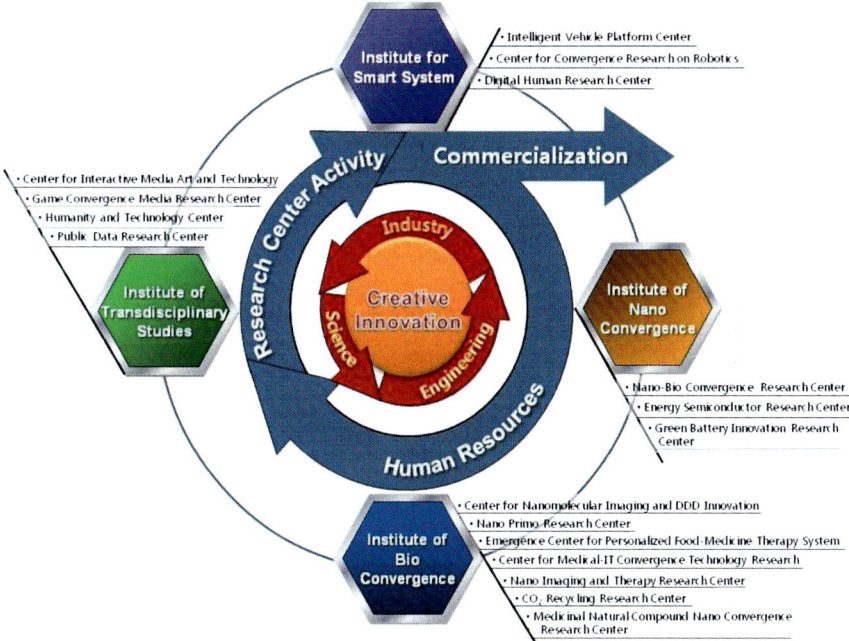

Fig. 8.16 Diagram illustrating the concept of AICT (Courtesy of Y. E. Pak)

In order to encourage this type of success, AICT provides seed funding and infrastructures for research projects that have high convergence content. Another area of high-impact research is the future intelligent vehicle platforms program based on electric power for clean environment and IT and cloud computing for convenience and safety. Companies can participate in this process and be partners in establishing centers of excellence whose members include researchers from the various participating organizations. Overseas institutes are also welcome to join or invest in these Technology Centers of Excellence. For example, the Energy Semiconductor Research Center was formed with the support from a global LED (light-emitting diode) company to work together on key technological issues as well as to identify new applications that require convergence of disciplines, including industrial design, psychology, and human sensory perception.

AICT managers believe that convergence technology is the key to successful future innovation, and they see large South Korean companies well suited for this new phase of global competition. In order to strengthen the budding SMEs to be more globally competitive, an executive education program called the World Class Convergence Program (WCCP) was created to train CEOs and

CTOs to utilize new convergence technologies so that new business opportunities can be created. To strengthen its alliance with the business sector, AICT is in close collaboration with the Korea National Industrial Convergence Center, whose charter is to make strategic plans for the successful incubation and application of convergence technologies in industry.

Key to successful convergence is learning to define a clear end goal and identifying critical constituent disciplines or technologies so that an appropriate team of experts can be brought together to meet the end goal. AICT has invited professionals from industry to join in this team-building effort with professors, researchers, and students. AICT also tries to create an environment to allow free exchange of ideas that encourage new and innovative concepts to develop from diverse backgrounds. To this end, a very flexible organizational structure is put in place so that new centers can readily be created and merged.

Technology convergence will continue to play an important role for innovation as technologies become more complex and diverse. Networking of innovation centers with the world-class production or manufacturing clusters will also be important in bringing up the speed to meet future market demands. AICT aims to play this important role in fostering international cooperation of Innovation Centers

8.9 International Perspectives

The following are summaries relevant to this chapter of discussions at the international regional WTEC NBIC2 workshops held in Leuven, Belgium, September 20–21, 2012; in Seoul, Korea, October 15–16, 2012; and in Beijing, China, October 18–19, 2012. Further details of those workshops are provided in Appendix A.

8.9.1 United States–European Union NlBIC2 Workshop (Leuven, Belgium)

The EU workshop on NBIC2 had three working groups: Human Development; Sustainable Development; and Co-evolution of Human Development and Technology. Each identified an important education component—personalized education, melding of STEM and social sciences/humanities in education, and life-long learning. Workshop participants are listed in Appendix B.

The Human Development Working Group discussed the opportunity for a paradigm shift by morphing formal and informal education toward adaptive life-long learning systems, a goal especially important as lifespan increases. Converging technologies will enable learning platforms that can be tailored to personal learning modes (print, video, oral, haptic, motion, interactive). To make these platforms successful,

8 Implications: People and Physical Infrastructure

it will be necessary to better understand cognition and learning processes, and the human–machine relationship. Because of resistance to change by the existing education systems, there will be challenges to develop the market for personalized education. In addition to the new personalized education platforms, it will be necessary to promote the convergence of traditional disciplines, preferably by improving awareness among those disciplines of the benefits to such a change. At the university level, this will likely require changes to traditional academic organization and reward systems. There needs to be a vetted portal that can provide access to education assets, perhaps an Amazon.edu.

The Sustainable Development Working Group principally addressed sustainability. While converging technologies can provide some partial solutions to a sustainable future, the human factor in education will be critical to changing societal behavior. Better integration of social sciences and humanities with STEM education will be important to realizing that goal. Change will be necessary at all levels, with special attention to integrated, progressive content and to the use of information technology in innovative teaching aides.

The Co-evolution of Human Development and Technology Working Group explored the interplay between machine and human capacities to augment productivity while retaining jobs. Our present education structures are already outdated, and the pace of converging technological change will exacerbate the situation. Education will be key to jobs in the coming economy. To enable the opportunities enabled by converging technology, changes will be necessary at all levels, but especially for vocational, undergraduate, graduate, and continuing education. Converging technology implies innovation, and this must be explicitly incorporated in the educational process. Because of increased use of automation by all organizations, high school education needs to be rethought to include vocational training in the machine–human interface. At the Ph.D. level, converging technologies will require teaming and multidisciplinary perspectives; this will require a different approach than those presently used by most universities.

8.9.2 United States–Korea–Japan NBIC2 Workshop (Seoul, Korea)

Panel members/discussants:

Kwyro Lee, National NanoFab Center (Korea)
James Murday, University of Southern California (U.S.)
Masahiro Takemura, National Institute for Materials Science (Japan)
Mark Lundstrom, Purdue University (U.S.)
Yoon-Hwae Hwang, Pusan University (Korea)
Chul Gi Ko, Korea Advanced Nano Fab Center (KANC, Korea)
Y. Eugene Pak, Seoul National University (Korea)

Physical Infrastructure

In the past 10 years, extensive nanoscale user facilities have been established and are being well used by academia. There remains the challenge to increase industrial usage and to adapt these facilities to converging technologies. For biology this poses a particular challenge since semiconductor-processing cleanrooms are over-pressured to reduce particulate contamination, while biology cleanrooms are under-pressured to reduce the risk of dissemination of any biological materials posing health risks. There is also the challenge to achieve self-sustainability by identifying sources of funding for continuing operation, maintenance, upgraded instrumentation, and availability of skilled operators. This challenge is exacerbated by the need for better coordination among the several government ministries involved in funding such operations.

As the science and engineering basis for converging technologies continues to mature, it will be important to transition into a technology and business development perspective, including test beds for manufacturing that are compatible with large-scale implementation.

The continuing installation of high-speed, real-time Internet linkages to individuals potentially enables the capability for them to gain access to center-based high-cost instruments, and to expand the application of citizen science beyond such topics as astronomy.

Education

The last decade saw increasing acceptance of interdisciplinarity as an essential component for advances in science and technology; this should evolve into a convergence of those disciplines into a transdisciplinary approach, reflected by new university organizational structures and faculty reward systems. As an example, Seoul National University opened its Graduate School of Convergence Science and Technology in March 2009 with the following missions: (1) act as a knowledge-producing base for global standards, (2) lead the development of new technologies in the future industry, (3) cultivate creative experts with international competitive power, (4) develop new technologies for industries through promotion of academic–industrial cooperation, (5) train field-oriented experts, (6) nourish experts with both interdisciplinary integration of knowledge and practical professionalism, and (7) foster creativity as well as field experts of the new convergence technology consisting of IT, BT, and NT, thereby promoting the creation and the development of new industries (see case studies, Sects. 8.8.9 and 8.8.10).

The maturation of digital information, including the development of three-dimensional displays and the replacement of desktop consoles with mobile devices, will instigate major changes in education. However, credentialing online courses is still a subject of concern; one approach could be the use of local testing centers for proctored exams.

8 Implications: People and Physical Infrastructure

Korea and Japan have government-funded programs providing world-class research capabilities:

1. In Korea, the World Class University (WCU) project is a higher education subsidy program of the Korean government, which invites international scholars who possess advanced research capacities to collaborate with Korean faculty members and establish new academic programs in key growth-generating fields. With a vision to enhance the competence of Korean universities and nurture high-quality human resources, the WCU project seeks to achieve two goals:

   - Enhance national, higher educational, and industrial competitiveness in interdisciplinary fields
   - Transform Korean universities into world-class research institutions

2. In Japan, the World Premier International Research Center Initiative (WPI) was launched in 2007 by the Ministry of Education, Culture, Sports, Science and Technology (MEXT) in a drive to build within Japan "globally visible" research centers that boast a very high research standard and outstanding research environment, sufficiently attractive to prompt frontline researchers from around the world to want to work in them. These centers are given a high degree of autonomy, allowing them to virtually revolutionize conventional modes of research operation and administration in Japan.

There is a need for a high-impact journal and a professional society addressing convergent technologies.

8.9.3 United States–China–Australia–India NBIC2 Workshop (Beijing, China)

Panel members/discussants:

James Murday, University of Southern California (U.S.)
Chennupati Jagadish, Australian National University (Australia)
Xiaomin Luo, BGI Healthcare (China)
Mark Lundsgtrom, Purdue University (U.S.)

Physical Infrastructure

The existing nanoscale user facilities need to be generalized to address converging technologies; these facilities are important in that they enable better communication between researchers as well as provide access to specialized instrumentation. While many modern electron microscopes can be remotely controlled, this capability needs to be instituted in other high-cost, user-facility equipment. Self-sustained funding for facilities is also a problem in this region of the world.

Both China and Australia have instituted efforts to facilitate the scale-up of nano-enabled technologies. The Cooperative Research Centres (CRC) program is an Australian Government Initiative administered by the Department of Industry, Innovation, Science, Research and Tertiary Education. The CRC program supports end-user-driven research collaborations to address major challenges facing Australia. CRCs pursue solutions to these challenges that are innovative, of high impact, and capable of being effectively deployed by the end users.

China's Suzhou Industrial Park held a joint groundbreaking and opening ceremony for 24 new projects in January 2011. Nanopolis, one of the industrial park's communities, has an estimated total investment of USD 1 billion. Its mission is to focus on micro-nano manufacturing engineering, especially the R&D of universal technologies and standard processes in the fields of nanomechanical-electrics and photo-electronics. By centering on pilot lines, the institute will offer support for technological commercialization and project incubation, as well as provide extensive services of talent training, quality control, safety assessment, and intellectual property rights protection. Suzhou Nanotech is a state-owned company under the Suzhou Industrial Park, which is home to the China International Nanotech Innovation Cluster jointly supported by the Ministry of National Science and Technology (MOST), the Ministry of Commerce, and Jiangsu Province.

Education

University programs for the nanoscale continue to evolve, presently mostly as concentrations within a traditional discipline. Some universities, such as Flinders University in Australia, are experimenting with full degrees.

With the maturation of digital information technology, the growing capability to biosense an individual's state of alertness, and growing insights into cognition, the vision of an digital tutor—a personal, lifelong educator—is in reach. This workshop affirmed the personalized digital education goals stated in the visionary priorities of Sect. 8.6. STEM education is the logical initial focus since it has lesser cultural ramifications.

References

ACS: Advancing graduate education in the chemical sciences. Full report of the ACS Presidential Commission on Graduate Education in the Chemical Sciences, 6 December 2012. Available online: http://portal.acs.org/portal/PublicWebSite/about/governance/CNBP_031603

Adobe Education: The silent transformation: evolution and impact of digital communication skills development in post secondary education (White Paper). Available online: http://www.adobe.com/education (2011)

Alexander, J.M., Svarovsky, G., Goss, J., Rosino, L., Mesiti, L.A., LeComte-Hinley, J., Reich, C.: A study of communication in the Nanoscale Informal Science Education Network. NISE Net. Online: http://nisenet.org/ncs (2012)

Atkins, D.E., Droegemeier, K.K., Feldman, S.I., Garcia-Molina, H., Klein, M.L., Messerschmitt, D.G., Messina, P., Ostriker, J.P., Wright, M.H.: Revolutionizing Science and Engineering Through Cyberinfrastructure: Report of the National Science Foundation Blue-Ribbon Advisory Panel on Cyberinfrastructure. NSF, Arlington (2003)

Balassone, M.: USC embraces online graduate education. USC Chron. **32**(04), 1+ (2012)

Barben, D., Fisher, R., Selin, C., Guston, D.: Anticipatory governance of nanotechnology: foresight, engagement, and integration. In: Hackett, E., Amsterdamska, O., Lynch, M., Wajcman, J. (eds.) The Handbook of Science and Technology Studies, pp. 979–1000. MIT Press, Cambridge, MA (2008)

Bell, L.: Engaging the public in technology policy: a new role for science museums. Sci. Commun. **29**(3), 386–398 (2008)

Bell, P., Lewenstein, B., Shouse, A.W., Feder, M.A. (eds.): Learning Science in Informal Environments: People, Places, and Pursuits. National Academies Press, Washington, DC (2009)

Blackwell, A.F., Wilson, L., Street, A., Boulton, C., Knell, J.: Radical Innovation: Crossing Knowledge Boundaries with Interdisciplinary Teams. University of Cambridge Computer Laboratory, Cambridge (2009). http://www.cl.cam.ac.uk/techreports/UCAM-CL-TR-760.pdf. Accessed 11 October 2012

Blumenstyk, G.: Carnegie leader calls for presidential commission to guide higher education future. The Chronicle of Higher Education 12 November 2012

Board on Science Education, The National Academies: Commissioned Papers. Available online: http://www7.nationalacademies.org/bose/PP_Commissioned_Papers.html (2012)

Brooks, D.: The campus tsunami. The New York Times, 3 May 2012

CAISE (Center for Advancement of Informal Science Education): Conference advances the field of public participation in scientific research. CAISE Newsletter 29 (2012)

Cole, J.R.: The Great American University: Its Rise to Preeminence, its Indispensable National Role, and Why it Must be Protected. Public Affairs, New York (2009)

Collins, H.M.: Tacit and Explicit Knowledge. University of Chicago Press, Chicago (2010)

Cook, G.: How crowdsourcing in changing science. The Boston Globe: Ideas, 11 November 2011

Davidson, C.: A core curriculum to create engaged entrepreneurs. Co. EXIST. Available online: http://www.fastcoexist.com/1680124/a-core-curriculum-to-create-engaged-entrepreneurs (n.d.)

ED (U.S. Department of Education): 10 Facts About K–12 Education Funding. U.S. Department of Education, Washington, DC (2005)

Engardio, P.: A Mexican technology park in Monterrey. Business Week, 1 June 2009. Available online: http://www.businessweek.com/innovate/content/jun2009/id2009061_243746.htm

Engler, J.: STEM education is the key to the U.S.'s economic future. U.S. News & World Report (June 15, Opinion) (2012)

Evans, M.: A Guide to Personalizing Learning: Suggestions for the Race to the Top–District Competition (White paper E-WP-005). Innosight Institute, San Mateo (2012). Available online: http://www.innosightinstitute.org/

Flagg, B.N.: Compilation of Nanoscale Communication Projects. Part 2A (Exhibits) of Front-end Analysis in Support of Nanoscale Informal Science Education Network. Multimedia Research, Bellport (2005). Available online: http://www.nisenet.org/catalog

Flagg, B., Knight-Williams, V.: Summative Evaluation of NISE Network's Public Forum: Nanotechnology in Health Care. Multimedia Research, Bellport (2008). Available online: http://informalscience.org/reports/0000/0391/NISEForumSummativeEval.pdf

Fonash, S.J.: Nanotechnology and economic resiliency. Nano Today **4**(4), 290–291 (2009)

Gai, T., Dadayan, L.: Federal, State, and Local Education Finances. Nelson A. Rockefeller Institute of Government, University at Albany, SUNY, Albany, (2012). Available online: http://www.rockinst.org/government_finance/

Gorman, M.E.: Mind in the world: cognition and practice in the invention of the telephone. Soc. Stud. Sci. **27**(4), 583–624 (1997)

Gorman, M.E.: Types of knowledge and their roles in technology transfer. J. Technol. Tran. **27**(3), 219–231 (2002)

Gorman, M.E.: A framework for anticipatory governance and adaptive management of synthetic biology. Int. J. Soc. Ecol. Sustain. Dev. **3**(2), 65–69 (2012)

Gorman, M.E., Spohrer, J.: Service science: a new expertise for managing sociotechnical systems. In: Gorman, M.E. (ed.) Trading Zones and Interactional Expertise: Creating new Kinds of Collaboration, pp. 75–105. MIT Press, Cambridge, MA (2010)

Gorman, M.E., Swami, N., Cohoon, J.M., Groves, J., Squibbs, K., Werhane, P.: Integrating ethics and policy into nanotechnology education. *Journal of Nano Education*, **4**(1–2), 25–32 (2013). http://www.ingentaconnect.com/content/asp/jne/2013/00000004/F0020001/art00004

Hallacher, P.M., Fonash, S.J., Fenwick, D.E.: The Pennsylvania Nanofabrication Manufacturing Technology (NMT) partnership: resource sharing for nanotechnology workforce development. Int. J. Eng. Educ. **18**(5) (2002)

Hamlett, P., Cobb, M., Guston, D.: National citizens' technology forum: Nanotechnologies and human enhancement. CNS-ASU Report #R08-0003. http://cns.asu.edu/cns-library/type/?action=getfile&file=88 (2008)

Hart, R., Karthigasu, K.: The benefits of virtual reality simulator training for laparoscopic surgery. Curr. Opin. Obstet. Gynecol. **19**(4), 297–302 (2007)

Hays, S., Miller, C., Cobb, M.: Public attitudes toward nanotechnology-enabled cognitive enhancement in the United States. In: Hays, S., Robert, J., Miller, C., Bennett, I. (eds.) Nanotechnology, the Brain, and the Future: Yearbook of Nanotechnology in Society. Springer, Dordrecht (2013)

Holt, L., Colburn, D., Leverty, L.: Innovation and STEM education. University of Florida Bureau of Economic and Business Research. Available online: http://www.bebr.ufl.edu/articles/innovation-and-stem-education (n.d.)

Isaacson, W.: Steve Jobs. Simon & Schuster, New York (2011)

ITU (International Telecommunications Union): The World in 2011: ICT facts and figures. http://www.itu.int/ITU-D/ict/facts/2011/material/ICTFactsFigures2011.pdf (2011)

Jasanoff, S.: Technologies of humility: citizen participation in governing science. Minerva **41**(3), 223–244 (2003)

Jones, B.F.: The burden of knowledge and the 'death of the renaissance man': is innovation getting harder? NBER Working Paper No. 11360. May Issue (2005)

Kahan, D.: Fixing the communications failure. Nature **463**, 296–297 (2010)

Katsouleas, T.: Who says online courseware will cause the death of universities? Forbes Singularity University, 20 November 2012

Langdon, D., McKittrick, G., Beede, D., Kahn, B., Doms, M.: STEM: Good Jobs Now and for the Future. ESA Issue Brief #03-11. Education and Statistics Administration, U.S. Department of Commerce, Washington DC (2011)

Lerner, L.S., Goodenough, U., Lynch, J., Schwartz, M., Schwartz, R.: The State of State Science Standards. The Thomas B Fordham Institute, Washington, DC (2012)

Lécuyer, C.: Making Silicon Valley: Innovation and the Growth of High Tech, 1930–1970. MIT Press, Cambridge, MA (2006)

Leshner, A.I.: Public engagement with science. Science **299**, 977 (2003)

Luks, F.: Post-normal science and the rhetoric of inquiry: deconstructing normal science? Futures **31**(7), 705–719 (1999)

McCallie, E.L., Bell, L., Lohwater, T., Falk, J.H., Lehr, J.L., Lewenstein, B., Needham, C., Wiehe, B.: Many Experts, Many Audiences: Public Engagement with Science and Informal Science Education. A CAISE Inquiry Group Report. Center for Advancement of Informal Science Education, Washington, DC (2009). http://caise.insci.org/uploads/docs/public_engagement_with_science.pdf

Melguizo, T., Wolniak, G.C.: The earnings benefits of majoring in STEM fields among high achieving minority students. Res. High. Educ. **53**, 383–405 (2012)

Molebash, P.: Technology and Education: Current and Future Trends. INDUS Training and Research Institute, Bangalore (1998)

Murday, J.S. (ed.): Partnership for nanotechnology education. NSF Workshop report under NSF award EEC 0805207. Available online: http://nsf.gov/crssprgm/nano/reports/educ09_murdyworkshop.pdf (2009)

Murday, J.S. (ed.): Workshop Report: International Benchmark Workshop on K–12 Nanoscale Science and Engineering Education (NSEE), Washington, DC, 6–7 December 2010. National Science Foundation, Arlington (2011). Available online: http://nsf.gov/crssprgm/nano/reports/educ09_murdyworkshop.pdf

Nielsen, M.: Reinventing Discovery: The New Era of Networked Science. Princeton University Press, Princeton (2011a)

Nielsen, J.: Lessons learned from graduated nanoeducation efforts. Presentation at 2011 NSF Nanoscale Science and Engineering Grantee Conference. Manhattan Strategy Group. http://www.nseresearch.org/2011/presentations/Day2_Jennifer_Nielson~FINAL_NSF_NSE_Grantees_Conference_Presentation.pdf (2011b)

NIH: Biomedical Research Workforce Working Group Report. National Institutes of Health, Washington, DC (2012). June 14

NRC (National Research Council): Preparing for the Revolution: Information Technology and the Future of the Research University. The National Academies Press, Washington, DC (2002)

NRC (National Research Council): Is America Falling off the Flat Earth? The National Academies Press, Washington, DC (2007a)

NRC (National Research Council): Rising Above the Gathering Storm: Energizing and Employing America for a Brighter Economic Future. The National Academies Press, Washington, DC (2007b). Initial report release in 2005

NRC (National Research Council): Engineering in K–12 Education: Understanding the Status and Improving the Prospects. The National Academies Press, Washington, DC (2009)

NRC (National Research Council): Successful K–12 STEM Education: Identifying Effective Approaches in Science, Technology, Engineering, and Mathematics. The National Academies Press, Washington, DC (2011)

NRC (National Research Council): A Framework for K–12 Science Education: Practices, Crosscutting Concepts, and Core Ideas. The National Academies Press, Washington, DC (2012a)

NRC (National Research Council): Community Colleges in the Evolving STEM Education Landscape: Summary of a Summit. The National Academies Press, Washington, DC (2012b)

NRC (National Research Council): Research Universities and the Future of America: Ten Breakthrough Actions Vital to our Nation's Prosperity and Security. The National Academies Press, Washington, DC (2012c)

NSB (National Science Board of the National Science Foundation): A National Action Plan for Addressing the Critical Needs of the U.S. Science, Technology, Engineering, and Mathematics Education System. National Science Foundation, Arlington (2007). http://www.nsf.gov/nsb/documents/2007/stem_action.pdf

NSB (National Science Board of the National Science Foundation): Preparing the next Generation of STEM Innovators: Identifying and Developing our Nation's Human Capital. NSF, Arlington (2010). http://www.nsf.gov/nsb/publications/2010/nsb1033.pdf

NSF: Chapter 5: Learning and workforce development (2006–2010). In: Cyberinfrastructure Vision for 21st Century Discovery. National Science Foundation Cyberinfrastructure Council, Arlington (2007). Available online: http://www.nsf.gov/pubs/2007/nsf0728/nsf0728.pdf

NSF: A Report of the National Science Foundation Advisory Committee for Cyberinfrastructure, Task Force on Cyberlearning and Workforce Development. National Science Foundation, Arlington (2011). http://www.nsf.gov/od/oci/taskforces/TaskForceReport_Learning.pdf

NSF Cyberinfrastructure Council: Cyberinfrastructure Vision for the 21st Century. National Science Foundation (NSF 07-28), Arlington (2007)

NSTC (National Science and Technology Council Interagency Working Group on Advanced Manufacturing): A national Strategic Plan for Advanced Manufacturing. Executive Office of the President, Washington, DC (2012). http://www.whitehouse.gov/

Oblinger, D.G. (ed.): Game Changers: Education and Information Technologies. Educause, Washington, DC (2012)

OSTP (Office of Science and Technology Policy): 2010 Federal STEM Education Inventory Data Set. Executive Office of the President, Washington, DC (2012a). http://www.whitehouse.gov/blog/2012/04/18/ostp-releases-data-stem-education

OSTP (Office of Science and Technology Policy): Preparing a 21st Century Workforce, Science, Technology, Engineering, and Mathematics (STEM) Education in the 2013 Budget. Executive Office of the President, Washington, DC (2012b). http://www.whitehouse.gov/sites/default/files/microsites/ostp/fy2013rd_stem.pdf

PCAST (President's Council of Advisors on Science and Technology): Report to the President and Congress on the Third Assessment of the National Nanotechnology Initiative. Executive Office of the President, Washington, DC (2010). http://www.whitehouse.gov/sites/default/files/microsites/ostp/pcast-nano-report.pdf

PCAST (President's Council of Advisors on Science and Technology): Policy workstream report. Annex 4 in Capturing Domestic competitive advantage in advanced manufacturing: Report of the Advanced Manufacturing Partnership Steering Committee. Available online: http://www.whitehouse.gov/administration/eop/ostp/pcast/docsreports (2012a)

PCAST (President's Council of Advisors on Science and Technology): Engage to Excel: Producing One Million Additional College Graduates with Degrees in Science, Technology, Engineering, and Mathematics. Report to the President. Executive Office of the President, Washington, DC (2012b). http://www.whitehouse.gov/administration/eop/ostp/pcast/docsreports

Pearson, G., Young, A.T. (eds.): Technically Speaking: Why all Americans Need to Know more About Technology. The National Academies Press, Washington, DC (2002)

Reich, C., Chin, E., Kunz, E.: Museums as forum: engaging science center visitors in dialogues with scientists and one another. The Informal Learning Review (July-August):1–8 (2006)

Reich, C., Bell, L., Kollmann, E.K., Chin, E.: Fostering civic dialogue: a new role for science museums? Mus. Soc. Issues **2**(2), 207–220 (2007)

Rizzolatti, G., Fogassi, L., Gallese, V.: Neurophysiological mechanisms underlying the understanding and imitation of action. Nat. Rev. Neurosci. **2**(9), 661–670 (2001)

Robinett, W.: Imagine the consequences of fully understanding the brain. In: Roco, M.C., Bainbridge, W.S. (eds.) Converging Technologies for Improving Human Performance: Nanotechnology, Biotechnology, Information Technology and Cognitive Science, pp. 3–27. Kluwer Academic, Dordrecht (2003)

Roco, M.C., Bainbridge, W.S.: Overview: converging technologies for improving human performance. In: Roco, M.C., Bainbridge, W.S. (eds.) Converging Technologies for Improving Human Performance: Nanotechnology, Biotechnology, Information Technology and Cognitive Science, pp. 166–170. Kluwer, Dordrecht (2003)

Sacramento Bee.: McGraw-Hill Education launches Digital Learning Partnership Program to increase affordability and access to e-books and digital solutions among colleges nationwide, 5 November 2012

Schiavelli, M.: STEM jobs outlook strong, but collaboration needed to fill jobs. U.S. News & World Report (November 3, News) (2011)

Schmidt, W.H., McKnight, C.C.: Inequality for All: The Challenge of Unequal Opportunity in American Schools. Teachers College Press, New York (2012)

Sen, A.: Development as Freedom. Anchor/Random House, New York (2000)

Sharp, P.A., Langer, R.: Promoting convergence in biomedical science. Science **333**(29), 527 (2011)

Sitzmann, T., Kraiger, K., Stewart, D., Wisher, R.: The comparative effectiveness of web-based and classroom instruction: a meta-analysis. Department of Defense contract number DASW01-03-C-0010. Blackwell Publishing (2006)

Sjoberg, S., Schreiner, C.: The next generation of citizens: attitudes to science among youngsters. In: Bauer, M., Shukla, R. (eds.) The Culture of Science: How Does the Public Relate to Science Across the Globe? Routledge, New York (2010)

Spohrer, J.: WorldBoard: What comes after the WWW? Research Report: Learning Communities Group32, ATG Apple Computer. Available at http://service-science.info/archives/2060 (1997)

Spohrer, J.C.: The meaning of learning from the perspective of rapid technological change. Educ. Technol. **42**(2), 31–34 (2002)

Spohrer, J.C., Engelbart, D.C.: Converging technologies for enhancing human performance: Science and business perspectives. In: Roco, M.C., Montemagno, C.D. (eds.) The Coevolution of Human Potential and Converging Technologies, pp. 50–82. The New York Academy of Sciences, New York (2004)

Spohrer, J., Giuiusa, A.: Exploring the future of cities and universities: a tentative first step. In: Proceedings of Workshop on Social Design: Contribution of Engineering to Social Resilience, 12 May 2012. System Innovation. University of Tokyo, Tokyo, Japan. Available at http://www.sys.t.u-tokyo.ac.jp (2012)

St. John, M., Helms, J.V., Castori, P., Hirabayashi, J., Lopez, L., Phillips, M.: NISE Network: Overview of the NISE Network Evaluation. Inverness Research, Inverness (2009)

The Foundation Center.: Top 50 U.S. foundations awarding grants for education circa 2012 (table). http://foundationcenter.org/findfunders/statistics/pdf/04_fund_sub/2010/50_found_sub/f_sub_b_10.pdf (2012)

The opportunity equation: Transforming mathematics and science education for citizenship and the global economy. Carnegie Corporation of New York, New York. Available online: http://carnegie.org/fileadmin/Media/Publications/PDF/OpportunityEquation.pdf (2009)

The Third Revolution: The Convergence of the Life Sciences, Physical Sciences, and Engineering. MIT White Paper, January 2011. Massachusetts Institute of Technology Washington DC Office, Washington, DC (2011). http://dc.mit.edu/. Accessed 11 October 2012

Trani, E.P., Holsworth, R.D.: The Indispensable University: Higher Education, Economic Development, and the Knowledge Economy. Rowman & Littlefield, New York (2010)

Turnpenny, J., Jones, M., Lorenzoni, I.: Where now for post-normal science? A critical review of its development, definitions, and uses. Sci. Technol. Hum. Values **36**(3), 287–306 (2011). [Electronic version] http://dx.doi.org/10.1177/0162243910385789

Vastag, B.: U.S. pushes for more scientists, but the jobs aren't there. Washington Post (July 7, Health and Science) (2012)

Wetzel, D.R.: Science and technology adult education: continuing education programs designed for nontraditional learners. Suite101, Aug 28, http://suite101.com/continuingeducation (2009)

Wiley, D., Green, C., Soares, L.: Dramatically Bringing Down the Cost of Education with OER. Center for American Progress, Washington, DC (2012). Issue Brief. Feb 7

Wladawsky-Berger, I.: Why CIOs desperately need a technology-literate society. CIO J, 23 September 2012

Wong, J.A., Matsumoto, E.D.: Primer: cognitive motor learning for teaching surgical skill—how are surgical skills taught and assessed? Nat. Clin. Pract. Urol. **5**(1), 47–54 (2008)

Yankelovich, D.: Winning greater influence for science. Issues Sci. Technol. http://issues.org/19.4/yankelovich.html (2003). Accessed 15 November 2009

Young, E.: Engineering your degree. Inside Higher Ed, 10 July 2012. http://www.insidehighered.com/

Recommended General Reading

Bell, L., Kollmann, E.K.: Strategic directions for PES. http://dimensionsofpes.wikispaces.com/Strategic+Directions+for+PES (2011). Accessed 20 December 2011

Bennet, G., Gilman, N., Stavrianakis, A., Rainbow, P.: From synthetic biology to biohacking: are we prepared? Nat. Biotechnol. **27**, 1109–1111 (2009)

Murday, J.S. (ed.): Partnership for nanotechnology education. NSF Workshop report under NSF award EEC 0805207. http://nsf.gov/crssprgm/nano/reports/educ09_murdyworkshop.pdf (2009). Accessed 11 October 2012

Murday, J.S. (ed.): Workshop Report: International Benchmark Workshop on K–12 Nanoscale Science and Engineering Education (NSEE), Washington, DC, 6–7 December 2010. National Science Foundation, Arlington. Available online: http://www.nsf.gov/crssprgm/nano/reports/nsfnnireports.jsp (2011)

Murday, J., Hersam, M., Chang, R., Fonash, S., Bell, L.: Developing the human and physical infrastructure for nanoscale science and engineering. In: Roco, M.C., Mirkin, C.A., Hersam, M.C. (eds.) Nanotechnology Research Directions for Societal Needs in 2020. Springer, Dordrecht (2011)

NRC (National Research Council): A Framework for K–12 Science Education: Practices, Crosscutting Concepts, and Core Ideas. The National Academies Press, Washington, DC (2012a)

NRC (National Research Council): Research Universities and the Future of America: Ten Breakthrough Actions Vital to our Nation's Prosperity and Security. The National Academies Press, Washington, DC (2012b)

NRC (National Research Council): Community Colleges in the Evolving STEM Education Landscape: Summary of a Summit. The National Academies Press, Washington, DC (2012c)

Sarasvathy, S.D.: Effectuation: Elements of Entrepreneurial Expertise. Edward Elgar Publishing, Cheltenham (2008)

Chapter 9
Implications: Convergence of Knowledge and Technology for a Sustainable Society

Mamadou Diallo, Bruce Tonn, Pedro Alvarez, Philippe Bardet,
Ken Chong, David Feldman, Roop Mahajan, Norman Scott,
Robert G. Urban, and Eli Yablonovitch

In October 2011, the world population reached seven billion (Coleman 2011). Several scenarios of world population growth based on projections by the United Nations (UN) and the United States Census Bureau (Wikipedia Commons 2012) estimate that the world population will reach eight billion to ten billion by 2050. The global challenges facing the world are complex and involve multiple

---

Corresponding editors M.C. Roco (mroco@nsf.gov) and W.S. Bainbridge (wbainbri@nsf.gov).

M. Diallo
California Institute of Technology, Pasadena, CA, USA

Korea Advanced Institute of Science and Technology, Daejeon, South Korea

B. Tonn
University of Tennessee, Knoxville, TN, USA

P. Alvarez
Rice University, Houston, TX, USA

P. Bardet
George Washington University, Washington, DC, USA

K. Chong
National Institute of Standards and Technology, Gaithersburg, MD, USA

D. Feldman
University of California, Irvine, CA, USA

R. Mahajan
Virginia Polytechnic Institute and State University, Blacksburg, USA

N. Scott
Cornell University, Ithaca, NY, USA

R.G. Urban
Johnson & Johnson Boston Innovation Center, Cambridge, MA, USA

E. Yablonovitch
University of California, Berkeley, Berkeley, CA, USA

interdependent areas. Every human being requires food, water, energy, shelter, transportation, healthcare, and employment to live and prosper on Earth. A great challenge facing the word in the twenty-first century is to continue to provide better living conditions to all people while minimizing the impact of human activities on the global environment. The United Nations 1987 World Commission on Environment and Development, commonly referred to as the Brundtland Commission, defined "sustainable development" as that "which meets the needs of the present without compromising the ability of future generations to meet their own needs" (Brundtland 1987). The commission called the three pillars of sustainability social, economic, and environment—suggesting that sustainability requires convergence of all three.

This chapter discusses the convergence of knowledge, technology and society (CKTS) for ensuring a sustainable society and a healthy, secure, and peaceful world. The focus is on sustainable solutions to our global problems, as the world's population continues to grow and as demands for basic commodities (e.g., food, water, and energy); finished goods (e.g., cars, airplanes, and cell phones); services (e.g., shelter, transportation, and healthcare); employment, and better living standards continue to increase in the next 20 years.

9.1 Vision

9.1.1 Changes in the Vision over the Past Decade

During the last 10 years, (1) population growth; (2) the demand and sustainable supply of energy, clean water, food, and critical materials; and (3) global climate change have emerged as being among the most critical problems facing society and the global economy in the twenty-first century (Diallo and Brinker 2011).

During the last decade, significant progress has been made in information and communication technologies (ICT). Several new ICT devices such as smart phones, e-pads, and other computing and communication devices have been developed. Such devices are using an increasing amount of energy and critical metals, and new methods are underway to reduce consumption of energy in electronic devices and to utilize less critical metals or replace them with nanostructured materials based on earth-abundant elements.

The emergence of industrial ecology as a well-established science is another important development of the last 10–20 years. Industrial ecology has provided new tools for assessing and minimizing the impact of industrial activities on the environment; these tools include (1) *material flow analysis* (MFA), (2) *life-cycle analysis* (LCA) and (3) *design for environment* (DFE) (den Hond 2000). Advances in earth-systems science are also providing new tools to probe the impact of human activities on Earth's climate and ecosystems (see Chap. 3 of this book). During the

9 Implications: Convergence of Knowledge and Technology...

last 20 years, a consensus has gradually emerged that human activities have become the main drivers of global environmental change (IPCC 2007a; Rapport 2007; Rockström et al. 2009a, b). Although Earth has experienced many cycles of significant environmental change, the planet's environment has been stable during the past 10,000 years, commonly referred to as the Holocene period (Zalasiewicz et al. 2010). This stability is now threatened as the world's population and demands and competition for basic commodities, finished goods, and services continue to increase in the foreseeable future (Moyo 2012).

A group of investigators from the Stockholm Resilience Center led by Johan Rockström have argued that human activities could put the "Earth System" outside a stable state, with significant or catastrophic consequences. Rockström et al. (2009a, b) subsequently proposed the concept of "planetary boundaries" (PB) as a unifying framework for (1) describing the impact of human activities on the environment and (2) determining a "safe operating space for humanity with respect to the functioning of the Earth System" (Rockström et al. 2009b). They suggested planetary boundaries for nine global, regional, and local environmental processes, as shown in Table 9.1.

9.1.2 The Vision for the Next Decade

The Brundtland Commission's definition of sustainability, the planetary boundaries concept, and the MFA, LCA, and DFE tools, taken together, provide an integrated and coherent framework for addressing our global needs for basic commodities, finished goods, services, decent employment, and better living standards, while minimizing the impact of human activities on Earth's climate and ecosystems as the world population reaches eight to ten billion by 2050. This chapter focuses on several critical problems in global sustainability where CKTS-based solutions will likely have the greatest impacts over the next decade, including energy and water, food and agriculture, human health and well-being, rural and urban communities, and materials supplies and utilization. Here we articulate and discuss the implementation of a vision of societal sustainability based on the following premises (Fig. 9.1):

- Sustainability is determined by the coupled interactions between: (1) population growth and human needs, (2) societal and cultural values, (3) the human-built environment, and (4) Earth system's boundaries.
- Sustainability is enabled by the convergence of knowledge and technologies developed by (and for) society (CKTS).

The overarching theme of this chapter is that CKTS will be critical to *"realizing the future we want for all"* (UN System Task Team 2012), that is, a sustainable, healthy, secure, and peaceful world.

Table 9.1 Planetary boundaries with proposed boundary and current values of the control variables (Source: Rockström et al. 2009b)

Earth-system process	Control variable	Proposed boundary	Current status	Preindustrial value
Climate change	1. Atmospheric carbon dioxide concentration (parts per million by volume)	350	387	280
	2. Change in radiative forcing (watts per m^2)	1	1.5	0
Rate of biodiversity loss	Extinction rate (number of species per million species per year)	10	>100	0.1–1
Nitrogen cycle (part of a boundary with the phosphorus cycle)	Amount of N_2 removed from the atmosphere for human use (millions of tons per year)	35	121	0
Phosphorus cycle (part of a boundary with the nitrogen cycle)	Quantity of P flowing into the oceans (millions of tons per year)	11	8.5–9.5	~1
Stratospheric ozone depletion	Concentration of ozone (Dobson unit)	276	283	290
Ocean acidification	Global mean saturation state of aragonite in surface sea water	2.75	2.90	3.44
Global freshwater use	Consumption of freshwater by humans (km^3 per year)	4,000	2,600	415
Changes in land use	Percentage of global land cover converted to cropland	15	11.7	Low
Atmospheric aerosol loading	Overall particulate concentration in the atmosphere, on a regional basis	To be determined		
Chemical pollution	For example, amount emitted to, or concentration of, persistent organic pollutants, plastics, endocrine disrupters, heavy metals, and nuclear waste in the global environment, or the effects thereof on ecosystem and functioning of Earth system	To be determined		

9.2 Advances in the Last Decade and Current Status

To continue to survive and thrive on Earth, Rockström and colleagues (2009a, b), as noted above, suggest that humanity must stay within defined planetary boundaries for a range of key ecosystem processes so as to avoid catastrophic environmental changes (Table 9.1). More to the point, they believe that we have already transgressed three of these nine boundaries: (1) atmospheric carbon dioxide (CO_2) concentration, (2) rate of biodiversity loss, and (3) input of nitrogen into the biosphere.

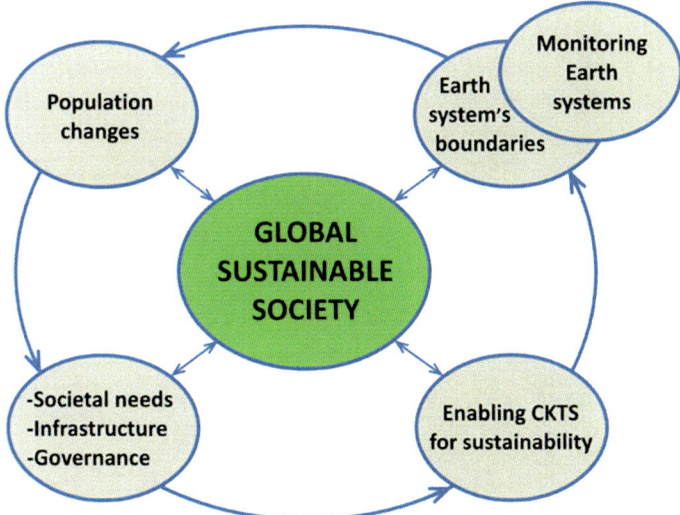

Fig. 9.1 Vision for achieving societal sustainability in the next 10 years and beyond. Four main factors will be considered as key determinants of societal sustainability in this chapter: (1) Earth system's boundaries, (2) population changes, (3) societal needs, infrastructure, and governance and (4) enabling converging knowledge and technologies (Courtesy of M. Diallo)

In the case of global freshwater, Rockström et al. (2009a, b) argue that "the remaining safe operating space for water may be largely committed already to cover necessary human water demands in the future."

To address the interrelated questions of societal sustainability, population growth and human needs, social and cultural values, and built and natural environments, the following sections provide a background discussion (the status) for each of the target sustainability goals to set the stage for a subsequent assessment of how CKTS will advance these goals in the next 10 years and beyond.

9.2.1 Population Growth and Structure

A scientific consensus about long-term demographic trends has not emerged, and it is unclear how many people the Earth can support, even assuming continued technological progress (Cohen 1995). More than two centuries ago, Thomas Malthus (1798) argued that human population growth would always consume resources faster than they could be produced, leaving most people living in abject poverty or dying from disease, starvation, and warfare. However, technological growth seemed to prove Malthus wrong, as wealth grew faster than population, lifting ever greater fractions of humanity into prosperity. By the middle of the twentieth century, a new perspective emerged, called demographic transition theory, that was far more optimistic.

A background assumption was that the institutions of society naturally adjust to changing conditions, for the improved functioning of society as a whole, and thus for most individuals who dwell within it (Parsons and Shils 1951). In the distant past, high childhood death rates required women to give birth to many children, so that enough would live to adulthood and produce children themselves. Thus, high mortality was balanced by high fertility. Then, technological advances reduced childhood deaths, but for a time the fertility rate remained high because societies require time to adjust to changing conditions—what is called cultural lag (Ogburn 1922). Eventually, the fertility rate will come down, according to demographic transition theory, until it is again in balance with the mortality rate, and the human population will be stable (Davis 1963). During the period of cultural lag, there will be a population explosion, but it will naturally subside as society adjusts.

In the 1960s, doubts began to be voiced from many quarters, both about the demographic transition theory and about the general idea that societal institutions like the family naturally adjust in a manner to maximize their functionality. Some critics focused on the continuing population explosion itself (Ehrlich 1968), while others argued that the theory failed really to explain how any kind of new balance could naturally emerge (Knodel and van de Walle 1979). Although countries like China could for a while forcibly limit fertility in order to promote economic growth, most countries do not have that option (Baochang et al. 2007). Now a new public debate is emerging, around two demographic issues that will impact the ability of converging technology to provide its anticipated benefits to humanity, and which may be addressed through societal convergence: humanity's changing age structure and disparate fertility rates.

The first issue is the changing age structure. Low fertility coupled with an increasing life span shifts the age distribution toward the later years of life. This trend is slow but apparently inexorable. For example, the fraction of the United States population under age 18 declined from 25.7 % in 2000 to 24.0 % in 2010, as the percent age 62 and older increased from 14.7 % to 16.2 % (Howden and Meyer 2011). The aging of the population is a major challenge to the healthcare system, as increasing fractions of the population suffer chronic disorders that perhaps can be managed but not currently cured. More broadly, huge economic challenges result from the fact that increasing fractions of the population are retired, including many people who lose jobs in late middle age and are unable to reenter the workforce because no employer will hire them.

Specific converging technologies will offer many partial solutions for one or another age-related problem, but general societal convergence must provide a general solution. One major possibility is instituting a new custom in developed countries, such that most workers at a certain age in a given profession are assisted with training and placement to enter entirely new fields of work, ones better suited for their advancing years. For example, someone in a job that requires heavy lifting or manual labor may no longer be able to perform the tasks, whether from fatigue or arthritis, but could very well do other kinds of work if properly trained, given suitable technologies to work with. Instituting this new system would require many changes in how education and business operate, and could be successful only in the context of general societal convergence.

The second issue is the disparity in fertility rates across different populations around the globe. In developed countries, on average, each woman must bear 2.1 children to sustain a stable total population, but already all countries of the European Union have dropped below that replacement rate (Bainbridge 2009). As of 2007, even before the changing age distribution had its full effect, in 14 European countries, deaths outnumbered births. Although the fertility rate for the United States in the period 2005–2010 was almost exactly the 2.1 replacement rate, 26 countries have fertility rates of 5.0 or greater, none of them in Europe or the Americas (United Nations 2012).

In order to achieve population stability in a democratic context, societal convergence will need to achieve three primary goals: (1) universal prosperity, so that no society is underdeveloped with the consequent pre-transition high fertility rate, (2) a cooperative economy, so that potential parents are not forced to compete so hard at work that they are discouraged from having children, and (3) gender fairness, so that women are not penalized for their reproductive function, and those who choose to have several children are appropriately assisted by the wider community. Exactly how converging technologies can contribute to these goals remains to be worked out, through research and innovation that combines social with technological elements (Cáceres-Delpiano 2012; Staff and Mortimer 2012). A related issue is how demographic convergence can be fine-tuned to achieve precise population stability at a level that maximizes human well-being and free choice.

9.2.2 Energy and Water

Energy and water supplies are strongly coupled. Power plants require abundant amounts of water to produce electricity, from 38 L per 1,000 kWh for natural gas to 180,900–969,000 L per 1,000 kWh for biodiesel. Conversely, the production and delivery of clean water requires a lot of energy (Webber 2008).

The development of hydraulic fracturing (commonly referred to as "fracking") has vastly increased the amounts of natural gas that can be extracted from shale gas formations during the last decade (DOE 2009b). Because natural gas is considered to be the "cleanest" among fossil fuels, it has become the transition fuel toward a post-fossil-energy society powered by renewable energies (DOE 2009b). However, the production of shale gas by fracking requires large amounts of water (2–4 millions of gallons per well) and is often carried out in proximity to valuable sources of surface water and groundwater (DOE 2009b). Although the increased production of natural gas from shale oil formations has resulted in tangible economic benefits for the United States, including both lower gas prices and less dependence on oil imports, substantial concerns remain about the adverse impact of fracking and shale gas production on water resources (both quantity and quality) and greenhouse gas (GHG) emissions.

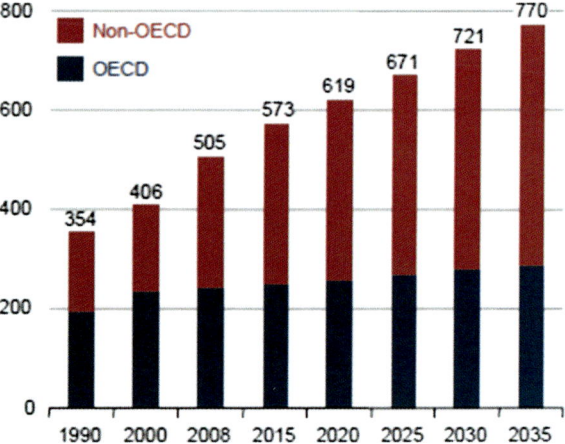

Fig. 9.2 World energy consumption in quadrillion Btu from 1990 to 2035 (EIA 2011; ©EIA, used by permission)

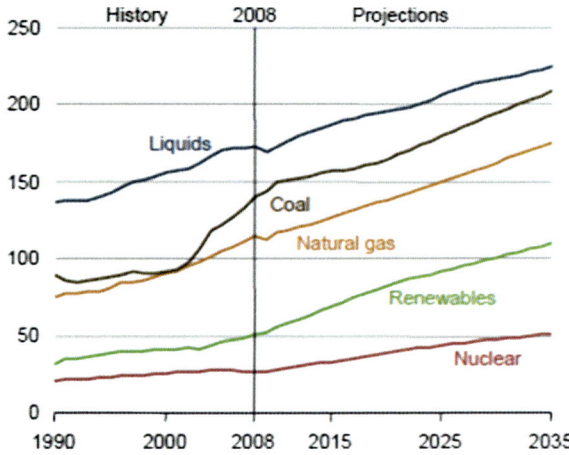

Fig. 9.3 World energy consumption in quadrillion Btu from 1990 to 2035 (EIA 2011; ©EIA, used by permission)

Energy

The availability of abundant, low-cost, and carbon-neutral energy is arguably the greatest challenge of the twenty-first century. The worldwide demand for energy is expected to increase by 40 % in the next 20 years (Fig. 9.2) (EIA 2011).

Figure 9.3 shows that fossil fuels will continue to provide a significant amount of the energy used worldwide in the foreseeable future. During the last 20 years, a consensus has gradually emerged that increasing emissions of GHG such as carbon dioxide from the combustion of fossil fuels (coal, petroleum, and natural gas) are among the key drivers of global climate change (IPCC 2007a).

Although significant progress is being made toward developing more efficient and safer nuclear energy fuels and reactors (see Sect. 9.8.2 below), the utilization of low-cost and non-GHG- emitting renewable energies might be the most sustainable solution to global climate change. During the last 10 years, significant advances have been made in the development and deployment of renewable energy sources, including biomass. Biomass can be burned directly to generate electricity, or it can be converted to liquid biofuels (i.e., ethanol and biodiesel). Because transportation accounts for approximately 33 % of CO_2 emissions in the United States and 66 % of oil consumption (Davis et al. 2008), significant research efforts are being devoted to the conversion of non-food plants (e.g., grasses) to liquid biofuels. In 2007, BP formed the Energy Biosciences Institute (EBI) as a way to initiate a major R&D program in cellulosic biofuels and fossil fuel microbiology (see http://www.energybiosciencesinstitute.org/). BP funded the EBI at the level of $50 million per year for 10 years. Its core R&D programs are being implemented by a team of scientists and engineers from the University of California at Berkeley, the Lawrence Berkeley National Laboratory, the University of Illinois at Urbana-Champaign, and BP.

During the last 10 years, significant progress also has been made in solar energy generation technologies. Solar energy has become the most attractive source of renewable energy due to its abundance, versatility, and ease of implementation with minimum environmental impact in terms of water consumption and land usage (Lewis 2007; Brinker and Ginger 2011). Key advances in solar energy generation include the development of (1) thin-film gallium arsenide (GaAs) photovoltaics (PV) cells with high efficiency (~23 %) (see http://altadevices-blog.com/) and (2) more efficient silicon PV cells. This has led to a significant drop in the cost of electricity produced from solar PV to $1 per Watt. The new GaAs solar cells are being scaled-up for commercialization by the California start-up company Alta Devices (http://altadevices.com/). Because renewable energy sources such as solar PV and wind power produce electricity intermittently, their large-scale implementation will require more efficient energy storage devices and electric grids. Thus, significant progress has also been made toward development of (1) low-cost batteries and supercapacitors with high energy and high power density, and (2) smart grids for improved delivery of solar-derived electricity (Brinker and Ginger 2011; WWF 2011).

As renewable energy sources become more affordable and efficient, energy demand continues to rise in many sectors, including in information and communications technologies. The invention of the smart phone, for example, is among the most important advances in ICT hardware during the last 20 years. In 2010, the number of cell/smart phones in use worldwide was estimated to be 5.3 billion (Moyo 2012), causing a significant energy demand. Large amounts of critical materials will also be needed as the worldwide demands for mobile/smart phones continue to increase. There is a growing consensus that the development and large-scale implementation of renewable energy generation and storage technologies will require sizeable amounts of valuable and critical metals such as copper, lithium, titanium, gallium, rare-earth elements (e.g., neodymium and europium), platinum group metals (e.g., platinum and palladium), and precious metals (e.g., gold and silver) (Diallo and Brinker 2011; Fromer et al. 2011).

Water

The availability of clean water is a critical problem facing society and the global economy in the twenty-first century. Already there is insufficient availability of

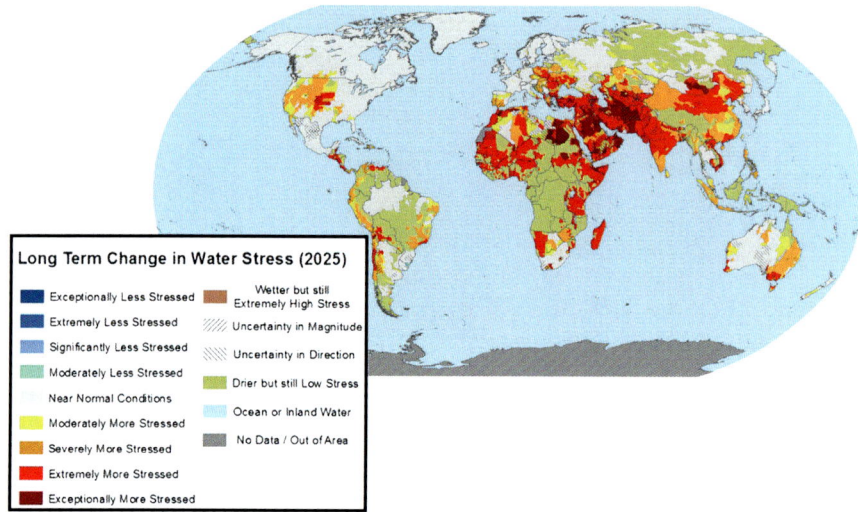

Fig. 9.4 Predicted long-term changes in water stress (2025). Water stress is expected to grow 2–8 times worse in many places by 2025 (Figure: World Resources Institute; Iceland 2011; data provided courtesy of the Coca Cola Company). The A1B scenario (In assessing the physical science bases of climate change, IPCC Working Group 1 developed a series of emissions scenarios for the future: "The A1 scenario family develops into three groups that describe alternative directions of technological change in the energy system. The three A1 groups are distinguished by their technological emphasis: fossil intensive (A1FI), non-fossil energy sources (A1T), or a balance across all sources (A1B) (where balanced is defined as not relying too heavily on one particular energy source, on the assumption that similar improvement rates apply to all energy supply and end-use technologies)"; see http://www.ipcc.ch/ipccreports/tar/wg1/029.htm#storya1) of the Intergovernmental Panel on Climate Change (IPCC) assumes a "future world of very rapid economic growth, global population that peaks in mid-century and declines thereafter, and the rapid introduction of new and more efficient technologies" with a "balanced" portfolio of fossil fuels and renewable energy sources (Bates et al. 2008, Figure 9.4; ©IPCC, used by permission)

clean water for human consumption, agriculture, and industry. Many regions of the world are experiencing higher demands for clean water while freshwater supplies are being stressed (Shannon et al. 2008; Diallo and Brinker 2011). The number of people living in water-stressed areas will continue to increase in the next decades as the amount of available freshwater decreases due to global climate change (Bates et al. 2008). By 2025 (Fig. 9.4), it is estimated that approximately 50 % of the world population will live in water-stressed areas (Iceland 2011).

To alleviate future water shortages and stresses, considerable research efforts have been devoted during the last decade to develop more efficient and cost-effective technologies to reclaim and reuse wastewater (Shannon et al. 2008; McCarty et al. 2011; Pennisi 2012; Logan and Rabaey 2012). Approximately 70–90 % of the water used in agriculture, industry, and human consumption is returned to the environment as wastewater (Shannon et al. 2008). Wastewaters contain organics, including nutrients (nitrogen and phosphorous) and compounds with fuel values. A recent study suggests that the fuel values of organics in wastewater per gram of solids (dry weight) range from 17.7 to 28.7 kJ (Heidrich et al. 2011). Current wastewater treatment plants are not sustainable; they spend significant amounts energy to destroy (i.e., mineralize) valuable organic compounds in wastewater. In the United States, the treatment of organic-rich wastewater consumes approximately 3 % of the electrical power produced (McCarty et al. 2011).

Significant progress has been made in the development of technologies to extract clean water, energy, and valuable compounds from wastewater (McCarty et al. 2011; Pennisi 2012; Logan and Rabaey 2012). Microbial fuel cells (MFCs) have emerged as promising platforms for converting organic compounds in wastewater to electricity (Logan and Rabaey 2012). However, McCarty et al. (2011) argue that integrated and anaerobic wastewater treatment systems would be more efficient and cost-effective than MFCs at extracting both clean water and energy from wastewater. A proposed integrated anaerobic wastewater treatment system would combine conventional solid–liquid separation with the latest developments in microbiology and membrane technology—anaerobic membrane bioreactors (McCarty et al. 2011).

In addition to wastewater reclamation and water reuse, low-energy desalination technologies with reduced environmental footprint (e.g., brine generation) also will be required to alleviate future water shortages and stresses (Shannon et al. 2008; Diallo and Brinker 2011; Elimelech and Phillip 2011). Saline water (including brackish water and seawater) constitutes ~97 % of the water on Earth. Key advances in desalination during the last 10 years include the following:

- Discovery of fast water transport through carbon nanotube membranes (Holt et al. 2006).
- Preparation of thin-film composite reverse-osmosis (RO) membranes with embedded nanoparticles (e.g., zeolite) that achieve higher flux (~1.2 times) than commercial RO membranes (Jeong et al. 2007).
- Preparation of low-pressure and ion-selective nanofibrous composite membranes with high water flux (Park et al. 2012).
- Progress in the preparation of biomimetic desalination membranes with embedded aquaporin channels (Tang et al. 2012).
- Emergence of forward osmosis (FO) as a potential low-energy desalination technology (Wang et al. 2012).

9.2.3 Agriculture and Food

While productivity has been a consistent and important goal in agricultural R&D, there has been significant increased emphasis on sustainability over the last 10 years. Indeed, the concept of sustainability of agriculture has been much discussed and was addressed in the National Research Council (NRC) report on sustainable agriculture (2010), wherein four goals are used to define sustainable agriculture:

- Satisfy human food, feed [for animals], and fiber needs, and contribute to biofuels needs
- Enhance environmental quality and the resource base
- Sustain the economic viability of agriculture
- Enhance the quality of life of farmers, farm workers, and society as a whole

Modern agriculture has had an impressive history of increasing productivity that has led to abundant, safe, and affordable food, fiber, and recently, biofuels. Farmers today are meeting expanding demands for both domestic and international markets on the same acreage as a century ago as a result of technological innovations, economies of scale, consolidation of food processing and distribution, and advanced retailing. During the decade, U.S. agriculture has continued to become increasingly dependent on large-scale, high-input farms that specialize in a few crops and concentrated animal production practices; for example, 2 % of U.S. farms are responsible for 59 % of U.S. farm products (NRC 2010). In contrast, small- and medium-sized farms, which represent more than 90 % of the total farmers, manage about half of U.S. farmland. Advances in biotechnology, automation and precision agriculture, conservation tillage, and livestock systems have maintained and increased yields.

Biotechnology crops as of 2011 (Clive 2011) are utilized in 29 countries and are reaching 160 million hectares. The use of crops derived from biotechnology has yielded several benefits:

- Contribution to food, feed, and fiber security, and self-sufficiency, including more affordable food, by increasing productivity and economic benefits sustainably at the farm level.
- Conservation of biodiversity through land-saving technologies. Higher productivity on the current 1.5 billion hectares of arable land can preclude deforestation and protect biodiversity.
- Contribution to alleviation of poverty and hunger. For example, biotech has made significant contribution to incomes of approximately 15 million small resource-poor farmers in developing countries in 2011, primarily in cotton, maize, and rice.
- Reduction of agriculture's environmental footprint by reducing pesticide use, saving on fossil fuels, decreasing CO_2 emissions, and increasing efficiency of water usage.

Precision agriculture is a systems approach for site-specific management of crop production systems. The foundation of precision agriculture rests on geospatial data techniques for improving the management of inputs and documenting production outputs (Reid 2011). A key technology enabler for precision farming resulted from the public availability of the U.S. global navigation satellite system (GNSS). GNSS has provided highly accurate and high-resolution data for mapping yields and moisture content. Advances in precision agriculture benefited greatly from the design of new machinery, including precision planters, sprayers, fertilizer applicators, and tillage instruments.

Conservation tillage systems are methods of soil tillage that can have both environmental and economic benefits. Conservation tillage leaves a minimum of 30 % of crop residue, or at least 1,100 kg/ha of small grain residue, on the soil surface during the critical soil erosion period (NRCS 2012). The most significant advantage is less soil erosion due to wind and water. Conservation tillage systems also benefit farmers by reducing fuel consumption and soil compaction. Conservation tillage, although developed a number of years ago, has gained limited adoption in the United States during the last decade and was used on 44 million hectares, which constituted approximately 38 % of all U.S. cropland by 2004.

Livestock systems. Positive environmental gains along with considerable gains in livestock systems have been achieved during the past decade. In the United States, advances in animal nutrition, management systems, and genetics have resulted in a large increase in annual milk yield of dairy cattle. Capper et al. (2009) report a fourfold increase in milk yield in 2007 compared to that of 1944, with 84.3 billion kg produced in 2007 compared to 53 billion kg in 1944 with 64 % fewer cows. Carbon emissions and total emissions per unit of milk were reduced by 66 % and 41 %, respectively. Similar results for emissions per unit of product have been achieved in the beef and poultry industries.

9.2.4 Human Health and Wellness

Equitable health and wellness policies are critical to a sustainable society. Over the course of the last century, tremendous medical advances have consistently increased life span across the globe. Early in the 1900s, the discovery and use of antibiotics, followed in the middle of the century by widespread dissemination of effective vaccines, dramatically reduced the morbidity and mortality associated with infections (Oeppen and Vaupel 2002; Christensen et al. 2009). Later in the century, biochemistry-based advances led to new drugs for patients with elevated blood lipids (Tobert 2003), and for cases where lipid deposition had already resulted in clinically relevant occlusions, coronary stents were developed that sufficiently restored cardiac function—both life-saving developments (Tobert 2003). Advances in surgical techniques and molecular tissue typing/matching allowed, along with the development of optimized immunosuppressive treatments and the ability to surgically transplant whole organs from one person (alive or deceased) to another, thus giving "a second life" to those with whole-organ failure (King and Meier 2000). Over the course of this century, steady advances in life span are expected, and CKTS will be enabling many of these new life-extending discoveries.

Interrelated with advances in medicine, wellness knowledge, and technologies are issues of medical accessibility and affordability. Life span expectations are influenced by the gross domestic product of a given population, but factors such as obesity, diet, exercise, tobacco smoking, and drug and alcohol use can confound

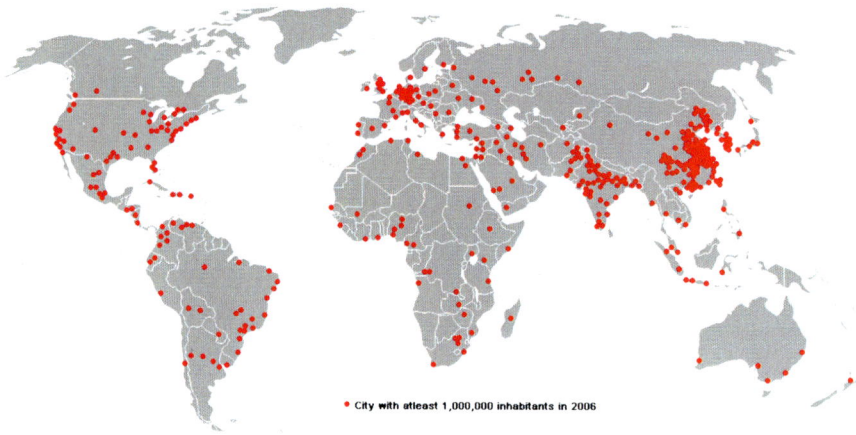

Fig. 9.5 Map showing urban areas with at least one million inhabitants in 2006. Only 3 % of the world's population lived in cities in 1800; this proportion had risen to 47 % by 2000, and reached 50.5 % by 2010. By 2050, the proportion may reach 70 % (Source: Wikimedia Commons 2007)

positive health outcomes for wealthier populations. A more detailed and in-depth discussion of CKTS and human health and wellness can be found in Chap. 5 of this report.

9.2.5 Urban and Rural Communities

Vibrant and healthy communities are the backbones of a sustainable society. Rural communities in many parts of the world face an array of poverty-related challenges, including lack of infrastructure, housing, access to quality healthcare, education, and a steady means of income, as discussed in Sect. 9.8.4 of this chapter. One of the most significant challenges to global sustainability is the emergence of large urban communities comprised of millions of people. The rapid growth of megacities with tens of millions of people is a global phenomenon that began to accelerate at the close of the last century and continues today. Over two thirds of the world's urban residents live in cities in Africa, Asia, and Latin America (Fig. 9.5; Wikimedia Commons 2007). Further, since 1950, the urban population of these regions has grown fivefold, and in Africa and Asia, urban population is expected to double by 2030 (UNPF 2007). These are also the regions where megacities—such as Mumbai, Shanghai, Beijing, Calcutta, Lagos, Mexico City, Rio de Janeiro, and Cairo—have become most common.

Megacities impose numerous environmental and resource impacts that until a generation ago were insufficiently appreciated. They challenge the achievement of a sustainable society in three major ways. First, megacities impose huge stresses upon regional freshwater supplies and water quality. In part, this is because most large cities are often located some distance from the water sources needed to support them. As a result, they must divert water from outlying rural areas. Large cities also generate huge volumes of wastewater, which is costly to treat and, if left untreated, can contaminate local wells and streams—a growing problem, especially in many developing nations. Wastewater generation is itself exacerbated by urban sprawl, which leads to increased paving of city streets and commercial districts, contributing to pollutant runoff and diminished groundwater recharge, as well as increased consumption of water for parks and outdoor residential use (increasing evapo-transpiration and taxing local supplies). While greater concentrations of people in cities may lower unit costs for many forms of urban infrastructure (Satterthwaite 2000), the need to expand water supply and treatment networks over vast distances increases the likelihood of distribution system leaks.

Second, megacities impose huge burdens on energy systems, infrastructure, and human health and hygiene. The quest to modernize growing urban areas has led to a desire to afford them with a higher quality of life, adding to demands for additional energy. Energy demands are expected to grow by some 60 % over present needs by 2030 in developing countries. While virtually every source of energy depends on water, one rapidly growing energy source in much of the developing world is biofuels such as ethanol, which can be produced by converting biomass, including trees, grasses, left-over agricultural residue from harvesting crops, and even algae. Biofuels firmly link agriculture and energy production with dramatic impacts upon fresh water. The need to expand water supply and treatment and energy infrastructure can also enormously strain existing capacity and outrace development capacity—as well as place inordinate pressure on the outlying regions from whence in-migration occurs (UN 2009).

Third is another energy–water interconnection: as cropland is diverted from food to energy-crop production, and as developing countries restrict imports of food to encourage more domestic farming and greater food security, food prices are likely to rise. We have already seen this occur as the result of corn-based ethanol production in many countries. As food prices rise, the cost of water used for production of both food and biofuel crops, and the cost to treat the new sources of water supply contamination resulting from this higher production also rise because of higher demand for water (IPCC 2007b).

9.2.6 Materials Supply and Utilization

Minerals and Critical Metals

Minerals are key building blocks of the sustainable economy of the twenty-first century. They have become the primary sources of the variety of metals used to fabricate the critical components of numerous products and finished goods, including airplanes, automobiles, cell/smart phones, and biomedical devices (NRC 2008). The application of CKTS to critical metal supply/utilization has thus far received little attention (Diallo and Brinker 2011). However, recent stresses in the global market of rare-earth elements (REEs) have brought the sustainable supply of critical metals to the forefront in the United States and other industrialized countries (Diallo and Brinker 2011; Fromer et al. 2011). In addition to REEs (e.g., neodymium and dysprosium) and platinum-group metals (PGMs) (e.g., platinum and rhodium), significant amounts of copper, silver, indium, lithium, and gallium will be needed to build the renewable energy technologies of the twenty-first century.

Recently, the uses of critical materials in renewable energy technologies have been the subject of extensive discussions (Fromer et al. 2011; DOE 2011). To ensure that the availability of critical metals will not adversely impact the development and deployment of renewable energy systems, the U.S. Department of Energy (DOE) through the Advanced Research Projects Agency Energy (ARPA-E) has recently initiated a broad range of research programs to develop rare-earth alternatives in critical technologies (REACT) (ARPA-E 2011). The initial focus of the REACT program is on building new magnets for electric vehicles and wind turbines using earth-abundant elements such as iron, nickel, manganese, and aluminum as building blocks. In 2013, the DOE established an "Energy Innovation Hub" led by Ames National Laboratory to address challenges in critical materials for energy generation, conversion, and storage, including (1) mineral processing/purification, (2) manufacturing, (3) substitution, and (4) recycling (DOE 2013). It is worth mentioning that the end-of-of-life recycling rate (EOL-RR) is very low for most of the critical materials that will be used to build the sustainable products, processes, and industries of the future (Reck and Graedel 2012; UNEP 2011b; Fig. 9.6).

Carbon-Based Materials and Biomass

Carbon-based materials derived from petroleum are the building blocks of a broad range of essential products and finished goods, including plastics, solvents, adhesives, fibers, resins, gels, and pharmaceuticals. Steep increases in the cost of petroleum and geopolitical concerns over the sustainable supply of oil and GHG emissions have made biomass an attractive feedstock for the chemical industry (Gallezot 2012; Tuck et al. 2012). During the last 10 years, significant research efforts have been devoted to the conversion of biomass to valuable chemicals. Three main conversion strategies that are being explored include:

- Chemical and/or biological conversion of biomass into small and intermediate molecules that are subsequently utilized to prepare specialty and fine chemicals
- Chemical conversion of biomass into a mixture of small molecules with similar functionalities that are subsequently employed to produce commodity and performance chemicals
- Chemical functionalization of biomass to produce new functional polymers that are converted to performance chemicals

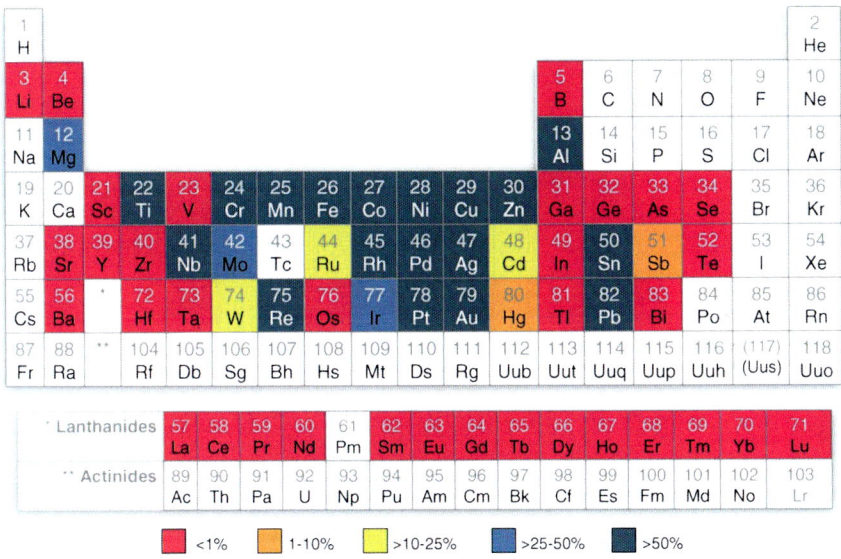

Fig. 9.6 Global estimates of end-of-life recycling rates for 60 metals and metalloids: 30 are <1 %; 2 are 1–10 %; 3 are 10–25 %; 3 are 25–50 %; only 18 are over 50 % (Reck and Graedel 2012, Adapted from a figure in Graedel et al. 2011; ©AAS, Reprinted with permission)

9.2.7 Climate Change and Clean Environment

Climate Change Mitigation

Global climate change is a formidable challenge facing human society and the environment in the twenty-first century. During the last two decades, a scientific consensus has gradually emerged that increasing emissions of GHG such as CO_2 from the combustion of fossil fuels are the key drivers of global climate change (IPCC 2007a). Currently, fossil fuels provide approximately 80 % of the energy used worldwide (IPCC 2005). Thus, the utilization of non-CO_2-emitting renewable energy sources might be the most effective means of reducing GHG emissions (IPCC 2011). Although various non-GHG-emitting and carbon-neutral renewable energy sources are being developed/deployed (see Sect. 9.2.2), the world will continue to burn significant amounts of fossil fuels in the foreseeable future. During the last decade, significant progress has been made in the development of novel GHG mitigation strategies, including carbon capture and storage (CCS) (IPCC 2005) and CO_2 conversion to fuels and useful products (Diallo and Brinker 2011). Relevant examples include:

- More efficient and selective sorbents for CO_2 capture, release, and storage, including metal organic frameworks (MOFs) (Britt et al. 2009) and zeolitic imidazolate frameworks (ZIFs) (Phan et al. 2010)

- Conversion of CO_2 to fuels and organic chemicals using microbial electrosynthesis (Logan and Rabaey 2012)
- Conversion of CO_2 to cement using seawater and low-cost sources of alkalinity (Service 2012)

In addition to CCS and CO_2 conversion, geoengineering is being considered as a potential climate mitigation technology. To alleviate public and regulatory concern over the global/regional environmental impact of climate mitigation by "solar radiation management" using stratospheric aerosols (Keith et al. 2010), less risky and "soft" geoengineering technologies that "touch gently on biological and social systems" are being developed (Olson 2012). Recent progress includes the invention and field evaluations of the Ice911 technology to slow ice melting in Arctic regions (Field 2012). The Ice911 concept is predicated on the use of light-colored and lightweight materials to cover and protect "water or areas in danger of melting" (Field 2012).

Clean Environment

Industrial manufacturing has a heavy environmental footprint. First, it requires significant amounts of land, energy, water, and materials. Second, it generates a lot of wastes (gaseous, liquid, and solid) and toxic by-products that need to be disposed of or converted into harmless products. "Green manufacturing" might be the most efficient means to reduce and (eventually) eliminate the release of toxic pollutants into the soil, water, and air. Green manufacturing encompasses a broad range of approaches that are being used to:

- Design and synthesize environmentally benign chemical compounds and processes (green chemistry)
- Develop and commercialize environmentally benign industrial processes and products (green engineering)

During the last 10 years, significant advances have been made in green engineering and green chemistry (Schmidt 2007). The SRC/SEMATECH Engineering Research Center for Environmentally Benign Semiconductor Manufacturing led by the University of Arizona is exploring the utilization of "Environment, Safety & Health (ESH) factors as design parameters in the development of new processes, tools, and protocols for semiconductor manufacturing" (see http://www.erc.arizona.edu/vision.htm). The NSF Center for Sustainable Materials Chemistry (CSM) led by Oregon State University is exploring the utilization of water-based chemistries for "producing very high-quality thin films and patterns" for the next generation of semiconductor devices (see http://sustainablematerialschemistry.org/). The CSM is also employing green chemistry to develop new materials for energy applications, including electrocatalysts, solid-state ionic conductors, and high dielectric constant materials.

Although green manufacturing could reduce and (eventually) eliminate the release of toxic pollutants into the soil, water, and air, its large-scale implementation by industry will take decades. Thus, more efficient and cost-effective technologies are still needed to (1) detect and monitor pollutants (environmental monitoring), (2) reduce the release of industrial pollutants (waste treatment), and (3) clean polluted sites (environmental remediation). During the last 10 years, the application of engineered nanomaterials to sensing and detection devices has enabled the development of a new generation of advanced monitoring and detection concepts, devices, and systems for various environmental contaminants (Fan et al. 2004; Vaseashta and Dimova-Malinovska 2005; Rickerby and Morisson 2007; Wang et al. 2008; Aravinda et al. 2009).

Advances in nanotechnology have also enabled development of more efficient and cost-effective environmental remediation technologies (Savage and Diallo 2005; Tratnyek and Johnson 2006). Nanoscale zero valent iron (NZVI) particles have proven to be very efficient redox-active media for the degradation of organic contaminants, especially chlorinated hydrocarbons (Lowry and Johnson 2004; Liu et al. 2005; Song and Carraway 2005). Dendritic nanomaterials, which consist of highly branched nanoscale polymers, have been successfully employed as supramolecular hosts and ligands to extract environmental pollutants from aqueous solutions; these toxins include inorganic pollutants, heavy metals, and biological and radiological compounds (Crooks et al. 2001; Diallo et al. 2004; Birnbaum et al. 2003; Balogh et al. 2001). More recently, Diallo and coworkers successfully utilized low-cost hyperbranched polymers to develop a new generation of high-capacity and selective anion exchange and chelating resins to remove pollutants such as perchlorate and boron/borate from contaminated water (Chen et al. 2012; Mishra et al. 2012).

9.2.8 *Human Progress, Societal Values, and Economic Development*

Two important goals of global sustainability are to achieve "inclusive" economic and societal development, which benefits all people while minimizing the impact of human activities on Earth's climate and ecosystems (UN System Task Team 2012). Currently, there are wide disparities in living standards and income between people living in developed and developing countries (UN System Task Team 2012). According to the International Labor Organization (ILO), approximately one billion people in the world are malnourished, and more than 200 million are unemployed (ILO 2011). In 2008, about 1.3 billion people in the world live on less than $1.25 a day (ILO 2010). Nearly one third (~2.4 billion people) of the world's population in 2015 will lack access to improved water supply and sanitation (UNICEF and WHO 2009.) More to the point, approximately 900 million people live in slum-like conditions (UN Millennium Project 2005). In Sub-Saharan Africa, 60–70 % of the people live in slums (Fig. 9.7).

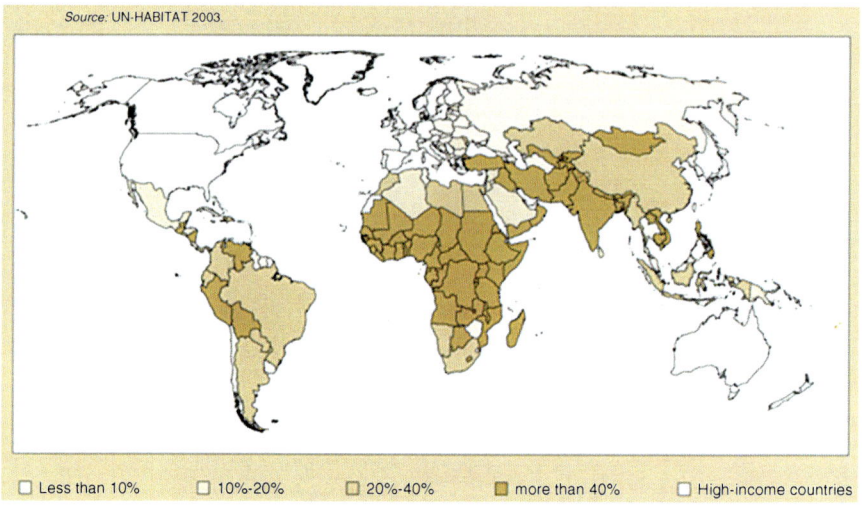

Fig. 9.7 Share of urban populations living in slums, *percent* (UN Millennium Project 2005, Chapter 2, p. 27, Map 2.4; ©UN-Habitat 2003, used by permission)

Since the adoption of the Millennium Development Goals in 2000 by the United Nations (UN Millennium Project 2005), significant advancements have been made toward achieving inclusive economic and societal development (UN System Task Team 2012). Strong economic growth has occurred in many developing countries, thus lifting millions of people out of poverty. One of the most significant developments in the last 10 years is the convergence between ICT (e.g., mobile phones), social media (e.g., Facebook), and social entrepreneurship (Rifkin 2011). This convergence has radically transformed the "way people communicate, organize, network, learn, and participate as national and global citizens" (UN System Task Team 2012, 14). This in turn is enabling major and beneficial changes:

- Participatory sustainable development
- Increased transparency
- Environmental and intergenerational equity/justice
- Public scrutiny and governance of emerging/transformative technologies

Although tangible progress has been made in the implementation of Millennium Development Goals, a great deal remains to be done. Costanza et al. (2012) recently published their report "*Building a Sustainable and Desirable Economy-in-Society-in-Nature*". In this report, the authors argue that we need a "new vision of the economy" to achieve "inclusive" economic and societal development with minimum environmental impact. More to the point, they argue that such new economic models should be based on the principles of "Ecological Economics," including:

1. Respect of "planetary boundaries," i.e., the carrying capacity of the natural environment
2. Recognition of the critical relationships between human development and well-being on societal/cultural values and "fairness"

9 Implications: Convergence of Knowledge and Technology...

Fig. 9.8 An integrated framework for realizing the "future we want for all" in the post-2015 UN development agenda (UN System Task Team 2012, Figure 9.1, p. 24; reuse under stock image license)

3. Acceptance that the ultimate goal of societal development is "real and sustainable human well-being, not merely growth of material consumption"

Costanza et al. (2012) believe that implementation of this new economic vision will be critical to realizing a sustainable, healthy, secure, and peaceful world. Figure 9.8 provides the UN System Task Team's integrated framework for achieving this goal.

9.3 Goals for the Next Decade

The overarching theme of this chapter is that in the next 10 years, progress in CKTS will be critical to achievement of sustainable development in consideration of the factors listed in Fig. 9.1. The UN study *"Realizing the Future We Want for All"* (UN System Task Team 2012) provides an integrated framework for achieving "inclusive" economic and societal development with minimum environmental impact (Fig. 9.8). Given the global environmental challenges and expected increase in population, realizing the UN vision will require transformative advances and the convergence of many fields, including physical/earth sciences, biological sciences, engineering, and social sciences, as outlined below.

9.3.1 Sustainable Supply of Energy for All

As previously discussed, solar energy is the most attractive source of non-GHG-emitting source of energy. Solar energy is both abundant and versatile. Concentrating solar power (CSP) can convert solar energy into thermal energy that can be subsequently stored and utilized to produce electricity via a steam turbine or a heat engine. Photovoltaics (PV) convert solar radiation directly into electricity. Both CSP and PV are flexible and modular. They can be incorporated into large centralized power plants as well as into smaller, distributed systems (e.g., rooftop units). Solar radiation could ultimately be combined with CO_2 and water to produce solar fuels (e.g., hydrogen and methanol). Solar energy has the potential to provide "*Sustainable Energy for All*" by 2050 (WWF 2011).

In the next 10 years, we envision that the convergence between physical sciences (e.g., chemistry, physics, materials sciences), nanotechnology, information technology, engineering (e.g., mechanical engineering, electrical engineering, chemical engineering, and systems engineering) and scalable manufacturing (e.g., 3D printing) will lead to transformative advances in solar energy generation, conversion, storage, and distribution. The most significant advances will include:

- High-efficiency solar cells and modules (30–35 %) that produce electricity at a cost of 0.35 per Watt
- Third-generation solar cells that are able to overcome the Shockley-Queisser limit (31–41 % power efficiency) for single band gap PV cells
- Scalable Li-S and Li-Air batteries with high energy density (2,000–3,000 Wh/kg) for storage of solar PV electricity
- More efficient grid-scale energy storage devices for electricity derived from solar energy
- First prototype of a scalable solar fuel generator that produces H_2 by splitting water
- More efficient solid-state materials for storage of H_2 derived from solar power

9.3.2 Sustainable Supply of Clean Water for All

As previously stated, water scarcity is expected to grow more acute as the world population increases and the amount of available freshwater decreases due to climate change. Global climate change will adversely impact the world's freshwater resources in several ways: (1) increase the frequency of droughts and floods; (2) decrease the amount of water stored in snowpack and glaciers; and (3) decrease the overall water quality due to salinity increase and enhanced sediment, nutrient, and pollutant transport in many watersheds throughout the world (Bates et al. 2008). Consequently, transformative advances in water purification technology and major behavioral changes in the ways people use water will be required to alleviate future water shortages. Section 9.8.3 provides a case study showing how the convergence between nanotechnology and microbiology could lead to more efficient and safer water disinfection strategies. In the next 10 years, we envision that the convergence between earth-systems science, physical sciences (e.g., chemistry and materials sciences), biological sciences (e.g., microbiology and biotechnology), nanotechnology, engineering (e.g., chemical/environmental process engineering), and social sciences will lead to transformative advances in water science, technology, management, and utilization. We expect that these major advances will enable us to:

- Understand, model, and manage the impact of global climate change on freshwater resources
- Develop more sustainable technologies to (1) reuse water, (2) reclaim wastewater and (3) desalinate brackish water and seawater while recovering energy, nutrients (nitrogen and phosphorous), and valuable elements (e.g., lithium and magnesium)
- Develop more efficient and cost-effective decentralized water treatment systems to meet the drinking and clean water needs of population clusters (e.g., residential buildings, villages, and private homes) in developed and developing countries
- Develop more efficient water treatment and reuse technologies to address the special needs and growing water stresses of industry worldwide
- Educate the public and regulators about the value of water, the current water crisis, and the necessity to save and reuse water

9.3.3 Sustainable Supply of Food for All

Modern agriculture has a heavy environmental footprint. Firstly, it requires significant amounts of land, energy, water, fertilizers, and pesticides. Secondly, it generates significant amounts of GHG and liquid wastes (e.g., run-off) containing nutrients and toxic by-products (e.g., excess pesticides). In the next decades, the world will face the daunting challenging of doubling the amount of food it currently produces to feed around nine billion people in 2050 while reducing energy usage and GHG emissions, as the amount of freshwater available decreases due to global climate change. In the next 10–20 years, we expect that the convergence between nanotechnology, biotechnology, information technology, and social sciences will produce the transformative advances required to meet this formidable challenge. Broad goals for the next decade are to:

- Satisfy human food, feed, and fiber needs, and contribute to biofuel needs
- Enhance environmental quality and the resource base
- Sustain the economic viability of agriculture
- Enhance the quality of life for farmers, farm workers, and society as a whole
- Reduce hunger, malnutrition, and poverty worldwide

    More specific goals will be to:

- Freeze agriculture's carbon footprint by slowing agricultural land expansion, in particular, loss of tropical forests

- Reduce yield (production) gaps between existing growth (production) levels and the genetic potential for both plants and animals
- Improve efficiencies of agriculture and natural resources (more output/input resource)
- Reduce "diet" gaps, changing the mix of food products to enhance food availability and reduce environmental impacts
- Reduce food wastes at every level in the agriculture and food system
- Integrate agriculture and food systems into sustainable community thinking, possibly via opportunities for renewable, distributed energy generation, including "vertical" farms in the urban areas

9.3.4 Health and Wellness for All

The emergence of large urban communities and megacities with more than ten million inhabitants is a formidable challenge to societal sustainability in the twenty-first century (See Sect. 9.2.5 of this report). In addition to their heavy environmental impact (e.g., high GHG emissions, increased air pollution, and huge stresses on freshwater resources), megacities pose significant challenges to achieving health and wellness for the world's population. Over the course of history "medical access" has been inextricably linked with "distance to doctor". As we enter into a world with a myriad of advanced/instant communication tools in which satellites and ground-based high-speed fiber optics-based information systems are all connected, the convergence of information and communication technology (e.g., smart phones) with biotechnology (e.g., low-cost and rapid DNA sequencing technologies), and medicine (e.g., molecular diagnostic tools) will lead to transformative advances in the 10–20 years. Relevant and important advances will include:

- Access to best-in-world medical and wellness information, instantly, anywhere, and anytime.
- More effective disease avoidance options/information.
- More personalized and cost-effective medical diagnostics tools.
- More personalized and cost-effective treatment and drugs.

In addition, we expect that some of the new molecular diagnostic tools will be integrated into advanced sensors and detection systems to monitor the presence of flu viruses or emerging pathogens in air, water, and urban living environments. This will provide city managers and first responders with more effective tools to protect public health by early detection of the presence of infectious agents and/or poisonous, toxic, or radioactive substances into urban air, water, soil, and transportation systems.

9.3.5 Sustainable Communities for All

A key societal goal in the 21st center is to build/establish sustainable communities for all; that is communities that meet the basic needs of food, clean drinking water, energy, healthcare, education, and providing meaningful jobs, affordable homes, and economical transportation for all people, with minimum impact on Earth's climate and ecosystems. Major and transformative advances and the convergence between many fields (e.g., physical sciences, engineering, information/communication technology, architecture, and social sciences) will be required to achieve this colossal undertaking in the next 10–20 years as global climate change accelerates and the world's population continues to grow. Section 9.8.4 discusses the development of CKTS-based solutions for building sustainable rural communities. Relevant and significant advances for building the sustainable urban communities of the twenty-first century will include smarter cities with:

- More affordable/resilient housing and distributed energy generation, storage, and distribution systems
- Resilient and distributed water/wastewater infrastructure
- More energy-efficient transportation systems
- More efficient food production and delivery systems
- Distributed healthcare infrastructures, employment opportunities for all, and more livable environments

9.3.6 Sustainable Materials Supply and Utilization

Innovations in materials science/engineering and the sustainable supply/utilization of materials will be critical to developing the next generation of sustainable technologies and products, including high-efficiency solar cells, high-efficiency magnets for electrical cars and wind turbines, high-capacity batteries, low-energy desalination membranes, energy-efficient transportation systems, and high-performance carbon capture and storage technologies. Because our current consumption of critical metals is at an all-time high (Reck and Graedel 2012), there is an urgent need for novel strategies to decrease the utilization of critical materials in energy generation, conversion, and storage, and/or to replace them with earth-abundant elements. New technologies and strategies are also needed to augment the supply of critical materials through recycling and reuse. In the next 10–20 years, we expect that the convergence between nanotechnology, materials science, separations science, and engineering (e.g., chemical, mechanical, and systems engineering) will lead to the transformative advances required to ensure global materials sustainability including:

- Nanocrystalline-silicon thin films for high-efficiency photovoltaics cells
- Nanostructured materials based on earth-abundant elements (e.g., silicon nanowires) for high-capacity energy storage systems (e.g., batteries)
- More efficient magnets for electric vehicles and wind turbines using earth-abundant elements (e.g., iron, nickel, and manganese) as alternatives to rare-earth elements
- Nanostructured catalysts based on earth-abundant elements (e.g., iron, copper, nickel, and manganese) for fuel cells and solar fuel generation

- More efficient and cost-efficient separation materials (e.g., sorbents and membranes) and systems for extracting and recovering critical materials from nontraditional sources, including mine tailings, industrial wastewater, seawater, and brines

9.3.7 Sustainable Climate and Clean Environment

Sustaining Earth's Climate

Mitigation of global climate change is among the most urgent tasks facing the world. As previously discussed, the switch to renewable and non-GHG emitting energy sources (e.g., solar energy) is the most effective mean for stabilizing Earth's climate. However, the deployment of terawatt-scale non-GHG-emitting energy sources will require decades. Thus, a consensus has emerged that more efficient mitigation and adaptation strategies will also be needed to stabilize the level of CO_2 in the atmosphere (see Table 9.1) and minimize/manage the impact. In the next 10–20 years, we expect that the convergence between nanotechnology, separations science, catalysis, engineering, earth-systems science, architecture urban/planning, and social sciences will lead to transformative advances in climate mitigation and adaptation technologies/strategies, including:

- Nanoscale sorbents containing functionalized size- and shape-selective molecular cages that can selectively capture carbon dioxide (CO_2) from flue gases and convert it to useable products
- More efficient and cost-effective systems that can directly extract CO_2 from air and convert it to useable fuels and useable products
- More robust strategies and solutions to increase the resilience and decrease the vulnerability of people, urban/rural communities, and ecosystems to extreme weather events (e.g., drought and flooding) caused by global climate change

Clean Environment for All

Pollution prevention through green manufacturing is arguably the most efficient way to reduce and/or prevent the release of many toxic pollutants into the soil, water, and air. In the next 10 years, a convergence between nanotechnology, materials science, green chemistry, and green engineering will enable society to build the sustainable products, processes, and industries of the twenty-first century while maintaining a clean environment. Relevant advances will include:

- Environmentally benign building blocks and manufacturing processes for the semiconductor, chemical, petroleum, metal/mineral, and pharmaceutical industries
- High-performance nanocatalysts for the chemical, petroleum, and pharmaceutical industries

9 Implications: Convergence of Knowledge and Technology...

- More efficient and environmentally acceptable industrial separation and purification process for the chemical, petroleum, metal/mineral, and pharmaceutical industries

In addition to pollution prevention and green manufacturing, more efficient and cost-effective technologies will continue to be needed to detect and monitor pollutants, reduce the release of industrial pollutants, and clean polluted sites and accidental releases of hazardous materials (e.g., oil spills). Thus, we envision that in the next 10 years, the convergence between nanotechnology, information communication/technology, and engineering will lead to major advances, including:

- Small-scale and ubiquitous sensors capable of performing real-time monitoring of environmental systems, including air, water, and soils
- More cost-effective environmental cleanup and remediation technologies for emerging contaminants, including pharmaceuticals, household products, and nanomaterials
- More efficient and cost-effective oil spill clean-up technologies

Human Progress, Societal Values, and Economic Development

As previously discussed, significant progress has been made toward achieving economic and human development worldwide since the adoption of the Millennium Development Goals by the United Nations (UN System Task Team 2012; UNEP 2011a). However, this progress has been uneven, with great disparities in incomes and access to basic service between the "haves" and "have-nots" (Milanovic 2010). Following the June 2012 Rio Conference on Sustainable Development, the United Nations began developing its post-2015 development agenda to promote and facilitate the implementation of a global development agenda that is inclusive, "people-centered," and sustainable (UN System Task Team 2012). Three guiding principles of the post-2015 UN development agenda are (1) human rights, (2) equality, and (3) sustainability (Fig. 9.8). The implementation of this development agenda will require transformative changes in the ways we produce and consume goods, manage our natural resources, and govern our society. In addition, the convergence of knowledge and technology from numerous fields (physical/biological sciences, engineering, social sciences) will also be required to produce the transformative advances critical to realizing a sustainable, healthy, secure, and peaceful world. Key goals for the next 10–30 years will include:

- Sustainable energy for all
- Universal access to clean water and sanitation
- Sustainable food and nutrition for all
- Universal access to quality healthcare
- Stable climate and clean environment
- Sustainable urban and rural communities
- Sustainable transportation for all
- Decent and productive employment for all

9.4 Infrastructure Needs

To realize the post-2015 UN development agenda (Fig. 9.8), it will also be necessary to merge knowledge with technology and entrepreneurship to develop, scale-up, and commercialize the next generation of sustainable products, processes, and technologies. Key science and technological infrastructure needs in the next 10 years include:

- Holistic investigations of all interdependent aspects of sustainable development, including the strong couplings between energy and water, water and agriculture, and energy and materials
- Dedicated facilities for scale-up and testing of materials/systems for sustainability applications
- Dedicated material characterization facilities for sustainability applications
- Computer-aided modeling and process design tools for sustainability applications
- Supercomputers or supercomputing nodes dedicated to solving sustainability-related grand challenges (e.g., catalyst design for solar water splitting and large-scale modeling of urban systems)
- Test-beds for benchmarking and rapid prototyping of sustainability-enabling products, processes, and technologies

9.5 R&D Strategies

The critical Grand Challenge for the post-2015 UN development agenda (Fig. 9.8) *is to stabilize Earth's climate while (1) achieving a more equitable and sustainable management of our natural resources and (2) enabling inclusive economic and societal development* (UNDPUN System Task Team 2012). Because sustainability entails considering social, economic, and environmental factors, it is critical in all cases to converge knowledge (e.g., materials science, nanotechnology, and multiscale modeling) with engineering (e.g., system design, fabrication, and testing), commercialization (e.g., scale-up and new products/solutions), and societal benefits (e.g., new jobs and cleaner environment) as we address the R&D priorities outlined in this report. Thus, sustainable development cannot simply be addressed at the level of small- and single-investigator research grants. Sustainability R&D needs to be integrated with broader research goals and included from the beginning in large interdisciplinary programs to be carried out by teams of investigators and/or dedicated Federally funded transdisciplinary research and development centers. In addition, partnerships between academia, government (Federal, state, and local), industry, and the venture capital community will be critical to translating R&D advances into innovative products and solutions. To achieve these objectives, it will be necessary to:

- Establish focused centers and networks to develop and implement solutions to our critical sustainability challenges in energy, water, food, materials, climate, and environment

- Develop new funding mechanisms to advance promising early-stage research projects, e.g., automatic supplemental funding for projects with commercial potential
- Involve industry at the outset of programs
- Establish focused innovation hubs and ecosystems to bring sustainable solutions/products into the marketplace for broader societal benefit

9.6 Conclusions and Priorities

The convergence of knowledge and technology for societal benefit (CKTS) will be critical to achieving the UN goals discussed above. The global sustainability challenges facing the world are complex and involve multiple interdependent and strongly coupled global problems. We expect CKTS to provide breakthrough and scalable solutions for sustainable development, particularly in the areas of renewable energy (generation, conversion, and storage), clean water resources, food/agriculture resources, materials resources, climate stabilization, and clean environment. The following key priorities have been identified for the next decade:

- Solar energy generation, storage, and distribution. Two priorities are:
  - Development and deployment of terawatt-scale solar energy at a cost comparable to or lower than that of energy derived from fossil fuels
  - Development of efficient and cost-effective technologies to convert solar energy into hydrogen and liquid fuels
- Water science, technology, management, and utilization. Two priorities are:
  - Understanding, modeling, and managing the impact of global climate change on freshwater resources
  - Development of sustainable technologies to (1) reuse water, (2) reclaim wastewater, and (3) desalinate brackish water and seawater while recovering energy, nutrients (nitrogen and phosphorous), and valuable metals (e.g., lithium and magnesium)
- *Sustainable agriculture and food security.* A key priority is to develop and deploy more sustainable agricultural and food production systems with reduced (or zero) greenhouse gas emissions, reduced water/land usage, and increased application of conservation tillage, biotechnology, and information technologies throughout the agricultural and food production chain.
- *Sustainable supply and utilization of critical materials.* Three key priorities are to develop novel technologies/strategies to:
  - Decrease the utilization of critical materials in energy generation, conversion, and storage, and/or to
  - Replace them with earth-abundant alternatives
  - Augment the supply of critical materials through recycling, reuse, and extraction from nontraditional sources such as mine tailings, industrial wastewater, seawater, and brines

- *Sustainable urban communities.* A key priority is to reconfigure existing megacities and configure future ones into smarter cities with:
  - Resilient and distributed energy and water infrastructures
  - Sustainable agriculture/food production and supply systems
  - Energy-efficient transportation systems
  - Distributed healthcare infrastructure, more livable environments, and employment opportunities for all

9.7 R&D Impact on Society

CKTS offers the potential to extend the limits of sustainability and enable inclusive human, economic, and societal development on Planet Earth. However, we must ensure that any potentially adverse effect on humans and the environment are effectively assessed and addressed before large-scale deployment of CKTS-enabled sustainability products and solutions.

9.8 Examples of Achievements and Convergence Paradigm Shifts

9.8.1 R&D Programs to Support Sustainable Development in the United States

Contact persons: *Jessica Robin and Bruce Hamilton, NSF*

A number of U.S. Federal agencies are making very substantial investments to support R&D for sustainable development. Examples include the Department of Energy (DOE) for clean energy (see http://www.eere.energy.gov/); the U.S. Department of Agriculture National Institute of Food and Agriculture (USDA-NIFA) for sustainable agriculture (see http://www.sare.org/Grants), and in collaboration with DOE, fuels from biomass (see http://www.usbiomassboard.gov/); and the U.S. Environmental Protection Agency (EPA) for sustainable materials management (see http://www.epa.gov/epawaste/conserve/smm/vision.htm). Correspondingly, the President's 2013 budget proposed $2.3 billion for DOE's Energy Efficiency and Renewable Energy Office and $292 million for USDA's bio-energy research (OSTP 2012). Concurrently, NSF formulated and implemented an agency-wide investment area, Science, Engineering, and Education for Sustainability (SEES).

SEES is a 10-year initiative with the overall mission to advance science, engineering, and education to inform the societal actions needed for environmental and economic sustainability and sustainable human well-being. Now in its fourth year, the SEES portfolio includes 17 interdisciplinary programs that represent a significant and growing cross-NSF investment. The first set of SEES programs focused on the environment and the portfolio has expanded to include programs in energy, materials, cyber, and resilience (see SEES program page, http://www.nsf.gov/sees/). NSF investment in SEES for FY10 was $70 million, for FY11 was $93 million, for FY12 was $170 million, and the request for FY13 was $203 million (NSF's budget requests to Congress are available online at http://www.nsf.gov/about/budget/). SEES programs give strong attention to incorporating the human sciences in addition to integrating education, partnerships, and networks both domestically and internationally.

A 2012 National Academy of Sciences sustainability symposium stated there lacks a commonly accepted operational framework for how to move forward in addressing sustainability challenges (Saunders 2012). Sustainability efforts must integrate and coordinate government resources and funding. Furthermore, the interface between science and policy needs a more adaptive management structure that allows agencies to learn from their management experiences and incorporate lessons learned into their management structures. In turn, sustainability research requires new knowledge, technologies, and approaches; education and workforce development; more integrative science; partnerships; and linking knowledge to action.

Sustainability science includes a broad range of research domains spanning climate, biodiversity, agriculture, fishery, forestry, energy, water, economic development, health, and lifestyles (Kajikawa 2008; Kajikawa et al. 2007). While the field of sustainability science has grown rapidly since the 1980s, the integration across disciplines into a new field only started in recent years (Bettencourt and Kaur 2011). Collaborative links between researchers accelerated after 1989, but it was not until 2000 that the field became dominated by a unified group of collaborations to which most authors in this field now belong. Bettencourt et al. (2009) argue that such unification is the hallmark of a true field of science. Correspondingly, there are numerous Federal programs that support new knowledge, technologies, and approaches, as well as education and workforce development, in the research domains of sustainability. By and large these programs coincide with the corresponding agencies' missions (e.g., NSF supporting fundamental science and engineering, National Institutes of Health [NIH] enhancing human health, the Department of the Interior (DOI) protecting natural resources). Some agencies' programs focus on generating new knowledge across a spectrum of domains, such as NSF's SEES portfolio, while others such as the NIH Clean Cookstoves initiative (NIH Fogarty International Center 2010) and DOI's WaterSMART Clearinghouse (http://www.doi.gov/watersmart/) focus on specific technologies and approaches.

The degree to which the sciences are integrative across these programs also depends on the agencies' mission and their partnerships. The National Science Foundation, which funds basic research across all nonmedical science and engineering disciplines from mathematics to geosciences and social sciences, provides opportunities for supporting integrative science across disciplines through initiatives such as SEES. USDA's mission to provide leadership on food, agriculture, natural resources, rural development, nutrition, and related issues allows for integrative research, extension, and education programs that focus on sustaining all components of agriculture, including renewable energy, rural communities, and human nutrition. Additionally, partnerships across Federal agencies integrate the sciences even further by coordinating resources and broadening the range of research and services that can be supported. NSF supports agriculture research in its Water Sustainability and Climate (WSC) and Decadal and Regional Climate Prediction using Earth System Models (EaSM) programs through a partnership with USDA. DOE also partners on the latter program. Concurrently, NSF's Arctic SEES program provides space for engaging community stakeholder groups and local, state, and regional governments in research through a partnership with the EPA. Additionally, partnerships with DOI (the Bureau of Ocean Energy Management, U.S. Geological Survey, U.S. Fish and Wildlife Service) and a consortium of French agencies broaden the scope of this program even further.

Such partnerships extend internationally as well. NSF, NIH, and the U.S. Agency for International Development (USAID) collaborate on the Partnerships for Enhanced Engagement in Research (PEER), which provides support to researchers from developing countries in a wide range of sustainability topics from disaster mitigation to renewable energy to child survival. USAID and NSF partner on the PEER Science component, and USAID and NIH partner on the PEER Health component, both managed by The National Academies. Funding resources are also leveraged by partnering with scientific funding agencies from other countries. NASA, NSF-China, and the São Paulo Research Foundation of Brazil (FAPESP) partner with NSF on their Dimensions of Biodiversity program (DoB). Additionally, NSF's Partnerships for International Research and Education (PIRE) program partners with funding agencies from the UK, Japan, and Russia, in addition to the Inter-American Institute for Global Change Research (IAI), USAID, and EPA. In this program, the partner agencies indicate priority sustainability research areas and provide support for their respective researchers collaborating with U.S. researchers funded through the PIRE program in those topics. Priority research areas include materials & engineering (UK), nanotechnology (Russia), energy (Japan and Russia), information technology (Russia), water (Japan and EPA), and global change (IAI and Japan).

Sustainability science, like agricultural and health sciences, is defined by the problems it addresses rather than by the disciplines it employs; it is neither basic nor applied but use-inspired science with a commitment to moving such knowledge into societal action (Clark 2007; Kates 2011). While use-inspired research aligns more closely with USDA and the U.S. Department of Human & Health Services, their mandates limits them to supporting only certain domains of sustainability (e.g., agriculture and health). DOE broadens the portfolio by supporting energy-related research, and NSF, with an even broader research mandate, covers multiple domains. But they lack the expertise, resources, and mandate to link the scientific research they support to action. EPA and DOI are much better suited for that role through their regional offices and partnerships with local and state governments. However, they need the research support of these other agencies. In short, no U.S. Federal agency has the capability, or mandate, to support and manage all the needed requirements for sustainability research.

Partnerships, domestically and internationally, are essential, and the ones above are moving in the right direction but opportunity-specific. What is needed is a comprehensive and coordinated approach that spans across all Federal agencies and research domains of sustainability science. The President's National Science and Technology Council, through its Committee on Environment, Natural Resources, and Sustainability, has taken preliminary steps in that direction. Additionally, an ad hoc National Research Council committee under its Science and Technology for Sustainability Program has been convened to identify priority areas for interagency cooperation among domains such as energy, water, and health that are not routinely considered in decision-making. This project is sponsored by multiple Federal agencies, private foundations, and industry. The task is not an easy one, and much can be learned from the experiences of the U.S. Global Change Research Program (USGCRP; http://globalchange.gov/), a consortium of U.S. agencies that coordinate Federal research on changes in the global environment, and the Belmont Forum (http://igfagcr.org/index.php/belmont-forum), a scientific council for global change research comprised of 14 countries and the European Union (EU), as well as the International Council for Science (ICSU) and the International Social Science Council (ISSC). In closing, sustainability research inherently requires converging knowledge and technologies. A comprehensive and coordinated U.S. Federal research program would provide the needed space for that to happen and to address the sustainability challenges we now face.

9.8.2 Nuclear Energy as a CKTS Solution for Sustainability

Contact persons: *Ken Chong, NIST and George Washington University, and Philippe M. Bardet, George Washington University*

Like wind and hydroelectricity, nuclear power has a very low carbon footprint. Nuclear power plant capital costs are about three times more than conventional coal power plant costs. However, the production cost of electricity from nuclear power is 50 % less than from coal. President Obama called for "a new generation of safe, clean nuclear power plants" in his 2010 State of the Union address.[1] As of November 2010, 29 countries worldwide were operating 441 nuclear reactors for electricity generation. After Japan's Fukushima Daiichi nuclear power plant disaster due to the earthquake and following tsunamis on March 11, 2012, new attention has been given to nuclear safety worldwide. Nevertheless by May 2012, 66 new nuclear plants were under construction in 14 countries.

Introduction and Historical Background Since the 1940s

The development of nuclear energy marks a great accomplishment of human ingenuity and collaboration (Marcus 2010). Driven initially by fundamental physicists

---

[1] http://www.whitehouse.gov/the-press-office/remarks-president-state-union-address

who uncovered the structure of the atom and how to extract energy from it, it took the twentieth century to transform nuclear energy into a safe and reliable source of electricity. Atomic energy is the densest known source of energy, and it has one of the smallest carbon footprints for electricity production. It is also one the least expensive sources of baseload electricity. However, high costs of building new plants, the handling of nuclear wastes, and negative public perceptions have all hindered its further deployment in some countries. In 2010, nuclear energy represented 13.5 % of the world electricity production. Despite severe accidents such as at Three-Mile Island, Chernobyl, and Fukushima, this energy source continues to grow: besides the 66 reactors being built now, another 250 are being planned within the next 25 years. Other countries are considering investing in this technology.

The development of nuclear technology is the result of multidisciplinary collaboration among physicists and chemists, as well as chemical, electrical, mechanical, structural, and materials science engineers, working together to create and build revolutionary technologies, ranging from electricity production to medical applications. The introduction of nuclear engineering departments at U.S. universities in the 1950s is an illustration of this phenomenon; professors from established and diverse academic backgrounds created and defined this new field of study.

Radioactive Sources for Medical, Energy, Explosive, and Other Purposes

Many isotopes of chemical elements are radioactive, and we are surrounded by low-level radioactivity. Depending on the application, various isotopes are used in treating cancers, medical imaging, space power systems, food sterilization, and nuclear power, among many other fields.

For nuclear energy production, both fissionable nuclides, i.e., capable of sustaining a chain reaction, and fertile nuclides, which first need to be converted to fissile ones by neutron absorption, have been demonstrated and used. Uranium-235 is the most commonly used fissile nuclide; plutonium-239 is also used in fast reactor and mixed oxide (MOX) fuels. Fertile nuclides such as thorium-232 and uranium-238 are all employed.

Many synthetic isotopes, such as cesium-137 or cobalt-60, are produced in nuclear reactors and employed for medical radiation therapy, food irradiation, or industrial radiography. Plutonium-238 is used for space radioisotope thermoelectric generators. Another alpha-emitter, Americium-241, a by-product of fission, is commonly employed in smoke detectors.

Research, Education, and Safety of Nuclear Power Plants

Continued public and private research has led to better understanding of reactor conditions. As a result, plant safety, operation, and predictability have improved significantly. However, there are still some major challenges and research areas with respect to nuclear power plants, and CKTS will have a transformative effect on them (Chong 2011), As an example, the broad field of mechanics plays a critical role in nuclear technology, through such subdisciplines as fracture mechanics, multiscale mechanics, new materials, nano-mechanics for high-temperature and extreme-environment applications and assured safety (Glade et al. 2006), nuclear penetration mechanics, and multiphase flow.

Simulation refers to the application of computational models to the study and prediction of physical events or the behavior of engineered systems (Oden et al. 2006). The development of computer simulation has drawn from a deep pool of scientific, mathematical, computational, and engineering knowledge and methodologies. Improvements in performance for large-scale simulations of turbulent magneto hydrodynamics show that advances in simulation algorithms over a period of a couple of decades have tripled the effective performance compared to performance improvements due to advances in processor speed alone. Similar results have been documented in many other domains, such as turbulent combustion and radiation transport.

There has been a significant increase in U.S. DOE R&D funding since 2000 (see section below on R&D investment). Nuclear engineering education has had contributions from many branches of engineering and fundamental sciences, such as chemistry, biology, and physics. Only a fraction of graduating nuclear engineers ends up working in the energy field. Enrollment in nuclear engineering degrees has increased since 2006, from a low in the early 2000s (ORISE 2011). In 2010, the number of undergraduate students getting a degree in nuclear engineering at 32 U.S. academic institutions was 443. This represents a 167 % increase compared to 2003.

Next-Generation Nuclear Energy Systems: GEN IV

The current operating fleet of nuclear power plants is comprised of generation II nuclear reactors, and most of the reactors being currently deployed worldwide are generation II+ systems or above, with nearly half being generation III and III+. The main evolution for generation III+ systems is the introduction of enhanced passive safety systems. Passive safety systems do not require electrical power for the reactor to be cooled down in case of accident.

The next-generation nuclear energy systems, or GEN IV, incorporate transformative technology with new coolants and materials. By design, this next generation of nuclear systems will produce reactors that are sustainable, economical, safe and reliable, and proliferation-resistant. To study the feasibility of alternative designs and initiate R&D, the Generation IV International Forum (GIF) was formed with 13 Members (Argentina, Brazil, Canada, China, Euratom [European Atomic Energy Community], France, Japan, Republic of Korea, Republic of South Africa, Russian Federation, Switzerland, the United Kingdom, and the United States). GIF selected six nuclear energy systems as most promising for future collaborative development and able to meet GIF goals: gas-cooled fast reactors, lead-cooled fast reactors, sodium-cooled fast reactors, molten-salt-cooled reactors, supercritical-water-cooled reactors, and very-high-temperature reactors.

Transmutation Science and Engineering

Nuclear wastes, or spent fuels, are significantly denser than those of any other energy sources; however, if untreated, they need to be stored for millenniums before their activity reaches non-radiotoxic levels. This makes it a very controversial issue that hinders public acceptance of atomic energy. By bombarding nuclides with neutrons, the nuclear structure can be artificially manipulated, i.e., the number and type of nucleons in a nucleus can be altered, changing "lead into gold" or fissioning large transuranic isotopes (actinides). Actinides are long-lived radioactive isotopes in spent fuels; by transmutation their activity can be reduced from millennia to a few decades. Nuclear wastes thus can become more manageable on a human scale.

A roadmap for the third- and fourth generation nuclear energy systems (NERAC/GIF 2002, 5) anticipates that these will be technically available after 2010 and 2030, respectively.

To artificially transmute used nuclear fuels, high-energy particles are necessary. They can be produced by neutrons in fast reactors (in the deep burn fuel cycle), in particle accelerators, or in fusion devices. Accelerator technology coupled with better predictability of composition of spent fuels rods has led to the advancement of the "reactor-driven reactors," which are becoming a very promising technology.

Public Investment in R&D Since 1950 for Civilian Nuclear Energy

Energy R&D support has a distinct share between government and industry. Government financing of R&D tends to be focused on long-term new technology development. On the other hand, private R&D financing is focused on further developing and improving existing technology. In the United States from 1950 till 2010, the government provided $837 billion for all energy development, in 2010 dollars (NEI 2011). Of this, the largest fraction was tax concessions, which mostly benefitted the oil, gas, and coal industries (47 % of the total); the nuclear industry did not receive any tax incentives. Nuclear energy, excluding fusion, received nearly half ($74 billion of $153 billion) of all public R&D funding—$42 billion from 1950 to 1976 and $31.4 billion from 1976 to 2010—and 9 % of all energy incentives.

Evolution of Investment in Civilian Nuclear RD&D in the U.S. in Comparison to Other Countries

Since 1950 in the United States, light water reactor technology accounted for only 8 % ($5.3 billion in 2010 dollars) of the nuclear RD&D (research, development, and demonstration) incentives, though it now provides almost 20 % of U.S. electricity. Many of the alternative reactor concepts studied in the United States served as the baseline for GEN-IV development. Public funding was significantly reduced in 1988 (Fig. 9.9) when the breeder reactor program was cancelled (EIA 2011). Overall,

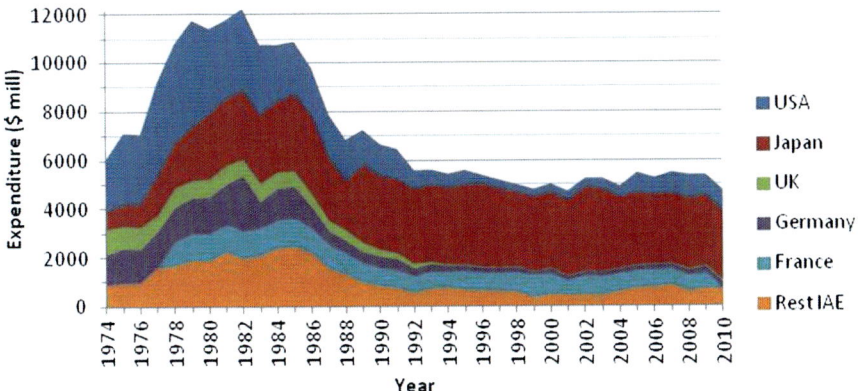

Fig. 9.9 IEA countries' expenditure in 2010 dollars from 1974 to 2010 for research development & demonstration on nuclear energy, including fusion (Courtesy of Philippe Barder; created using data from EIA 2011). Data are missing for France prior to 1976 and in 2010, and for South Korea prior to 2002. The main contributors to the "rest of IEA" category of countries have been Canada, Italy, and South Korea

nuclear R&D peaked at $4.4 billion in 1979: $0.04 billion for LWR, $0.90 billion for fuel cycle, $2.3 billion for breeder and other converter reactors, $1.2 billion for fusion, $0.2 billion for nuclear supporting technology. Public R&D funding declined sharply to about $550 million in 1987 then fell steadily to a low of $310 million in 1999 (with $281 million for fusion and only $29 million for fission). In 2010, funding for fission R&D increased back to $470 million. Since 1988, public spending on nuclear R&D has been less than for coal, and since 1994 it has been less than that for renewable energies as well.

The peak in public expenditure for nuclear energy R&D in countries associated with the International Energy Agency (IEA; http://www.iea.org/countries/, notably excluding China, Russian Federation, and India) was reached around 1979 ($12 billion, with more than a third of that in the United States) (Fig. 9.9). IEA-reported RD&D expenditures decreased sharply after the Chernobyl accident in 1986. From the late 1980s till 2010, while the United States, the United Kingdom, and Germany significantly decreased their investment in nuclear energy R&D, only Japan, France, and Canada have maintained a steady R&D program at an annual funding, respectively, of about $3 billion, $700 million, and $300 million. In 2010, the U.S. Government-supported nuclear energy R&D represented 0.3 % of total Federal R&D support; the comparative investments were 12 % for Japan and 4 % for France. In 2010, total national R&D funding was $402 billion for the United States, $138 billion for Japan, and $48 billion for France; government contributions were, respectively, 31 %, 18 %, and 39 % of total R&D funding.

9.8.3 Convergence of Nanotechnology and Microbiology: Emerging Opportunities for Water Disinfection, Microbial Control, and Integrated Urban Water Management

Contact person: *Pedro Alvarez, Rice University*

No other resource is as universally necessary for life as is water; its safety and availability is a grand challenge inextricably linked to global health and economic competitiveness. While a myriad of water contaminants can cause disease, by far the greatest waterborne threat arises from pathogenic microorganisms. Overcoming this challenge is becoming increasingly difficult as the demand for safer water grows with the world's population and climate change threatens to take away a large fraction of already scarce freshwater. Furthermore, aging water treatment and distribution systems in many cities cannot ensure reliable disinfection; in fact, some systems serve as incidental sources of microbial diseases. Traditional solutions to microbial control have not been able to keep up with the increasing complexity, new barriers, and renewed relevance of this problem. For example, 106 outbreaks and 5,024 recent cases of illness in the United States were attributed to waterborne pathogens in public water systems (Yoder et al. 2008), while each year 39 million Americans suffer infections from waterborne pathogens, leading to productivity losses on the order of $20 billion (Reynolds et al. 2008). The problem is more pronounced in developing nations. Almost one billion people still lack access to safe water (Hutton and Haller 2004), and diarrhea causes about two million infant deaths every year (UNICEF and WHO 2009). Overall, the importance of enhancing water disinfection and microbial control cannot be overstated. The challenge to efficiently disinfect without forming harmful disinfection byproducts (DBPs) and the growing demand for retrofitting aging water infrastructure and developing distributed point of use (POU) water treatment and reuse systems, underscore the needs for new technologies and water management approaches that provide practical solutions for clean water.

Convergence of Nanotechnology and Microbiology as a Synergistic Interdisciplinary Area

The most intellectually stimulating and technologically productive areas of research often occur at the interfaces between disciplines. Such interdisciplinary research has great potential to generate new products and services; enhance human capacity, economic competitiveness, and social achievements; and enable sustainable development (Roco and Bainbridge 2003). Nanotechnology has had a transformative impact on numerous disciplines, including surface science, organic chemistry, molecular biology, semiconductor physics, and microfabrication. Similarly, the convergence of nanotechnology with environmental microbiology would likely result in an interdisciplinary field with great potential for meaningful disruptive

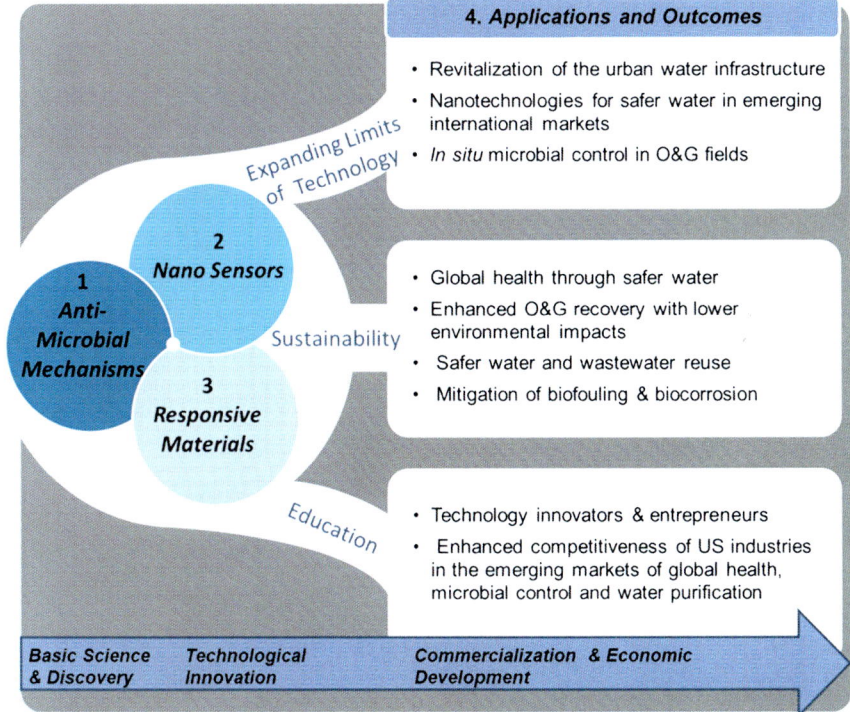

Fig. 9.10 Operational vision and potential outcomes of nanotechnology-enabled water disinfection and microbial control (Courtesy of Pedro Alvarez, Rice University)

innovation. This convergence could expand the limits of water technologies and enhance industrial competitiveness in the emerging markets of global health, microbial control, and water purification, as well as contribute to water security (and thus, energy and food security) and sustainable development (Fig. 9.10).

How Can Nanotechnology Make a Difference in Water Treatment?

Previous research suggests a great potential for nanotechnology-enabled microbial control (Li et al. 2008). Some nanoparticles interact directly with microbial cells to disrupt the integrity of the cell membrane (e.g., carbon nanotubes) and interrupt respiration and energy transduction (e.g., fullerenes and ceric oxide [CeO_2]). Other nanoparticles act indirectly by producing secondary products that serve as disinfection agents, e.g., reactive oxygen species generated by titanium dioxide (TiO_2) or dissolved metal ions released by silver nanoparticles (AgNPs). Nanotechnology can also contribute to integrated urban water and wastewater management by enabling a distributed and differential water treatment and reuse paradigm where water and wastewater are treated locally to the level required by the intended use. This would minimize water quality degradation within aging distribution networks, alleviate dependence on large and centralized system infrastructure (e.g., use only basic treatment near the source water to enhance distribution, and complement it by tailored POU treatment), exploit alternative water sources (e.g., recycled wastewater or storm water) for potable, agricultural, or industrial use, and decrease energy requirements for treating and moving water (Qu et al. 2012).

Implementation Barriers

Cost and performance are critical factors for the broad acceptance of novel water treatment nanotechnologies. In developing countries, water treatment often only covers the most basic needs, such as disinfection, when available. In contrast, industrialized nations tend to use more advanced technologies to remove a wider spectrum of emerging pollutants. However, in both scenarios, there is a need to treat increasingly complex pollutant mixtures and supply higher-quality water at lower cost, which is pushing the boundaries of current treatment approaches. The proposed nanotechnology-based treatment options are high performance—enabling more efficient treatment. However, the relatively high costs of nanomaterials represent a significant (but, perhaps, only temporary) implementation barrier.

Nanomaterial costs are unlikely to decrease significantly without an increase in demand to favor economy of scales. Nevertheless, the cost of nanomaterial synthesis is generally small compared to that incurred by separation and purification steps, due to the high energy and chemical requirements of the latter. This suggests an opportunity to decrease costs by using nanomaterials of lower purity. Another cost-reduction strategy is to facilitate nanomaterial reuse, such as immobilizing photocatalysts that retain high activity after multiple reuse cycles, and iron-containing nano-adsorbents (e.g., nano-magnetite for arsenic removal), which can be separated magnetically and regenerated. Reuse decreases nanomaterial costs per volume of water treated, which is a more relevant feasibility metric than the price of nanomaterials per gram. Potential impacts to human or ecosystem health associated with incidental or accidental releases of nanomaterials represent another important barrier from both regulatory and public acceptance perspectives (Alvarez et al. 2009). Therefore, it is important to take a proactive approach to assess the fate and mitigate potential risks associated with nanomaterials used in (or flowing into) water and wastewater treatment processes.

Social acceptability is also a critical consideration that requires balancing economic and human dimensions while adopting a position of proactive responsibility and inclusiveness. An unbalanced focus on technical innovation may pose risks to the human dimension and jeopardize the sustainability of the technology, whereas focusing too much on responsible development may generate too-restrictive regulations and approaches that delay economic and societal benefits (Roco and Bainbridge 2003). Similarly, disregarding the need to include all stakeholders represents a wasted opportunity to integrate social and ethical issues that intersect with pertinent governmental functions (e.g., funding and regulation), and to establish mechanisms to inform and involve the public about potential impacts of nanotechnology and dispel common misconceptions. This could lead to slower technology implementation and dissemination (and even isolationism). Finally, it is important to resist a tendency to focus on short-term economic feasibility, and rather prioritize a longer-term vision for water security (e.g., nanotechnology-enabled integrated water management) for current and future generations.

Outlook for Nanotechnology in the Water Sector

Despite the aforementioned potential barriers, nanotechnology will likely be increasingly relied upon for needed innovations in water treatment and reuse. The benefits of incorporating nanomaterials have a clear overall benefit when one or more of the following conditions prevails:

- Current processes fail to meet existing or upcoming requirements
- Wastewater reuse is hindered by hazardous micropollutants that break through the treatment process
- POU approaches are needed because of insufficient infrastructure
- Nanomaterials can improve the cost-effectiveness of the treatment process at low additive ratios

Near-term applications include upgrading and enhancing treatment capabilities without major alterations to existing infrastructure (e.g., more efficient disinfection of resistant microbes, and lower potential for DBP formation, microbial-induced corrosion, and membrane fouling) while possibly enabling the use of nonconventional water sources for different reuse scenarios (Qu et al. 2012). Nanomaterials could also be incorporated in POU systems to differentially treat drinking water to higher standards, obviating concerns about secondary contamination through the distribution system. Distributed and differential treatment approaches enabled by POU devices would also be attractive for rural areas and expanding cities in developing countries that lack extensive water infrastructure, where capital investment for new infrastructure may not be feasible. In such cases, nanotechnology could help develop POU systems that are tailored to site-specific needs with minimal use of electricity or imported chemicals.

Conclusions

The extraordinary size-dependent properties of some nanomaterials (e.g., high specific surface area, photosensitivity, catalytic and antimicrobial activity, electrochemical, optical, and magnetic properties, and/or tunable pore size and surface chemistry) offer leapfrogging opportunities to develop next-generation applications for drinking water disinfection and safer wastewater reuse (e.g., photocatalytically enhanced disinfection, biofouling-resistant membranes, biofilm- and corrosion-resistant surfaces, and sensors for pathogen detection). The convergence of nanotechnology

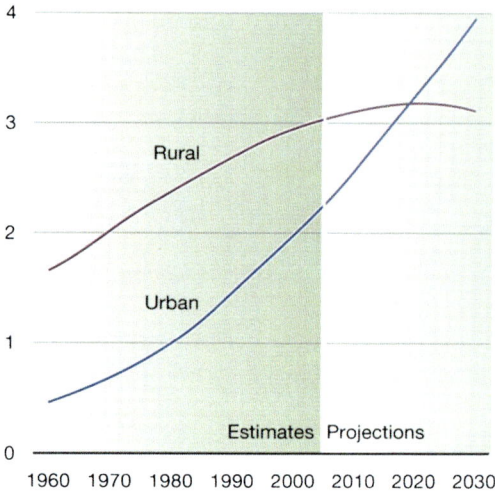

Fig. 9.11 Urban and rural populations in less developed regions (billions), estimated projections (Ahlenius 2009, ©UNEP/GRID-Arendal, used by permission)

with environmental microbiology is a fertile interdisciplinary research area that could expand the limits of technology, enhance global health through safer water reuse, serve as an innovation ecosystem to nurture intellectual entrepreneurs, and contribute towards sustainable and integrated urban water management.

9.8.4 Sustainable Rural Communities

Contact person: *Roop Mahajan, Virginia Polytechnic Institute and State University*

Rural communities in large swaths of the developing world face an array of poverty-related challenges, including lack of housing, infrastructure, access to quality healthcare, education, and a steady means of income. Despite an historic population shift towards urban areas (Fig. 9.11), global poverty remains a massive and predominantly rural phenomenon. According to the Rural Poverty Report 2011, "of the 1.4 billion people living in extreme poverty [less than US$1.25/day] in 2005, approximately one billion—around 70 %—lived in rural areas." With an estimated global population of 8.3 billion in 2030, the rural population in less developed countries, after accounting for the expected migration from rural to urban areas, is estimated to be over three billion (Ahlenius 2009; Cohen 2006).

Lifting this large population out of poverty is a mammoth undertaking, and not surprisingly, one of the top priorities of most governments. However, the implementation of rural development programs over the years has had mixed results, at best, due to political instability, lack of resources, and most important, the absence of a planned approach for sustainable long-term solutions. In many cases where non-profits, philanthropists, and businesses have intervened to supplement the scant resources and fill the gap, many of these efforts remain uncoordinated. Needed are innovative solutions that integrate the best available technologies and practices,

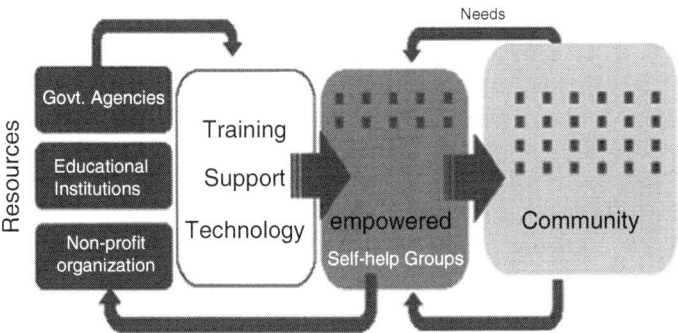

Fig. 9.12 Role of self-help groups as drivers of positive change at the grass-roots level (Courtesy of Roop Mahajan)

and seek the active involvement of the affected population to achieve sustainable development. The goal is to establish *sustainable rural communities* that meet the basic needs of food, clean drinking water, energy, healthcare, education, and provide meaningful jobs, affordable homes, and economical transportation.

To reach these goals, regions much rely as much as possible on their capabilities and the natural resources of the area. Such self-reliant and vibrant rural communities will help retain and attract young people and thus stem the migration to increasingly unsustainable urban centers that are characterized by sprawl where people live in slums larger than the cities themselves, inefficient energy use resulting in heavy smog, uncontrolled wastewater, and overburdened systems that impede rather than enable human ability. A possible model to identify, develop, and successfully implement solutions for creating sustainable rural communities is to form partnerships among self-help groups (SHGs)—*informal clubs or associations of people who choose to come together to find ways to improve their life situations*, government agencies at the local level, privately supported nonprofits, and local colleges and communities (Fig. 9.12).

At the core of these partnerships are SHGs, which have been shown to be critical in formulating the needs of the community and gaining acceptance of the proposed solutions in education, micro-financing and technology adoption, to name a few (Mayoux 1997). In the words of Kanayo F. Nwanze, the president of the International Fund for Agricultural Development (IFAD), "It is time to look at poor smallholder farmers and rural entrepreneurs in a completely new way—not as charity cases but as people whose innovation, dynamism and hard work will bring prosperity to their communities and greater food security to the world in the decades ahead" (http://www.un.org/en/globalissues/briefingpapers/ruralpov/progress.shtml). In these partnerships, institutions of higher learning play multifaceted roles. They serve as interlocutors between the SHGs and other stakeholders. Through their extension and outreach programs, they can provide the curriculum, structure, and in some cases, workforce to implement solutions. Lastly, their student bodies can act as role models for the young people in the community. The other stakeholders—the nonprofits and government agencies—can bring the resources and the best practices for the partnerships to succeed. Organic in nature, such partnerships can serve as the testing grounds for an appropriate mix of converging knowledge and technology, which could possibly migrate to neighboring communities and beyond.

An example of a portfolio of technologies and practices that can be deployed to marshal a rural community in a developing country to reach self-reliance is shown in Table 9.2. The solutions represent a holistic combination of technological solutions, practices, and principles of self-empowerment targeted specifically for rural communities.

Although not specifically listed in Table 9.2, it is implied that the latest converging and emerging technologies, where feasible, will be part of the solution. In large-scale farming, for example, new farm equipment with GPS and radio signals can be used to tailor planting, fertilizing, and insect-control to the specific strengths and needs of the soil, allowing plants to better withstand severe conditions. Other examples include smart sensors and delivery systems that can help the agriculture industry in early disease detection, combating viruses and other crop pathogens, monitoring environmental conditions, and delivering nutrients or pesticides as appropriate (ETC Group 2004). Nanotechnology can also improve our understanding of the biology of different crops and thus potentially enhance yields or nutritional value. In addition, it can offer routes to added-value crops or environmental remediation (Nanoforum 2006). Nanofibers, electrospun from biodegradable cellulose and scrap material, can be used for air filtration, protective clothing, and possibly as mats to absorb and release fertilizers and pesticides for targeted application (Johnson 2005).

As one of the emerging technologies, additive manufacturing (AM) may offer a novel means for the incorporation of converging technologies into prototype and finished products (Ivanova et al. 2013). Since AM allows products to be designed and printed that are more appropriate for local consumption using local materials, including recycled materials, the rural community could reduce reliance on expensive imports by designing and making their own products and reaping the profits from this production.

Finally, the need for sustainable communities, including rural, is also receiving attention from the developed world. For example, in the United States, the interagency Partnership for Sustainable Communities was established on June 16, 2009, among the Department of Transportation (DOT), Department of Housing and Urban Development (HUD), and U.S. Environmental Protection Agency (EPA) to coordinate investments and policies to support sustainable communities—urban, rural, and suburban. The partnership provided six guiding "living principles": (1) safe, reliable, and economical transportation choices; (2) equitable and affordable housing; (3) enhanced economic competitiveness; (4) support of existing communities; (5) healthy, safe, and walkable neighborhoods; and (5) coordination of Federal policies and resources. The U.S. Department of Agriculture (USDA) joined this Partnership to form a Rural Work Group to reinforce the initiatives and ensure that the four agencies' efforts lead to economically vibrant and environmentally sustainable rural communities. In 2003, the British government launched the "Communities

Table 9.2 Convergence of knowledge and technology to build sustainable communities

Need	Enabling technology/ service	Comments
Energy	Solar technology	Electricity, water purification, and food processing at a community level
	Wind	Low cost, locally built windmills for power generation and/or pump water for irrigation
	Biomass	A user-friendly biogas plant that can be linked to a community kitchen as well as linked to organic farming
	Oil	Oil-based power generation unit that supplements a solar + wind power plant
	Natural gas	Natural gas to supplement the bio gas system to provide fuel for the community
Water for household	Rain water harvesting	Harvested rain water linked to a point of use (POU) purification system to supplement existing drinking water supply
	Waste water treatment	Use of sustainable filtration system for grey water
	POU systems	POU systems installed and run by the community, for safe drinking water (Heierli 2008)
Agriculture	Small-scale farming	Promoted by self-help groups, it can result in jobs in agriculture, food processing and transportation sectors
	Large-scale farming	Industrial-scale farming equipment, a catalyst for subsidiary services, training centers, and fabrication facilities
	Smart sensors	Better communication & efficiency among various system components
	Drip irrigation	A service offered by a local self-help group to small-scale farmers. It can energize food production, jobs
Healthcare	Immunization	Effective immunization to reduce child mortality, promote women's health and community productivity; a cell-phone-based immunization system will simplify the process for the government agencies and users
	Mobile clinics	Mobile clinics with a link up to a regional hospital can serve multiple communities
	Telemedicine	Access to a physician at remote locations using hi-tech communication devices
Women empowerment	Micro entrepreneurship	Micro-finance self-help groups provide opportunities for community-level solutions (Mayoux 1997)
	Skills training	SHGs can be trained in the areas of small scale manufacturing, food processing, and providing support services for education
	Local governance	Role of SHGs expanded to include women in local governments

(continued)

Table 9.2 (continued)

Need	Enabling technology/service	Comments
Education	Education	Innovative ways to provide non-formal access to education
	Secondary education	Secondary level education program that is in tune with the local needs and culture
	Community centers	To be run by SHGs after they have been trained in critical areas like small-scale manufacturing
	Cultural/folk art centers	To preserve and promote the artistic expressions of the local community and provide much needed encouragement
Employment	IT centers	Easy access to information in a user-friendly way can catalyze a community in a revolutionary way
	Job opportunities	To educate and inform the community about various employment opportunities

Plan" (Sustainable Communities: Building for the Future) and set out plan of actions for "delivering sustainable communities in both urban and rural areas" (Smith 2008). As expected, there is an overlap between the aspirations of rural communities in the developed and the developing economies. The concept of participatory SHGs and the deployment of CKTS can therefore be easily adapted for rural communities across the world.

9.8.5 Convergence of Knowledge and Technology for Extracting Valuable and Critical Metals from Waste Streams

Contact person: *David Allen, The University of Texas at Austin*

Industrialized economies use large quantities of fuels, minerals, biomass, and other materials. Although material use varies among developed economies, on average, total material use in all industrial economies is greater than 100 lb per person per day, not including water use (Matthews et al. 2000). Each of these materials has a life cycle: they are extracted from the lithosphere or biosphere, processed into commodity materials and products, then recycled or disposed of.

Most industrial systems use materials once, with no engineered recycling systems. It could be argued that the low rates of material reuse in industrialized economies are due to the inherently low value of materials in wastes; however, empirical evidence suggests that much more extensive mining of materials from wastes could be done economically. Doing so will require a convergence of knowledge concerning the flows of materials in industrialized economies and the development of technologies suitable for recovering critical materials.

Mapping Material Flows

As noted above, per capita material flows in industrialized economies are significant. Characterizing the flows and emissions of these materials over the course of their life cycles requires data on material flows entering the economy, manufacturing processes, and information on wastes, emissions, and recycling structures. Data that enable tracking and optimization of national and global material flows are just emerging, and terminology and data analysis frameworks are still evolving (NRC 2003). Material flow analyses are performed on systems with well-defined boundaries. The system boundary might be the geopolitical boundaries of a nation, the natural boundaries of a river's drainage basin, or the technological boundaries of a cluster of industries.

Consider, as an example of material flow analyses, the element lead (Pb). Pb is a neurotoxin, and Pb exposure is associated with developmental delays (http://www.epa.gov/iris), so human exposure should be minimized. Historically, some of the principal uses of lead have been as an octane enhancer in gasoline, in batteries, and in paint. The material flows of Pb in the United States in 1970 and in the mid-1990s are shown in a USGS publication (2000, 14). Such a figure can provide details such as in 1970 the fraction of virgin material was 36 % (450 tons recycled/1,250 tons total usage), while in the mid-1990s the fraction had increased to 65 % (910 tons recycled/1,400 tons total usage). The Pb was incorporated into a variety of products. Some products, even when used as designed (such as lead paint applied outdoors or lead additives in gasoline), result in the release of Pb into the environment (dissipative uses). Other products, when used as designed (lead acid batteries), allow Pb to be effectively recovered and recycled at the end of the product's life (nondissipative uses).

Figure 9.13 illustrates another potential structure for the mapping of material flows (Graedel et al. 2011). The figure includes additional types of flows, using global use of iron as a case study. It separates iron flows into production, fabrication/manufacturing, use, and waste management categories. There are flows between these stages of materials processing, for example, as "home" scrap within production operations is reprocessed. The material flow mapping of Fig. 9.13 also shows flows of material from different processing stages into anthropogenic repositories (such as landfills) and the flows of material into durable goods (stock).

Detailed material flow mappings are not available for most materials. Yet, combinations of materials scarcity and geopolitical factors will put increasing pressure on material flows, and this will necessitate better knowledge of material flow patterns in industrialized economies, so that critical materials can more readily be recovered and recycled.

The Role of Separation Technologies

If materials are to be mined from waste streams, there must not only be the knowledge of the stocks and flows of the materials, but technologies to recover critical materials from wastes. Currently, the range of technologies used to manage waste

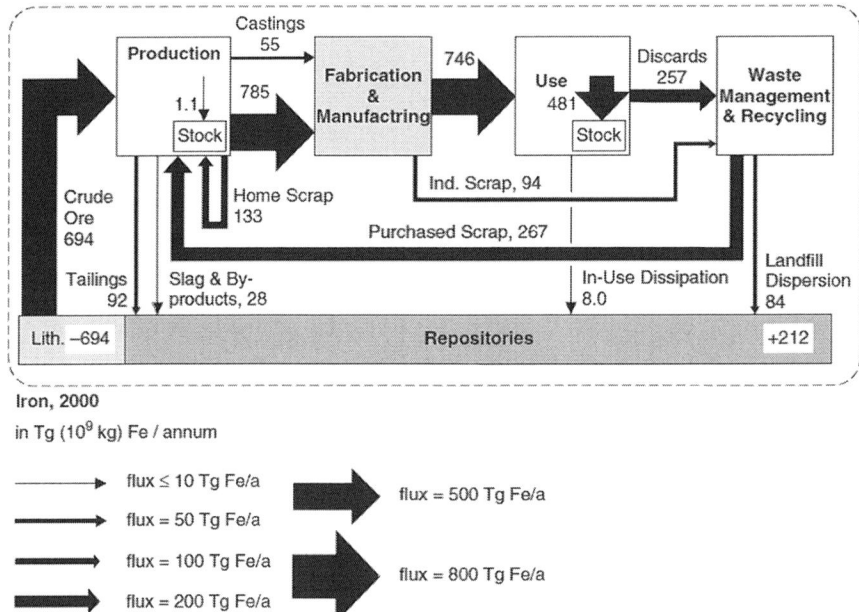

Fig. 9.13 Global flows of iron in 2000 (Graedel et al. 2011; STAF Project, Yale University Center for Industrial Ecology; © *Journal of Industrial Ecology* 2007, http://dx.doi.org/10.1111/j.1530-9290.2011.00342.x, used by permission)

streams is relatively limited. Combustion, wastewater treatment, and land disposal dominate, leaving substantial opportunity for technological progress (Allen 1992; Baker et al. 1992; Allen and Behmanesh 1992). A sense of the critical role that new technologies may play in mining materials from wastes is provided by free energies of mixing. A simple calculation reveals that the entropy that needs to be overcome in separating a pound of a critical material from a million pounds of ore or waste is on the order of 1,000 BTU (a gallon of gasoline has a heating value in excess of 100,000 BTU). This amount of energy is negligible relative to the amounts of energy required by technologies employed to perform such separations. Therefore, fundamental physical limitations are not the key limitation in separating critical materials from wastes. Effective and energy-efficient separation technologies are required.

The potential value of critical materials that might be recovered from waste streams can also be illustrated through a simple case study. Allen and Behmanesh (1994) examined the extent to which hazardous wastes in the United States might be mined (cost-effectively recycled) by comparing the degree of dilution of metals in hazardous waste streams to the concentrations at which the materials were mined. Their original analysis, summarized in Table 9.3, found that many hazardous waste streams in the United States had high enough concentrations of metals to merit additional recycling. Hazardous wastes were chosen because detailed data existed on their compositions, flow rates, and fates.

9 Implications: Convergence of Knowledge and Technology...

Table 9.3 Percentage of metals in hazardous wastes in the U.S. that can be recovered economically (as estimated by Allen and Behmanesh 1994)

Metal	Percent theoretically recoverable (%)	Percent recycled in 1986 (%)
Antimony (Sb)	74–87	32
Arsenic (As)	98–99	3
Barium (Ba)	95–98	4
Beryllium (Be)	54–84	31
Cadmium (Cd)	82–97	7
Chromium (Cr)	68–89	8
Cooper (Cu)	85–92	10
Lead (Pb)	84–95	56
Mercury (Hg)	99	41
Nickel (Ni)	100	0.1
Selenium (Se)	93–95	16
Silver (Ag)	99–100	1
Thallium (Tl)	97–99	1
Vanadium (V)	74–98	1
Zinc (Zn)	96–98	13

Surprisingly, many hazardous waste streams contained relatively high concentrations of metals. Approximately 90 % of the copper, 95 % of the zinc, and 100 % of the nickel found in hazardous wastes was, at the time, at a concentration high enough to recover. For every metal for which data existed, recovery occurred at rates well below rates that would be expected to be economically viable. The economic value of these waste streams is significant. For example, the roughly 3.6 million tons per year of Ni identified as unrecovered in the 1994 analysis has a current market value of over $50 billion ($7.00/lb).

This focused analysis, initially performed in 1994, led to the conclusion that many opportunities exist for recovering materials from wastes. There are limitations to the analysis, however. The analysis focused only on hazardous wastes, where legal liability concerns may limit the desire to recycle. The identification of "recycleable" streams was simplistic. It ignored issues related to economies of scale (processing geographically dispersed, heterogeneous waste streams may be more expensive than extracting relatively homogeneous ore from a single mine). Nevertheless, the analysis indicated that resources are not effectively recovered from many waste streams.

The United Nations has assembled more recent, global data on metals recycling. That analysis concludes that global recycling rates can continue to be improved. Metals like Pb (along with Fe, Cr, Co, Ni, Cu, Zn, and many precious metals) have post-consumer recycling rates that exceed 50 % globally, but many other metals (e.g., rare earth elements) have post-consumer recycling rates of less than 1 % (UNEP 2011b). Key factors in changing this situation will be knowledge of material flow patterns and separation technologies required to recover critical materials.

9.9 International Perspectives

The following are summaries relevant to this chapter of discussions during the international regional WTEC NBIC2 workshops held in Leuven, Belgium, September 20–21, 2012; in Seoul, Korea, October 15–16, 2012; and in Beijing, China, October 18–19, 2012. Further details of those workshops are provided in Appendix A.

9.9.1 United States–European Union NIBIC2 Workshop (Leuven, Belgium)

Panel members/discussants

Daan Schuurbiers (Chair), Utrecht University (EU)
Mamadou Diallo, Caltech (U.S.) and Korea Advanced Institute of Science and Technology (KAIST)
Albert Duschl, University of Salzburg (EU)
Barbara Harthorn, University of California–Santa Barbara (U.S.)
Todd Kuiken, Woodrow Wilson International Center for Scholars (U.S.)
Andy Miah, University of the West of Scotland (EU)
Alfred Nordmann, Darmstadt University of Technology (EU)
Anders Sandberg, Oxford University (EU)
Bruce Tonn, University of Tennessee (U.S.)

The U.S.–EU Working Group 2 consisted of a diverse group of physical scientists, engineers, and social scientists. The group devoted its discussion to the broad areas of sustainable development and human development. The discussion focused on four topics of global sustainability: water (Topic 1), energy and environment (Topic 2), megacities and urban communities (Topic 3), and agriculture and food (Topic 4). For each topic, the U.S.-EU Working Group 2 outlined broad research questions/objectives and discussed how the convergence of knowledge and technology for the benefit of society can help address these questions to achieve concrete outcomes in the next 10, 20, and 50 years. The group also discussed governance-related issues for each topic. Below we give a summary of the discussion for each topic.

Topic 1: Water

Core Idea: Use wastewater and saline water as sources of clean water, nutrients, and materials.
Concrete outcomes:

- Beyond reverse osmosis—develop biomimetic desalination membranes that work like ion channels and aquaporins
- Solutions of major energy and water problems

Governance:

- Competing technological options: water reuse or desalination
- Public acceptance of water reuse: "purity" and "disgust" issues

Topic 2: Energy

Core Idea: Reduce energy imports; reduce GHG by 80 %: low or no carbon emissions by 2050.
Concrete outcomes:

- Solar energy-develop more efficient and advanced photovoltaic cells
- Fusion energy after 2050.

Governance:

- Energy security as a social concern
- Energy as international development issue
- Energy and society: how does the social system of energy production, distribution, and use have to change to accommodate a shift to a society that consumes less energy?

Topic 3: Megacities and Urban Communities

Core Idea: Design and build the sustainable cities and urban communities of the twenty-first century.
Concrete Outcomes:

- Urban planning for megacities – maximizing positive effects
- Sustainable energy infrastructure ("on-the-spot" generation of energy and smart grids)
- Sustainable materials usages (recycling, urban mining and 3D printing)

Governance:

- Intelligent management systems: information/communication technologies to address social planning questions, harness creativity and tacit knowledge to identify, improve, and share solutions for a sustainable lifestyle in mega cities and urban communities.

Topic 4: Agriculture and Food

Core idea: Build sustainable agriculture and food production systems
Concrete Outcomes:

- Non-genetically modified drought-resistant crops
- More efficient hydroponic agricultural and vertical farming systems
- Precision agriculture systems to optimize the delivery of water, nutrients (nitrogen and phosphorus), and pesticides to crops.

Governance:

- Food quality and taste: "Keep things as they are"
- Necessary condition for the EU debate, otherwise it would jeopardize the acceptance of CKTS as tool in building sustainable agriculture and food production systems

9.9.2 United States–Korea–Japan NBIC2 Workshop (Seoul, Korea)

Panel members/discussants

Mamadou Diallo (co-chair), Caltech (U.S.) and KAIST (Korea)
Kazuyo Matsubae (co-chair), Tohoku University, Japan
Young Hyun Cho, Dongbu Hitek Co., Ltd. (Korea)
Bruce Tonn, University of Tennessee (U.S.)
Robert Urban, MIT (U.S.)

The U.S.-Korea-Japan Working Group S9 devoted its discussion to the broad area of sustainable development. The group focused on three topics: renewable energy sources (Topic 1), agriculture, food and natural resources (Topic 2), megacities and urban communities (Topic 3). For each topic, the U.S.-Korea-Japan Working Group S9 outlined broad research questions/objectives and discussed how the convergence of knowledge and technology (CKTS) can help address these questions in the next 10 years. Below we give a summary of the discussion for each topic.

Topic 1: Renewable Energy Sources (Solar and Biomass)

Core Idea: Develop and deploy more efficient renewable energy sources from solar and biomass.
CKTS Tools

A. Solar energy

- World's best Si-based solar cell modules have an efficiency of 24.7 %
- Multi-junction concentrators are very expensive but can get up to 42 %.
- Next generation of high-efficiency solar cells: nanocrystalline silicon-quantum dots, thin film solar cells using quantum dots

B. Biomass

- Korea established the Advanced Biomass R&D Center at KAIST to convert algae biomass to fuels. The program is funded at a level of $12–15 million per year for 9 years.

Topic 2: Agriculture, Food, and Natural Resources

Core Idea: Recovery of nutrients (phosphorous) from non-traditional sources

CKTS Tools
- Monitor the global flow of phosphorous using advanced sensor networks
- Develop more efficient separation technologies to recover phosphorous from non-traditional sources including wastewater and slags from steels plant
- Increase public/regulatory acceptance of using recycled phosphorus to grow food crops

Topic 3: Megacities and Urban Communities

Core Idea: Design and build smart and sustainable cities

CKTS Tools
- New forms of transportation that anticipates destinations and builds routing, timing, clustering, and redistribution
- Balance power usages through on and off grid energy bi-directional transfers
- Wellness-based living: walkable, breathable, community health-centric, safety, stress-minimized, family/education-directed
- Ubiquitous monitoring/sensing of urban systems (environment, diseases, hazards)
- On-demand and local manufacturing (3D printers, auto assemblers, carbon legos)
- Urban agriculture/food production systems (e.g., high-efficiency vertical farms)
- Design efficient closed-loop urban water and wastewater systems
- Develop strategies to retrofit existing megacities into smarter and more sustainable cities

9.9.3 United States–China–Australia–India NBIC2 Workshop (Beijing, China)

Panel members/discussants

Mamadou Diallo (co-chair), Caltech (U.S.) and KAIST (Korea)
Ming Liu (co-chair), Chinese Academy of Sciences (China)
Ian Lowe (co-chair), Australian Conservation Foundation (Australia)
Yajun Guo, National Natural Science Foundation (China)
Xiangyu Jiang, National Center for NanoScience and Technology (China)
Bruce Tonn, University of Tennessee (U.S.)
Robert Urban, MIT (U.S.)

The U.S.-China-Australia-India Working Group S9 devoted its discussion to the broad area of sustainable development. The group focused on three topics: minerals and materials (Topic 1), global climate change and environment (Topic 2) and rural and low-income communities (Topic 3). For each topic, the U.S.–China–Australia Working Group S9 outlined broad research questions/objectives and discussed how the convergence of knowledge and technology can help address these questions to achieve concrete outcomes in the next 10 years. Below we give a summary of the discussion for each topic.

Topic 1: Minerals and Materials

Core Idea: Optimize materials usage and supply to reduce the reliance on critical minerals/materials in sustainable energy generation and storage.

CKTS Tools
- Materials are not the limit, if energy is unlimited, non-critical materials (earth-abundant materials) can be found/used in lower amounts (supply, utilization, substitution)
- Recycling advances (extraction of industrial waste, mine tailings)
- Energy efficient separation technologies using high-capacity and selective ligands/hosts
- High-efficiency supercapacitors using nanostructured materials
- Integrated battery-supercapacitor systems

Topic 2: Environment and Global Climate Change

Core Idea: Develop more efficient strategies to mitigate CO_2 emission and utilize CO_2 as feedstock to produce liquid fuels and useful products
  CKTS Tools

A. Mitigation

- Develop and deploy low-CO_2 emitting sources of energy (e.g., solar and wind)
- Develop and deploy carbon capture and storage technologies (e.g., flue gases and ambient air)
- Improve the energy efficiency of existing products and processes

B. Utilization

- Conversion of CO_2 to fuels and chemicals by artificial photosynthesis
- Use CO_2 as a raw material in chemical manufacturing to produce polymers and cement

Topic 3: Rural and Low-Income Communities

Core Idea: Sustainable technologies for supporting rural and low income communities

CKTS Tools
- Connect every family to "the cloud" to enable access to health and wellness information
- Effective information/communication (ICT) tools to support and empower rural and low-income communities
- Affordable healthcare products (e.g., vaccines, paper-based diagnostics, nano-patches)
- Low-cost energy generation and storage technologies
- Sustainable and distributed water purification and sanitation technologies
- Open source building materials/plans
- Sustainable and affordable food production and storage technologies.

References

Ahlenius, H.: Chart in Nellemann, C, MacDevette, M, Manders, T, Eickhout, B, Svihus, B, Prins, AG, Kaltenborn, BP (eds.) The environmental food crisis—the environment's role in averting future food crises UNEP/GRID-Arendal. Available online: http://www.grida.no/graphicslib/detail/trends-in-urban-and-rural-populations-less-developed-regions-1960-2030-estimates-and-projections_bd04 (2009)

Allen, DT: An overview of industrial waste generation and management practices. MRS Bull. **17**(3), 30–33 (1992)

Allen, DT, Behmanesh, N: Non hazardous waste generation. Hazard. Waste Hazard. Mater. **9**, 91–96 (1992)

Allen, DT, Behmanesh, N: Wastes as raw materials. In: Allenby, BR, Richards, DJ (eds.) The Greening of Industrial Ecosystems, pp. 69–89. The National Academies Press, Washington, DC (1994)

Alvarez, PJ, Colvin, V, Lead, J, Stone, V: Research priorities to advance eco-responsible nanotechnology. ACS Nano **3**(7), 1616–1619 (2009)

Aravinda, CL, Cosnier, S, Chen, W, Myung, NV, Mulchandani, A: Label-free detection of cupric ions and histidine-tagged proteins using single poly(pyrrole)-NTA chelator conducting polymer nanotube chemiresistive sensor. Biosens. Bioelectron. **24**, 1451–1455 (2009)

ARPA-E: Rare earth alternatives in critical technologies (REACT). Available online: http://arpa-e.energy.gov/programsprojects/react.aspx (2011)

Bainbridge, WS: Demographic collapse. Future **41**, 738–745 (2009)

Baker, RD, Warren, JL, Behmanesh, N, Allen, DT: Management of hazardous waste in the United States. Hazard. Waste Hazard. Mater. **9**, 37–60 (1992)

Balogh, L, Swanson, DR, Tomalia, DA, Hagnauer, GL, McManus, ET: Dendrimer-silver complexes and nanocomposites as antimicrobial agents. Nano Lett. **1**(1), 18–21 (2001)

Baochang, G, Feng, W, Zhigang, G, Erli, Z: China's local and national fertility policies at the end of the twentieth century. Popul. Dev. Rev. **33**(1), 129–147 (2007)

Bates, BC, Kundzewicz, ZW, Wu, S, Palutikof, JP (eds.): Climate Change and Water. Technical Paper of the Intergovernmental Panel on Climate Change. IPCC Secretariat, Geneva (2008). Available online: http://www.ipcc.ch/pdf/technical-papers/climate-change-water-en.pdf

Bettencourt, LMA, Kaur, J: Evolution and structure of sustainability science. PNAS **108**(49), 19540–19545 (2011)

Bettencourt, LMA, Kaiser, DI, Kaur, J: Scientific discovery and topological transitions in collaboration networks. J. Informetr. **3**(3), 210–221 (2009). http://dx.doi.org/10.1016/j.joi.2009.03.001

Birnbaum, ER, Rau, KC, Sauer, NN: Selective anion binding from water using soluble polymers. Sep. Sci. Technol. **38**(2), 389–404 (2003)

Brinker, JC, Ginger, D: Nanotechnology for sustainability: energy conversion, storage, and conservation. In: Roco, MC, Mirkin, C, Hersham, M (eds.) Nanotechnology Research Directions for Societal Needs in 2020: Retrospective and Outlook, pp. 261–303. Springer, Dordrecht (2011). Science Policy Reports

Britt, D, Furukawa, H, Wang, B, Glover, TG, Yaghi, OM: Highly efficient separation of carbon dioxide by a metal-organic framework with open metal sites. PNAS **106**, 20637–20640 (2009)

Brundtland, GH: Annex A/42/427 to Report of the World Commission on Environment and Development: Our Common Future. United Nations World Commission on Environment and Development, New York (1987). Available online: http://www.channelingreality.com/Documents/Brundtland_Searchable.pdf

Cáceres-Delpiano, J: Can we still learn something from the relationship between fertility and mother's employment? Evidence from developing countries. Demography **49**(1), 151–174 (2012)

Capper, JL, Cady, RA, Bauman, DE: The environmental impact of dairy production: 1944 compared with 2007. J. Anim. Sci. **87**, 2160–2167 (2009)

Chen, DP, Yu, CJ, Chang, C-Y, Wan, Y, Goddard, WA, Diallo, MS: Branched polymeric media: perchlorate-selective resins from hyperbranched polyethyleneimine. Environ. Sci. Technol. **46**, 10718–10726 (2012)

Chong, K.P.: Nuclear energy: safety, production, mechanics research and challenges. Plenary Lecture IMECE2011-65987, ASME Congress, Denver, CO, November (2011)

Christensen, K, Doblhammer, G, Rau, R, Vaupel, JW: Aging populations: the challenges ahead. Lancet **374**(9696), 1196–1208 (2009)

Clark, WC: Sustainability science: a room of its own. PNAS **104**, 1737–1738 (2007)

Clive, J.: Global status of commercialized Biotech/GM crops: 2011. ISAAA Brief No. 43. ISAAA, Ithaca: International Service for the Acquisition of Agri-Biotech Applications. Available online: http://www.isaaa.org/resources/publications/briefs/43/default.asp (2011)

Cohen, JE: How Many People Can the Earth Support? W. W. Norton, New York (1995)

Cohen, JE.: Human population: the next half century. In Kennedy, D. (ed.) Science Magazine's State of the Planet 2006–7. Island Press, London (2006). Joseph, T., Morrison, M.: Nanotechnology in Food and Agriculture. Institute of Nanotechnology. http://www.nanoforum.org/dateien/temp/nanotechnology%20in%20agriculture%20and%20food.pdf (2006)

Coleman, J.: World's 'seven billionth baby' is born. The Guardian. Sunday 30 Oct 2011. Available online: http://www.guardian.co.uk/world/2011/oct/31/seven-billionth-baby-born-philippines?intcmp=122 (2011)

Costanza, R., Alperovitz, G., Daly, H., Farley, J., Franco, C., Jackson, T., Kubiszewski, I., Schor, J., Victor, P.: Building a Sustainable and Desirable Economy-in-Society-in-Nature. Report to the United Nations for the 2012 Rio+20 Conference. Division for Sustainable Development, United Nations Department of Economic and Social Affairs, New York (2012)

Crooks, RM, Zhao, M, Sun, L, Chechik, V, Yeung, LK: Dendrimer-encapsulated metal nanoparticles: synthesis, characterization, and application to catalysis. Acc. Chem. Res. **34**, 181–190 (2001)

Davis, K: The theory of change and response in modern demographic history. Popul. Index **29**, 345–366 (1963)

Davis, SC, Diegel, SW, Boundy, RG: Transportation Energy Data Book. ORNL-6981, 27th edn. Oak Ridge National Laboratory, Oak Ridge (2008)

den Hond, F: Industrial ecology: a review. Reg. Environ. Chang. **1**(2), 60–69 (2000)

Diallo, MS, Brinker, JC: Nanotechnology for sustainability: environment, water, food, minerals and climate. In: Roco, MC, Mirkin, C, Hersham, M (eds.) Nanotechnology Research Directions for Societal Needs in 2020: Retrospective and Outlook, pp. 221–259. Springer, Dordrecht (2011). Science Policy Reports

Diallo, MS, Christie, S, Swaminathan, P, Balogh, L, Shi, X, Um, W, Papelis, C, Goddard, WA, Johnson, JH: Dendritic chelating agents. 1. Cu(II) binding to ethylene diamine core poly(amidoamine) dendrimers in aqueous solutions. Langmuir **20**, 2640–2651 (2004)

DOE (Department of Energy): Buildings energy data book. Available online: http://buildingsdatabook.eere.energy.gov (2009a)

DOE (Department of Energy): Modern shale gas development in the United States: A primer. Available online: http://fossil.energy.gov/news/techlines/2009/09024-Shale_Gas_Primer_Released.html (2009b)

DOE (Department of Energy): Critical materials strategy. Available online: http://energy.gov/pi/office-policy-and-international-affairs/downloads/2010-critical-materials-strategy (2011)

DOE (Department of Energy): Critical materials hub. Available online: http://energy.gov/articles/ames-laboratory-lead-new-research-effort-address-shortages-rare-earth-and-other-critical (2013)

Ehrlich, PR: The Population Bomb. Ballantine, New York (1968)

EIA (U.S. Energy Information Administration): Annual energy outlook 2011. http://www.eia.gov/oiaf/aeo/gas.html (2011)

Elimelech, M, Phillip, WA: The future of seawater desalination: energy, technology and environment. Science **333**, 712–717 (2011)

ETC Group. Down on the Farm: The Impact of Nano-scale Technologies on Food and Agriculture. ETC Group, Ottawa. Available online at http://www.etcgroup.org/content/down-farm-impact-nano-scale-technologies-food-and-agriculture (2004)

Fan, ZY, Wang, DW, Chang, PC, Tseng, WY, Lu, JG: ZnO nanowire field-effect transistor and oxygen sensing property. Appl. Phys. Lett. **85**, 5923–5925 (2004)

Field, L.: Soft geoengineering: Ice 911. Available online: http://www.wilsoncenter.org/event/considering-soft-geoengineering#field_files (2012)

Fromer, N., Eggert, R.G., Lifton, J. (ed.): Critical Materials for Sustainable Energy Applications. Caltech Resnick Institute Report. Available online: http://resnick.caltech.edu/programs/critical-materials/index.html (2011)

Gallezot, P: Conversion of biomass to selected chemical products. Chem. Soc. Rev. **41**, 1538–1558 (2012)

Glade, SC, Wirth, BD, Odette, GR, Asoka-Kumar, P: Positron annihilation spectroscopy and small angle neutron scattering characterization of nanostructural features in high-nickel model reactor pressure vessel steels. J. Nucl. Mater. **351**, 197–208 (2006)

Graedel, TE, Allwood, J, Birat, J-P, Buchert, M, Hagelüken, C, Reck, BK, Sibley, SF, Sonnemann, G: What do we know about metal recycling rates? J. Ind. Ecol. **15**(3), 355–366 (2011). http://dx.doi.org/10.1111/j.1530-9290.2011.00342.x

Heidrich, ES, Curtis, TP, Dolfing, J: Determination of the internal chemical energy of wastewater. Environ. Sci. Technol. **45**(2), 827–832 (2011)

Holt, JK, Park, HG, Wang, Y, Stadermann, M, Artyukhin, AB, Grigoropoulos, CP, Noy, A, Bakajin, O: Fast mass transport through sub-2-nanometer carbon nanotubes. Science **312**, 1034–1037 (2006)

Howden, LM, Meyer, JA: Age and Sex Composition: 2010. 2010 Census briefs. U.S. Census Bureau, Washington, DC (2011)

Hutton, G, Haller, L: Evaluation of the Costs and Benefits of Water and Sanitation Improvements at the Global Level. World Health Organization WHO/SDE/WSH/04.04, Geneva (2004)

Iceland, C.: Aqueduct and the water–energy–food nexus. Presentation, Bonn Nexus Conference, Bonn, Germany, 17 Nov 2011. Interactive water map available online: http://insights.wri.org/aqueduct (2011)

EIA (International Energy Agency): Energy technology RD&D budgets. Available online: http://www.iea.org/stats/rd.asp (2011)

ILO (International Labour Organization): World Social Security Report 2010/11: Providing Coverage in Times of Crisis and Beyond. ILO, Geneva (2010)

ILO (International Labour Organization): Global Employment Trends 2011 (Geneva). Available from http://www.ilo.org/wcmsp5/groups/public/@dgreports/@dcomm/@publ/documents/publication/wcms_150440.pdf (2011)

IPCC (Intergovernmental Panel on Climate Change): Carbon Dioxide Capture and Storage. A Special Report of Working Group III of the Intergovernmental Panel on Climate Change, Eds.

Metz, B, Davidson, O, de Coninck, H, Loos, M, Meyer, L. Cambridge University Press, Cambridge (2005)

IPCC (Intergovernmental Panel on Climate Change): Climate change 2007: the physical science basis. In: Solomon, S, Qin, D, Manning, M, Chen, Z, Marquis, M, Averyt, KB, Tignor, M, Miller, HL (eds.) Contribution of Working Group I to the Fourth Assessment Report of the Intergovernmental Panel on Climate Change. Cambridge University Press, Cambridge/New York (2007a)

IPCC (Intergovernmental Panel on Climate Change): Impacts, adaptation, and vulnerability. In: Parry, ML, Canziani, OF, Palutikof, JP, van der Linden, PJ, Hanson, CE (eds.) Contribution of Working Group II to the Fourth Assessment Report of the Intergovernmental Panel on Climate Change. Cambridge University Press, Cambridge, New York (2007b)

IPCC: Summary for policymakers. In: Edenhofer, O., Pichs-Madruga, R., Sokona, Y., Seyboth, K., Matschoss, P., Kadner, S., Zwickel, T., Eickemeier, P., Hansen, G., SchloÅNmer, S., von Stechow, C. (eds.) IPCC Special Report on Renewable Energy Sources and Climate Change Mitigation. Cambridge University Press, Cambridge/New York (2011)

Ivanova, O., Williams, C., Campbell, T.: Additive manufacturing (AM) and nanotechnology: promises and challenges. Rapid Prototyp. J. **19**(5), 353–364 (2013)

Jeong, BH, Hoek, EMV, Yan, Y, Huang, X, Subramani, A, Hurwitz, G, Ghosh, AK, Jawor, A: Interfacial polymerization of thin film nanocomposites: a new concept for reverse osmosis membranes. J. Membr. Sci. **294**, 1–7 (2007)

Johnson, A.: Agriculture and Nanotechnology (University of Wisconsin–Madison Student-Produced Report). Available online at http://www.tahan.com/charlie/nanosociety/course201/nanos/AJ.pdf (2005)

Kajikawa, Y: Research core and framework of sustainability science. Sustain. Sci. **3**(2), 215–239 (2008). http://dx.doi.org/10.1007/s11625-008-0053-1

Kajikawa, Y, Ohno, J, Takeda, Y, Matsushima, K, Komiyama, H: Creating an academic landscape of sustainability science: an analysis of the citation network. Sustain. Sci. **2**(2), 221–231 (2007). http://dx.doi.org/10.1007/s11625-007-0027-8

Kates, RW: What kind of a science is sustainability science? PNAS **108**(49, 99), 19449–19450 (2011). http://dx.doi.org/10.1073/pnas.1116097108

Keith, DW, Parson, E, Morgan, MG: Research on global sun block needed now. Nature **463**, 426–427 (2010)

King III, SB, Meier, B: Interventional treatment of coronary heart disease and peripheral vascular disease. Circulation **102**, Iv-81–Iv-86 (2000)

Knodel, J, van de Walle, E: Lessons from the past: policy implications of historical fertility studies. Popul. Dev. Rev. **5**, 217–245 (1979)

Lewis, NS: Toward cost-effective solar energy use. Science **315**, 798–801 (2007)

Li, Q, Mahendra, S, Lyon, DY, Brunet, L, Liga, MV, Li, D, Alvarez, PJ: Antimicrobial nanomaterials for water disinfection and microbial control: potential applications and implications. Water Res. **42**(18), 4591–4602 (2008)

Liu, Y, Majetich, SA, Tilton, RD, Sholl, DS, Lowry, GV: TCE dechlorination rates, pathways, and efficiency of nanoscale iron particles with different properties. Environ. Sci. Technol. **39**, 1338–1345 (2005)

Logan, BE, Rabaey, K: Conversion of wastes into bioelectricity and chemicals by using microbial electrochemical technologies. Science **337**, 686–690 (2012)

Lowry, GV, Johnson, KM: Congener-specific dechlorination of dissolved PCBs by microscale and nanoscale zero-valent iron in a water/methanol solution. Environ. Sci. Technol. **38**(19), 5208–5216 (2004)

Malthus, TR: An Essay on the Principle of Population. J. Johnson, London (1798)

Marcus, GH: Nuclear Firsts: Milestones on the Road to the Nuclear Power Development. American Nuclear Society, La Grange Park (2010)

Matthews, E, Amann, C, Bringezu, S, Fischer-Kowalski, M, Hüttler, W, Kleijn, R, Moriguchi, Y, Ottke, C, Rodenburg, E, Rogich, D, Schandl, H, Schütz, H, van der Voet, E, Weisz, H: The Weight of Nations. World Resources Institute, Washington, DC (2000)

Mayoux, L.: Participatory learning for women's empowerment in micro-finance programmes: negotiating complexity, conflict and change, (October 1996), 1–21 (1997)

McCarty, PL, Bae, J, Kim, J: Domestic wastewater treatment as a net energy producer—can this be achieved? Environ. Sci. Technol. **45**(17), 7100–7106 (2011)

Milanovic, B: The Haves and the Have-Nots: A Brief and Idiosyncratic History of Global Inequality. Basic Books, New York (2010)

Mishra, H, Yu, C, Chen, DP, Dalleska, NF, Hoffmann, MR, Goddard III, WA, Diallo, MS: Branched polymeric media: boron-chelating resins from hyperbranched polyethyleneimine. Environ. Sci. Technol. **46**, 8998–9004 (2012)

Moyo, DF: Winner Take All. China's Race for Resources and What It Means for the World. Basic Books, New York (2012)

Nanoforum Report. Nanotechnology in Agriculture and Food. Available online at ftp://ftp.cordis.europa.eu/pub/nanotechnology/docs/nanotechnology_in_agriculture_and_food.pdf (2006)

NEI (Nuclear Energy Institute), prepared by Management Information Services, Inc.: 60 years of energy incentives: analysis of federal expenditures for energy development. NEI, Washington, DC. Available online: http://www.misi-net.com/publications/NEI-1011.pdf (2011)

NERAC (U.S. Nuclear Energy Research Advisory Committee) and the Generation IV International Forum (GIF): A technology roadmap for Generation IV nuclear energy systems. Available online: http://www.gen-4.org/PDFs/GenIVRoadmap.pdf (2002)

NIH Fogarty International Center: NIH partners on clean cookstoves initiative. Adv. Sci. Glob. Health 9(5). http://www.fic.nih.gov/News/GlobalHealthMatters/Oct2010/Pages/cookstove.aspx (2010)

NRC (National Research Council): Materials Count: The Case for Material Flows Analysis. The National Academies Press, Washington, DC (2003)

NRC (National Research Council): Minerals, Critical Minerals, and the U.S. Economy. The National Academies Press, Washington, DC (2008)

NRC (National Research Council): Toward Sustainable Agricultural Systems in the 21st Century. Committee on Twenty First Century Systems Agriculture, Board on Agriculture and Natural Resources, Division on Earth and Life Sciences. The National Academies Press, Washington, DC (2010)

NRCS (National Resources Conservation Service): National Resource Conservation Handbook. NRCS, Washington, DC (2012). Available online: http://www.nrcs.usda.gov/wps/portal/nrcs/main/national/home

Oden, J.T., Belytschko, T., Fish, J., Hughes, T.J.R., Johnson, C., Keyes, D, Laub, A., Petzold, L., Srolovitz, D., Yip, S.: Simulation-Based Engineering Science: Revolutionizing Engineering Science Through Simulation. Report of the National Science Foundation Blue Ribbon Panel on Simulation-Based Engineering Science. Available online: http://www.nsf.gov/pubs/reports/sbes_final_report.pdf (2006)

Oeppen, J, Vaupel, JW: Broken limits to life expectancy. Science **296**, 1029–1031 (2002)

Ogburn, WF: Social Change with Respect to Culture and Original Nature. B. W. Huebsch, New York (1922)

Olson, RL: Soft geoengineering: a gentler approach to addressing climate change. Environ. Sci. Policy Sustain. Dev. **54**(5), 29–39 (2012)

ORISE (Oak Ridge Institute for Science and Education): Nuclear engineering enrollments and degrees survey, 2010 Data, June 2011 Update. Available online: http://orise.orau.gov/files/sep/NE-Brief-68.1-2011-data.pdf (2011)

OSTP (Office of Science and Technology of the Executive Office of the President): Innovation for America's Economy, America's Energy, and American Skills: Science, Technology, Innovation, and STEM Education in the 2013 Budget. OSTP, Washington, DC (2012). Summary available online: http://www.whitehouse.gov/sites/default/files/microsites/ostp/fy2013rd_summary.pdf

Park, S, Cheedrala, RK, Diallo, MS, Kim, CH, Kim, IS, Goddard III, WA: Nanofiltration membranes based on polyvinyldene fluoride nanofibrous scaffolds and crosslinked polyethyleimine networks. J. Nanopar. Res. **14**, 884 (2012)

Parsons, T, Shils, EA (eds.): Toward a General Theory of Action. Harvard University Press, Cambridge (1951)

Pennisi, E: Water reclamation going green. Science **337**, 674–676 (2012)

Phan, A, Doonan, CJ, Uribe-Romo, FJ, Knobler, CB, O'Keeffe, M, Yaghi, OM: Synthesis, structure, and carbon dioxide capture properties of zeolitic imidazolate frameworks. Acc. Chem. Res. **43**, 58–67 (2010)

Qu, X, Brame, J, Li, Q, Alvarez, PJ: Nanotechnology for a safe and sustainable water supply: enabling integrated water treatment and reuse. Acc. Chem. Res. (Jun 27), 834–843 (2012)

Rapport, DJ: Sustainability science: an ecohealth perspective. Sustain. Sci. **2**, 77–84 (2007)

Reck, BK, Graedel, TE: Challenges in metal recycling. Science **337**, 690–695 (2012). http://dx.doi.org/10.1126/science.1217501

Reid, J.F: The impact of mechanization on agriculture. In Fall issue of The Bridge on Agriculture and Information Technology. National Academy of Engineering (NAE) **41**(3), 22–29. Available online: http://www.nae.edu/Publications/Bridge/52548/52645.aspx (2011)

Reynolds, KA, Mena, KD, Gerba, CP: Risk of waterborne illness via drinking water in the United States. Rev. Environ. Contam. Toxicol. **192**, 117–158 (2008)

Rickerby, DG, Morisson, M: Nanotechnology and the environment: a European perspective. Sci. Technol. Adv. Mater. **8**, 19–24 (2007)

Rifkin, J: The Third Industrial Revolution: How Lateral Power is Transforming Energy, the Economy, and the World. Palgrave Macmillan, New York (2011)

Rockström, J, Steffen, W, Noone, K, Persson, Å, Chapin III, FS, Lambin, EF, Lenton, TM, Scheffer, M, Folke, C, Schellnhuber, HJ, Nykvist, B, de Wit, CA, Hughes, T, van der Leeuw, S, Rodhe, H, Sörlin, S, Snyder, PK, Costanza, R, Svedin, U, Falkenmark, M, Karlberg, L, Corell, RW, Fabry, VJ, Hansen, J, Walker, B, Liverman, D, Richardson, K, Crutzen, P, Foley, JA: A safe operating space for humanity. Nature **461**, 472–475 (2009a)

Rockström, J, Steffen, W, Noone, K, Persson, Å, Chapin III, FS, Lambin, EF, Lenton, TM, Scheffer, M, Folke, C, Schellnhuber, HJ, Nykvist, B, de Wit, CA, Hughes, T, van der Leeuw, S, Rodhe, H, Sörlin, S, Snyder, PK, Costanza, R, Svedin, U, Falkenmark, M, Karlberg, L, Corell, RW, Fabry, VJ, Hansen, J, Walker, B, Liverman, D, Richardson, K, Crutzen, P, Foley, JA: Planetary boundaries: exploring the safe operating space for humanity. Ecol. Soc. **14**(2), art.32 (2009b)

Roco, MC, Bainbridge, WS: Converging Technologies for Improving Human Performance: Nanotechnology, Biotechnology, Information Technology, and Cognitive Science, p. 467. Kluwer Academic Publishers, Dordrecht (2003). xiii

Satterthwaite, D: Will most people live in cities? Br. Med. J. **321**(7269), 1143–1145 (2000). Available online: http://www.pubmedcentral.nih.gov/articlerender.fcgi?artid+1118907

Saunders, J: Meeting Summary: Partnerships, Science, and Innovation for Sustainability Solutions, A Symposium, May 16–18, 2012. The National Academies Press, Washington, DC (2012). Available online: http://sites.nationalacademies.org/PGA/sustainability/SustainabilitySymposium/

Savage, N, Diallo, M: Nanomaterials and water purification: opportunities and challenges. J. Nanopart. Res. **7**, 331–342 (2005)

Schmidt, KF: Green Nanotechnology: It is Easier Than you Think. Project on Emerging Nanotechnologies (Pen 8). Woodrow Wilson International Center for Scholars, Washington, DC (2007)

Service, RF: Save pave the world. Science **337**, 676–678 (2012)

Shannon, MA, Bohn, PW, Elimelech, M, Georgiadis, JG, Mariñas, MJ, Mayes, AM: Science and technology for water purification in the coming decades. Nature **452**, 301–310 (2008)

Smith, M.K.: Sustainable Communities and Neighbourhoods. Theory, Policy and Practice. In the Encyclopaedia of Informal Education. Available online: http://www.infed.org/community/sustainable_communities_and_neighbourhoods.htm (2008)

Song, H, Carraway, ER: Reduction of chlorinated ethanes by nanosized zero-valent iron kinetics, pathways, and effects of reaction conditions. Environ. Sci. Technol. **39**, 6237–6254 (2005)

Staff, J, Mortimer, JT: Explaining the motherhood wage penalty during the early occupational career. Demography **49**(1), 1–21 (2012)

Tang, CY, Zhao, Y, Wang, R, Hélix-Nielsen, C, Fane, AG: Desalination by biomimetic aquaporin membranes: review of status and prospects. Desalination **308**, 34–40 (2012). http://dx.doi.org/10.1016/j.desal.2012.07.007

Tobert, JA: Lovastatin and beyond: the history of the HMG-CoA reductase inhibitors. Nat. Rev. Drug Discov. **2**, 517–526 (2003)

Tratnyek, PG, Johnson, RL: Nanotechnologies for environmental cleanup. Nano Today **1**(2), 44–48 (2006)

Tuck, CO, Pérez, E, Horváth, I, Sheldon, RA, Poliakoff, M: Valorization of biomass: deriving more value from waste. Science **337**, 695–699 (2012)

UN (United Nations): 2010 Statistical Yearbook. United Nations, New York (2012)

UN (United Nations): Demographic, economic, and social drivers. Chapter 2 in World Water Development Report 3. Paris: World Water Assessment Program, UNESCO Division of Water Sciences, UNESCO Publishing (2009)

UN Millennium Project: A Practical Plan to Achieve the Millennium Development Goals. United Nations Development Group, New York (2005). Available online (historic site): http://www.unmillenniumproject.org/reports/fullreport.htm

UN System Task Team on the Post-2015 UN Development Agenda: Realizing the Future We Want for All: Report to the Secretary-General. UN System Task Team, New York (2012). Available online: http://www.un.org/millenniumgoals/pdf/Post_2015_UNTTreport.pdf

UNEP (United National Environment Programme): Towards a Green Economy: Pathways to Sustainable Development and Poverty Eradication. Part I, Investing in Natural Capital: Chapter 1, Agriculture. UNEP, Nairobi (2011a). Available online: http://www.unep.org/greeneconomy/greeneconomyreport/tabid/29846/default.aspx

UNEP (United National Environment Programme): Recycling Rates of Metals: A Status Report, 2011. Available online: http://www.unep.org/resourcepanel/Publications/Recyclingratesofmetals/tabid/56073/Default.aspx (2011b). Accessed July 2011

UNICEF/WHO: Diarrhoea: Why Children Are Still Dying and What Can Be Done, Ed. Jensen. The United Nations Children's Fund (UNICEF), New York. World Health Organization (WHO), Geneva (2009)

UNPF (United Nations Population Fund): State of World Population 2007: Unleashing the Potential of Urban Growth. United Nations Population Fund, New York (2007)

USGS (United States Geological Survey): Materials and Energy Flows in the Earth Science Century: A Summary of a Workshop Held by the USGS in November 1998, USGS Circular 1194. Available online: http://pubs.usgs.gov/circ/2000/c1194/c1194po.pdf (2000)

Vaseashta, A, Dimova-Malinovska, D: Nanostructured and nanoscale devices, sensors, and detectors. Sci. Technol. Adv. Mater. **6**, 312–318 (2005)

Wang, B, Cote, AP, Furukawa, H, O'Keeffe, M, Yaghi, OM: Colossal cages in zeolitic imidazolate frameworks as selective carbon dioxide reservoirs. Nature **453**, 207–211 (2008)

Wang, R.,Setiawan, L., Fane, A.G. (eds.): Forward osmosis: Current status and perspectives. J. Membr. Sci. Virtual Special Issue. Available online: http://www.journals.elsevier.com/journal-of-membrane-science/virtual-special-issues/forward-osmosis-current-status-and-perspectives/ (2012)

Webber, ME: Energy versus water: solving both crises together. Sci. Am. Earth **3**, 34–41 (2008)

Wikimedia Commons: Top 400 "urban areas" with at least 1,000,000 inhabitants in 2006. Available online at http://commons.wikimedia.org/wiki/File: 2006megacities.PNG (2007)

Wikipedia Commons: World Population. Available online at http://en.wikipedia.org/wiki/File:World-Population-1800-2100.svg (2012)

WWF (World Wildlife Fund): The Energy Report–100% Renewable Energy by 2050. Available online: http://wwf.panda.org/what_we_do/footprint/climate_carbon_energy/energy_solutions/renewable_energy/sustainable_energy_report/ (2011)

Yoder, J, Roberts, V, Craun, GF, Hill, V, Hicks, LA, Alexander, NT, Radke, V, Calderon, RL, Hlavsa, MC, Beach, MJ, Roy, SL: Surveillance for waterborne disease and outbreaks associated with drinking water and water not intended for drinking-United States, 2005–2006. MMWR Surveill. Summ. **57**(9), 39–62 (2008)

Zalasiewicz, J, Williams, M, Steffen, W, Crutzen, P: The new world of the anthropocene. Environ. Sci. Technol. **44**(7), 2228–2231 (2010)

Chapter 10
Innovative and Responsible Governance of Converging Technologies

Mihail C. Roco, David Rejeski, George Whitesides, Jake Dunagan,
Alexander MacDonald, Erik Fisher, George Thompson,
Robert Mason, Rosalyn Berne, Richard Appelbaum,
David Feldman, and Mark Suchman

10.1 Vision

10.1.1 Changes in the Vision over the Past Decade

In the last 10 years, the following developments have taken place that impact the governance of converging technologies:

- The viability of integrating nanotechnology, biotechnology, information science, and cognitive science (NBIC) has been confirmed in multiple settings since 2001, mostly in a "reactive approach" responding to collaborative opportunities. This NBIC concept has been extended in this study to other converging platforms and to a holistic systematic approach called Converging Knowledge, Technologies, and Society (CKTS).
- There is an increasing emphasis on the roles of innovation (for societal benefit, jobs, and economic competitiveness), sustainability (in energy, health, food, climate change, etc.) and realizing human potential (in education, workforce, aging with

---

Corresponding editors M.C. Roco (mroco@nsf.gov) and W.S. Bainbridge (wbainbri@nsf.gov).

M.C. Roco
National Science Foundation, Arlington, VA, USA

D. Rejeski
Woodrow Wilson International Center for Scholars, Washington, DC, USA

G. Whitesides
Harvard University, Cambridge, MA, USA

J. Dunagan
The Institute for the Future, Palo Alto, CA, USA

A. MacDonald
National Aeronautics and Space Administration, Washington, DC, USA

dignity, economic and legal support of family, etc.) in evaluating emerging new technologies. The combination of multiple emerging fields has enhanced both expectations and capabilities, exemplified by simultaneous advances in imaging, electronics, genetics, brain research, and other NBIC-based technologies; CKTS offers increased means for pursuing such goals. Examples of convergence ecosystems are the Semiconductor Research Corporation, Silicon Valley, and State and regional science and technology (S&T) initiatives.

- Concerns about the impact of new technologies have grown. Three main concerns have been raised: (1) that powerful new technologies such as synthetic biology and quantum information systems pose uncertain transformational impacts for society; (2) that the ethical challenges will be daunting of balancing new capabilities to improve human healthcare against defining limits to human enhancement; and (3) greater recognition by an increasing number and variety of actors of the role that nanotechnology-based environmental, health, and safety (EHS) and ethical, legal, and social issues (ELSI) play in research, regulatory challenges, and governance under conditions of uncertainty and/or knowledge gaps, voluntary codes, and accepted practices.

- Societal implications of emerging and converging technologies are tied to key contextual factors, including changes in demographics, the multiple S&T poles, knowledge and technology transfer from the Western to Eastern hemisphere, and both the nature and the human awareness of the rapid developments that science and society are experiencing. Smaller countries tend to be able to adjust governance faster, e.g., see the South Korea case study, Sect. 10.8.9.

- Two regulatory approaches are developing in parallel: one that is probing the extendibility of regulatory schemes ("developing the science" approach), and one that is developing exploratory (soft) regulatory and governance models that work reasonably well even with insufficient knowledge for full risk assessment.

E. Fisher
Arizona State University, Phoenix, AZ, USA

G. Thompson
Intel, Santa Clara, CA, USA

R. Mason
University of Washington, Seattle, WA, USA

R. Berne
University of Virginia, Charlottesville, VA, USA

R. Appelbaum
University of California–Santa Barbara, Santa Barbara, CA, USA

D. Feldman
University of California, Irvine, Irvine, CA, USA

M. Suchman
Brown University, Providence, RI, USA

- Two international organizations have emerged: the Converging Technologies Bar Association (CTBA 2003, http://www.convergingtechnologies.org/defalt2.asp) and the Society for the Study of Nanoscience and Emerging Technologies (S.Net 2009, http://www.thesnet.net/). There is also an "International Journal of Emerging Technologies and Society" based in Australia: http://www.swinburne.edu.au/hosting/ijets/ijets/. Such relatively modest beginnings for specialized audiences nevertheless have underlined the need for new approaches to converging knowledge and technology, going beyond current methods that are focused on individual disciplines, individual players, or coincidental collaborations.
- Social media are providing new methods for collaboration enabled by converging technologies, e.g., social-media-enabled innovation and development of "leaderless" movements/networks.

10.1.2 The Vision for the Next Decade

CKTS implications are expected to increase significantly in the next decade, with the following characteristics:

- Convergence will contribute to major changes in science, technology, and society, and will become a condition for national but also industrial and human competitiveness. The convergence approach will become increasingly proactive and systemic (decisions taken considering the system as a whole, as compared to a reactive approach based on interaction of system components). New sociotechnical convergence platforms of various sizes will emerge. Systematic convergence in knowledge and technology promises to increase the rate of scientific breakthroughs, lead to the establishment of new S&T domains and support growing expectations for human progress, including improved productivity, education, and quality of life.
- A virtual spiral of creativity and innovation evolving in time (see schematic in Fig. 4.1 in Chap. 4) between and within CKTS platforms will be created, along with an increase in the speed of circulation (transfer) of ideas from one field to another. This will have a significant effect on innovation, productivity, and commercialization. It will be a condition of competitiveness for sectors or regions.
- It is expected that society will increasingly guide and authorize the CKTS research and investment agenda through various governance means. This role will increase as scientific knowledge increases quasi-exponentially, technologies enable more powerful tools, population growth coincides with increased expectations for better quality of life, and global competition intensifies. Governance will increasingly shape CKTS for societal progress.
- The roles of individuals in public groups (e.g., entrepreneur/inventor Elon Musk and his company SpaceX) and of public–private partnerships will increasingly

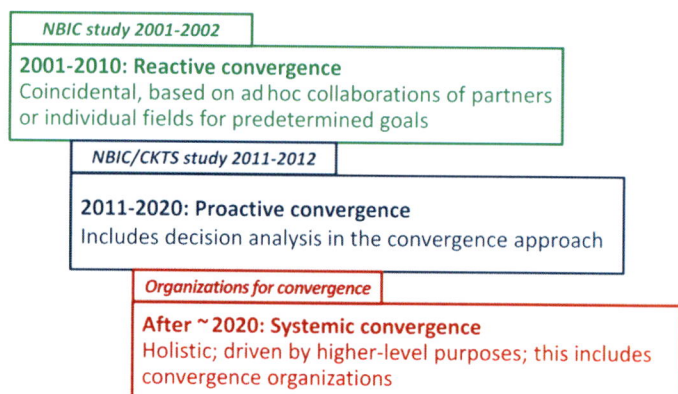

Fig. 10.1 Estimated timeline for progress in converging knowledge and technologies for society. The 2001–2002 study report is Roco and Bainbridge (2003) (Figure courtesy of M. C. Roco)

push the development of new converging technologies, separate from the roles of governments. New tools will emerge for participatory governance, e.g., games, collaborative design, and social media.

- An international community dedicated to these governance considerations is expected to be established, extending efforts already underway within individual emerging technologies. Co-evolution between science, technology, and societal norms and values will become increasingly evident to a larger number of actors.
- CKTS advances will provide improved methods and databases in support of governance.
- The successes of the National Nanotechnology Initiative (NNI) in integrating social science and governance considerations with technology development will be relevant to CKTS.
- The success of CKTS will depend on whether risk governance methods in the short and long terms are addressed from the beginning of each project and whether social scientists and public participants are meaningfully involved.
- A horizontal, vertical, and system-integrated infrastructure will be developed for the essential convergence platforms. Existing and new S&T domains are envisioned to be enabled by CKTS, such as distributed and digital manufacturing, cognitive and neuromorphic engineering, synthetic biology, and quantum information systems. Similar to the case of nanotechnology development, it will become imperative over the next decade to focus not only on how CKTS can generate economic and medical value ("material progress") and enable *cognitive, social, and environmental value* ("moral progress"), but also on how to contribute to quality of life and how to foster international collaboration.
- Three successive and overlapping steps for convergence of knowledge and technology seem to be emerging on the approximate timeline shown in Fig. 10.1.

10.2 Advances in the Last Decade and Current Status

10.2.1 Innovation and Converging Technologies

This chapter discusses specific aspects of innovation and risk governance of CKTS. By applying the holistic system deductive approach, higher-level language approach, convergence–divergence evolutionary approach proposed in Chap. 4 (see Fig. 4.1), and vision-inspired basic research approach (Fig. 4.5), the CKTS innovation opportunities increase due to the multidisciplinary discoveries within the general convergence platforms, an accelerated innovation process facilitated by the convergence methods, and multisector applications available in the general convergence platforms (see Fig. 10.2). An index of innovation rate (4.1) has been defined by correlation.

The concept of differentiation has been identified as a principle of action in evolutionary society (Parsons and Shils 1951; Parsons 1964) as well as in biological systems. The divergence phase in the convergence–divergence cycle that initially was proposed for the coherent evolution of megatrends in science and technology (Roco 2002) and extended in this volume to evolution of knowledge-centered human activities (R&D, design, production, etc.) has qualitative similarities with the above social science and biological concept.

At the same time, convergence of emerging technologies brings specific kinds of risk challenges:

- Increased technology complexity, uncertainty, and ambivalence in comparison with traditional technologies.

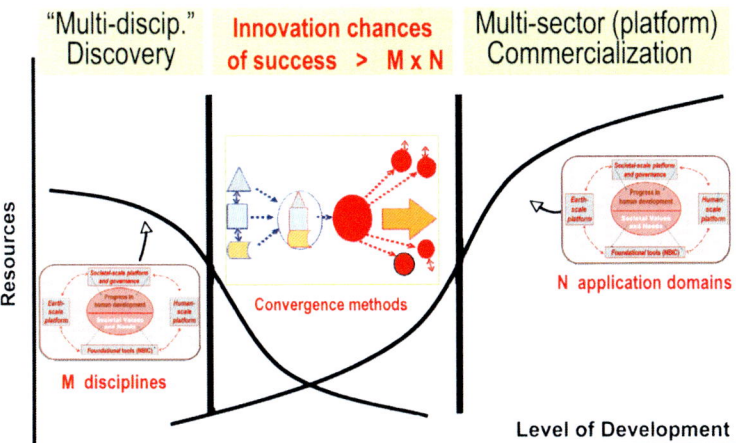

Fig. 10.2 CKTS innovation opportunities increase in proportion with the M number of disciplines supporting the R&D, the N number of application domains, and improvement by applying the convergence methods (overall > M×N). The convergence–divergence approach offers multiple possible pathways from discovery to commercialization (Courtesy of M. C. Roco)

- Interdependency with wide-ranging effects throughout scientific, industrial, and social systems, including convergence and integration trends.
- Increased importance of societal implications, which cannot be fully known at the release of the technology. It will be essential to reduce the time lag between development of scientific knowledge and evaluation of societal implications. Similarly, it will be crucial to integrate anticipation of societal implications with research and development, commercialization, and regulation.

The NBIC technologies have been defined as a multidisciplinary foundational platform for improving the benefits of emerging and converging technologies in society, offering new approaches for education, innovation, learning, and governance (Roco and Bainbridge 2003). As the definition of converging knowledge and technology expands, so do its transforming and risk implications in society. CKTS builds on the convergence process defined within the general platforms of human activities (see Chaps. 1, 2, 3, and 4). Governance of CKTS has many facets, from fostering research to increasing innovation and productivity, to addressing ethical concerns and long-term human development issues (IRGC 2006, 2008). It includes "transformational, responsible, inclusive, and visionary" development, as described below (Roco 2008). The converging technologies approach has been brought to the attention of legislative institutions such as the German Bundestag (Coenen 2008).

10.2.2 Societal Dimensions of Converging Knowledge and Technologies

Principles of holistic interdependence, connectivity, and co-evolution have long-time roots in human civilization (e.g., in ancient oriental cultures, the European Renaissance, and indigenous Indian culture in the Americas). As noted in Chap. 4, in the decade since the first NBIC study was completed, NBIC research programs have been undertaken in the United States, the European Union, China, Russia, Japan, and other countries, as well as by international organizations such as the International Risk Governance Council (IRGC). Converging knowledge and technology bring new opportunities to address societal needs with increasingly more coherent knowledge and technologies.

An international community of scholars who specifically address ethics and societal dimensions of emerging technologies, S.NET, was created in 2009, in part through extension of various national nanotechnology networks (e.g., the National Science Foundation's Nanotechnology in Society Network in the United States, since 2005). Journals such as *NanoEthics* and the S.NET have diagnosed a range of science–society disconnects, from "science leaps ahead/ethics lags behind" (Mind the gap [Mnyuisiwalla et al. 2003]) in about 2000, to "ethics leaps ahead/science lags behind" in 2010. A European Community "Code of Conduct for Research Integrity" has been proposed, but globally, a common terminology and shared levels of national commitment to ethics in R&D are still to be reached. In 2008, the German government evaluated the risks of NBIC (Bundestag 2008).

Public concern over the societal implications of innovative technologies in general is illustrated by ongoing resistance to genetically modified (GM) foods, due in part to the lack of attention to addressing public concerns at the time GM foods were first put on the market in 1996. Critics have continued to object to GM foods on several grounds, including safety issues, ecological concerns, and economic concerns raised by the fact that organisms capable of reproduction are subject to intellectual property law (http://en.wikipedia.org/wiki/Genetically_modified_food). The NNI has explicitly incorporated investment in environmental, health, and safety (EHS) and ethical, legal, and other social issues (ELSI) research to consider and address these kinds of existing and potential public concerns concerning the nation's investment in nanotechnology research.

Development of "leaderless" or "multicentered" movements/networks, in part enabled by social media and converging technologies, is a new phenomenon that has emerged in the last few years.

10.2.3 Governance of Converging Knowledge and Technologies

Challenges to governance of CKTS have included developing the multidisciplinary knowledge foundation; strengthening the innovation chain from priority-setting and discovery to societal use; establishing a common language in nomenclature and patents; addressing broader societal implications; and overall, creating the tools, people, and organizations to responsibly develop and distribute the benefits of the new technologies.

To address those challenges in nanotechnology R&D, *four simultaneous functions of governance* were proposed and have been applied in the United States since 2001 (Roco 2008; IRGC 2008), that it should be

- *Transformative*, including having a results-oriented, project-oriented focus and advancing multidisciplinary and/or multisector innovation
- *Responsible*, including addressing EHS, ELSI, and equity concerns
- *Inclusive*, having all-inclusive, all-agency, and all-stakeholder participation
- *Visionary*, including long-term planning and anticipatory, adaptive policies

Table 10.1 gives U.S. examples of these functions, which have international counterparts.

Growth of converging and emerging technologies research is expected to exceed the average rates of growth in scientific R&D worldwide in the next decade, particularly because of its importance for improving economic efficiency in Western countries and the focus in Asian countries on emerging technologies. Emerging technology areas have been identified in previous chapters of this report, and examples of major emerging technologies in the United States are given in Table 10.2. Other topics such as biofuels, solar energy (photovoltaics), aeolian energy, electric cars, vaccines, cognitive, prosthetics, the game industry, and mass media (networks, twitter, etc.) have more dispersed R&D programs.

Table 10.1 Examples of U.S. governance functions supporting CKTS (2001–2010)

Governance aspect	Example 1	Example 2
TRANSFORMATIVE function		
National and regional investment policies	Support for R&D programs and emerging industries with high economic return and societal relevance (see Table 10.2)	NIH grand challenges set up, and CKTS R&D and facilities developed to address priorities in human health
	Support technologies for renewable natural resources (water, food, energy, environment)	
Science, technology, and business policies	Support for S&T integration in long-term programs (such as Apollo space, nuclear energy, ITR, NNI) through competitive peer-reviewed, multidisciplinary R&D programs	NSF, DOD, NASA support for innovation in converging technologies (nano-bio-info-…)
Education and training	Interdisciplinary Grant for Education, Research and Training (IGERT) program at NSF and NIH	NSF's Science of Learning Centers (addressing topics from brain research to teaching methods)
Technology and economic transformation tools	Support for clusters of manufacturing capabilities for integrated industrial platforms (NRC 2011), such as cell phone technology	Establish research clusters for various emerging areas, such as synthetic biology
RESPONSIBLE function		
Environmental, health, and safety (EHS) issues	The December 2003 Nanotechnology R&D Act includes EHS guidance; OSTP, PCAST, and NRC make EHS recommendations; NNI publishes national nanoEHS strategy in 2008 and 2011	Establish NSTC Emerging Technologies Interagency Policy Committee (2010–) and NSET Nanotechnology Environmental and Health Implications working group (NEHI, 2005–)
Ethical, legal, and social issues and other issues (ELSI+)	Ethics of converging technologies addressed in publications (Roco and Bainbridge 2001, 2007; NGO reports, and UNESCO reports 2006a, b, c)	Program announcements for nano-ELSI (NSF, 2004–); Equitable benefits for developing countries (ETC/NGO 2005; CNS-UCSB 2009)
Methods for risk governance	Risk analysis, including the social context, supported by NSF and U.S. EPA and applied in EPA, FDA, and OSHA policies	Governance for multilevel risk in converging technologies in the global ecological system (IRGC 2008)
Regulations and reinforcement	IT, genome, bio-ethics, and nanotechnology-focused regulatory groups created at NSTC, EPA, FDA, and NIOSH	Voluntary measures for nanotechnology EHS at EPA, 2008

Communication and participation	Increased interactions among experts, users, and public at large via public hearings	Public and professional society participation in legislative processes for IT, genome, and NNI funding
INCLUSIVE function		
Partnerships to build national capacity	NSTC investment in IT, global climate change, robotics, and nanotechnology; e.g., fostering interagency partnerships (27 agencies); industry–academia–state–Federal government partnerships (NNI support for 4 regional-state-local nanotech. workshops)	Partnering among research funding and regulatory agencies in the NSTC for dealing with nanotechnology implications; e.g., the NSET Subcommittee and its NEHI Working Group
Global capacity	International Dialogue Series on Responsible Nanotechnology (2004, 2006, 2008); OECD and ISO working groups on various emerging technologies	International Risk Governance Council reports on emerging technologies; e.g., on nanotechnology, and on food and cosmetics (IRGC 2009)
Public participation	Public debates on human enhancement capabilities and transhumanism caused by human-technology co-evolution (ELSI issues) (Hamlett et al. 2008)	Combined public and expert surveys; public deliberations; informal science education (e.g., NSF programs beginning in ~2001)
VISIONARY function		
Long-term, global view	U.S. strategic plans for space, nanotechnology, information technology, health research, and neurosciences	Long-term effects of technology on human development (*Humanity and the Biosphere*, Foundation for the Future, UNESCO 2007)
Support human development, incl. sustainability	Research on energy and water resources using nanotechnology (DOE, NSF, EPA, others)	Research connecting brain functions, mind, and education (NSF, NIH)
Long-term planning	Ten-year vision statements published for 2001–2010 (published in 2000) and 2011–2020 (Roco et al. 2011) (Similar long-term R&D planning in EU, Korea, Singapore, Japan, China, and other countries)	NNI strategic plans every 3 years (last three in 2004, 2007, and 2010), followed by PCAST and NRC evaluations

Acronyms: *ISO* International Standards Organization, *ITR* Information Technology Research, *NEHI* Nanotechnology Environmental and Health Implications Working Group, *NIOSH* National Institute of Occupational Safety & Health, *NRC* National Research Council, *NSET* Nanoscale Science, Engineering and Technology (Subcommittee of OSTP's Committee on Technology), *NSTC* National Science and Technology Council (U.S.), *OECD* Organization for Economic Co-operation and Development, *OSHA* Occupational Safety and Health Administration, *OSTP* Office of Science and Technology Policy of the White House, *PCAST* President's Council of Advisors in Science and Technology, *UNESCO* United Nations Educational, Scientific and Cultural Organization

Table 10.2 Examples of emerging technology programs in the U.S. since 1950 with timelines

Emerging technology	First commercial prototypes	Expected to reach large-scale application	Related chapters/sections in this report
Nuclear energy	1957	1965	Section 9.8.2
Space: NASA program; unmanned autonomous systems in aerospace	1961	1970	Section 10.8.1
Supercomputers	1980	2000	Chapters 1 and 7
Large databases	1985	2010	Chapter 1
Genetics: genome program	2000	2020	Chapter 5
Nanotechnology	2000	2020	Chapter 1
Smart mobile phones (iPhone)	2005	2010	Sects. 1.1.1 and 4.3.5
Systems and synthetic biology	2010	2025	Chapter 1, Sect. 1.8.3
Robotics for personal service	2010	2040	Chapters 2 and 7, Sects. 2.3.1, 2.8.5, 6.3.1, 7.8.2
Neurotechnology	2010	2060	Chapter 6

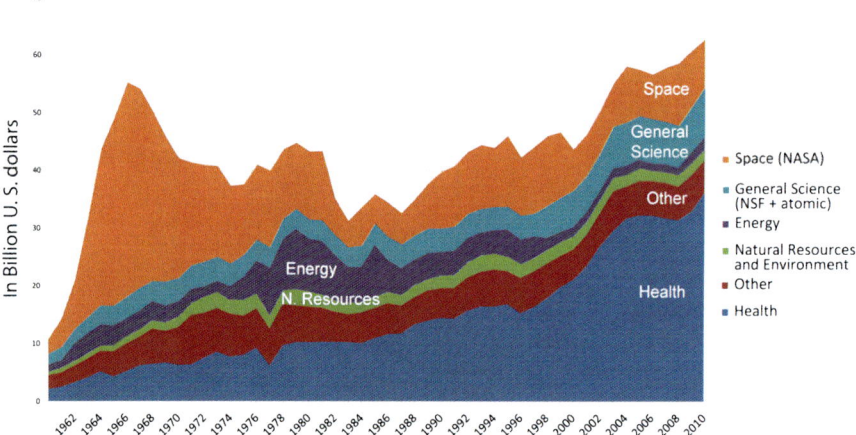

Fig. 10.3 U.S. Federal Government science R&D investments in major areas of focus in the last 50 years, in fixed 2010 U.S. dollars (Figure by M. C. Roco; data source: http://www.whitehouse.gov/sites/default/files/omb/budget/fy2014/assets/hist09z8.xls)

The proportions for R&D investment have changed over time and by country. Such changes in time are relevant to the convergence–divergence process in S&T. Figure 10.3 provides figures for U.S. Government R&D investment between 1960 and 2003. The proportion of health R&D has increased, while the funds remaining for other areas, including emerging technologies in space, energy, and general sciences, have decreased.

Ethical (ELSI), EHS, and public engagement aspects are still being defined for CKTS, but the attention paid to these aspects has increased. Specific aspects related

Table 10.3 Websites with ELSI content related to CKTS

Center	Website
Center for Nanotechnology in Society at Arizona State University	http://cns.asu.edu
Center for Nanotechnology in Society at University of California, Santa Barbara	http://www.cns.ucsb.edu/
Socio-Technical Integration Research (STIR), Arizona State University	http://cns.asu.edu/stir/
Museum of Science, Boston	http://www.mos.org/
Wikipedia	http://en.wikipedia.org/wiki/

to the role of equity in a connected society have been raised (Chinni and Gimpel 2010; Sunstein 2009). Prospects for regulation and legislation are low for the entire range of fields, even if regulations may be developed for various sectors of emerging technologies as the science and applications are better defined. Despite their importance, aspects related to nomenclature, standards, legislation, and policies yet have to receive adequate attention. International interactions are essential in this exploratory phase of the development of CKTS.

Addressing human development and global grand challenges are important goals of CKTS. Several reference websites are given in Table 10.3.

10.2.4 Return on Investment from Emerging Technologies

Emerging technologies have the promise to bring higher than normal returns on public and private investment because of their transforming and disruptive nature. Such returns also depend on the general socio-economic, private sector–public partnerships, governance methods, and international context. The return on material progress has to accompany efforts to advance moral progress. Below are three illustrations of government investment, in information technology, the Human Genome Project, and the National Nanotechnology Initiative. Space (Sect. 10.8.1) and nuclear energy (Sect. 9.8.2) programs are discussed elsewhere in this report.

The IT-intensive "information-communications-technology-producing" industries grew a total of 16.3 % and contributed nearly 5 % to the overall U.S. gross domestic product (GDP), according to estimates for 2010 by the U.S. Bureau of Economic Analysis (2011). A 2011 study by the McKinsey Global Institute (Du Rausas et al. 2011) found that in 2009, Internet-related activities alone contributed an average of 3.8 % to the U.S. GDP. In comparison to this return on private sector and public investment, the total Federal funding in fiscal year 2010 for networking and IT research and development programs was approximately $4.3 billion, just under 0.03 % of GDP. The return on investment in knowledge areas broadly speaking (nanotechnology, genome, space, nuclear, etc.) is broad and difficult to estimate in product returns only. The direct economic benefits listed above generally do not capture the societal benefits realized from the enabled application of other emerging and converging technologies. The investment of the Federal Government in basic research has been recognized as essential (NRC 2012).

The Human Genome Project (HGP)'s $3.8 billion U.S. Government investment from 1988 to 2003 helped, together with private sector investment, to drive $796 billion in economic impact ($796 billion/ $3.8 billion = 210 times) and the generation of $244 billion in total personal income, according to a study released by Battelle (2011). In 2010 alone, human genome sequencing projects and associated genomics research and industry activity directly and indirectly generated $67 billion in U.S. economic output ($67 billion/$3.8 billion = 18 times in 2010) and supported 310,000 jobs that produced $20 billion in personal income. Genomics-enabled industry also provided $3.7 billion in Federal taxes during 2010 (Battelle 2011).

The U.S. National Nanotechnology Initiative (NNI) has stimulated considerable research in nanotechnology, but its broader economic impacts elude calculation because the field is broad and it is difficult to separate the impact of Government-only R&D investment. With a worldwide public and private investment in 2010 of $18 billion (of which $11 billion was public), the market for nanotechnology products reached $300 billion (Roco et al. 2011). In the United States alone, an investment of $4.1 billion (of which $1.8 billion is from Federal/NNI funds) has led to the production of $110 billion of nanotechnology-enabled products representing over 220,000 jobs (Roco et al. 2011). Without including the operation expenses in production, the annual return is about 110/4.1 or ~25 times that for the U.S. nanotechnology R&D investment and 300/18 or ~17 times that for the world. The average growth of markets and numbers of people involved in nanotechnology from 2001 to 2010 has been about 25 % worldwide (Roco et al. 2011; Roco 2011; PCAST 2012).

The goal of nurturing science discoveries into innovative technologies and products has been pursued by the U.S. Federal Government for decades. The President's Council on Bioethics was established under the National Science and Technology Council (NSTC) of the White House Office of Science and Technology Policy (OSTP) in 2001, replacing an earlier group focused on science. An interagency program on Network and Information Technology Research and Development (NITRD) was created in 1991. The year 2001 saw the creation of the NNI, a prototypical experiment in integrating and synergizing a broad-scale program that engaged multiple academic disciplines and multiple government funding agencies and envisioned widespread technology impacts. While the NNI has had its limitations, it is considered by many to have been successful in fostering the incorporation of multidisciplinary perspectives as well as in making strides in dealing with societal concerns. Cognitive neuroscience is the focus of the U.S. BRAIN initiative (Brain Research through Advancing Innovative Neurotechnologies) which began formal interagency collaboration in April 2013.

Previous evaluations of the impact of science and technology generally have concluded that 50–85 % (an average of 67 %) of increase in productivity and economic progress in the United States and other developed countries is due to science and technology (Solow 1957; PCAST 2012). Better integration of knowledge, technology, and society are expected to improve such results.

Because NSF's traditional panel review of proposals may underestimate interdisciplinary, innovative ideas (high-return, high-risk proposals), in 2011 NSF began experimenting with new approaches to proposal evaluation under the umbrella of Integrated Support Promoting Interdisciplinary Research and Education (INSPIRE); the initial program in this effort is the Creative Research Award for Transformative Interdisciplinary Ventures (CREATIV; http://www.nsf.gov/od/oia/creativ/). NSF has also introduced expanded programs to foster international collaborations.

A growing number of universities have created innovation programs and centers that simultaneously address the scientific and technical transition of a university's discoveries into commercial products and preparing students with the business skill sets necessary to accomplish commercialization. Examples include the Martin Trust Center for MIT Entrepreneurship (1958; http://entrepreneurship.mit.edu/), Stanford's Epicenter (http://epicenter.stanford.edu/; "creating a nation of entrepreneurial engineers"), Berkeley's Lester Center for Entrepreneurship (http://entrepreneurship.berkeley.edu/), Carnegie Mellon's Robert Mehrabian Collaborative Innovation Center (http://www.cmu.edu/corporate/partnerships/cic/), and the University of Southern California's Stevens Center for Innovation (http://stevens.usc.edu/).

10.3 Goals for the Next Decade

10.3.1 Advance Innovation for Economic Productivity

Key ideas are:

- *Supporting innovation* enabled by convergence. New models of facilitating innovation will have to develop that are suitable to a diversity of S&T converging inputs and divergent application outputs.
- *Developing common languages* for communication and standards across general convergence platforms for enabling innovation across areas.
- *Creating capacity to address EHS and ELSI* concerns related to rapidly evolving converging knowledge and technology, and to do so in an integrated way that enables enhancements in both innovation and public value.
- *Placing special emphasis on advancing investment in R&D for emerging technologies* as compared to investing in classical S&T fields. Several Asian countries in the last two decades have increased their CKTS investments—partially using basic research from the West.

Gordon (2012) wrote about the "end of progress" thesis (or the average slowdown of innovation in society), which does need to be considered in the context of current debates about economic policy and public investments. His paper questions the assumption that economic growth is a continuous process that will persist forever (Gordon 2012, 1). There was virtually no growth before 1750 (see Fig. 10.4), and there is no guarantee that growth will continue indefinitely. If indeed the current

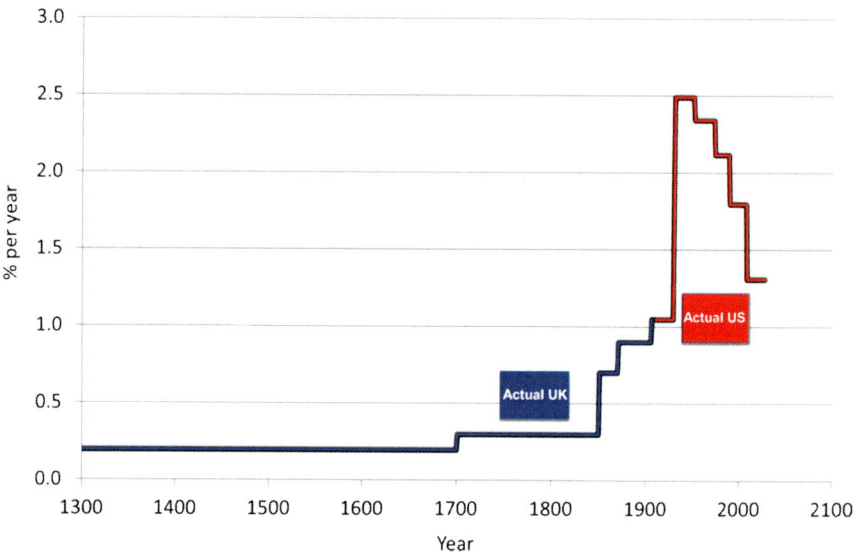

Fig. 10.4 Growth rate in real GDP per capita in the last eight centuries showing "Actual UK" and "Actual US" growth (Gordon 2012, Figure 1; Courtesy of R. J. Gordon)

economic problems are "structural" or the indirect result of decelerating scientific progress, then many of the very expensive proposed solutions will fail. Rather, we must vigorously support work in two kinds of scientific research and engineering innovation: (1) transformative emerging areas like nanotechnology where the possibilities for progress have not yet been exhausted, and (2) efforts to understand the socio-economic consequences and integrate the socio-economic projects aligned with CKTS.

The Gordon thesis is challenged by the position expressed in the *Economist* (2013) that sees the increase in productivity in waves connected to the introduction of a major technology. As an illustration, introduction of electricity after 1890 and of information technology after 1970 correlate with doubling the U.S. labor productivity in about 40 years (*Economist* 2013, Figure 4). In our opinion, convergence would provide the next wave in the conditions of limited natural and investment resources.

10.3.2 Support Converging Technologies to Advance Human Potential and Quality of Life

Key ideas are:

- *Creating new governance arrangements and organizations dealing with CKTS and global issues long-term* is an essential goal for advancing human capacity

(enhanced productivity, learning, active aging) and life security (sustainability, health, security). This is aligned with the general CKTS goal of increasing the focus on people.
- *Extending governance of emerging technologies to include broader public(s)* by expanding public participation and adding suitable governance criteria such as sustainability at local and global levels; access and equity; evaluation of causal relations within the societal system; and managing Earth systems.
- *Redesigning foundational governance systems dealing with long-term and global issues.*
- *Building governance structures (personal and social) that account for the neurological and cognitive biases of the brain.*
- *Addressing ethical issues specific to CKTS* related to human enhancement, human–machine interactions, risk of transhumanism, dislocations in the work force, cost of healthcare, and equal rights and opportunities.

Governing the impact of CKTS on quality-of-life indicators in order to limit the risks and maximize the benefits for people and the environment can take numerous forms, including institutional guidelines, regulations, rules, codes, and laws to monitor and manage the development of converging technologies. Meanwhile the deployment of converging knowledge and technologies can also shape the task of (re)designing foundational governance systems.

With the rapid pace of change and innovation occurring in science and technical applications, government and industry officials face an ever-growing challenge to develop effective standards and regulations in the rapid time frames staked out by today's innovation environment. Legal standards and codes are being ported over from legacy models of science and technology and uncomfortably applied to new business and research models, creating a Procrustean bed of confusion that can impede progress. Given the challenges that governments face in dealing effectively with the pace and complexity of scientific innovation at present, much less the long-term impacts, global issues such as climate change suggest that the redesign of governance may be the most critical endeavor for the long-term survival of human civilization. While they are largely in dispersed and inchoate states, converging technologies and scientific approaches both underscore the importance of attending to governance issues and arrangements and also in some cases point in potentially hopeful directions. In fact, novel governance ideas are emerging from research labs and from ad hoc experiments in governance, especially those coming from virtual communities on the Internet.[1]

Converging technologies are demonstrating potential applicable insights for governance in many ways. Cognitive neuroscience is providing rich and sometimes counterintuitive data on human behavior, decision-making processes, and even the definition of the mind and personhood. These insights can be combined with emerging sciences around organizing systemic structures and bio-inspired processes

---

[1] See http://www.dailykos.com/story/2012/01/24/1057803/-Democracy-Technology for a list of labs, initiatives, and experiments in "democracy technology."

to provide governance designers' ideas for new ways of ordering society at a global level. In addition, new abilities to process vast amounts of data about human behavior are making robust behavior modeling more accurate and useful. Leveraging this data and making it meaningful will be crucially important. Global modeling and modeling of complex systems will bring about new decision-making tools for businesses and governments.

Finally, while there is much overreach in the many imaging studies claiming to locate certain behaviors and traits in the brain, finding the neural correlates of critical human behaviors could help us build metacognitive extensions in order to create better outcomes. For example, understanding our brains' capacities and limitations for foresight will allow us to build governance structures (both personal and social) that account for neurological and cognitive biases that tend to favor short-term and deemphasize long-term thinking. Insights in brain communication may open new possibilities in governance (Chi and Snyder 2011).

The governance of converging technologies and the use of converging technologies to redesign governance will critically influence how we address the grand challenges we face in the "Anthropocene," an era defined by human activities and a future made by human decisions.

10.3.3 Set Grand Challenges in the Vision-Inspired Quadrant

There are two ways of thinking about Pasteur's Quadrant.[2] One way is the application of known science to achieve incremental improvements in existing technology. Another way, recommended for CKTS, is "vision-inspired grand challenges" (see Sect. 4.3.8 and Fig. 4.5), that is, the identification of big problems—ones recognized by society as vital—for which there is no current solution and that require fundamentally new science and technology to solve. Governance has a critical role to play in setting the goals for fundamental research using society's big challenges, besides the approach of curiosity-driven research.

Grand challenges in the area of governance that we will address in the next decade include the following:

- Providing governance for improving productive efficiency by combining emerging technology tools (NBIC) and human-scale platform (such as human–robot interfaces, brain-to-brain and human–machine interactions).
- Addressing global issues such as space exploration, sustainability, and a global communications system in an integrated Earth-scale approach to development. Providing CKTS solutions for a global *sustainable society* and improving overall life security on Earth. Breakthrough technical solutions for protecting and preserving natural resources will help extend the limits of sustainable development.

---

[2] A concept introduced by Donald Stokes (1997) to describe scientific research that is both fundamental and "use-inspired," that takes a systems engineering approach to extending basic S&T understanding but applies it to solution of problems.

- Expanding the role of the mind (imagination, expectation, etc.) in CKTS.
- Reconstructing public healthcare based on nano-/biosensing advances and info/cognitive pattern recognition to create observational capacity for emerging diseases and an ability to anticipate public health problems emerging in large complex systems (e.g., Fisher et al. 2012). Instituting governance of theranostic medicine to integrate diagnostics, therapy, and monitoring within one system.
- Establishing anticipatory governance, including forecasting (Martin 1996). Extending governance of emerging technologies to broaden public participation. Rewriting the social contract between academic research and society; new structures are needed to handle very large problems. Adding criteria to governance that support sustainability at local and global levels.

10.3.4 Design Integrated Socio-Technical Systems

Key ideas are to create:

- Methodology and policy centers for converging technologies research and development.
- Converging knowledge and technologies platform research for emerging technologies and visionary ideas.
- Societal convergence databases and systemics (logical and mathematical paradigms to study systems from a holistic point of view; Bunge 1979; Vester 2008). This is an attempt to develop logical, mathematical, engineering and philosophical paradigms, and information research.

Human activities are increasingly embedded in complex systems that mix social with technical components interacting in complex and dynamic ways (Chap. 7). Two longstanding examples are manufacturing industries and urban transportation systems. A manufacturing enterprise not only manages the complex system that constitutes a factory assembly line but also the supply chain and financial investment systems that make manufacturing possible and the distribution systems that make it profitable. Increasingly, public transportation in urban areas is linking its components through information technology, as riders pay through smart cards (channeling money via the Internet), the location of every bus and train is provided in real time, and coordination centers are responsible for responding effectively to emergencies of many kinds. Given that some sectors of society are already organized as complex socio-technical systems, the questions arise, which others are evolving in the same direction, what innovations are required to make each different system function well, and do we need to manage all of society as single technically convergent system?

Mathematical methods, typically computer-based, can model complex systems, but they do so in a somewhat abstract manner and must work in partnership with several other approaches. Socio-technical systems tend to be assembled from components, for reason of cost, efficiency, and intelligibility, and much innovation involves improvements in a single component. For example, a fleet of busses in an urban transportation system may include a mixture of designs and over time may shift from traditional gasoline engines to natural gas. However, once a system is functioning well, it can be exceedingly difficult to improve it substantially merely by improving components. The most familiar example is a computer operating system, on which plug-and-play software programs may be designed to run; sometimes many of those programs will not run well on a substantially redesigned version of that operating system, and radical improvement of the system could require redesign of each and every piece of software. Thus, conceptual and mathematical approaches to complex systems that serve well to optimize an already-defined system must be supplemented by such other approaches as social science research, evolutionary computing that applies principles from biology to engineering design, and socio-technical laboratories for prototyping entirely novel forms of systems.

10.3.5 Address Deficits in Risk Governance for Next-Generation CKTS Products

Risk governance is an essential component for anticipatory, participatory, and adaptive CKTS. A model to address deficits in risk governance of converging technologies is the Risk Management Escalator and Stakeholder Involvement model (IRGC 2006, Figure 4) that has four progressing stages of application that require distinct risk management and public discourse processes:

For simple systems, where statistical risk analysis can be applied

1. For component-complexity–induced risk, where epistemological discourse and probabilistic risk modeling are necessary
2. For system-uncertainty–induced risk, where reflexive discourse and risk balancing are necessary
3. For ambiguity-induced risk, where participative discourse and risk trade-off analysis and deliberations are necessary

The governance of science and engineering currently involves very complex economic, legal, and management systems; it seems unlikely that the institutions developed decades ago are perfectly suited to today's rapidly changing circumstances. It would not be wise to recommend reforms without first using rigorous methods to assess the current institutions and develop appropriate alternatives. Yet there are clear signs that problems are endemic to the current systems, as illustrated by the public controversies about climate change, healthcare, and national defense investments, all of which have substantial science and engineering components. One oversimplification that dominated past thinking was that public involvement in decisions about technological development was required only when unusual harm might be done, such as health hazards if certain potentially harmful particulate matter were released into the environment in substantial quantities. Otherwise, it was assumed that the financial investment industry and the free market would make the right decisions. Clearly, rapid scientific and technological progress is absolutely essential for the wellbeing of humanity, but a better sense seems to be needed and deliberated of which directions such progress should take.

Two approaches for the social sciences to take can be identified now, and perhaps others can be invented. First, modern methods based on traditional opinion polling and conducted online could measure the changing values of the general public to identify the goals that are important to people, and then experts in the relevant fields of science and technology can determine how to achieve those public priorities. Second, in each area of technical decision-making, ordinary citizens may select professionals, such as academics or leaders in industry, who will serve as their representatives in deliberations in the given area; in making their selections, citizens could consult online blogs by the candidate technical representatives or rely upon advice from opinion leaders in their own community whom they personally trust. The fact that we cannot specify now the exact way these challenges need to be addressed only reinforces the need for imaginative research and active experimentation.

10.3.6 Improve the Cultural Balance Between Collaboration/Harmony and Confrontation, as Informed by the Global Context and by CKTS

There are different interpersonal levels to address when considering governance and CKTS:

- Collaboration and conflict resolution among people from different countries or communities
- Maintaining harmony between country or community groups
- Promoting better interaction between scientists and the general population
- Facilitating convergence between the real and virtual worlds (Bainbridge 2010)
- Enabling better work and personal satisfaction for individuals

The contexts show the urgency and the possibilities: an increasingly crowded planet with smaller buffers between countries or communities; the increased benefits from collaboration facilitated by global information systems and science and technology resources; and the increased destructive power of new tools and technologies.

10.4 Infrastructure Needs

The main infrastructure needs for CKTS are in the following areas:

- Preparing people and tools for convergence, including in formal and informal education settings; establishing multidisciplinary physical infrastructure with measurement and manufacturing capabilities; implementing scientific results in

Table 10.4 Examples of regional and local partnerships

Partnership model	Main location	Specific
Silicon Valley (see Sect. 10.8.2 below)	California	Venture funds
Oregon Nanoscience and Microtechnologies Institute	Oregon	Technology cluster
Albany NanoTech Complex	New York	CKTS industry-government-education
Research Triangle	North Carolina	CKTS university-industry-government
Huntsville Aeronautics	Alabama	CKTS industry lead
Pharmaceutical industry in NE United States	New Jersey	CKTS industry lead
Grenoble Center	France	Nano-bio-electronics
Aachen-IMEC-Eindhoven	Germany-NL	Nano-bio-electronics
Dresden platform	Germany	Nano-bio-electronics
Tsukuba platform	Japan	University-industry-government
Samsung platform	Korea	CKTS industry lead
Nanopolis	China	CKTS industry lead

multiple areas; and advancing a creative, integrative, and innovative culture. An illustration is regional convergence, where geographically grouped partnerships and initiatives are supported. Several examples are shown in Table 10.4.
- New communication methods (Internet, various telecommunications, etc.) and social media as infrastructure for governance.
- Development of common nomenclature and informatics for CKTS, as well as neutral professional or societal "observatories" for dialogue.
- A support infrastructure for public participation.
- Investment into research on methods of convergence, creativity, and innovations in governance.

10.5 R&D Strategies

Several possibilities exist for improving CKTS governance in the global ecosystem:

- Use open-source (including social media) and incentive-based models in governance.
- Implement long-term planning with systemic and international perspectives.
- Build in flexibility in investments to adapt to technological developments.
- Harmonize regulations in high-technology areas, and institute voluntary measures for risk management when the regulations for emerging technologies are not in place.
- Adopt anticipatory, participatory, real-time technology assessment and adaptive governance of nanotechnology. The shift to new generations of CKTS products needs to focus on higher productivity and products not available before, uncertainty in risk management, and making decisions with incomplete information.

- Harmonize global R&D by standardizing principles for merit review and research integrity, sharing resources to increase the scope and global impact of scientific experimentation, exploring new options to share the research output of major scientific infrastructure projects, developing ways to guide the collection, analysis, and distribution of scientific information and "big data," and enhancing transborder mobility of researchers (Suresh 2012).

10.6 Conclusions and Priorities

Increased recognition of the dynamic interactions between scientific, technological, and societal developments associated with converging technologies points to the overall importance of governance as a critical component for maintaining national and regional competitiveness and cooperation in this area. Converging technologies are potentially transformational and therefore offer immense societal benefits but also raise social and ethical concerns. Converging technologies promise an array of innovative products, skills, and solutions, but these must be developed in socially responsive ways in order to ensure that public investments contribute to advancement of the key policy goals of economic strength, societal benefits, and national competitiveness.

These goals cannot be achieved in isolation from one another, and they require developing institutional capacities for facilitating and enhancing the interactions between science and society. Governance refers to the collective capacity for achieving socially desired benefits under complex and changing conditions. This capacity is most robust to the extent that it is distributed across multiple stakeholder groups and consists of multiple instruments, both voluntary (organic) and enforced (hierarchical). The twofold emphasis on innovation and responsibility represents a new frontier in science policy, as is evident in numerous nanotechnology and synthetic biology programs throughout the industrialized world. Leadership in converging technologies will, in large part, depend on continuing to perfect this twofold governance capacity.

The emphasis on governance has evolved over the last decade due to a variety of factors, including the continued effects of regional and global integration, the increasing rate and scope of technological change, and enhanced stakeholder abilities to discriminate credible information. Policy experimentation with different modes of program design, coordination, and evaluation, including public engagement and interdisciplinary collaboration, has generated a variety of governance models and approaches. In order to promote social responsiveness and at the same time reward scientific creativity and innovation, traditional models of knowledge production, translation, and assessment need to be integrated with each other in novel and synergistic ways.

New thinking about governance also brings its own set of challenges. Traditional institutions have a reduced role, being bypassed by social-media-enabled movements. Increased social dependence on complex technological infrastructure and evolving knowledge systems, from national security to international finance, requires input from multiple sources of intelligence that span the expert–lay divide. The convergence of disciplinary techniques, technological platforms, and governance models will also need to maintain a productive balance among competing interests—public and private, economic and ecological, individual and collective, national and international.

To avoid counterproductive regulatory and bureaucratic burdens, effective governance will need to enable researchers and policymakers to act on inputs from a broad array of stakeholders, to resist over-simplistic models of science–technology–society interactions, and to maintain flexibility in the absence of scientific certainty and normative consensus. Rather than asking scientists to supply and citizens to consume a predetermined set of technological outcomes that may fail to be productive, governance should tap individual aspirations that are the hallmark of modern democratic society, allowing these to influence the direction of evolving trajectories. Governance approaches that facilitate social learning and develop fundamental leadership skills can strengthen existing connections among scientific, entrepreneurial, and democratic competencies—connections that have shown signs of strain in recent years. In this way, convergence represents an opportunity not only to develop the technological bases, but also to build the societal foundations, for continued national competitiveness.

In order to strengthen existing linkages among national policy goals, science and innovation programs, and broad-based societal norms, governance of converging technologies will need to be guided by prominent policy criteria. In addition to supporting competitiveness technologically, economically, and strategically, these policy criteria include:

- Sustainable development—respecting the integrity of social, natural, and technological systems
- Individual privacy, especially in light of increased nanotechnology-enabled and digitally transmitted molecular diagnostics
- Human dignity and autonomy in the face of ever more powerful performance enhancements
- International coordination to avoid duplicate research targets in areas such as toxicology and to harmonize emerging regulatory frameworks

Governance arrangements should be informed by tested models that allow for a diversity of approaches, respect multiple sources of insight and innovation, and seek to strengthen rather than bypass underlying connections between scientific expertise, technological innovation, and public norms and values. Nanotechnology has helped test and evolve recent models of anticipatory and participatory governance, in which the humanities and social sciences have been brought to bear both directly and indirectly on science and innovation—from policy formulation to laboratory research—along with expanded roles for expert deliberation and citizen engagement.

For instance, Arizona State University's Center for Nanotechnology in Society has demonstrated the viability of a large-scale social scientific research program to productively engage members of the public at the national scale (Cobb 2011), synergistically integrate innovation and responsibility in the laboratory (Fisher and Mahajan 2010; Schuurbiers 2011), and develop long-term scenarios (Selin 2011) in the areas of nanotechnology, synthetic biology, and other converging technologies. The coordination of the three capacities of foresight, engagement, and integration into a research ensemble provides proof of concept for the transformative role of social science in enabling anticipatory governance of new technologies (Barben et al. 2008; Guston 2008).

In short, the innovative and responsible governance of converging technologies has emerged as a key condition for realizing societal benefits internationally and for advancing regional and national competiveness, even as it presents its own set of challenges and opportunities. As is evident from the early years of the National Nanotechnology Initiative, the United States has already provided an international leadership role in the responsible development of nanotechnology, and it is poised to take its experience into the area of converging technologies. The main priorities are to:

- Create *national centers for societal convergence* (to address knowledge-technology-society policy issues and methodologies such as convergent–divergent cycles, systems approaches, evolutionary approaches, responsible innovation, etc.) with application in areas such as research planning and evaluation, investment policies, healthcare, Earth systems, space, and other areas of national interest.
- Develop data, standards, methods, organizations, systemics, and informatics research to enable societal convergence platforms, and advance their rigorous evaluation, benefits, risks, and governance.
- Guide convergence by higher-purpose criteria such as improvement of economic productivity, human potential, and life security, including sustainable development. Both international coordination and international competitiveness are necessary components.
- Revise rules and regulations to advance individual and group creativity and innovation in convergent processes in the economy as a critical condition for competitiveness.
- Involve a broad array of publics and experts in setting priorities for large-scale CKTS initiatives.
- Adapt traditional institutions that have diminished roles as they are bypassed by social media-enabled movements. Address the opportunities and threats arising from changes in technology and governance roles.
- Support innovative and responsible governance of visionary ideas (see list in Appendix D).

10.7 R&D Impact on Society

- Proactive CKTS governance is essential for obtaining the benefits of the new technologies, limiting their negative implications, and fostering global collaboration. More specific implications are discussed in Chaps. 5, 6, 7, 8, and 9. CKTS governance affects public society at large and also international interactions.

- CKTS will support emerging technologies, general education programs, and social and even philosophical aspects of S&T development.
- EHS and ELSI are determinant factors in the success of emerging technologies and their governance.
- Responsible governance will increase creativity and innovation by offering multiple (multidisciplinary) science paths, application areas, and societal targets.

10.8 Examples of Achievements and Convergence Paradigm Shifts

10.8.1 Spaceflight: Lessons from an Earlier Convergence Platform

Contact person: *Alexander MacDonald, NASA*

The convergence of many technologies was required to make spaceflight possible, including electrical, mechanical, chemical, and computational engineering, and sciences like astronomy and physics. Thus, the history of spaceflight development is a case study in convergence—one that highlights different governance strategies. The process of spaceflight development has extended over centuries, driven largely by the intrinsic motivations of individuals to realize a specific potential future. NBIC convergence is similar in that it also is driven by a specific vision of a potential future that individuals and technical communities hope to see realized. We can divide the historical process of spaceflight development into six stages that may also be found in many other major technology developments:

- *Articulation of the Vision.* In the 1830s, three Americans independently articulated visions for a future that included human spaceflight: Edgar Allan Poe, Edward Everett Hale, and John Leonard Riddell. These visions were expressed in the form of fictional narratives that served as vehicles for the transmission of the idea of spaceflight to others. Late in the nineteenth century, individual scientists and engineers like Robert Goddard foresaw themselves personally enacting the convergence of emerging technologies to create spaceflight systems.
- *Community Formation.* As the convergence of the requisite technologies of spaceflight became possible, communities began to form around the vision as a shared goal that might be achieved through the specific technology of liquid-fuel rockets. As in the case of the American Rocket Society in the United States or the *Verein für Raumschiffahrt* (Society for Space Travel) in Germany, the first significant resources were applied to the problem, through which the physical process of technological convergence began to produce prototype systems.
- *Exchange with Other Interests.* Next, to obtain further resources in order to build operational spaceflight systems, leading groups entered into exchanges with entities that did not necessarily share their intrinsic desire for spaceflight but which saw the value of the technologies for achieving their own objectives—such

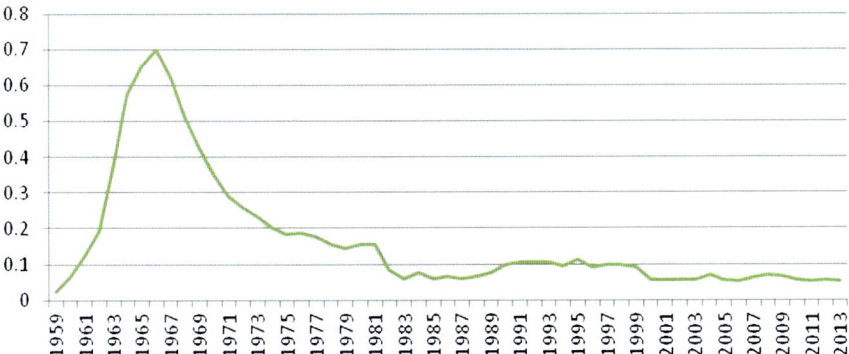

Fig. 10.5 NASA (Space) budget share of U.S. GDP 1959–2013 (Figure by M. C. Roco; data sources: http://www.whitehouse.gov/sites/default/files/omb/budget/fy2014/assets/hist09z8.xls, http://www.whitehouse.gov/sites/default/files/omb/budget/fy2014/assets/hist10z1.xls)

as the desire of German, American, and Soviet militaries for new long-range bombardment capabilities. This convergence for mutual advantage increased funding by orders of magnitude (Fig. 10.5), but also introduced the first governance strategies, which were focused on rapid development and on limiting the transfer of spaceflight technology outside the nation's military-industrial system.

- *Grand Challenge*. In 1961, President John F. Kennedy issued a grand challenge to land a man on the Moon by the end of the decade that greatly accelerated the development of spaceflight, motivated largely by geopolitical competition with the Soviet Union. This required further technological convergence, notably the integration of the digital computer into the Command Module and Lunar Module for guidance, navigation, and control.
- *Maturation*. After completion of the Apollo program, the political value of further spaceflight development waned. Nonetheless, human spaceflight retained enough political value to ensure its continuance, and the spaceflight industry entered a period of maturation where further technological convergence and development was often incremental rather than transformational.
- *Institutional Transformation*. Although there was significant cumulative progress in spaceflight systems during the maturation stage, development was nonetheless considered too slow by some members of the spaceflight community, who were motivated by an expansive vision that included near-term human settlement on other worlds. Beginning in the 1980s, a number of wealthy individuals began to invest significant private-sector capital in the development of commercial spaceflight capabilities to realize that vision.

Although mature, spaceflight technology has achieved only a small fraction of its developers' original hopes. The question faced today by spaceflight and many other mature technologies concerns whether private efforts could be sufficient to spark a renaissance of progress, or whether fundamental Government-led reformation, based on dynamic convergence across many fields of science and technology, would be required.

10.8.2 Convergence Case Study on Innovation: Silicon Valley (The Rainforest)

Contact persons: Greg Horowitt and Victor Hwang, T2 Venture Capital (*Interview*)

The book *The Rainforest: The Secret to Building the Next Silicon Valley* (Hwang and Horowitt 2012) proposes various tools for designing, building, and sustaining rainforests. People learn culture not from top–down instruction, but through actual practice, role modeling, peer-to-peer interaction with diverse partners, feedback mechanisms that penalize bad behavior, and making social contracts explicit. Leaders who can bridge between social networks to bind greater communities together for common action are essential to building and maintaining rainforests. Public subsidies of venture capital are ineffective when fund managers are not culturally attuned to foster symbiotic relationships between investors and investees.

The Rainforest model is more than a metaphor. Innovation ecosystems are not merely like biological systems; they are biological systems. And talent, ideas, and capital are the nutrients moving through this system. Certain social behaviors are essential to allowing the movement of those nutrients to be even freer—as they are in rainforests. It is these human networks, properly formed, that are the key to generating sustainable innovation.

10.8.3 SRC and the Semiconductor Industry as a Model for Multidimensional Convergence

Contact person: Celia Merzbacher and Ralph Cavin, Semiconductor Research Corporation

The Semiconductor Research Corporation (SRC) was established 30 years ago to help the U.S. semiconductor industry to better compete in the global market. The model that was created by its visionary founders called for SRC participants—at first only from industry and later also from government—to pool resources to fund basic university research that addresses the semiconductor industry's long-term technology challenges.[3] Since its inception, SRC has had three overarching objectives: (1) define relevant research directions, (2) explore potentially important new technologies, and (3) generate a pool of experienced faculty and relevantly educated students.

In the process of achieving its goals, SRC enhances and expedites convergence—the coming together of two or more distinct entities or phenomena—*in multiple dimensions*. Figure 10.6 shows the dimensions of convergence, described in more detail below, among SRC stakeholders, including member companies, State and Federal government participants, universities, and society at large. Arrows indicating the flow of information and intellectual property (IP), people, and funding illustrate

---

[3] Cavin et al. (1989) provide an overview of SRC's early organization, operation, and research results.

Fig. 10.6 Schematic diagram showing the relationships among the key stakeholders in SRC research: SRC and its member companies, government partners, universities, and society at large. *Arrows* indicate the flow of information in various forms and forums, people, and funding (Courtesy of Ralph Cavin, SRC)

the strength and two-way nature of the relationship between SRC and its members and the university community. The knowledge, educated students, and IP flow to and impact not only the member companies, but also other companies and sectors of society.

The convergence dimensions of SRC's model are:

- *Convergence of discovery and innovation.* The semiconductor industry has been driven for decades to continue the trend known as Moore's Law, which states that the number of transistors per computer chip doubles roughly every 2 years. SRC works with its member companies to define fundamental research that addresses technical barriers and thereby enables continued technological progress. The portfolio of selected research is guided by industry experts, and results are extracted in near real time and delivered to the members. Many current technologies (strained silicon, high-k gate dielectrics, copper interconnects, lead-free packaging, etc.) were the subject of SRC research years in advance of their use by industry. The connection between basic research and practical applications is at the heart of SRC.

Much of the research funded by SRC addresses technology challenges described in the International Technology Roadmap for Semiconductors (http://www.itrs.net). Since the 1990s, this regularly updated document has defined challenges facing the industry in a 15-year timeframe. Many of the longest-term challenges have no known solution and require fundamental "out of the box" research. These are the areas where SRC is focused.

- *Convergence of research and education.* SRC funds leading-edge research aimed at important industry problems. Such real-world challenges attract high-quality faculty researchers, who in turn attract outstanding students. Each year, SRC supports approximately 1,500 students, who not only participate in the research as part of their education, but who also regularly interact with industry engineers and scientists. Upon graduation, most SRC students work in semiconductor-related fields in industry or academia and are able to "hit the ground running."
- *Convergence of science and engineering.* Addressing many of the challenges facing industry requires multidisciplinary and multiscale approaches. SRC supports research that ranges from materials and device science to advanced manufacturing and design tools, involving researchers from chemistry, physics, and engineering. SRC seeks ideas that address precompetitive industry challenges from the broad science and engineering university community. After research is underway, periodic research reviews bring together investigators working on related projects to share results and to obtain industry feedback. Disciplinary boundaries are unimportant in SRC decision-making.
- *Convergence of industry and academia.* Industry and academia have distinct missions and modes of operation. The mission of industry is to add value and create wealth, ultimately growing the economy; the mission of academia is education and advancing knowledge. Industry rewards profits and growth; academia rewards good research as measured by publication, recognition by peers, and ability to obtain research funding. Industry tends to operate on a short time frame and generally treats information as proprietary. Academics typically share research results openly and operate on the time scale of a graduate student's Ph.D. research project (~3 years). SRC serves as a bridge between these two worlds, working with industry to identify longer-term research problems that are suited to university research, providing academics the right to publish their results, ensuring SRC members have necessary IP rights, and providing in-person and electronic mechanisms for industry–university interaction.
- *Convergence of U.S. and international research(ers).* Today, research expertise and excellence is distributed around the globe, and companies, including SRC members, have operations worldwide. Although restricted to U.S. universities at the outset, since 2000 SRC has accepted and funded proposals from institutions worldwide, thereby expanding the academic research enterprise focused on semiconductor research. By end of 2012, SRC had funded 86 projects at universities in 26 countries outside the United States. Researchers at universities outside the United States are integrated into the overall program, interacting and collaborating with other SRC-funded academics and with member company technologists.

- *Convergence of students and mentors.* A critical element—and benefit—of SRC research is the enhanced experience of students. At the task or project level, at least one industry expert serves as a liaison, guiding the research in near real time and mentoring the students. Industry mentors help students appreciate the industry perspective and needs motivating their research and open their eyes to careers in industry, arranging for internships and other opportunities. Students interact with many more industry representatives at research reviews and an annual technical conference that includes students from across all SRC research programs.

SRC Outcomes (Divergence of Results)

The impact and outcomes of SRC after 30 years are both quantitative and qualitative. Since 1982, SRC has directed more than $1.4 billion in university research. In addition, SRC staff and industry members have participated in proposal reviews, workshops, and other activities that influence how Federal funds are spent. Through its investments, SRC has built a substantial network of university researchers focused on semiconductor-related research. When SRC was created in 1982, fewer than 100 academic researchers were doing research relevant to the semiconductor industry. Today, SRC supports approximately 2,000 faculty and student researchers annually. To date, a total of more than 9,000 graduate students and 2,000 faculty members have been supported at over 200 universities worldwide. A number of faculty researchers have received SRC funding for many years, in some cases going back to their own graduate student research. The numerous relationships between industry technologists and faculty researchers are a less readily quantified yet extremely valuable asset, providing industry with access to leading experts and providing academia with access to real world experience, and on occasion, to samples and facilities.

Of high value to industry is the pipeline of relevantly educated scientists and engineers. Upon graduation, approximately half of SRC students take a first position at a member company. About 30 % go to other semiconductor-related industry positions, and 15 % go to academia or government labs.

SRC-funded research also produces numerous technical publications in peer-reviewed journals. A measure of the impact of such papers is the number of citations by others, and a measure of potential commercial value is the fraction of citations from papers by authors from industry. At least 210 papers reporting SRC-funded research have received more than 100 citations, and some have surpassed 1,000 citations. Of those receiving more than 100 citations, almost two thirds received at least 15 % of the citations from industry-authored papers.

Another measure of research output is patents. SRC's IP policy is primarily defensive, allowing universities to own the rights to any resulting intellectual property, including patents, while retaining a nonexclusive, paid-up, royalty-free license for all members. This policy insures members have the freedom to operate. When an invention is made, SRC, in consultation with the members, decides whether a patent application should be filed, and if so, it pays the associated costs. By 2012, SRC research has resulted in nearly 400 patents.

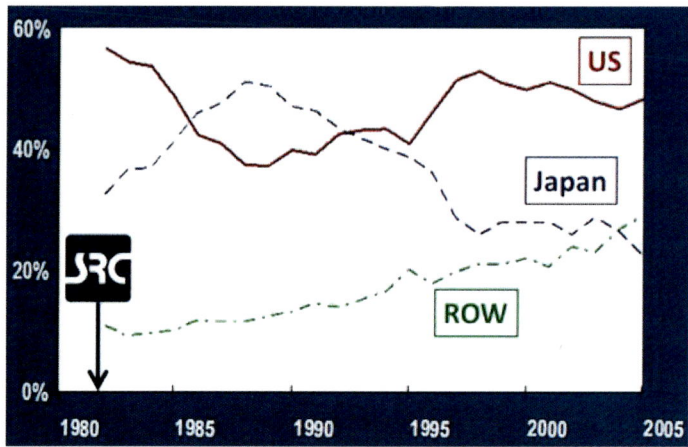

Fig. 10.7 Global semiconductor market share held by companies headquartered in the United States, Japan, and the rest of the world (ROW) in 1980–2005 (Courtesy of Semiconductor Industry Association)

Despite the open publication policy and the IP rights provided by SRC contracts, at least 25 startup companies have roots in SRC research performed at universities across the country. The companies provide products and services that range from software for chip design to specialized metrology tools. Several have been acquired, in some cases by an SRC member company. These new businesses are built upon innovative solutions and create both value and jobs.

The direct economic impact of SRC on the semiconductor industry is difficult to quantify. Despite notoriously short product lifecycles, as in other sectors it is typically on the order of 12 years from the time a new material or technology is discovered (i.e., the basic research phase) to when it is incorporated in a commercial semiconductor product or process. Even SRC research shows a similar research-to-product transition time (Herr and Zhirnov 2004). SRC was established in part to help address the decline in the world semiconductor market share held by U.S. companies. Within 10 years after the creation of SRC, the trend reversed, and since the late 1990s the U.S. semiconductor industry has held about 50 % of the global market (Fig. 10.7). During this time, SRC has continuously produced research results and a relevantly educated workforce to which SRC members—which until 2000 were limited to U.S. companies—have enhanced access.

SRC Impact Beyond the Semiconductor Industry

Although established by and for the semiconductor industry, SRC has had broader impact. SRC produces research results that are disseminated broadly in scientific and technical publications and presentations at conferences and workshops. It provides top-notch graduates who follow careers not only in the semiconductor industry, but also in other sectors and in academia and government. Today, SRC alumni are science and engineering leaders at top universities and in businesses ranging from large technology companies to startups.

To the extent that SRC has helped the semiconductor industry to continue along the trend defined by Moore's Law, it has contributed to the broad impact of the industry across the economic landscape. Semiconductors are at the heart of information technology, making the Internet and the "knowledge-based economy" possible. Semiconductors provide the "intelligence" in devices and systems used in medicine, energy generation and use, and security. Moreover, the use of semiconductors in a host of products and services has enabled increased productivity and growth in virtually every sector. Based on analysis of data collected by the Department of Commerce from 1960 to 2007, semiconductors accounted for nearly 30 % of economic innovation overall and correlated with 37 % growth in labor productivity in the Communications Equipment industry, 25 % growth in the Other Electronic Products sector, and 14 % growth in Educational Services (Samuels 2012).

Finally, the success of SRC and its various programs has been evaluated and held up as an example for public–private partnership and research management for other sectors to consider. The President's Council of Advisors on Science and Technology commended in particular the Nanoelectronics Research Initiative, a subsidiary of SRC, in its 2010 review of the National Nanotechnology Initiative (PCAST 2010). More recently, a Harvard Kennedy School report on energy innovation concludes that the strong connection between industry and academia created and managed by SRC contributes to technology relevance and private sector uptake (Diaz Anadon et al. 2011). These reports acknowledge the accomplishments of SRC and suggest that a similar approach could improve the efficiency and effectiveness of other (government-funded) research programs.

Where Is Convergence Taking the Semiconductor Industry?

In the near term, convergence is creating opportunities for the semiconductor industry. Computing power that is smaller, faster, and cheaper makes possible new products and new capabilities. Examples include medical diagnostic tools that allow physicians to see inside the body and automobile systems that detect when an accident is imminent and take action. Such advanced electronics require much greater convergence during the design process between the semiconductor and biomedical or automotive engineers. As Moore's Law and the ability to increase performance by scaling to smaller dimensions reaches physical limits, SRC seeks possible new approaches for advancing the power and utility of what today is referred to as "semiconductor technology" by supporting research to explore new materials and physics to store and transmit information.

In the longer term, convergence among scientific and engineering disciplines points the way to the future of the semiconductor industry. Many have noted the efficiency of biological systems, comparing the human brain that consumes 20 W

Table 10.5 Share of total global R&D spending, by country or region

	2011 (%)	2012 (%)	2013 estimated (%)	Difference 2012 vs. 2011 (%)
Americas	34.8	34.3	33.8	−0.5
U.S.	29.6	29.0	28.3	−0.6
Asia	34.9	36.0	37.1	+1.1
Japan	11.2	11.1	10.8	−0.1
China	12.7	13.71	14.7	+1.01
India	2.8	2.8	3.0	+0.
Europe	24.8	24.5		−0.3
Rest of World	5.7	5.7	5.7	+0.0

and a high-performance computer that requires nearly 10 MW of power. Of course humans and computers are optimized for different types of tasks, e.g., facial recognition vs. numerical computation. But can computation take advantage of some of Nature's strategies for signal processing and communication, as well as for extracting energy from the environment? Advances at the convergence of biology, electronics, and information technology offer prospects for novel cyber-physical systems that connect people with each other, the environment, and information in order to provide on a global scale everything from improved security, energy, and transportation systems to accessible and affordable healthcare.

Conclusion

The semiconductor industry provides tools that grow the economy. It has continued to improve those tools over time, in part by its investment in research for the future; the U.S. semiconductor industry invests on average 17 % of sales in research annually. A component of that investment is through SRC, which serves as the most forward-looking arm of industry research. SRC plays a key role in bringing industry together to define, support, and guide basic precompetitive university research, in coordinating multidisciplinary academic researchers, and in delivering results and outstanding graduates to its members. In the process, SRC has helped the semiconductor industry to continue advancing the power of computers and intelligent systems. With convergence among disciplines and along the innovation pathway becoming increasingly crucial to economic and societal progress, consortia like SRC can play a vital role in bridging boundaries in multiple dimensions.

10.8.4 Global Trends in R&D Investment

Contact person: *Martin Grueber, Battelle*

Table 10.5 shows the percentage contributions of various countries and regions to the annual global R&D spending, and their respective changes between 2011 and 2012 (based on Battelle 2012; Grueber and Studt 2012).

A separate study shows that of all students who studied abroad in 2000, 23 % chose to study in the United States. By 2009, this had declined to 18 % (OECD 2011).

These trends are expected to be significantly affected by the adoption of emerging technologies and convergence (CKTS) in the respective economies.

10.8.5 Examples of NSF-Funded Research Projects on NBIC (Converging Technologies) 2001–2012

Contact persons: M. C. Roco, National Science Foundation

Table 10.6 illustrates various modes of support of NBIC awards at the U.S. National Science Foundation.

Since 2010, the NBIC awards with at least two components per award have represented about 5 % of total NSF awards.

10.8.6 Public Participation and Innovation Ecosystems for Convergence

Contact person: David M. Berube, North Carolina State University

Public participation in science and technology policy decision-making has been asserted by many contemporary critics to be a tenet of public sphere theory. Contemporary participation exercises come in many forms: planned and unplanned. Planned exercises run from scheduled elections, which might include visionary technological issues to consensus conferences and citizen juries. Unplanned exercises are instantaneous discussions and dialogues at work stations or in church basements. And somewhere in the middle are consumption patterns (we buy what we want and exercise our choices in that fashion). Such exercises can be inorganic or organic (Gehrke 2008; Breau and Brook 2007). Inorganic exercises involve formal meetings for which participants are recruited and often paid. Organic exercises involve reaching out to the public, where they already meet, to join in on their agenda. An interesting variant of this is the science café, which mixes both—participants gather to listen, discuss, and argue in an untraditional venue such as a restaurant or a pub. While inorganic participation has a clear and often rigid agenda, the organic form is fluid, accommodating, and "messy".

The convergence of knowledge and technology could have profound impacts on who we become and how we fit in the general order of nature and the cosmos. The impact sets include a range of effects, such as developments in computation and improvements in health and well-being. On a different scale, the sets might also include off-planet development, significant life prolongation, and transhumanist evolution. The first set is less problematic for the public, whereas the second set can conflict with beliefs and attitudes, hence, confirmation conflicts (see above). This potential for substantive impacts of a defining order demands participation by

Table 10.6 Examples of awards made by NSF since 2001

Year	Award number	Title	Award	Principal investigator	Institution
2006	550169	Developing Ontological Schema Training Methods to Help Students Develop Scientifically Accurate Mental Models of Engineering Concepts	$755,163	Miller, Ronald	Colorado School of Mines
2010	1002410	Nanotechnology from Basic Science to Emerging Applications: Institute for Functional Nanomaterials (IFN)	$20,000,000	Weiner, Brad	University of Puerto Rico
2012	1160483	NSF Nanosystems Engineering Research Center for Advanced Self-Powered Systems of Integrated Sensors and Technologies (ASSIST)	$18,499,997	Misra, Veena	North Carolina State University
2012	1241701	Center for NanoBiotechnology Research	$4,995,710	Singh, Shree	Alabama State University
2004	335765	NNIN: National Nanotechnology Infrastructure Network	$152,762,150	Tiwari, Sandip	Cornell University
2010	936384	Operation of the Cornell High Energy Synchrotron Source (CHESS)	$77,586,038	Gruner, Sol	Cornell University
2003	310717	Engineering Research Center for Extreme Ultraviolet Science and Technology	$40,726,563	Rocca, Jorge	Colorado State University
2006	618647	Collaborative Research: Institutionalizing a Reform Curriculum in Large Universities	$599,961	Haugan, Mark	Purdue University
2001	122419	ITR/SY: Center for Bits and Atoms	$25,326,990	Gershenfeld, Neil	Massachusetts Inst. of Technology
2010	939514	Center for Energy Efficient Electronics Science (Center for E3S)	$25,000,000	Yablonovitch, Eli	University of California–Berkeley
2011	1125565	Institute for Quantum Information and Matter (IQIM)	$11,400,000	Kimble, H.	California Institute of Technology
2004	0425780	NSEC Nano-bio	$11,426,000	Bonnell, Dawn	University of Pennsylvania

as many stakeholders as possible. Members of the public will need to reexamine their sensibilities, if not their beliefs and attitudes sets, to accommodate life and definitional changes. If the investments in time, psyche, and funding toward technological convergence are to become manifest, public support, or at least acquiescence, is needed, especially if products of convergence end up as market goods that need to be consumed.

The assumption that the convergence of knowledge and technology is affected by societal/public needs is partially true. There are calls for unprecedented solutions to especially challenging or sticky (von Hippel 1994) problems, such as climate change (Nisbet 2009) and infrastructure shortfalls (Spence 2011). When a social problem surfaces that cannot be resolved by nature or humanity's plodding evolution of solutions, there are demands for "new" ways to resolve issues, and especially those steeped in uncertainty and speculation, such as synthetic biology (Armstrong 2012; Keck Futures Initiative 2009).

While it might be true that the causes of some of society's most vexing problems may be society itself, one reaches a point where that argument does not provide solace. How we got where we are is less important than how we get out of the predicaments we may have created. We must accept responsibility for these problems we created and try to come together to find ways to solve them. While technological fixes are not necessarily the best answer to all technological problems (Tenner 1997), they do tend to be appropriate for many. In an effort to resolve societal conundrums, we attempt to merge what we know and what we can do, in turn spurring technological evolution. While we may produce another set of problems as well as solutions, the costs of inaction and retrogression are simply too great. Thus, if the best option at hand is to rise to the occasion and move forward, human society remains an important partner in the process, and its members deserve to know what is at stake and to help to choose the directions we take (Sarewitz and Nelson 2008).

On the other hand, critics contend that convergence is the product of a few stakeholders attempting to control resources and wealth to forward a more self-serving agenda, rule by mega-elites (ETC Group 2003). It is true that those who can afford to move ahead, shedding externalities like financial resources that might impede their momentum because they often can afford to aggrandize both power and wealth. Whether this case can withstand scrutiny or not (Baumol 1986; De Long 1988; Baumol and Wolff 1988), it may be prudent to find as many ways as we can to enable the whole set of stakeholders to participate in understanding as well as advancing a public convergence agenda that benefits as many stakeholders as possible, as much as possible, and as equally as possible (ETAG 2005). While irregularities and inconsistencies may still erupt, participation may offer the best solution to minimizing public losses and maximizing public gains.

Understandably, the public will demand that immediate problems are not forgotten as we advance an agenda involving long-term problems, and this is an important cost to development of all sorts. Participating publics want solutions to immediate issues, like housing, employment, and other basic needs. While this mindset does not preclude longer-term planning, it limits it. Anyone perpetrating in a long-term convergent technological set of inquiry as well as policies will need to deal with this reality. Members of the public are reluctant to sacrifice their present for someone else's contrivance of their future. As such, public participation activities involving convergence are prickly.

Consider the convergence platform associated with computation and digital media. We have discovered that the personal computer in every home has risen to the necessity level of home heating and cooling systems. This phenomenon is evident throughout much of the West and extends to the developed cultures of the East and the South as well. While we hear voices calling out to warn us of its effects in defining who we are (Carr 2010) and bemoan the birth of a "search engine society" (Burgess and Green 2009; Halavais 2008; Hassan 2008), except for students in digital critical theory and some technology cognoscenti, these voices are seldom heard by the public. The public seems more or less to have acquiesced. Whether they will respond similarly to convergence is unclear.

Finally, and importantly, convergence should contribute new platforms for participation (Mossberger et al. 2007; West 2011). Innovation networks characterize much of the literature on technological convergence. They are overlapping and intersecting visions of how knowledge and its products iteratively interact (Burke 2011). These ecosystems have nodes dedicated to financial inputs, sustainability values, technological breakthroughs, and educational values, among others. These are authentic axiologies (Rescher 2005), or value systems. The business world has learned how to tap the public via tools like the Internet to help them decide what products the public needs and wants as well as how to improve or even remove products that are in their inventory and on the market. The feedback loops work best when the range of participation is very broad. Convergence affects a plethora of human beliefs, attitudes, and values. How public participation interacts with it can be described as an innovation ecosystem (Adner 2006; Anthony et al. 2006).

In general, innovation is driven by a battery of motivations, competing innovators, and a general awareness of separating rediscoveries from truly innovative acts. In turn, innovation enters the overall agenda of many different stakeholders who comment and interact by specifying how the innovative act interacts with their needs and wants. Responding to stakeholders' input, innovators assess, adapt, and adjust the innovative processes and products to improve their interactions with as many stakeholders as practical. Next, stakeholders reassess and reevaluate the processes and products to determine exigency (importance) and salience (relevance) (Bitzer 1968). In response, innovators reassess, readapt, and readjust the innovation before it becomes marketable. Further developments associated with the innovative act, process, and product renter the cycle, which is iterative (Fig. 10.8).

Innovative ecosystems cannot be ignored, and those who ignore them do so at their own peril. "In a world where the rate of initiative and change continues to accelerate, where expertise and ideas are more distributed than ever, enterprises that do not feel in their day-to-day activities the expansion of their ecosystem are probably falling behind on innovation and growth opportunities" (Cramer 2011). Furthermore, "an innovation ecosystem could be envisaged as a system which supports the birth and growth of innovative activities in a self-sustaining manner" (Dasgupta 2010). Simply put, given investments, the system functions as

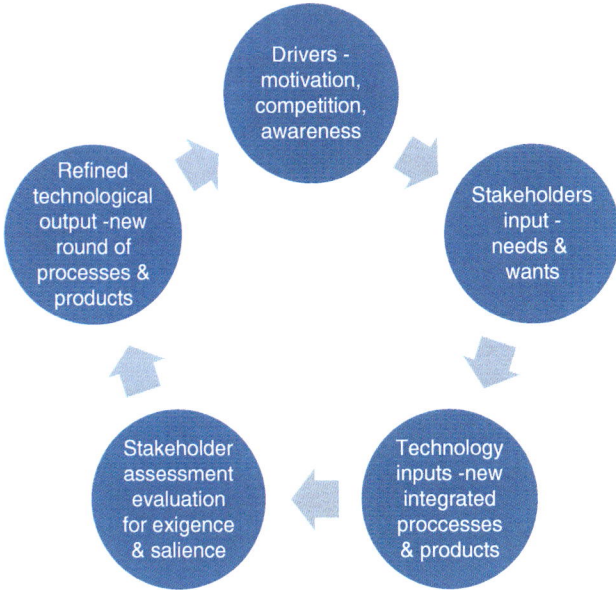

Fig. 10.8 Innovation ecosystem and stakeholders (Courtesy of D. M. Berube)

a working economy to perpetuate itself. Investments made today should percolate through the system, magnifying subsequent inputs and outputs. Consequently, participation in innovation ecosystems becomes essential to prevent the self-perpetuating nature of the ecosystem from sidelining much of the public from its benefits (Collier 2007).

10.8.7 Global Risk Governance Through Managing Resources at Multiple Scales

Contact person: *Marietta L. Baba, Michigan State University*

A new trend in risk governance is incorporating polycentric global commons. The risk governance framework for CKTS could be enhanced by incorporating the work of Nobel Laureate Elinor Ostrom on economic governance of the commons. She shows that groups of users can develop and implement mechanisms (e.g., rules) for managing common economic resources to "sustain tolerable outcomes" (Ostrom 2009). This work has been extended from the scale of the "local" to the "global" through the concept of "polycentric governance" (that is, managing resources at multiple scales; Ostrom 2010). Ostrom's design principles for governance of the commons have been elaborated to include emerging technologies (Stern 2011; nanotechnology is specifically named).

Different dimensions of the convergence that consider interactions at multiple scales must be included for this purpose. Cognitive science is a successful academic program that is broadly interdisciplinary, linking several different natural science and social science disciplines, driven by the need for the knowledge from those disciplines and making links in society. This field recognizes that cognition is a process that takes place not only in the (individual) brain but also between and amongst brains and minds; that is, cognition develops within and through social interactions between people—cognition is in part social; e.g., cultural schemas are shared (Quinn 2005). This is why CKTS must incorporate social science and humanities, because humans are fundamentally social beings.

10.8.8 Convergence Aspects of the EU Horizon 2020 Vision

Contact person: *Christos Tokamanis, European Commission*

The CKTS workshops findings and report has commonalities with the European Commission's proposal for a multi-annual 2014–2020 framework programme of research and innovation activities called "Horizon 2020" (H2020). The points of "convergence" between this CKTS study and the H2020 proposal (April 2013) can seen in the brief description below.

The priorities of H2020 are—in line with the Europe 2020 strategy and the Flagship Innovation Union—short-to-medium-term oriented, and they focus on growth, competitiveness, solutions for society's challenges, and employment. H2020 has three pillars: generator of knowledge (excellent science), technology developer (industrial leadership), and preparing the building blocks for answering the societal challenges. These pillars may correspond to the creative phase, integration phase, and application divergence phase in the convergence–divergence model (Fig. 4.1 in Chap. 4).

A. *The Pillar "Excellent Science"*

Its structure is made up of generating knowledge bottom-up (European Research Council, ERC) and top-down (Frontier Emerging Technologies, FET), as well as infrastructure and training.

The core of the generation of fundamentally new knowledge and technologies in H2020 is concentrated in the pillar "Excellent Science" (about 33 % of the H2020 budget[4]) which focuses on fundamental research, infrastructures, and the development of scientific talent. The ERC does not foresee any thematic orientation. The proposed FET scheme aims at fundamental research with the potential for long-term applications (it foresees a funding of about three billion Euros for 7 years). It comprises three main parts with different kinds of focus: (1) FET Open, very broad, with potential for break-through results; (2) FET

---

[4] The budget data are taken from the COM-Proposal for HORIZON 2020 (COM(2011)811), 30.11.2011, and can be strongly modified due to readjustment of the budget due to negotiations of MFF 2014–2020.

Proactive, which is thematically more focused on the goals for long-term applications in the topical field that are not yet ready for inclusion in industry research roadmaps; and (3) recent FET Flagships, preparing the ground by long-term, large-scale research on high-potential topics

B. *The Pillar "Industrial Leadership"*
The pillar "Industrial Leadership" (24 % of the proposed budget) means leadership in enabling and industrial technologies, the major component of which is key enabling technologies (KETs): nanotechnology, nanoelectronics, photonics, biotechnology, materials, and (cross-cutting) manufacturing. KETs (about € 5.8 billion in 7 years) aim at improving the connection between research and applications, with a strong accent on application and demonstration. Strong private sector involvement, including public–private partnerships, in such activities will be a prerequisite. The demand-oriented activities "Access to Risk Finance" and "Innovation in SMEs" complement the R&D&I [research, development, and innovation] activities and do not include long-term or fundamental research. One focus of this pillar is "cross-cutting key enabling technologies, which will enhance product competitiveness and impact and stimulate growth and jobs as well as provide new opportunities to tackle societal challenges" leading to a "joint work programme for cross-cutting KETs activities." In the cross-cutting activities, however, impetus can be expected for topics in fundamental research.

As examples of convergence, the ICT [information and communication technologies] section of this pillar foresees that "a number of activity lines will target ICT industrial and technological leadership challenges *along the whole value chain* and cover generic ICT research and innovation agendas." The biotechnology section states that, "The objective is to lay the foundations for the European industry to stay at the front line of innovation, *also in the medium and long term. It encompasses the development of emerging technology areas* such as synthetic biology, bioinformatics and systems biology, as well as exploiting the convergence with other enabling technologies such as nanotechnology (e.g., bionanotechnology) and ICT (e.g., bioelectronics), and engineering technology." In order to assure the inflow of fresh ideas with a breakthrough potential in the area of key technologies, however, a well-conceived link is established in both directions, i.e., the knowledge and information flow from Excellent Science to KET and vice versa would be beneficial, particularly in the medium-to-long term.

C. *The Pillar "Societal Challenges"*
The third pillar, "Societal Challenges" (43 % of the budget), is strongly oriented towards societal challenges. The broad lines of the activities and specific objectives are:

- Improving lifelong health and well-being
- Securing sufficient supplies of safe and high-quality food and other bio-based products, by developing productive and resource-efficient primary production systems, fostering related ecosystem services, alongside competitive and low-carbon supply chains

- Making the transition to a reliable, sustainable, and competitive energy system in the face of increasing resource scarcity, increasing energy needs, and climate change
- Achieving a European transport system that is resource-efficient, environmentally friendly, safe, and seamless for the benefit of citizens, the economy, and society
- Achieving a resource-efficient and climate-change-resilient economy and a sustainable supply of raw materials in order to meet the needs of a growing global population within the sustainable limits of the planet's natural resources
- Fostering inclusive, innovative, and secure European societies in a context of unprecedented transformations and growing global interdependencies

10.8.9 Innovative R&D System for Converging Technologies (Korea)

Contact: Heyoung Yang, Korea Institute of S&T Evaluation and Planning (KISTEP), Korea

The proposed organization for governance of converging technologies has been informed by the NBIC studies since 2001 and the 2012 NBIC2 workshop in Korea.

In recent years, the goals of science and technology innovation policy include not only contributing to economic development but also satisfying general societal needs. A recent study concerning future society shows that closely related keyword clusters can be used to draw an "issue-keyword network" (Fig. 10.9).[5] These clusters designate future society issues and needs and can be used to trigger the creation of new user-friendly converging systems, assembling a variety of technologies at the beginning stages.

These technologies are dynamically converging to the new technology and a system with shared goals, by breaking boundaries between them. In addition, the characteristics of converging technologies might differ between nations since their societies' issues and needs might vary depending on their national situations. At the same time, the time dependency of an issue keyword network makes us deduce the dynamic characteristics of converging technologies due to changes over time in the issues and needs that are identified.[6]

---

[5] Based on keywords from trend analysis, the frequencies for co-word pairs between keywords appearing in Google documents (2007–2009) have been counted. The number of counts designates the strength of connectivity between the keywords, and this can be shown as in Fig. 10.10 utilizing the network drawing program, "Netminer".

[6] As compared with the results of the same analysis for Google documents during 2004–2006, megatrend which is related to the formation of the clusters does not change, but the degree of the connectivity between the issues does change, meaning that there are emerging hot issues in recent years such as climate change–health, climate change–global governance and social network–healthcare connections.

10 Innovative and Responsible Governance of Converging Technologies

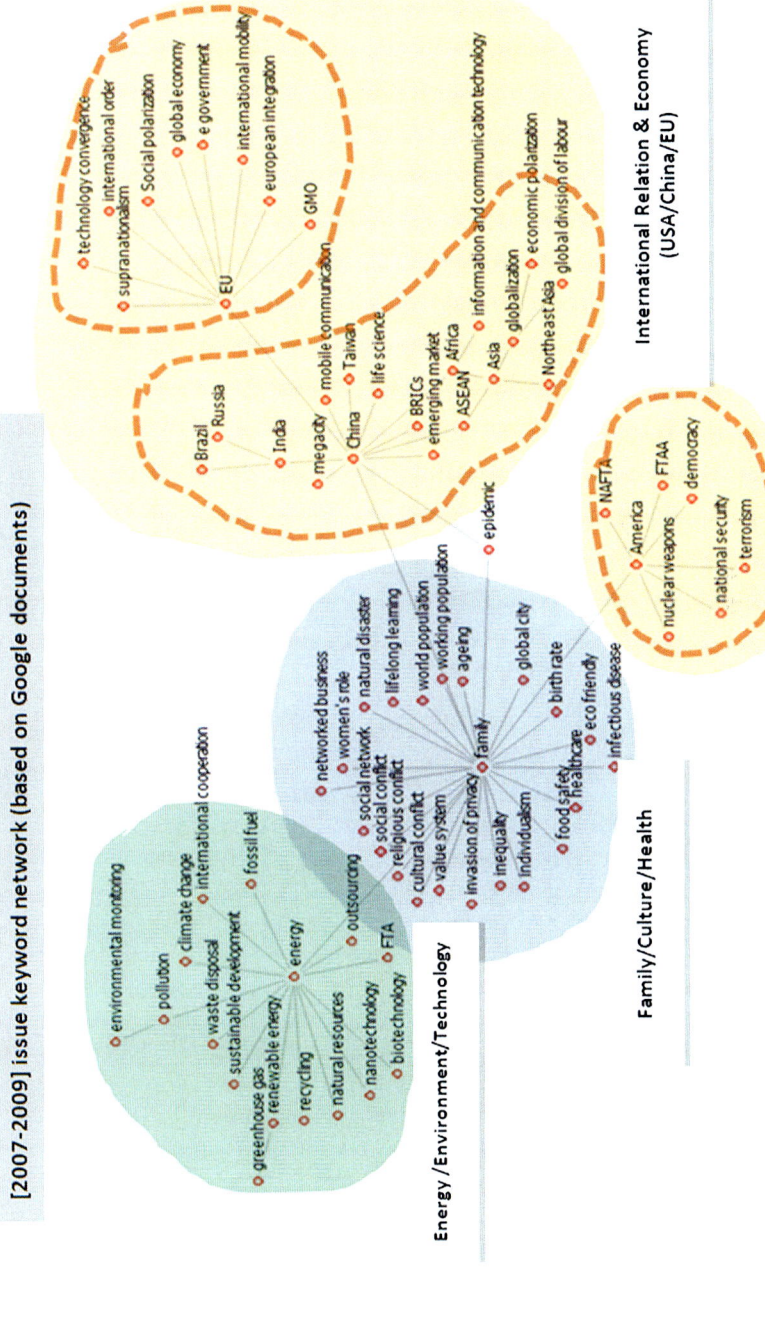

Fig. 10.9 Issues and needs for future society (Yang 2012: Courtesy of H. Yang, KISTEP)

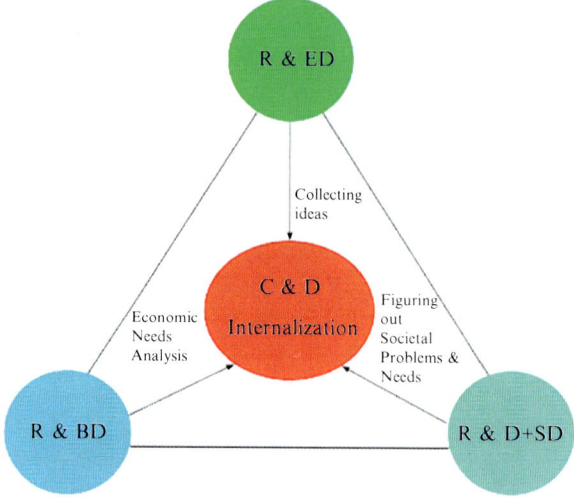

Fig. 10.10 Innovative "organization-friendly" open-cooperation R&D system for converging technologies (Based on Jeon and Jeong 2010, 23, ©ETRI; used by permission)

In this context, there must be a discussion on how to achieve an innovative R&D system that considers the characteristics and the trends of convergence. Especially, to attain policy goals such as economic development, improved quality of life, and sustainable development, a holistic innovation policy and integrated approach, such as multisectional policy plans, should be established that untangles the complicated and interrelated socio-economic needs and problems.

From the viewpoint of organizational theory, the collaboration system between government ministries and R&D institutes and organization units should be built to facilitate spontaneous and intimate cooperation, breaking up "sectionalism." Therefore, the top priority for R&D investment is to build a research collaboration program. Furthermore, the flexibility of research organizations has to be secured, which will enhance the mobility of human resources and creative environments.

Current regulations and laws related to converging technology need to be gradually reformed. The introduction of an "open-innovation R&D system" such as an organization-friendly R&D eco-system should be considered (Fig. 10.10). This ecosystem should integrate the R&ED (research and education development) system to emphasize collecting innovative and original ideas, the R&BD (research & business development) system to focus on the pragmatic business needs for commercialization, and the R&SD (research and solution development) system to address societal needs and problem-solving. Here, the main concern is how well we can construct a collaboration system based on a "connect and development" approach between creative human resources. Inside and outside networks and internalization of new outsourcing knowledge into the organization should be the main focus in this system.

In Korea, a bill for enhancing the competitiveness of converging technologies and facilitating commercialization was proposed to become law in 2012 after passage by the congress. The "Convergence Research Policy Development Centre

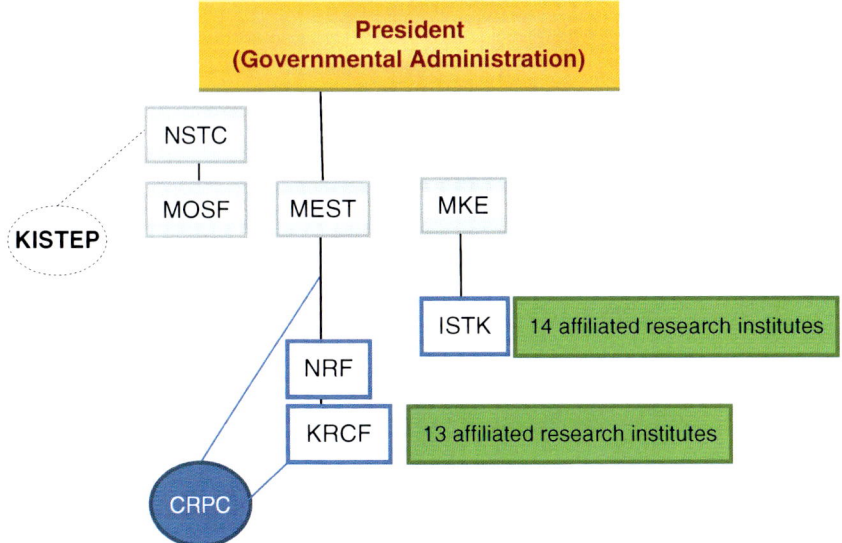

Fig. 10.11 Science and technology governance structure/hierarchy in S. Korea (Courtesy of H. Yang) (Agency Acronyms: *NSTC* National Science & Technology Commission, *KISTEP* Korea Institute of Science & Technology Evaluation and Planning, *MOSF* Ministry of Strategy & Finance, *MEST* Ministry of Education, Science and Technology, *MKE* Ministry of Knowledge Economy, *NRF* National Research Foundation, *ISTK* Korea Research Council for Industrial Science & Technology, *KRCF* Korea Research Council of Fundamental Science & Technology)

(CRPC[7])" (Fig. 10.11) has been set up very recently to implement this law for building an innovative R&D system. Centre staff has started to develop a national agenda and cross-ministerial planning of long-term convergence research for the coming decades with the support of the Ministry of Education, Science, and Technology (MEST) and Korea's National Research Foundation.

Simultaneously, several representative convergence research programs with their own creative R&D systems have been launched, seemingly considered as policy experiments. The first one is the interministry program called "Nano-Based-Convergence

---

[7] This is a tentative name for the organization, provided by the author, since the exact English name for the organization had been not decided at the time this description was written.

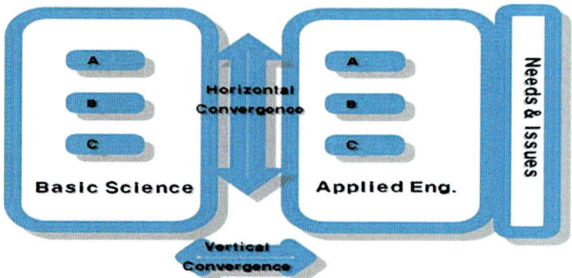

Fig. 10.12 Horizontal vs. vertical convergence in Korea's "Cutting-Edge Convergence Research Program" (Courtesy of H. Yang)

2020," sponsored by the Ministry of Knowledge and Economy and MEST. From the view of "breaking up sectionalism", this program pursues the synergy effects between two ministries and maximizing R&D efficiency by translational research, connecting basic R&D performance and ideas to business development.

The other program is the "Cutting-Edge Convergence Research Program", which emphasizes inter-disciplinarity with multi-organizational cooperation. The program collects ideas from researchers, and at the same time, planning managers conduct tests for compatibility with national R&D strategy and relevancy to societal needs through consultations with expert groups. This program pursues a "high-risk, high-return" strategy and therefore embraces the "sincere" research failure, thus supporting a creative R&D environment. The main differences between these two programs are the types of convergence characteristics, vertical or horizontal, they exhibit (Fig. 10.12).

The concept, "vertical convergence," which is well-explained for converging phenomena between basic science and applied engineering areas for specific societal needs, bridges the R&D stages in different phases. "Horizontal convergence" is a phenomena explaining converging activity within similar R&D phases, mainly creating new academic areas or ideas that can be applicable in the future.

10.8.10 Contribution of Knowledge and Technology to Sustainable Development in Emerging Economies

Contact person: *Richard Appelbaum, University of California, Santa Barbara*

The potential contribution of knowledge and technology to sustainable development in emerging economies is substantial. One recent study (Singer et al. 2005; Salamanca-Buentello et al. 2003) consulted a panel of 63 experts, 60 of whom were from developing countries, to rank the ten nanotechnology applications they felt would be of greatest benefit to developing countries over the next decade. In order of ranking (from top to bottom), these were (1) energy storage, production, and conversion; (2) agricultural productivity enhancement; (3) water treatment and remediation; (4) disease diagnosis and screening; (5) drug delivery systems;

(6) food processing and storage; (7) air pollution, prevention, and remediation; (8) construction; (9) health monitoring; and (10) vector and pest detection. Another study (Bürgi and Pradeep 2006, 648) concludes that nanotechnology "has the potential to become the flagship of the industrial production methods of the new millennium in developed as well as in the developing world."

Parker and Appelbaum (2012a, b) have identified four areas where nanotechnology and other emerging technologies (particularly at the intersection of biology and nanotechnology) have the potential to achieve breakthroughs that would improve the lives of many people in the Global South:

- *Energy/Environment*: Potential applications include more efficient appliances, solid-state lighting, enhanced energy storage with lithium ion batteries, and nanotechnology-enabled solar panels; biofuels (such as sugar cane in Brazil; Brito Cruz 2012); and potentially, solar cells.[8]
- *Water*: Potential applications include cheaper and faster diagnosis and removal of contaminants, disinfection or desalination of water through nanoporous membranes, and coatings that releases chlorine or other known disinfectants over extended periods of time.[9]
- *Food Security*: Potential applications include nanotechnology-enabled sensors (capable of detecting plant and crop disease, measuring pest and fertilizer levels, and ensuring food safety), nanotechnology-enhanced delivery of pesticide control, enhanced nutrition bioavailability, nanotechnology-enabled increased animal fertility and reproduction rates, greater protection against food-borne illness, and early detection of pathogens and disease.[10]
- *Health*: Potential applications include peptides for biopharmaceuticals, targeted and controlled drug delivery systems, labs-on-a-chip with sensors that contain thousands of nanowires able to detect proteins and biomarkers at the site of tumors, and gold nanoshells for dual imaging and cancer therapies.[11]

---

[8] First-generation single-crystalline silicon solar cells operate at 10–15 % conversion efficiency; cadmium telluride (CdTe) photovoltaics may reach 20 % efficiency; multijunction, thin-layer films promise 40 % efficiencies; quantum-dot structures hold high promise, with efficiencies approaching theoretical limits (Osman 2012, 23).

[9] Chinese engineers are developing a variety of nanoscale approaches to filtration that, if successful, will remove virtually all water and soil contaminants, whether they are bacterial in origin stemming from organic wastes, or industrial effluences such as toxic metals (Wang and Huang 2012). Osman (2012, 24) notes that "novel polymeric materials and nanofibrous media may enable high flux, low pressure membranes, thus reducing energy demand." (See also Hillie and Hlophe 2012, for examples from other developing countries.)

[10] "Nanotechnology is fast emerging as the new platform for the next wave of development and transformation of agri-food systems... projected to have the potential to provide foundation for large emerging agriculture centered economies like India" (Sastry et al. 2009, 91; see also Rogers and Zader 2012).

[11] Chinese scientists in particular are making advances in the areas of disease diagnosis and targeted drug delivery (Wang and Huang 2012).

How a Focus on Sustainable Development Affects Other Areas of Knowledge, Technology, and Global Outcomes

Increased cross-border collaboration—much of it across the advanced/developing country divide—furthers the ability of nanotechnology and other emerging technologies to play a key role in sustainable development (Wagner 2008; Wagner et al. 2001). This increase will ideally produce a truly global "open-source" approach to the development and regulation of nanotechnology and other emerging technologies, one that involves transparent and readily adaptable systems of governance and that speeds the advance of nanotechnology in support of more sustainable forms of development (Roco 2012). To take one example, China–U.S. scientific collaboration, after declining slightly during a "nationalist phase" (2000–2005), is again increasing (Mehta et al. 2012); Shapira and Wang (2010) found that U.S.–China collaborations are among the densest of all inter-country collaborations. There are, of course, vast differences between such emerging economies such as China, India, Brazil, Argentina, and Chile—which already have fairly advanced scientific institutions—and the low-income countries of Latin America, Asia, and Africa, where scientific infrastructure is weak or entirely absent.

Some Outcomes

While the NNI has had clear success in supporting basic research, its payoff in terms of sustainable development, particularly developments that would benefit the Global South, is less clear. China can offer some useful lessons in this regard.

China has invested not only in basic research (China's nanotechnology-related scientific publications now equal or exceed those of the United States, although quality and impact are not as high), but also in potential for commercialization. China's 15-year *National Medium- and Long-Term Plan for Science and Technology Development, 2006–2020* (MLP) picks numerous winners for government funding, including eleven "key areas," eight "frontier technologies," thirteen "engineering megaprojects," and four "science megaprojects," one of which is nanotechnology (the other three are development and reproductive biology, protein science, and quantum research; see Fig. 10.13).

China's 12th Five-Year Plan (2011–2015) further identifies seven "strategic areas" for funding, including biotechnology and new materials, and emphasizes the importance of green development (the plan calls for reducing carbon emissions per unit of GDP by 17 %). China's overarching goal is to develop an "indigenous innovation" capability that would enable it to become less dependent on foreign technology transfer, transitioning from "made in China" to "designed in China."

Our own research suggests that China's investment in nanotechnology and other emerging technologies, while not without its challenges, is likely to pay off (Parker and Appelbaum 2012c, forthcoming; Appelbaum et al. 2011; Cao et al. 2013).

> **Box 2. Areas and programs identified in China's 15-year science plan**
>
> **Key areas**
> Agriculture
> Energy
> Environment
> Information technology industry and modern services
> Manufacturing
> National defense
> Population and health
> Public securities
> Transportation
> Urbanization and urban development
> Water and mineral resources
>
> **Frontier technology**
> Advanced energy
> Advanced manufacturing
> Aerospace and aeronautics
> Biotechnology
> Information
> Laser
> New materials
> Ocean
>
> **Engineering megaprojects**
> Advanced numeric-controlled machinery and basic manufacturing technology
> Control and treatment of AIDS, hepatitis, and other major diseases
> Core electronic components, high-end generic chips, and basic software
> Drug innovation and development
> Extra large scale integrated circuit manufacturing and technique
> Genetically modified new-organism variety breeding
> High-definition Earth observation systems
> Large advanced nuclear reactors
> Large aircraft
> Large-scale oil and gas exploration
> Manned aerospace and Moon exploration
> New-generation broadband wireless mobile telecommunications
> Water pollution control and treatment
>
> **Science megaprojects**
> Development and reproductive biology
> Nanotechnology
> Protein science
> Quantum research

Fig. 10.13 Areas and programs identified in China's National Medium- and Long-term plan for science and technology development, 2006–2020 (Cao et al. 2006, 43; © American Institute of Physics, used by permission)

Its substantial investment in science research parks, such as Suzhou Industrial Park ("China's Silicon Valley") has proven attractive to foreign as well as local firms. (For a more detailed discussion, see Parker and Appelbaum 2012a, 148–149.) In a comparison of the U.S. and Chinese approach to the role of the state in driving high-tech research, development, and commercialization, we conclude, "the Chinese government [is] playing a facilitative role in providing the infrastructure, science parks, and greenfield university campuses that may eventually let a thousand nano-based products bloom" (Appelbaum et al. 2012, 127–128).

In my view, the U.S. should be investing more in bringing promising technologies to fruition, not just in terms of basic research, but also in terms of commercialization. The U.S. Government's support for small business innovation and commercialization (SBIR and STTR) programs should be significantly expanded. Additionally, U.S. visa policies need to be revised. Because U.S. universities remain among the best in the world for scientific innovation, a substantial percentage of science and engineering students enrolled in U.S. graduate programs come from other countries. According to an NSF study (2009), in 2006 foreign students earned a third of U.S. Ph.D. degrees in science and nearly two-thirds of U.S. Ph.D. degrees in engineering; one out of every three such students were from China (see also Matthews 2010). Current U.S. visa policies virtually assure that most of these students will return home after earning their degrees. China benefits from this short-sighted approach, enticing the best and brightest Chinese students and expatriates to return to China through highly attractive start-up packages (for example, the "Thousand Talents" and "Thousand Young Talents" Programs).

A Long-Term Perspective on Converging Technology, Innovation, and Sustainable Development in Emerging Economies

The world in the twenty-first century is facing extraordinary challenges. World population is projected to grow from the present seven billion to more than nine billion by 2050, with most of the growth occurring in developing countries. The probable detrimental impact of such growth on global warming and its associated climate change, potable water scarcity, food security, and public health issues are well known. The solutions to these challenges are both technological and political: while emerging technologies will clearly play a key role in their solution, it remains an open question whether the political (and economic) will exists to develop and utilize these technologies effectively.

The NNI should develop into an effective set of well-funded policies and programs in support of converging technologies, one that addresses basic research while providing the expanded support necessary to turn innovative ideas into practical results. Such an approach is necessary, I believe, given the central role that science and technology must play in achieving some degree of global economic, political, and environmental security in the twenty-first century.

Highlighted below are three important exploratory directions for converging knowledge and technology in the next 10–20 years:

- *"Open-source" scientific development.* Cross-border collaboration to solve the world's most pressing problems should be strongly encouraged. In the next 10–20 years, these problems will surely be in the four areas discussed above: energy/environment (sustainable development, as China, India, and other emerging economies expand); water purification/filtration; food security; and health. NSF could greatly expand its Partnerships for International Research and Education (PIRE) program, for example. Another approach would be to develop an open-source database to link experts and practitioners focused on the use of advanced technologies to address these issues. Such a database would be populated with the names of interested scientists and engineers, NGOs, sustainable development experts, potential funding sources, practitioners—all potential stakeholders in local sustainable development projects. It would be truly global, user-friendly, and designed in such a way that (for example) a rural Bolivian community with water purification problems could search out and identify a potential team whose expertise and practical experience would help develop viable solutions.
- *Participatory approaches to North/South collaboration.* This would involve developing appropriate technologies in full cooperation with developing countries, including learning from local practices (Lacy 2012). It is crucial to break down the traditional North/South relationship of technology transfer, where experts "parachute in" with an absence of local knowledge and participation, resulting in ill-suited—and hence ill-fated—solutions (Lewis 2012). The most advanced technologies, for example, may not always be the best suited answers to local needs. It is important to determine the most appropriate technology; simple water pumps may be more useful than advanced filtration technologies (Musaazi 2012). Should

the emphasis be on small-scale projects grounded in local communities, in which outside experts act in the service of local needs, or should the emphasis be on massive, government-led projects best suited to serving the needs of large numbers of people (Irvine-Halliday 2012; Wang and Huang 2012)?

- *Labor and EHS effects.* Expanded analysis of advanced technologies (such as nanotechnology) on labor markets and environmental health and safety is essential. Will more or fewer jobs result? What kinds of jobs will they be (Cozzens 2012; Foladori 2012; Invernizzi 2012; Invernizzi and Foladori 2005)? How can we better understand the largely unknown EHS issues of nano- and other emerging technologies—and how can we better communicate these effects to all stakeholders (Harthorn et al. 2012; Maynard et al. 2012)? While important work is now being done both on EHS issues (through the University of California and Duke University Centers for the Environmental Implications of Nanotechnology) and perceptions of EHS risk (by Harthorn and her collaborators at CNS-UCSB), very little work to date is being done on labor impacts.

10.9 International Perspectives

The following are summaries relevant to this chapter of discussions at the international regional WTEC NBIC2 workshops held in Leuven, Belgium, September 20–21, 2012; in Seoul, Korea, October 15–16, 2012; and in Beijing, China, October 18–19, 20110. Further details of those workshops are provided in Appendix A.

10.9.1 United States–European Union NBIC2 Workshop (Leuven, Belgium)

Panel members/discussants

Daan Schuurbiers, De Proeffabriek (Netherlands)
Albert Duschl, Paris Lodron University of Salzburg (EU)
Andy Miah, University of West Scotland (EU)
Alfred Nordmann, Darmstadt University of Technology (EU)
Anders Sandberg, Oxford University Future of Humanity Institute (EU)
Barbara Harthorn, University of California–Santa Barbara (U.S.)
Todd Kuiken, Woodrow Wilson International Center for Scholars (U.S.)
Jim Murday, University of Southern California (U.S.)
Mihail C. Roco, National Science Foundation and NNI (U.S.)

This group of scientists found consensus around three important themes: (1) human development, (2) sustainability and human development, and (3) co-evolution of human development and technology.

There is a difference of perspective on governance issues between the United States and the European Union. The former, as represented by George Whitesides' remarks in Leuven, seem to suggest that one should move ahead with converging technologies and implement governance processes when an opportunity develops. According to EU speakers, the EU position is that one needs to prove that some imagined advance in converging technologies will not be harmful before one moves on. This difference in perspective is certainly prevalent in the area of environmental risk management. A good middle ground is to implement adaptive management and governance, which really isn't brought out in great detail in the chapter.

For the first theme, the core idea was an emerging trend to use personalized molecular information to enhance both medical treatment and cognition. Other important trends were quality-of-life enhancements made possible by new prosthetic device technologies and regenerative medicine. Finally, the trend towards personalized innovative education was recognized. All of these trends reflect technological convergence.

For the second theme, the core idea was using convergent technologies to enhance human sustainability. One core idea was the use of waste and sea water as sources for human development. Another core idea was the reduction of energy imports while simultaneously reducing carbon footprints. The discussants imagined a future where megacities would be ubiquitous and materials would be infinitely renewable.

For the third theme, the central point was the accelerating co-evolution between human development and technology. The discussants agreed that the co-evolution would manifest in increased productivity, human capacity, and personalized lifelong education.

10.9.2 United States–Korea–Japan NBIC2 Workshop (Seoul, Korea)

Panel members/discussants

Sang-Ki Jeong (co-chair), Korea Institute of S&T Evaluation and Planning (KISTEP, Korea)
Mihail C. Roco (co-chair), National Science Foundation (U.S.)
Tomoji Kawai (co-chair), Osaka University (Japan)
Ji Woong Yoon (participant), Kyung Hee University (Korea)
Eugene Pak (participant), Seoul National University (Korea)

The viability of integrating NBIC technologies has been confirmed in multiple settings, but it was deliberately applied only occasionally (reactively). Concerns have grown about the impact of new technologies, particularly those affecting global environmental issues and energy, health, food, and human performance. There is an increasing emphasis on the roles of innovation, sustainability, and family wealth in evaluating new technologies. Innovative and societal implications are manifested in changing contexts: multiple S&T poles, transfer of wealth and initiative from West to East, demographic changes, and overall rapid developments.

By convergence, new outcomes are expected for user-friendly societal systems leading to advanced discovery, innovation, and entrepreneurial ecosystems. There is an increasing role for public/private partnerships in development of converging technologies, with emergence of new tools for participatory governance, e.g., games and social media. Studies have been performed in Korea on the main societal factors to effect societal trends. As a result, a government organization in the office of the Korean Prime Minister has been proposed to facilitate S&T convergence with a targeted implementation date at the end of 2012. Co-evolution between technology, values, and societal norms has been highlighted. The Japanese group provided as an example for converging knowledge and technology the work in robotics.

The main conclusions from the discussions are:

- Combined innovative and responsible governance is critical for competitiveness
- Governance should promote individual development
- Convergence should be guided by criteria such as sustainable development, individual privacy rights, international coordination, and international competitiveness
- Understanding opportunities/threats arising from changes in governance roles; traditional institutions have reduced role, being bypassed by social media-enabled movements

10.9.3 United States–China–Australia–India NBIC2 Workshop (Beijing, China)

Panel members/discussants

Xian-En Zhang (co-chair), Ministry of Science and Technology (China)
Mihail C. Roco (co-chair), National Science Foundation (U.S.)
Ron Johnston (co-chair), Australian Centre for Innovation

Others

Craig Johnson, University of Tasmania, Australia
Ke Xu, Suzhou Institute of Nano-Tech and Nano-Bionics (China)
Bo Tang (China)

Knowledge and converging technologies are progressing fast in China and Australia. Concerns about the impact of new technologies have grown, particularly those affecting health, food, and human performance.

There is an increasing emphasis on the roles of innovation and sustainability in evaluating new technologies. Also, there is an increasing emphasis on the roles of *innovation and sustainability* in evaluating new technologies.

Examples were presented of how investment policies in multidisciplinary/multi-application R&D centers have been developed for China. Three main drivers are global issues (energy, resources, etc.), science breakthroughs (in brain research, single cell biology, etc.), and human dimensions (aging, cognitive and communication capacity, etc.). Combining specialization with converging thinking (in time, across fields) is an essential aspect.

References

Adner, R.: Match your innovation strategy to your innovation ecosystem. Harvard Business Review. http://pds12.egloos.com/pds/200811/07/31/R0604Fp2.pdf (2006). Accessed 2 Aug 2012

Anthony, S.D., Eyring, M., Gibson, L.: Mapping your innovation strategy. Harvard Business Review. http://ww3.harvardbusiness.org/corporate/assets/content/SDAnthony.pdf (2006). Accessed 2 Aug 2012

Appelbaum, R., Parker, R., Cao, C.: Developmental state and innovation: nanotechnology in China. Glob. Netw. **11**(3), 298–314 (2011)

Appelbaum, R., Cao, C., Parker, R., Motoyama, Y.: Nanotechnology as industrial policy: China and the United States. In: Harthorn, B.H., Mohr, J.W. (eds.) The Social Life of Nanotechnology, pp. 111–133. Routledge, New York (2012)

Armstrong, R.: Living Architecture: How Synthetic Biology Can Remake Our Cities and Reshape Our Lives. TED Books. Amazon Digital Services, New York (2012)

Bainbridge, W.S.: Online Worlds: Convergence of the Real and the Virtual. Springer, Dordrecht (2010)

Barben, D., Fisher, E., Selin, C., Guston, D.H.: Anticipatory governance of nanotechnology: foresight, engagement, and integration. In: Edward, O.A., Hackett, J., Lynch, M., Wajcman, J. (eds.) The Handbook of Science and Technology Studies, 3rd edn, pp. 979–1000. MIT Press, Cambridge, MA (2008)

Battelle: Economic Impact of the Human Genome Project. Battelle Memorial Institute, Columbus (2011). http://battelle.org/docs/default-document-library/economic_impact_of_the_human_genome_project.pdf

Battelle: 2013 Global R&D Forecast. Available on the web http://www.battelle.org (2012)

Baumol, W.J.: Productivity, growth, convergence, and welfare: what the long-run data show. Am. Econ. Rev. **76**(5), 1072–1085 (1986)

Baumol, W.J., Wolff, E.N.: Productivity, growth, convergence, and welfare: reply. Am. Econ. Rev. **78**(5), 1155–1159 (1988)

Bitzer, L.: The rhetorical situation. Philos. Rhetor. **1**(1), 1–14 (1968)

Breau, D.L., Brook, B.: "Mock" mock juries: a field experiment on the ecological validity of jury simulations. Law Psychol. Rev. HighBeam Res. **31**, 77–92 (2007). http://www.highbeam.com/doc/1G1-170157388.html

Brito Cruz, C.H.: Energy for development: the case of bioenergy in Brazil. In: Parker, R., Appelbaum, R. (eds.) Can Emerging Technologies Make a Difference in Development? pp. 127–139. Routledge, New York (2012)

Bunge, M.: A World of Systems. Reidel, Dordrecht/Boston (1979)

Bureau of Economic Analysis: 2010 recovery widespread across industries. April 26, http://www.bea.gov/newsreleases/industry/gdpindustry/2011/pdf/gdpind10_adv_fax.pdf. See also interactive access to industry economic accounts data, http://www.bea.gov/iTable/iTable.cfm?ReqID=5&step=1.5 (2011)

Burgess, J., Green, J.: YouTube: Online Video and Participatory Culture. Polity, Cambridge (2009)

Bürgi, B.R., Pradeep, T.: Societal implications of nanoscience and nanotechnology in developing countries. Curr. Sci. **90**(5), 645–658 (2006)

Burke, S.J.: How to Build an Innovation Ecosystem. New York Academy of Sciences, New York (2011). http://www.nyas.org/publications/Detail.aspx?cid=da1b8e1d-ed2d-4da4-826d-00c987f63c82. Accessed 27 July 2012

Cao, C., Suttmeier, R.P., Simon, D.F.: China's 15-year science and technology plan. Phys. Today **59**, 38–43 (2006)

Cao, C., Appelbaum, R., Parker, R.: Research is high and the market is far away: commercialization of nanotechnology in China. Tech Soc. **35**(1), 55–64 (2013). http://dx.doi.org/10.1016/j.techsoc.2013.03.004

Carr, N.: The Shallows: What the Internet Is Doing to Our Brains. Norton, New York (2010)

Cavin III, R.K., Sumney, L.W., Burger, R.M.: The semiconductor research corporation: cooperative research. Proc. IEEE **77**, 1327–1344 (1989)

Chi, R., Snyder, A.: Facilitate insight by non-invasive brain stimulation. PLoS ONE **6**(2), e16655 (2011). doi:10.1371/journal.pone.0016655

Chinni, D., Gimpel, J.: Our Patchwork Nation: The Surprising Truth About the Real America. Gotham, New York (2010)

CNS-UCSB (Center for Nanotechnology in Society at University of California at Santa Barbara). Emerging Technologies/Emerging Economies [Nano]technology for Equitable Development International Workshop. Woodrow Wilson Center for International Scholars, Washington, DC, 4–6 Nov 2009. See http://www.cns.ucsb.edu/events/nanoequity2009 (2009)

Cobb, M.D.: Creating informed public opinion: citizen deliberation about nanotechnologies for human enhancements. J. Nanoparticle Res. **13**(4), 1533–1548 (2011)

Coenen, C.: Konvergierende technologien und wissenschaften (Converging technologies: the status of the debate and political activities). Office of Technology Assessment at the German Bundestag, Background paper #16. Available online: http://www.tab-beim-bundestag.de/en/publications/reports/hp016.html (2008)

Collier, P.: The Bottom Billion: Why the Poorest Countries Are Falling and What Can Be Done About It. Oxford, New York (2007)

Cozzens, S.: Emerging technologies and inequalities: beyond the technological transition. In: Parker, R., Appelbaum, R. (eds.) Can Emerging Technologies Make a Difference in Development? pp. 47–60. Routledge, New York (2012)

Cramer, J.: Innovation ecosystems: everyone willing is at the center of the universe. *InnovToday* online: http://innovtoday.wordpress.com/2011/04/25/innovation-ecosystems-everyone-willing-is-at-the-centre-of-the-universe/#comments (2011). Accessed 27 July 2012

Dasgupta, J.: How an innovation system may look like. Innovation & Change (blog). http://jibakdasgupta.blogspot.com/2010/08/how-innovation-ecosystem-may-look-like.html (2010). Accessed 27 July 2012

De Long, J.B.: Productivity, growth, convergence, and welfare: comment. Am. Econ. Rev. **78**(5), 1138–1154 (1988)

Du Rausas, M.P., Manyika, J., Hazan, E., Bughin, J., Chui, M., Said, R.: Internet Matters: The Net's Sweeping Impact on Growth, Jobs, and Prosperity. McKinsey Global Institute, San Francisco (2011). http://www.mckinsey.com/Insights/MGI/Research/Technology_and_Innovation/Internet_matters

Economist: Innovation pessimism: has the ideas machine broken down? Available online: http://www.economist.com/news/briefing/21569381-idea-innovation-and-new-technology-have-stopped-driving-growth-getting-increasing (2013, Jan. 12)

ETAG (European Technology Assessment Group): Technology assessment on converging technologies. IP/A/STOA/SC/2005-183. http://www.scribd.com/doc/59157531/23/The-doom-scenarios-of-the-ETC-Group#page=15 (2005). Accessed 2 Aug 2012

ETC Group: The big down: from genomes to atoms, Atomtech: technologies converging at the nanoscale. Available online: http://www.etcgroup.org/sites/www.etcgroup.org/files/thebigdown.pdf (2003). Accessed 2 Aug 2012

ETC/NGO. The Potential Impact of Nanoscale Technologies on Commodity Dependent Developing Counties. ETC Group, Ottawa. Online at http://www.etcgroup.org/upload/publications/45/01/southcentre.commodities.pdf (2005)

FFF (Foundation for the Future), UNESCO (United Nations Educational, Scientific and Cultural Organization): Humanity and the Biosphere, The Next Thousand Years. Seminar Proceedings 20–22 September 2006. Foundation for the Future, Bellevue (2007)

Fisher, E., Mahajan, R.L.: Embedding the humanities in engineering: art, dialogue, and a laboratory. In: Gorman, M.E. (ed.) Trading Zones and Interactional Expertise: Creating New Kinds of Collaboration, pp. 209–230. MIT Press, Cambridge, MA (2010)

Fisher, E., Boenink, M., van der Burg, S., Woodbury, N.: Responsible healthcare innovation: anticipatory governance of nanodiagnostics for theranostics medicine. Expert. Rev. Mol. Diagn. **12**(8), 857–870 (2012)

Foladori, G.: Achieving equitable outcomes through emerging technologies: a social empowerment approach. In: Parker, R., Appelbaum, R. (eds.) Can Emerging Technologies Make a Difference in Development? pp. 40–46. Routledge, New York (2012)

Gehrke, P.: Public Engagement and Practicable Democracy. A Presentation at the Communicating Health & Safety Risks of Emergent Technologies Workshop, 28–29 August. North Carolina State University, Raleigh (2008). http://www.youtube.com/playlist?list=PL096FE2F4436C3783&feature=plcp. Accessed 2 Aug 2012

German Bundestag: Konvergierende technologien und wissenschaften: Der stand der debatte und politischen aktivitäten zu »converging technologies. Online: http://www.tab-beim-bundestag.de/de/publikationen/berichte/hp016.html (2008)

Gordon, R.J.: Is U.S. economic growth over? Faltering innovation confronts the six headwinds. NBIR Working Paper 18315, http://www.nber.org/papers/w18315. National Bureau of Economic Research, Boston (2012)

Grueber, T., Studt, T.: R&D funding forecast. R&D Magazine December16. Available online: http://go.nature.com/v2xzms (2012)

Guston, D.H.: Innovation policy: not just a jumbo shrimp. Nature **454**(7207), 940–941 (2008)

Halavais, A.: The Search Engine Society. Polity, Cambridge (2008)

Hamlett, P., Cobb, M., Guston, D.H.: National Citizens Technology Forum: Nanotechnologies and Human Enhancement. CNS-ASU Report #R08-0003. Center for Nanotechnology in Society, Arizona State University, Tempe (2008)

Harthorn, B.H., Shearer, C., Rogers, J.: Risk perception, public participation, and sustainable global development of nanotechnologies. In: Parker, R., Appelbaum, R. (eds.) Can Emerging Technologies Make a Difference in Development? pp. 188–197. Routledge, New York (2012)

Hassan, R.: The Information Society: Cyber Dreams and Digital Nightmares. Polity, Cambridge (2008)

Herr, D., Zhirnov, V.: Developments in the strategic partnership between the U.S. Government and the semiconductor industry. Future Fab International issue 17. Available online: http://www.future-fab.com/documents.asp?d_ID=2592 (2004)

Hillie, T., Hlophe, M.: Nanotechnology for potable water and general consumption in developing countries. In: Parker, R., Appelbaum, R. (eds.) Can Emerging Technologies Make a Difference in Development? pp. 105–115. Routledge, New York (2012)

Hwang, V.H., Horowitt, G.: The Rainforest: The Secret to Building the Next Silicon Valley. Regenwald, Los Altos (2012)

Invernizzi, N.: Implications of nanotechnology for labor and employment: assessing nanotechnology products in Brazil. In: Parker, R., Appelbaum, R. (eds.) Can Emerging Technologies Make a Difference in Development? pp. 140–152. Routledge, New York (2012)

Invernizzi, N., Foladori, G.: Nanotechnology and the developing world: will nanotechnology overcome poverty or widen disparities? Nanotechnol. Law Bus. **2**(3), 101–110 (2005)

IRGC (International Risk Governance Council): White Paper on Risk Governance: Towards an Integrative Approach. IRGC, Geneva (2006). Online: http://www.irgc.org/IMG/pdf/IRGC_WP_No_1_Risk_Governance__reprinted_version_.pdf

IRGC: Global Risk Governance: Applying and Testing the IRGC Framework. Eds. Renn, O., Walker, K. Springer, Berlin (chapter on nanotechnology by M.C. Roco and O. Renn) (2008)

IRGC: Appropriate Risk Governance Strategies for Nanotechnology Applications in Food and Cosmetics. International Risk Governance Council, Geneva (2009). Available online: http://irgc.org/wp-content/uploads/2012/04/irgc_nanotechnologies_food_and_cosmetics_policy_brief1.pdf

Irvine-Halliday, D.: Solid state lighting: a market-based approach to escaping the 'poverty trap.'. In: Parker, R., Appelbaum, R. (eds.) Can Emerging Technologies Make a Difference in Development? pp. 116–126. Routledge, New York (2012)

Jeon, H.R., Jeong, S.Y.: Introduction of open innovation system for convergent technology R&D, electronics and telecommunications trends. ETRI J. **25**(1), 23–31 (2010)

Keck Futures Initiative: Synthetic Biology Building on Nature's Inspiration. Interdisciplinary Research Team Summaries. November 20–22. The National Academies Press, Washington, DC (2009)

Lacy, S.: (Nano)technology and food security: what scientists can learn from Malian farmers? In: Parker, R., Appelbaum, R. (eds.) Can Emerging Technologies Make a Difference in Development? pp. 86–98. Routledge, New York (2012)

Lewis, D.: Rural development, technology, and "policy memory": anthropological reflections from Bangladesh on technological change. In: Parker, R., Appelbaum, R. (eds.) Can Emerging Technologies Make a Difference in Development? pp. 28–39. Routledge, New York (2012)

Martin, B.: Foresight in science and technology. Technol. Anal. Strateg. Manage. **7**, 139–168 (1996)

Matthews, C.H.: Foreign Science and Engineering Presence in U.S. Institutions and the Labor Force. Congressional Research Service, Washington, D.C. (2010). March 23

Maynard, A., Grobe, A., Renn, O.: Responsible innovation, global governance, and emerging technologies. In: Parker, R., Appelbaum, R. (eds.) Can Emerging Technologies Make a Difference in Development? pp. 168–187. Routledge, New York (2012)

Mehta, A., Herron, P., Motoyama, Y., Appelbaum, R., Lenoir, T.: Globalization and de-globalization in nanotechnology research: the role of China. Scientometrics (March), 1–20 (2012)

Mnyuisiwalla, A., Daar, A.S., Singer, P.A.: Mind the gap: science and ethics in nanotechnology. Nanotechnology **13**, R9–R13 (2003)

Mossberger, K., Tolbert, C.J., McNeal, R.S.: Digital Citizenship: The Internet, Society, and Participation. The MIT Press, Cambridge, MA (2007)

Musaazi, M.K.: Innovations for development: the African challenge. In: Parker, R., Appelbaum, R. (eds.) Can Emerging Technologies Make a Difference in Development? pp. 99–104. Routledge, New York (2012)

Nisbet, M.: Communicating climate change: why frames matter for public engagement. Environ. Sci. Policy Sustain. Dev. **51**(2), 12–23 (2009)

NRC (National Research Council): Growing Innovation Clusters for American Prosperity: Summary of a Symposium. The National Academies Press, Washington, DC (2011)

NRC (National Research Council): Continuing Innovation in Information Technology. The National Academies Press, Washington, DC (2012)

NSF (U.S. National Science Foundation): Science and Engineering Doctorate Awards, 2006. Detailed Statistical Tables, Table 3. National Science Foundation, Arlington (2009 [March])

OECD: Where do students go to study abroad? In: Education at a Glance 2011: Highlights. Online at http://www.oecd-ilibrary.org/docserver/download/fulltext/9611051ec013.pdf?expires=1325523346&id=id&accname=guest&checksum=C0E2EC02146BFEDD86488DB643C474AB (2011)

Osman, T.: Creating the future: materials, innovation, and the scientific community. In: Parker, R., Appelbaum, R. (eds.) Can Emerging Technologies Make a Difference in Development? pp. 21–27. Routledge, New York (2012)

Ostrom, E.: Economic governance. Scientific background on the Sveriges Riksbank prize in economic sciences in memory of Alfred Nobel. Economic governance. Compiled by the Economic Sciences Prize Committee of the Royal Swedish Academy of Sciences. Available online: http://www.nobelprize.org/nobel_prizes/economics/laureates/2009/advanced-economicsciences2009.pdf (2009)

Ostrom, E.: Polycentric systems for coping with collective action and global environmental change. Glob. Environ. Chang. **20**, 55–557 (2010)

Parker, R., Appelbaum, R. (eds.): Can Emerging Technologies Make a Difference in Development? Routledge, New York (2012a)

Parker, R., Appelbaum, R.: Introduction: the promise and perils of high-tech approaches to development. In: Parker, R., Appelbaum, R. (eds.) Can Emerging Technologies Make a Difference in Development? pp. 1–20. Routledge, New York (2012b)

Parker, R., Appelbaum, R.: The Chinese century? Some implications of China's move to high-tech innovation for U.S. policy. In: Harthorn, B.H., Mohr, J.W. (eds.) The Social Life of Nanotechnology, pp. 134–165. Routledge, New York (2012c)

Parsons, T.: Evolutionary universals in society. Am. Sociol. Rev. **29**, 339–357 (1964)

Parsons, T., Shils, E.A. (eds.): Toward a General Theory of Action. Harvard University Press, Cambridge, MA (1951)

PCAST (President's Council of Advisors on Science and Technology): Report to the President and Congress on the Third Assessment of the National Nanotechnology Initiative. Executive Office of the President, Washington, DC (2010). Available online: http://www.whitehouse.gov/sites/default/files/microsites/ostp/pcast-nano-report.pdf

PCAST (President's Council of Advisors on Science and Technology): Transformation and Opportunity: The Future of the U.S. Research Enterprise. Executive Office of the President, Washington, DC (2012). Available online: http://www.whitehouse.gov/administration/eop/ostp/pcast/docsreports

Quinn, N.: Finding Culture in Talk: A Collection of Methods. Palgrave, New York (2005)

Rescher, N.: Value Matters: Studies in Axiology. Ontos Verlag, Frankfurt (2005)

Roco, M.C.: Coherence and divergence of megatrends in the science and engineering. J. Nanopart. Res. **4**, 9–19 (2002)

Roco, M.C.: Possibilities for global governance of converging technologies. J. Nanopart. Res. **10**, 11–29 (2008). http://dx.doi.org/10.1007/s11051-007-9269-8

Roco, M.C.: Nanotechnology: from discovery to innovation and socioeconomic projects. CEP **107**(5), 21–27 (2011)

Roco, M.C.: Global governance of emerging technologies: from science networking to coordinated oversight. In: Parker, R., Appelbaum, R. (eds.) Can Emerging Technologies Make a Difference in Development? pp. 198–201. Routledge, New York (2012)

Roco, M.C., Bainbridge, W.S.: Societal Implications of Nanoscience and Nanotechnology. Springer (previously Kluwer), Dordrecht (2001)

Roco, M.C., Bainbridge, W.S. (eds.): Converging Technologies for Improving Human Performance: Nanotechnology, Biotechnology, Information Technology, and Cognitive Science. Kluwer Academic Publishers (now Springer), Dordrecht (2003)

Roco, M.C., Bainbridge, W.S.: Nanotechnology: Societal Implications, I and II (2 volumes). Springer (2007)

Roco, M.C., Mirkin, C.A., Hersam, M.C. (eds.): Nanotechnology Research Directions for Societal Needs in 2020: Retrospective and Outlook. Springer, Berlin/Boston (2011)

Rogers, J., Zader, A.: Food security: from the green revolution to nanotechnology. In: Parker, R., Appelbaum, R. (eds.) Can Emerging Technologies Make a Difference in Development? pp. 75–85. Routledge, New York (2012)

Salamanca-Buentello, F., Persad, D.L., Court, E.B., Martin, D.K., Daar, A.S., Singer, P.A.: Nanotechnology and the developing world. PLoS Med. **2**, 5 (2003) (May)

Samuels, J.D.: Semiconductors and U.S. Economic Growth. Semiconductor Industry Association, Washington, DC (2012, April 1). http://www.sia-online.org/clientuploads/directory/DocumentSIA/ecoimpactsemidraft_Samuels.pdf

Sarewitz, D., Nelson, R.: Three rules for technological fixes. Nature **456**, 871–872 (2008)

Sastry, K., Rashmi, H.B., Rao, N.H., Ilyas, S.M.: Nanotechnology and agriculture in India: the second green revolution? OECD Conference on Potential Environmental Benefits of Nanotechnology: Fostering Safe Innovation-Led Growth. OECD Conference Center, Paris, July 15–17 (2009)

Schuurbiers, D.: What happens in the lab: applying midstream modulation to enhance critical reflection in the laboratory. Sci. Eng. Ethics **17**(4), 769–788 (2011)

Selin, C.: Negotiating plausibility: intervening in the future of nanotechnology. Sci. Eng. Ethics **17**(4), 723–737 (2011)

Shapira, P., Wang, J.: Follow the money. Nature **468**(December 2), 627–628 (2010)

Singer, P.A., Salamanca-Buentello, F., Daar, A.S.: Harnessing Nanotechnology to Produce Global Equity. Issues On-Line in Science and Technology. National Academy of Sciences, Washington, DC (2005, Summer). http://www.issues.org/21.4/singer.html. Accessed 18 June 2012

Solow, R.M.: Technical change and the aggregate production function. Rev. Econ. Stat. **39**, 312–320 (1957)

Spence, M.: The Next Convergence: The Future of Economic Growth in a Multi-speed World. Farrar, Straus and Giroux, New York (2011)

Stern, P.: Design principles for the global commons: natural resources and emerging technology. Int. J. Commons **5**(2), 213–232 (2011)

Stokes, D.E.: Pasteur's Quadrant: Basic Science and Technological Innovation. Brookings Institution, Washington, DC (1997)

Sunstein, C.: Going to Extreme: How Like Minds Unit and Divide. Oxford UP, New York (2009)

Suresh, S.: Global challenges need global solution. Nature **490**(7420), 337–338 (2012)

Tenner, E.: Why Things Bite Back: Technology and the Revenge of Unintended Consequences. Vintage, New York (1997)

UNESCO. The Ethics and Politics of Nanotechnology. UNESCO, Paris. Online at http://unesdoc.unesco.org/images/0014591e.pdf (2006a)

UNESCO. Global Ethics Observatory. UNESCO, Paris. Online at www.unesco.org/shs/ethics/geobs (2006b)

UNESCO. Humanity and the Biosphere: The Next Thousand Years. UNESCO and "The Foundation for the Future". Paris (2006c)

Vester, F.: The Art of Interconnected Thinking: Tools and Concepts for a New Approach to Tackling Complexity. MCB, Munich (2008)

von Hippel, E.: Sticky information and the locus of problem solving: implications for innovation. Manag. Sci. **40**(4), 429–439 (1994)

Wagner, C.: The New Invisible College: Science for Development. Brookings Institution Press, Washington, DC (2008)

Wagner, C.S., Brahmakulam, I., Jackson, B., Wong, A., Yoda, T.: Science and Technology Collaboration: Building Capacity in Developing Countries? Monograph Report-1357-WB. RAND Corporation, Santa Monica (2001)

Wang, C., Huang, C.: The progress of nanotechnology in China. In: Parker, R., Appelbaum, R. (eds.) Can Emerging Technologies Make a Difference in Development? pp. 61–74. Routledge, New York (2012)

West, D.M.: The Next Wave: Using Digital Technology to Further Social and Political Innovation. Brookings Institution Press, Washington, DC (2011)

Yang, H.: Summer conference. The Korean Society for Innovation Management & Economics (2012, June)

Appendices

Appendix A
List of U.S. and International Workshops

Americas Workshop

"Emerging and Converging Technologies"
São Paulo, Brazil, November 24–25, 2011
Hosted and co-sponsored by:

- Centro de Gestão e Estudos Estratégicos (CGEE: Center for Strategic Studies of the Government of Brazil)
- Rede Latino-Americana de Prospectiva em Convergência Tecnológica (RLPCT: Latin American Network for Technological Convergence Prospective)
- Programa Sul-Americano de Apoio às Atividades de Cooperação em Ciência e Tecnologia (PROSUL: S. American Program to Support Cooperation Activities in Science and Technology of the Brazilian National Council for Scientific and Technological Development, CNPq)

  40 participants
  Website: http://www.wtec.org/NBIC2-Americas

United States Workshop

"Transforming Tools of Emerging and Converging Technologies for Societal Benefit (Beyond Nano-Bio-Info-Cognitive Technologies, NBIC2)"
Arlington, Virginia, United States. June 25–26, 2012
Sponsored by NSF, NIH, NASA, EPA, DOD, and USDA
Hosted by WTEC
85 participants
Website: http://www.wtec.org/NBIC2-US/

United States–European Union Workshop

"Converging Nano-Bio-Info-Cognitive S&T for Responsible Innovation and Society"
Leuven, Belgium. September 20–21, 2012
Co-sponsored by the European Commission
Hosted by IMEC
50 Participants
Website: http://www.wtec.org/NBIC2Leuven

United States–Korea–Japan Workshop

"International Study of Converging Technologies for Societal Benefit"
Seoul, Korea. October 15–16, 2012
Co-sponsored by:

- National Research Foundation of Korea (NRFK, Korea)
- Ministry of Education, Science and Technology (MEST, Korea)
- Japan Science and Technology Agency (JST, Japan)

Hosted by the Ministry of Education, Science, and Technology (Korea)
105 Participants
Website: http://www.wtec.org/NBIC2Seoul

United States–China–Australia Workshop

"International Study of Converging Technologies for Societal Benefit"
Beijing, China. October 18–19, 2012
Hosted and co-sponsored by the Chinese Academy of Sciences (CAS, China); Australian Nanotechnology Network; Australian National University; and Department of Innovation, Industry, Science, Research and Tertiary Education, Australian Government
41 Participants
Website: http://www.wtec.org/NBIC2Beijing

Final Workshop: Summary and Conclusions of the Study

"Converging Knowledge and Technologies for Societal Benefit (Beyond Nano-Bio-Info-Cognitive Technologies, NBIC2)"
Arlington, Virginia, United States. December 11, 2012
Sponsored by NSF, NIH, NASA, EPA, DOD, and USDA

Hosted by WTEC
85 participants (63 in person and 26 via Internet webcast)
Website: http://www.wtec.org/NBIC2
Webcast: http://www.tvworldwide.com/events/nsf/121211/

Convergence Interviews

Selected interviews with experts on convergence topics have been conducted and provided online by the Woodrow Wilson International Center for Scholars: http://wilsoncenter.org/convergence.

Appendix B
List of Participants and Contributors

This section lists the participants of all six workshops conducted in the course of this study: *Sao Paolo*, Brazil (November 2011); ***Arlington-US*** (June 2012); ***Leuven***, Belgium (September 2012); ***Seoul***, Republic of Korea (October 2012); ***Beijing***, People's Republic of China (October 2012); and ***Arlington-final*** (final presentation of findings, December 2012). (Appendix A gives workshop details.) This section also indicates the individuals who contributed to the report. After each person's institution is a key to his or her role. For example, "[Arlington-US]" means the person attended the June 2012 workshop in Arlington, VA, US; "[contributor]" means the person has made a contribution to this report by writing sections of text; "[webcast]" indicates attendance at the final Arlington workshop via webcast. Institutional affiliations are as of the date of the workshop(s) attended.

Sumanta Acharya
National Science Foundation
USA [Arlington-final]

George B. Adams
Network for Computational
 Nanotechnology
USA [contributor]

Mike Adams
University of Birmingham
UNILEVER
UK [Leuven]

Sung-Hoon Ahn
Korea National University
Korea [Seoul]

Rosa Alegria
Pontifical Catholic University
 São Paulo
Brazil [São Paulo]

Jacqueline Allan
Organization for Economic
 Co-operation and Development
France [Leuven; webcast]

David Allen
The University of Texas at Austin
USA [contributor]

Pedro Alvarez
Rice University
USA [Arlington-US; contributor]

Richard Appelbaum
University of California–Santa Barbara
USA [Arlington-US; contributor]

Ramy Arnaout
Massachusetts Institute of Technology (MIT)
USA [contributor]

Amanda J. Arnold
Massachusetts Institute of Technology (Washington, DC)
USA [Arlington-US; Arlington-final; contributor]

Giorgio Ascoli
George Mason University
USA [Arlington-US; contributor]

Masafumi Ata
National Institute of Advanced Industrial Science and Technology
Japan [Seoul]

Dan Atkins
Michigan State University
USA [Arlington-US]

Marietta Baba
Michigan State University
USA [Arlington-US; webcast; contributor]

Toshio Baba
Japan Science and Technology Agency
Japan [Seoul; webcast]

Guillermina Baena
National Autonomous University of Mexico
Mexico [São Paulo]

Chunli Bai
Chinese Academy of Sciences
PRC [Beijing]

William S. Bainbridge
National Science Foundation
USA [Arlington-US; Arlington-final; contributor]

Laura Ballerini
University of Trieste
Italy [Leuven; webcast]

Paola Balletti
European Commission
[Leuven]

Roberto Barbero
Office of Science and Technology Policy
USA [webcast]

Philippe Bardet
George Washington University
USA [contributor]

Linda Barker [webcast]

Matthew E. Bates
U.S. Army Corps of Engineers
USA [contributor]

Marie-Ange Baucher
Organization for Economic Co-operation and Development
France [webcast]

Paulo Beirão
National Council for Scientific and Technological Development
Brazil [São Paulo]

Larry Bell
Museum of Science Boston
USA [Arlington-US; contributor]

Gregory Benford
California Institute of Technology
USA [Arlington-US]

Rosalyn Berne
University of Virginia
USA [Arlington-US; contributor]

David M. Berube
North Carolina State University
USA [contributor]

Appendix B

Clement Bezold
Institute for Alternative Futures
USA [Arlington-US; Arlington-final; contributor]

So'Nin Bio
National Science Foundation
USA [Arlington-final]

Matteo Bonazzi
European Commission
Belgium [Leuven]

Barry Bruce
University of Tennessee, Knoxville
USA [Arlington-US; contributor]

Eugeniy Bykov
Moscow State University
Russia [webcast]

Shushan Cai
Tsinghua University
PRC [Beijing]

Chris Cannizzaro
U.S. Department of State
USA [Arlington-US; Arlington-final]

Jian Cao
Northwestern University
USA [Arlington-US; Leuven; Seoul; Beijing; Arlington-final; contributor]

Altaf H. Carim
Office of Science and Technology Policy, EOP
USA [Arlington-US; Arlington-final]

Esper Cavalheiro
Center for Strategic Studies of the Government of Brazil
[São Paulo; contributor]

Ralph Cavin
Semiconductor Research Corporation
USA [contributor]

Erwin P. Gianchandani
Computing Research Association
USA [contributor]

Ni-Bin Chang
National Science Foundation
USA [Arlington-final]

Robert Chang
Northwestern University
USA [contributor]

Claudio Chauke
Center for Strategic Studies of the Government of Brazil [São Paulo]

Chen Chen
Soochow University
PRC [Beijing]

Dongyi Chen
University of Electronic Science and Technology
PRC [Beijing]

Hongda Chen
U.S. Department of Agriculture
USA [webcast]

Byung-ki Cheong
Korea Institute of Science and Technology
Korea [Seoul]

Hee Chan Cho
Korea National University
Korea [Seoul]

Young Hyun Cho
Dongbu Hitek Co. Ltd.
Korea [Seoul]

Changhwan Choi
Hanyang University
Korea [Seoul]

Eunsun Choi
GTC-Korea
Korea [Seoul]

Jong Suk Choi
Korea Institute of Science
  and Technology
Korea [Seoul]

Kuiwon Choi
Korea Institute of Science and
  Technology
Korea [Seoul]

Young Choi
GTC-Korea
Korea [Seoul]

Ken Chong
National Institute of Standards
  and Technology
USA [Arlington-US; Arlington-final;
  contributor]

Myung-Ae Chung
Electronics and Telecommunications
  Research Institute
Korea [Seoul]

Vanessa Clive [webcast]

Brooke Coley
National Science Foundation USA
  [Arlington-final]

Clark Cooper
National Science Foundation USA
  [Arlington-final]

Khershed Cooper
Naval Research Laboratory USA
  [Arlington-final]

Genya Dana
U.S. Department of State
USA [webcast]

Frederica Darema
Air Force Office of Scientific Research
USA [Arlington-US; contributor]

Heather Dean
AAAS/ National Science Foundation
USA [Arlington-final]

Jo De Boeck
IMEC-International
Belgium [Leuven]

Gilbert Declerck
IMEC-International
Belgium [Leuven]

Michael DeHaemer
World Technology Evaluation Center,
  Inc.
USA [Arlington-US; Arlington-final]

Roger De Keersmaecker
IMEC
Belgium [Leuven]

Ivan de la Vega
Universidad Simón Bolivar
Venezuela [São Paulo]

Mamadou Diallo
California Institute of Technology, and
Korea Advanced Institute of Science
  and Technology
USA [Arlington-US; Leuven; Seoul;
  Beijing; Arlington-final; contributor]

Alain C. Diebold
University at Albany
USA [contributor]

Eduardo do Couto
Center for Strategic Studies of the
  Government of Brazil [São Paulo]

Andreas Doenni
National Institute for Materials
  Science
Japan [Seoul]

Robert Doering
Semiconductor Research Corp.
USA [contributor]

Maurits Doorn
Netherlands Organization for Applied
  Scientific Research
Netherlands [São Paulo]

Appendix B

Dalci Maria dos Santos
Federal University of São Paulo
Brazil [São Paulo]

Judith Droitcou
Government Accountability Office
USA [Arlington-final]

Jake Dunagan
The Institute for the Future
USA [Arlington-US; contributor]

Albert Duschl
Paris-Lodron University
Germany [Leuven]

Leon Esterowitz
National Science Foundation
USA [Arlington-US]

Tarek Fadel
National Nanotechnology Coordination Office USA [Arlington-final]

David Feldman
University of California, Irvine
USA [Arlington-US; contributor]

Placid Matthew Ferreira
University of Illinois
    at Urbana–Champaign
USA [Arlington-US; contributor]

Lelio Fellows Filho
Center for Strategic Studies of the
    Government of Brazil
Brazil [São Paulo]

Joe Filvarof
Woodrow Wilson International Center
    for Scholars
USA [Arlington-final]

Erik Fisher
Arizona State University
USA [Arlington-US; contributor]

Patricia Foland
World Technology Evaluation Center, Inc.
USA [São Paulo, Arlington-US; Leuven; Seoul; Beijing; Arlington-final]

Stephen J. Fonash
Pennsylvania State University
USA [contributor]

Christy Foran
U.S. Army Engineer Research and
    Development Center
USA [Arlington-final]

Lisa Friedersdorf
National Nanotechnology Coordination
    Office
USA [Arlington-final]

Takahiro Fujita
National Institute for
    Materials Science
Japan [Seoul]

James Gavigan
European Commission
EU [Arlington-final]

Andre Gazso
Austrian Academy of Sciences
Austria [webcast]

Erik Goodman
Michigan State University
USA [contributor]

Lieve Goorden
University of Antwerp
Belgium [webcast]

Michael E. Gorman
University of Virginia
USA [webcast; contributor]

Aura Gimm
AAAS/Department of Defense
USA [webcast]

Piotr Grodzinski
National Institutes of Health
USA [Arlington-US; Arlington-final; contributor]

Martin Grueber
Battelle
USA [contributor]

Ram Gupta
National Science Foundation USA
[Arlington-final]

Bruce Hamilton
National Science Foundation
USA [contributor]

Hyejin Han
Korea Food and Drug Administration
Korea [Seoul]

Barbara Harthorn
University of California Santa Barbara
USA [Leuven; webcast; contributor]

Tsuyoshi Hasegawa
National Institute for Materials Science
Japan [Seoul]

James Heath
California Institute of Technology
USA [Arlington-US; contributor]

Matthew Henderson
World Technology Evaluation Center, Inc.
USA [Arlington-US; Leuven; Seoul; Beijing; Arlington-final]

Michael Hochella, Jr.
Virginia Polytechnic Institute and State University
USA [Arlington-US; webcast; contributor]

Geoffrey Holdridge
National Nanotechnology Coordination Office
USA [Arlington-US; Arlington-final]

Greg Horowitt
T2 Venture Capital
USA [contributor]

Bin Hu
Lanzhou University
PRC [Beijing]

Frank Huband
World Technology Evaluation Center, Inc.
USA [Arlington-US; Arlington-final]

Victor Hwang
T2 Venture Capital
USA [contributor]

Yoon-Hwae Hwang
Pusan University
Korea [Seoul]

Sujin Hyung
Korea Institute of Science and Technology
Korea [Seoul]

Takanori Ichiki
University of Tokyo
Japan [Seoul]

Ramjitti Indaraprasirt
NANOTEC
Thailand [webcast]

Chennupati Jagadish
Australian National University
Australia [Beijing; webcast]

Heejin Jang
Technovalue
Korea [Seoul]

In Soo Jeon
Ministry of Education, Science, and Technology
Korea [Seoul]

Sang-Ki Jeong
Korea Institute of S&T Evaluation and Planning
Korea [Seoul]

Tianzi Jiang
National Laboratory of Pattern Recognition
The Chinese Academy of Sciences
PRC [Beijing]

Xingyu Jiang
National Center for Nanoscience & Technology
PRC [Beijing]

Minwoong Joe
Korea Institute of Science and Technology
Korea [Seoul]

Patricia Johnson
Johnson Edits
USA [Arlington-final]

Lawrence Kabacoff
Office of Naval Research
USA [Arlington-US]

Tom Kalil
Office of Science and Technology Policy
USA [Arlington-US]

Hana Kang
Hanyang University
Korea [Seoul]

Keon Wook Kang
Seoul National University
Korea [Seoul; webcast]

Man Kyu Kang
Hanyang University
Korea [Seoul]

Sungchul Kang
Korea Institute of Science and Technology
Korea [Seoul]

Barbara Karn
National Science Foundation USA [Arlington-final]

Tomoji Kawai
Osaka University
Japan [Seoul]

Seiichiro Kawamura
Japan Science and Technology Agency
Japan [Seoul]

Takeshi Kawano
Toyohashi University of Technology
Japan [Seoul]

Mitsuo Kawato
Advanced Telecommunications Research Institute International
Japan [Seoul]

Pradeep Khosla
Carnegie Mellon University
USA [Arlington-US]

Chang Woo Kim
National Nanotechnology Policy Center
Korea [Seoul]

CheolGi Kim
Chungnam National University
Korea [Seoul]

Hak Min Kim
Korea Institute of Machinery and Materials
Korea [Seoul]

Hyun Chul Kim
GF-Chem
Korea [Seoul]

In-Suk Kim
Sungkyunkwan University
Korea [Seoul]

Ji-Hwan Kim
Korea National University
Korea [Seoul; webcast]

Jin Hyung Kim
Ministry of Education, Science, and Technology
Korea [Seoul]

Jiyoung Kim
Kookmin University
Korea [Seoul]

Junghwan Kim
Sejong University
Korea [Seoul]

Kyoung-Jae Kim
Korea Institute of Science and Technology
Korea [Seoul]

Myung Joon Kim
Electronics and Telecommunications Research Institute
Korea [Seoul]

Sangtae Kim
Mortgridge Institute
USA [Arlington-US; contributor]

Sukpil Kim
Korea Institute of S&T Evaluation and Planning
Korea [Seoul]

Yong Joo Kim
Korea Electrotechnology Research Institute
Korea [Seoul]

Young Keun Kim
Korea University
Korea [Seoul]

Younsun Kim
SungKyunKwan University
Korea [Seoul]

Julian Kinderlerer
University of Cape Town
South Africa [Leuven]

Chin Hua Kong
Indiana University
USA [Arlington-US; contributor]

Todd Kuiken
Woodrow Wilson International Center for Scholars
USA [Leuven; contributor]

Anastasia Kuusk
University of South Australia
Australia [webcast]

Duygu Kuzum
Stanford University
USA [contributor]

Ick Chan Kwon
Korea Institute of Science and Technology
Korea [Seoul; webcast]

Ojung Kwon
Hanyang University
Korea [Seoul]

Robert Langer
Massachusetts Institute of Technology
USA [Arlington-US; contributor]

Mariano Laplane
Center for Strategic Studies of the Government of Brazil [São Paulo]

Deukhee Lee
Korea Institute of Science and Technology
Korea [Seoul]

Eungsug Lee
Korea Institute of Machinery & Materials
Korea [Seoul]

Geun Jae Lee
Ministry of Education, Science, and Technology
Korea [Seoul]

Haiwon Lee
Hanyang University
Korea [Seoul]

Jong-Gu Lee
Ministry of Food and Drug Safety
Korea [Seoul]

Jo-Won Lee
Hanyang University
Korea [Seoul; contributor]

Appendix B 505

Kwang Ryeol Lee
Korea Institute of Science and
 Technology
Korea [Seoul]

Kwyro Lee
National NanoFab Center
Korea [Seoul]

Kyu Back Lee
Korea National University
Korea [Seoul]

Sang-Rok Lee
Korea Institute of Machinery &
 Materials
Korea [Seoul]

Soobum Lee
Hanyang University
Korea [Seoul]

Soo-Hong Lee
Sejong University
Korea [Seoul]

Young Jik Lee
Electronics and Telecommunications
 Research Institute
Korea [Seoul]

Errol Levy
European Union
[Arlington-final]

Wei Li
International Joint Cancer Institute, The
 Second Military Medical University
PRC [Beijing; webcast]

Chuck Liarakos
National Science Foundation
USA [Arlington-US; contributor]

Hanjo Lim
Ajou University
Korea [Seoul]

Stuart Lindsay
Arizona State University
USA [contributor]

Igor Linkov
U.S. Army Corps of Engineers
USA [contributor]

Minghua Liu
Institute of Chemistry, Chinese
 Academy of Sciences
PRC [Beijing]

Witold Lojkowski
Institute of High Pressure Physics
Poland [webcast]

Aaron Lovell
Woodrow Wilson International Center
 for Scholars
USA [Arlington-US; Arlington-final]

Ian Lowe
Griffith University
Australia [Beijing]

Max Lu
University of Queensland
Australia [Beijing]

Mark Lundstrom
Purdue University
USA [Arlington-US; Leuven; Seoul;
 Beijing; Arlington-final; contributor]

Xiaomin Luo
BGI (Genomics)
PRC [Beijing]

Elizabeth Lyons
U.S. Department of State
National Science Foundation USA
 [Arlington-final]

Alexander MacDonald
National Aeronautics and Space
 Administration
USA [Arlington-US; Arlington-final;
 contributor]

Donald MacGregor
MacGregor Bates, Inc.
USA [contributor]

Marc Madou
University of California, Irvine
USA [Arlington-US; contributor]

Roop Mahajan
Virginia Polytechnic Institute and State University
USA [Arlington-US; contributor]

Jonathan Manton
University of Melbourne
Australia [Beijing; contributor]

Mira Marcus-Kalish
Tel Aviv University
Israel [Leuven; webcast]

Manuel Mari
Ministry of Science and Technology
Argentina [São Paulo]

René Martins
European Commission
Belgium [Leuven; webcast]

Robert Mason
University of Washington
USA [Arlington-US; contributor]

Kazuyo Matsubae
Tohoku University
Japan [Seoul]

Anne McLaughlin
North Carolina State University
USA [Arlington-US; webcast; contributor]

Michael A. Meador
National Aeronautics and Space Administration
USA [Arlington-US; Arlington-final; contributor]

Celia Merzbacher
Semiconductor Research Corporation
USA [contributor]

Javier Medina Vasquez
Universidad de Cali
Colombia [São Paulo]

Paolo Milani
University of Milan
Italy [Leuven]

Evando Mirra de Paula e Silva
Center for Strategic Studies of the Government of Brazil [São Paulo]

Tanya Monro
University of Adelaide
Australia [Beijing]

Luc Mortelmans
Department of Nuclear Medicine
University Hospital "Gasthuisberg",
Leuven
Belgium [Leuven]

James Murday
University of Southern California
USA [Arlington-US; Leuven; Seoul; Beijing; Arlington-final; contributor]

Padraig Murphy
Dublin City University
Ireland [webcast]

Shinnya Nakamoto
Japan Science and Technology Agency
Japan [Seoul]

Milos Nesladek
Academy of Sciences of the Czech Republic
University of Hasselt
MINATEC
Czech Republic [Leuven]

Amy Newhart
National Science Foundation USA [Arlington-final]

Leah Nichols
National Science Foundation USA [Arlington-final]

Alfred Nordmann
Darmstadt Technical University
Germany [Leuven]

James Olds
George Mason University
USA [Arlington-US; Leuven; Seoul; Beijing; Arlington-final; contributor]

Aude Oliva
Massachusetts Institute of Technology
USA [Arlington-US; webcast; contributor]

Fernando Ortega de San Martin
National Council for Science, Technology and Innovation
Peru [São Paulo]

Halyna Paikoush
World Technology Evaluation Center, Inc.
USA [Arlington-US; Arlington-final]

Y. Eugene Pak
Korea National University
Korea [Seoul]

Choonsun Park
SK Hynix, Inc.
Korea [Seoul]

Dae-Hwan Park
Ewha Womans University
Korea [Seoul]

Gung won Park
Sungeun Tech Co., Ltd.
Korea [Seoul]

Migyoung Park
Hanyang University
Korea [Seoul]

Sung Ha Park
SungKyunKwan University
Korea [Seoul]

Sung Kee Park
Korea Institute of Science and Technology
Korea [Seoul]

Abani Patra
University of Buffalo
USA [Arlington-US; contributor]

Lianmao Peng
Peking University
PRC [Beijing]

Peter Peumans
IMEC
Belgium [Leuven]

Robert Pohanka
National Nanotechnology Coordination Office
USA [Arlington-US; Arlington-final]

Melur Ramasubramanian
National Science Foundation
USA [Arlington-US]

Sohi Rastegar
National Science Foundation
USA [Arlington-US]

Paulo Antonio Rebelo
PETROBRAS
Brazil [São Paulo]

David Rejeski
Woodrow Wilson International Center for Scholars
USA [Arlington-US; Arlington-final; contributor]

Fernando Rizzo
Center for Strategic Studies of the Government of Brazil [São Paulo]

Jessica Robin
National Science Foundation
USA [contributor]

Douglas Robinson
TEQNODE
France [webcast]

Mihail C. Roco
National Science Foundation
USA [São Paulo, Arlington-US; Leuven; Seoul; Beijing; Arlington-final; contributor]

Françoise Roure
Organization for Economic
 Co-operation and Development
France [Leuven; webcast]

Sylvie Rousset
National Center for Scientific Research
 (CNRS)
France [Leuven]

Philip Rubin
Office of Science and Technology
 Policy
USA [Arlington-US; Arlington-final;
 contributor]

Skip Rung
Oregon Nanoscience and Microtech-
 nologies Institute
USA [contributor]

Alvaro Saavedra
PETROBRAS
Brazil [São Paulo]

Paul Saffo
Discern Analytics
USA [Arlington-US]

Maurizio Salvi
European Commission
Belgium [Leuven]

Yves Samson
Alternative Energies and Atomic
 Energy Commission
France [Leuven; webcast]

Anders Sandberg
University of Oxford
UK [Leuven]

Nora Savage
U.S. Environmental Protection Agency
USA [Arlington-US; Arlington-final;
 contributor]

Walt Scacchi
University of California, Irvine
USA [Arlington-US; contributor]

Brian Scassellati
Yale University
USA [Arlington-US; contributor]

Daan Schuurbiers
De Proeffabriek
Centre for Society and the Life
 Sciences
The Netherlands [Leuven]

Carla Schwingel
University of Mackenzie-São Paulo
Brazil [São Paulo]

Norman Scott
Cornell University
USA [webcast; contributor]

Cynthia Selin
Arizona State University
USA [contributor]

Jeongeun Seo
Hanyang University
Korea [Seoul; webcast]

Robert Shelton
World Technology Evaluation Center,
 Inc.
USA [Arlington-US; Arlington-final]

Hiroki Shimazu
Japan Science and Technology Agency
Japan [Seoul]

Kwang Min Shin
National Nanotechnology Policy Center
 (NNPC)
Korea [Seoul]

Min Cheol Shin
Korea Advanced Institute of Science
 and Technology
Korea [Seoul]

Jaihoon Sim
SK Hynix, Inc.
Korea [Seoul]

Woo Young Sim
NanoEnTek, Inc.
Korea [Seoul]

Appendix B

Lew Sloter
Department of Defense
USA [Arlington-US]

Dae Sup So
National Nanotechnology Policy
    Center
Korea [Seoul]

Fernanda Sobral
Center for Strategic Studies of the
    Government of Brazil [São Paulo]

Keun Hyoung Sok
Sungkyunkwan University
Korea [Seoul]

Jong-Guk Song
Science & Technology Policy Institute
Korea [Seoul]

Yibin Song
Hanyang University
Korea [Seoul]

Eunezio Souza
Latin American Network for
    Technological Convergence
    Prospective
Brazil [São Paulo]

James C. Spohrer
IBM Corporation
USA [Arlington-US; contributor]

Mark Suchman
Brown University
USA [Arlington-US; contributor]

Sang-Hee Suh
Korea Nanotechnology Researchers
    Society
Korea [Seoul]

Jyrki Suominen
European Commission
Belgium [Leuven]

Kohyama Takashi
Hokkaido University
Japan [Seoul]

Masahiro Takemura
National Institute for Materials Science
Japan [Seoul]

Suk-Wah Tam-Chang
National Science Foundation USA
    [Arlington-final]

Kazunobu Tanaka
Japan Science and Technology Agency
Japan [Seoul]

Bo Tang
Shandong Normal University
PRC [Beijing]

Clayton Teague
Independent Nanotechnology
    Consultant
USA [Arlington-US; contributor]

George Thompson
Intel
USA [Arlington-US; contributor]

Sally Tinkle
National Nanotechnology Coordination
    Office
USA [Arlington-US]

Christos Tokamanis
European Commission
Belgium [Leuven; webcast; contributor]

Mario Tokoro
Keio University and Sony Computer
    Science Laboratories
Japan [contributor]

Bruce Tonn
University of Tennessee
USA [Arlington-US; Leuven; Seoul;
    Beijing; Arlington-final; contributor]

Kohei Uosaki
National Institute for Materials Science
Japan [Seoul]

Robert G. Urban
Johnson & Johnson
USA [Arlington-US; Leuven; Seoul;
    Beijing; contributor]

Cesar Villarroel
Universidad Privada Bolivariana
Bolivia [São Paulo]

Luc Van den hove
IMEC
Belgium [Leuven]

Johan Van Helleputte
IMEC
Belgium [Leuven]

Marc Van Rossum
IMEC
Belgium [Leuven]

Herbert von Bose
European Commission
Leuven; Belgium [Leuven]

Gordon Wallace
University of Wollongong
Australia [Beijing]

Xiaomin Wang
Capital Medical University
PRC [Beijing]

George Whitesides
Harvard University
USA [Arlington-US; Leuven; Arlington-final; contributor]

Philip Westmoreland
North Carolina State University
USA [Arlington-US; webcast; contributor]

R. Stanley Williams
Hewlett-Packard Laboratories
USA [Arlington-final; contributor]

Frances Wilson
Universidad de Chile
Chile [São Paulo]

H.H. Wito
National Science Foundation USA
[Arlington-final]

H.-S. Philip Wong
Stanford University
USA [Arlington-US; Leuven; Seoul; Beijing; Arlington-final; contributor]

Ke Xu
Suzhou Institute of Nano-Tech and Nano-Bionics (SINANO)
PRC [Beijing]

Eli Yablonovitch
University of California, Berkeley
USA [Arlington-US; contributor]

Heyoung Yang
KISTEP
Korea [contributor]

Fruma Yehiely
Northwestern University
USA [webcast]

Won Jong Yoo
SungKyunKwan University
Korea [Seoul]

Ji Woong Yoon
Kyung Hee University
Korea [Seoul]

Seungjoon Yoon
Korea Environmental Industry and Technology Institute
Korea [Seoul]

Hui-suk Yun
Korea Institute of Material Science
Korea [Seoul]

Moises Zavaleta
North American Free Trade Agreement
Mexico [webcast]

Byoung-Tak Zhang
Korea National University
Korea [Seoul]

Xian'en Zhang
Ministry of Science and Technology
 of China
PRC [Beijing]

Xiang Zhang
University of California, Berkeley
USA [Arlington-US; contributor]

Tingshao Zhu
Institute of Psychology
PRC [Beijing]

Xing Zhu
Peking University
PRC [Beijing]

Appendix C
Abstract of the Converging Technologies Workshop in São Paulo, Brazil, November 2011

Contact: *Esper Abrao Cavalheiro, University of Sao Paulo, Brazil, and Center of Strategic Studies of the Government of Brazil*

With the support of the CGEE (Center for Strategic Studies of the Government of Brazil) and of the Program PROSUL (Brazilian National Research Council, or CNPq), an international symposium on "Converging Technologies" was held in São Paulo, Brazil, in the period of 24–25 November 2011. The symposium was co-sponsored by the World Technology Evaluation Center (WTEC, USA) that completed a survey on converging technologies with the participants and established a private passworded website with translations in English, to be included in an international study. Participants were scientists and members of STI (science, technology, and innovation) agencies of several Latin America countries. Two keynote speakers were especially invited to this symposium: Mihail Roco from the National Science Foundation (NSF, USA) and Maurits Doorn from STCorp (Netherlands).

On the one hand, the participants gave an overview of the initiatives proposed in their countries in the four areas (NBIC), as strictly defined in the converging technologies, i.e., nanotechnology, biotechnology, information technology, and cognitive sciences. The presentations attempted to highlight not only the stage at which these areas are in the Latin America, but also point out the strategic direction of research conducted in their countries. Although, in general, no country has reported the presence of research groups specifically grouped to work in all four areas of NBIC, some of the representatives could, at least, show the existence of isolated initiatives of groups combining two or three areas of the converging technologies. However, it was noticeable that the reinforcement of strategies of CT&I (i.e., "STI"), linking research with the need for socio-economic growth, will conduct the close emerging of NBIC activities in the region, specially focused in researches related to agriculture and livestock, energy, and environment, just to name a few.

The key presentation made by Professor M. Roco focused initially on a review of the original concept and the grounds on which the term "converging technologies" was build up where nanotechnology is the field whose contribution enables and facilitates the interaction between living organisms and devices designed for human beings. In this context, the term "converging technologies" refers to trends in or expectations of synergy in the development of these four scientific areas in order to strengthen them and, by combining them, creating new fields of application. Moreover, it was instructive to review the possible areas of knowledge that could benefit from the use of converging technologies in the search for appropriate solutions to their most pressing issues. But he advised that this working strategy should include changes in the governance of universities, research centers, companies, and government agencies to implement a transforming vision that needs to be inclusive and collaborative in the long run.

Mauritz Doorn from the Science & Technology Experts Group (STCorp, Netherlands) during his presentation pointed out that over the years, various actors took up the promises about nanotechnology and converging technology and their possible impact; the resulting discussions resulting in a considerable amount of hype and counter-hype about negative impacts. In his opinion, the hype and counter-hype discussion has drawn the attention away from actual realities in the field, and this has hampered learning and progress. At the same time, important qualitative changes are occurring, and the question is how actors can anticipate the changes in their respective fields. This can provide a starting point for actors to analyze the dynamics of converging technologies in their own fields of practice, enabling them to better anticipate changes and shape developments.

The discussions that followed the various presentations were very rich and vibrant—and the conclusions were not always convergent. In conclusion, it was extremely important to note that all agreed that the current idea of converging research extends to all fields of knowledge and that this new way of working important human issues could facilitate finding different solutions or similar solutions more quickly. According to Roco, converging technologies implies starting a research project from the problems to be solved using integrative approaches, not starting with the subjects and disciplines involved. It also involves the pursuit of common goals, shared theories, and shared approaches to work, the appreciation of the capabilities and achievements of people, and anticipating and managing opportunities and risks.

Appendix D
Review of NBIC Visionary Goals

Contact: *William Sims Bainbridge, National Science Foundation*

The primary focus of the 2011–2012 series of NBIC2 workshops was to consider action plans for the next decade of convergence, both in fine detail and in consideration of a CKTS Initiative.[1] Longer-term visions have therefore been kept in the background, yet they are crucial for planning near-term research and development priorities. The 2001 NBIC study and report (Roco and Bainbridge 2003) was a bridge to a new frontier, but there are mountains in the distance that must be climbed after the next decade and that must be approached by the right path. This appendix summarizes some of the earlier NBIC visions for the longer-term future, and extracts representative visions for the future from the ten chapters of this report.

There have been several Converging Technologies conferences, studies, and reports beginning in 2001, which have offered many inspiring visions of the future, but not all of them were positive. Fear can be just as motivating as hope, so candid consideration of the most difficult problems facing humanity needs to be part of the discussion. Most participants agreed that the NBIC2 vision can be achieved, and doing so will create a better future. But it also seems the case that humanity faces grave dangers and may not be able even to sustain current levels of prosperity and peace without focusing on convergence. This appendix begins by emphasizing the reasonable but grand hopes of earlier conferences, then poses some questions

---

[1] Editor's note: The acronym NBIC refers to the convergence of nanotechnology, biotechnology, information technology, and cognitive science. CKTS refers to a key concept in this report, the convergence of knowledge and technology for (the benefit of) society. The first major NBIC conference was held Arlington, VA, in 2001; there were follow-on NBIC conferences in 2003 (Los Angeles, CA), 2004 (New York, NY), and 2005 (Kona, HI). This 2011–2012 workshop series and 2013 report revisit technology convergence issues. This study originally was termed NBIC2 (meaning "beyond NBIC"), but because it focuses its discussion and recommendations on how convergence can address societal needs—including its recommendation of a Federal CKTS initiative—the study and report now also take the acronym CKTS.

based on the discussions of the most recent conference and reflecting the tension between hope and fear.

Besides reviewing the 20 visionary ideas put forward in 2001, this appendix offers 80 other ideas drawn from the full range of applications considered at the five NBIC studies, including this one. Some are framed as predictions and some as problems. Each one could deserve a chapter in its own right, and 40 from the 2013 study are indeed discussed in one or more of the chapters of this report. The goal of this brief appendix is not to analyze any of the ideas in depth but to raise them for the reader's own consideration. This 2013 report is especially focused on relatively near-term challenges and applications, so this appendix is designed to place the discussion of convergence of knowledge and technology in a somewhat longer historical context, anchored in the past but looking toward the more distant future.

The 20 Ideas from the Original Report

The original NBIC report listed 20 applications of technological convergence that could benefit humanity in a time frame of 10–20 years (Roco and Bainbridge 2003, 5–6). In 2005, 26 participants in the NBIC meetings of that year completed a questionnaire that used the Delphi technique, asking each of them to predict the year in which each of the 20 applications would be "substantially achieved" and how positive or beneficial that would be on a scale from 0 to 10. Of course, no one can predict precisely, but the median predicted date for achievement of each application was a reasonable indicator of how much research and development time was likely to be required, based on the judgments of those experts on NBIC convergence. The median was the appropriate measure, because it takes account of those cases in which some respondents predicted the application would never be achieved, although in most cases respondents put a date within the current century. The 26 respondents to the 2005 questionnaire predicted on average that three of the 20 application goals would be substantially achieved within a decade, that is, by 2015:

1. Anywhere in the world, an individual will have instantaneous access to needed information, whether practical or scientific in nature, in a form tailored for most effective use by the particular individual.
2. New organizational structures and management principles based on fast, reliable communication of needed information will vastly increase the effectiveness of administrators in business, education, and government.
3. Comfortable, wearable sensors and computers will enhance every person's awareness of his or her health condition, environment, chemical pollutants, potential hazards, and information of interest about local businesses, natural resources, and the like.

Indeed, from the perspective of 2013, it seems the conference participants were right. The technology is in place for each of these 2015 predictions, and the remaining questions about them primarily concern implementation. The phrase in prediction one, "most effective use by the particular individual," suggests an open-ended goal, because personalization of information resources can always be improved. Yet for a very wide range of current information needs, desired resources can be consulted from any location that has Internet access. Search engines can be personalized, but it is also easy for individuals to personalize their access to many specific digital libraries and archives. Not many people really have the skills yet to gain maximum benefit, and we may be passing through a period in which a skilled minority temporarily has a great advantage over others, including scientists able to do research better and faster than colleagues whose information technology skills are weaker. However, increasing fractions of the population have the skills, and the quality of interfaces continues to improve, so the first prediction will apply to more and more individual people with each passing year.

Regarding prediction two, many new organizational structures, especially those called "virtual organizations" that employ information technologies to overcome geographic distance, have indeed been developed on the basis of a large number of support technologies. Many of these shift responsibilities from one kind of management position to another, and thus they may also shift power relations between groups within an organization and between organizations. Thus there can be both social resistance to organizational change and interest-based distortions in how change is implemented. These are topics for investigation by the social sciences, which NBIC always conceptualized as the next in line to join the convergence, probably via unification with cognitive science.

A very large fraction of the population currently carries computer-like smart phones or other mobile devices that are equivalent to computers, and many kinds of sensors have been developed to work with them, to monitor either health conditions of an individual or environmental conditions. Implementation of many specific applications implied by prediction three has lagged somewhat, probably due to cost considerations, both the trade-offs for individual money investments in a time of economic problems, and the cost of effort required to learn about and use new devices.

Conveniently, these three predictions suggest valid points about application predictions in general. As the due date for a prediction approaches, especially if it is a rather correct prediction, then the meanings of its terms are likely to change, becoming refined and sometimes significantly revised. Each of these three also highlights the importance of a particular context. First, satisfaction of individual needs requires people to articulate what they need and to learn skills that maximize their ability to exploit resources. Second, the organizational context may retard, accelerate, or direct change, as groups and other social forces compete and cooperate in making decisions about the implementation of new technologies. Third, even innovation has a cost, and in a world rich in innovations, economic markets can play crucial roles in determining which applications succeed. The respondents to the

2005 questionnaire placed 8 of the 20 predictions about 15 years into their future, becoming substantially achieved around the year 2020:

4. People from all backgrounds and of all ranges of ability will learn valuable new knowledge and skills more reliably and quickly, whether in school, on the job, or at home.
5. Individuals and teams will be able to communicate and cooperate profitably across traditional barriers of culture, language, distance, and professional specialization, thus greatly increasing the effectiveness of groups, organizations, and multinational partnerships.
6. National security will be greatly strengthened by lightweight, information-rich war fighting systems, capable uninhabited combat vehicles, adaptable smart materials, invulnerable data networks, superior intelligence-gathering systems, and effective measures against biological, chemical, radiological, and nuclear attacks.
7. Engineers, artists, architects, and designers will experience tremendously expanded creative abilities, both with a variety of new tools and through improved understanding of the wellsprings of human creativity.
8. Average persons, as well as policymakers, will have a vastly improved awareness of the cognitive, social, and biological forces operating in their lives, enabling far better adjustment, creativity, and daily decision-making.
9. Factories of tomorrow will be organized around converging technologies and increased human–machine capabilities as intelligent environments that achieve the maximum benefits of both mass production and custom design.
10. Agriculture and the food industry will greatly increase yields and reduce spoilage through networks of cheap, smart sensors that constantly monitor the condition and needs of plants, animals, and farm products.
11. The work of scientists will be revolutionized by importing approaches pioneered in other sciences, for example, genetic research employing principles from natural language processing and cultural research employing principles from genetics.

Aspects of all eight of these applications have been achieved, although by no means completely, and again, the factors determining implementation are important. Several of them express optimism that human beings will possess greater understanding and thus be more effective in achieving their personal and professional goals. This contrasts with the rather significant societal pessimism that began with the financial crises 2 or 3 years after the predictions were made, and the increasing dissatisfaction with their governments acutely felt by citizens of Europe, Japan, the United States, and other technologically advanced nations. This observation suggests that these predictions should be translated into prescriptions, things we need to achieve to get the world back on the track to progress.

The national security prediction was somewhat controversial, as indicated by the fact it received the lowest "good" score of any of the 20, 5.5 on the scale from 0 to 10. Most of the predictions got scores in the 7–9 range. Of course, what this suggests is that human beings—both scientists and engineers are humans—differ in their values and how they might vote in a referendum about society's defense policies. No one predicted, "Military technology will be unnecessary, because humanity will have found a reliable path to peace." Yet the respondents differed in whether they viewed some of the technology as security-enhancing or as ultimately dangerous. The following three predictions with median dates around 2025 are not close to achievement now:

12. Robots and software agents will be far more useful for human beings, because they will operate on principles compatible with human goals, awareness, and personality.
13. The human body will be more durable, healthier, more energetic, easier to repair, and more resistant to many kinds of stress, biological threats, and aging processes.
14. A combination of technologies and treatments will compensate for many physical and mental disabilities and will eradicate altogether some handicaps that have plagued the lives of millions of people.

Perhaps by pure chance they collectively identify a crucial question for mid-term thinking about progress related to technological convergence: To what extent will progress be achieved by improving human beings, versus improving human-centered machines? One can take either side of this argument, or try to split the difference by talking about human–machine collaboration—as the National Robotics Initiative does when referring to "co-robots" designed to work well with humans. It may be that a different answer will prove to be best for each application. But the question clearly has profound policy, ethical, and social implications. An additional five predictions were scheduled for substantial achievement around the year 2030, and they appear to cover five very separate areas of achievement rather than sharing one or two common principles:

15. Fast, broadband interfaces between the human brain and machines will transform work in factories, control automobiles, ensure military superiority, and enable new sports, art forms, and modes of interaction between people.
16. Machines and structures of all kinds, from homes to aircraft, will be constructed of materials that have exactly the desired properties, including the ability to adapt to changing situations, high energy efficiency, and environmental friendliness.
17. The ability to control the genetics of humans, animals, and agricultural plants will greatly benefit human welfare; widespread consensus about ethical, legal, and moral issues will be built in[to] the process.
18. Transportation will be safe, cheap, and fast, due to ubiquitous real-time information systems, extremely high-efficiency vehicle designs, and the use of synthetic materials and machines fabricated from the nanoscale for optimum performance.
19. Formal education will be transformed by a unified but diverse curriculum based on a comprehensive, hierarchical intellectual paradigm for understanding the architecture of the physical world from the nanoscale through the cosmic scale.

The final prediction, set around the year 2050, concerns what is popularly called the "Final Frontier," namely a practical goal for the space exploration program:

20. The vast promise of outer space will finally be realized by means of efficient launch vehicles, robotic construction of extraterrestrial bases, and profitable exploitation of the resources of the Moon, Mars, or near-Earth-approaching asteroids.

There is much room to debate what exactly the promise of outer space is, although this item suggests one particular view on this difficult question. Mention of near-Earth asteroid mining usually implies a new source of rare mineral resources for terrestrial industry. Discussions of exploiting Martian resources could have a very different goal, the colonization of Mars itself. Robotic construction of extraterrestrial bases could be an efficient way of preparing for human habitation without the initial cost of creating a viable ecology and economy prior to the point at which many extraterrestrial resources could be exploited. And efficient launch vehicles would be required to support any significant expansion of physical human activity across the solar system.

Given that nanotechnology was the historical starting-point for NBIC convergence, it is possible to raise a serious issue about exploitation of extraterrestrial resources for terrestrial benefit. The Earth itself has all the chemical elements required for the full range of technologies, and a variety of new means for concentrating rare elements could certainly be devised, from robotic mining of the seabed to scrubbing the wastes that flow from our civilization. Nanoscience stresses the importance of how matter is structured on the nanoscale, thereby reducing the need for vast quantities of rare elements. Considering NBIC further, information can be more valuable than gold, and some of the most valuable materials are created by geological or biological processes native to our own planet.

Thus, if the goal is to increase the power of industrial society, NBIC convergence is the way to achieve it, not mining distant objects in the solar system. If the goal is to increase human understanding of the universe, then indeed, activities in outer space will be important, but probably in the form of deep space probes, robot landers exploring Europa or Titan, and orbiting telescopes scanning the universe across the full electromagnetic spectrum. If the goal is to expand civilization, the colonization of Mars may be an essential step, but so too will be expansion of cognitive science into the realm of cultural diversity.

By placing the promise of outer space farthest into the future of the 20 applications, the NBIC participants explicitly recognized how difficult it will be. But they also imply another perhaps more important truth: As we progress into a bright future, we will come to see things in a different light.

20 Selected Ideas from the 2005 Predictions—2030 and Beyond

The 2005 questionnaire included an additional 50 applications (Bainbridge and Roco 2006a: 340–344) culled from the first three NBIC reports (Roco and Bainbridge 2003; Roco and Montemagno 2004; Bainbridge and Roco 2006b); 20 of these

Appendix D 521

were not expected to be fulfilled until after the first third of the twenty-first century. The median predicted dates for all 50 were within this century, although we can expect great uncertainty about those that the respondents placed rather far in the future. A quarter of this group were judged to be achievable late in the decade of the 2030s:

1. A fresh scientific approach to culture, based on concepts from evolutionary biology and classification techniques from information science, will greatly facilitate humanities scholarship, marketing of music or literature, and artistic innovation.
2. Three-dimensional printers will be widely used not only for rapid prototyping but also for economical, local, on-demand manufacture of art objects, machine parts, and a host of other things from a variety of materials.
3. It will be technically and economically possible to sequence the genetic code of each unique individual, so we will fully understand genetic variations in human performance.
4. Nano-enabled sensors, implanted inside the human body, will monitor metabolism and health, diagnosing any health problem before the person even notices the first symptom.
5. Assistive technologies will largely overcome disabilities such as blindness, deafness, and immobility.

In fact, the first three of these now look achievable much sooner than the respondents predicted back in 2005. The "fresh scientific approach to culture" is covered in Chap. 2 of this report, three-dimensional (3D) printers are central to the distributed manufacturing covered in Chap. 7, and personalized genetic sequencing appears in Chap. 6. The predictions about implanted sensors and assistive technologies were phrased in somewhat extravagant terms, but progress along these lines is already happening today. Another five predictions were placed around the year 2040:

6. Humane machines will adapt to and reflect the communication styles, social context, and personal needs of the people who use them.
7. A combination of techniques will largely nullify the constraints associated with a human's inherent ability to assimilate information.
8. Science will achieve great progress in understanding and predicting the behavior of complex systems, at multiple scales and between the system and the environment.
9. A new form of computing will emerge in which there is no distinction between hardware and software, and biological processes calculate the behavior of complex, adaptive systems.
10. Nanoscale molecular motors will be mass-produced to perform a variety of tasks, in fields as diverse as materials manufacturing and medical treatment.

Achievement of number 6 above, which researchers in human-centered computing call *personalization* and *context-aware systems*, is a matter of degree and thus not really capable of having a specific year of achievement. The following four would all seem to require rather major scientific and technological breakthroughs, conceivable and perhaps even capable of being described in some detail, but not achievable in the near term.

The next seven predictions have that same quality and were placed between 2045 and 2050:

11. Warfighters will have the ability to control vehicles, weapons, and other combat systems instantly, merely by thinking the commands or even before fully forming the commands in their minds.
12. New research tools will chart the structure and functions of the human mind, including a complete mapping of the connections in the human brain.
13. Molecular machines will solve a wide range of problems on a global scale.
14. Memory enhancement will improve human cognition, by such means as external electronic storage and infusion of nerve growth factors into the brain.
15. A predictive science of the behavior of societies will allow us to understand a wide range of socially disruptive events and allow us to put mitigating or preventive strategies in place before the harm occurs.
16. A nano-bio processor will be developed that can cheaply manufacture a variety of medicines that are tailored to the genetic makeup and health needs of an individual.
17. Nanorobots will perform surgery and administer treatments deep inside the human body, achieving great health benefits at minimum risk.

These last seven applications appear to reflect slightly greater doubts about feasibility and a sense that more work needed to be done to accomplish these seven goals. The statement about a "predictive science of the behavior of societies" is challenging in a different way from the others, because the emphasis in all the Converging Technologies reports on societal implications requires inputs from the best available social science. If we cannot predict the behavior of societies, then we cannot anticipate the second-order consequences of technological innovations.

Ability to predict and accept varying societal behavior sets the basis for contemplating two predictions not expected to be fulfilled for an additional generation, around the year 2070:

18. Scientists will be able to understand and describe human intentions, beliefs, desires, feelings, and motives in terms of well-defined computational processes.
19. Rather than stereotyping some people as disabled, or praising others as talented, society will grant everybody the right to decide for themselves what abilities they want to have.

Put together, these two predictions raise the specter of social control versus freedom. If scientists really understood their fellow human beings, they could have great power over them. But our job is not to force people to fulfill our hopes for them, but to give people the means to achieve their own goals, even if often those goals may be modest in scope.

The final prediction in this set, with a median fulfillment date of 2085, is far out, both technically and ethically:

20. The computing power and scientific knowledge will exist to build machines that are functionally equivalent to the human brain.

If that capability comes into being, should we use it? Countless science fiction authors have imagined possible consequences of such a development, but their aim has been to provide an interesting focus of conflict to motivate an exciting story, not to develop serious design principles for technologies that would enhance human life. A more subtle issue concerns whether "functionally equivalent" means that the machines would operate on the same principles as the human brain, rather than using very different principles to simulate the human mind. In either case, developing such machines would require much deeper understanding of the nature of intelligence than we currently posses, and thus there could be great consequences for our conceptions of ourselves, even if the machines never left the research laboratories.

Ideas, Questions, and Tradeoffs Suggested by the 2013 CKTS Study

This section and the following one together contain 40 ideas about the future culled from this report and presented more as challenges than as predictions. They are not discussed here in the context of when each might come to closure, but rather are offered as issues for the reader to contemplate and to evaluate on the basis of the reader's own professional expertise and personal values. This first set of 20 are phrased explicitly as questions, perhaps to be answered by future research, or dilemmas to be resolved through logical analysis. After each question, the chapter number is given in which it is discussed or implied, although information from all parts of this report may be relevant.

1. Can transition to a knowledge-based economy provide new routes to upward social mobility for talented individuals without rendering many unskilled people totally unemployable? (Chap. 1)
2. Will the pervasive use of information technology bring social groups together or push them further apart? (Chap. 1)
3. Can the National Robotics Initiative realize its "co-robot" principle of designing beneficial collaborations between humans and robots, thereby becoming an example for other fields of engineering to follow? (Chap. 2)
4. Will it prove beneficial to expand convergence far beyond its NBIC origins, for example reaching through information and cognitive science to social science, and beyond that to incorporate the arts and humanities? (Chap. 2)
5. Is it possible to achieve environmental sustainability, for example, in use of energy resources, without causing serious negative political, social, and economic externalities? (Chap. 3)
6. Will the governments and other institutions of the world cooperate sufficiently to establish a comprehensive global system for monitoring the natural environment and the components of society that affect it? (Chap. 3)
7. How rapidly will a science of convergence arise, especially if it proves difficult to discover overarching principles to understand varied traditional fields in terms of new shared concepts? (Chap. 4)

8. What systems of evaluation and governance will make it possible to find the right dynamic balance between convergence and divergence, properly integrating the two in the convergence–divergence cycle? (Chap. 4)
9. Will it prove possible to mobilize scientists, medical professionals, and the general public to transform medicine from a reactive into a proactive practice, preventing illness, detecting illnesses early, and thereby reducing the costs of healthcare? (Chap. 5)
10. Will governments, insurance companies, and healthcare delivery systems find ways to mitigate significantly the demographic challenge of an aging population, in the context of extreme and possibly increasing economic inequality? (Chap. 5)
11. What principles of design and manufacture will make it possible for robots to serve diverse populations, from early childhood education to home living assistance for elderly persons? (Chap. 6)
12. In what ways would human culture be forever changed if a unified theory-based rule set for understanding human cognition were developed? (Chap. 6)
13. What organizational principle, such as franchises or guilds, will prove effective to establish distributed manufacturing systems that simultaneously maximize efficiency and local autonomy? (Chap. 7)
14. Is there any way to revive science-and-technology studies, which proper implementation of technological innovation requires, given that social scientists are now largely barred from doing research inside scientific laboratories, industrial corporations, and government agencies? (Chap. 7)
15. Given the many different kinds of infrastructure required for research, education, and manufacturing demonstration projects, will it be possible to prioritize them in a way that is affordable and does not unbalance the convergence process? (Chap. 8)
16. What methods of data collection and criteria of success should be used to evaluate programs to reform education along convergence lines? (Chap. 8)
17. What combination of energy sources and public policies could move the world away from the use of fossil fuels, without plunging humanity back into the limited resource situation experienced in history? (Chap. 9)
18. Given the many disappointments over the years, how can the nations of the world come together to institute effective environmental protection and resource sustainability? (Chap. 9)
19. What measures will encourage a virtuous spiral of creativity and innovation, versus a vicious circle of degradation? (Chap. 10)
20. If the pessimistic argument that we are reaching the end of scientific and technological progress is correct, would there be any way to transition smoothly to an enduring form of benevolent civilization? (Chap. 10)

Superficially, some of these 20 questions might be answered with a simple "yes" or "no," and others by a single sentence. Yet, really, each of them could be the topic of an extended debate today, and could be answered confidently only after years of convergence experience. They tend to emphasize evolution of the technological system based on convergence principles, with secondary consequences for human wellbeing, whereas the final 20 questions explicitly focus on social impact dilemmas that may be created by technology convergence.

20 Potential Social and Ethical Tradeoffs

The point of presenting the following 20 issues as tradeoffs is not to imply each one is an inescapable dilemma, but rather to highlight the ways in which multiple goals are often interrelated. To say that achieving A will prevent achieving B, can be an exhortation to seek a way of achieving both, perhaps through some previously unimagined synergy that removes the conflict between A and B. Some examples are presented as conditionals, rather than tradeoffs, such that A can be achieved, but only if another perhaps even more difficult goal, B, is achieved. Thus, the phraseology in this section is meant to stimulate creative thinking rather than pessimism, with an understanding that many important decisions have implications for each other.

1. Direct stimulation of the human brain will counteract many diseases and disabilities, but in so doing, it raises huge questions about the propriety of technological control of the individual mind. (Chap. 1)
2. Extensive use of open source software will greatly increase the functionality and customizability of information systems, at the cost of innumerable security breaches and cyber attacks. (Chap. 1)
3. The governance of technological innovation could be centralized and authoritarian, or decentralized and democratic. (Chap. 2)
4. Recruitment of nonprofessionals to research teams, in what is often called "citizen science," promises to advance science through their volunteer efforts and to integrate science better into the wider culture, but at the risk of encouraging pseudoscience and politically tainted science. (Chap. 2)
5. General policies based on global risk assessment could benefit humanity as a whole, while damaging the prospects of people living in some specific geographic areas. (Chap. 3)
6. Extensive public debate about the full range of environmental and social implications of nanotechnology could ensure that it achieved maximum human benefit for humanity, or could ensnare progress in popular misconceptions based on fear and ignorance. (Chap. 3)
7. A remarkable range and number of new strategies will be developed for research and development investment and implementation, but one result could be diminished support for tried-and-true older strategies that have not really outlived their usefulness. (Chap. 4)
8. Creation of higher-level, multidomain technical languages could facilitate convergence, but at the cost of a whole new program of education for young scientists and engineers and the obsolescence of older people whose technical training is finished. (Chap. 4)

9. Regenerative medicine and advanced prosthetics will be among the new treatments improving the quality of survivors' lives, while less emphasis may be given to prolonging the lives of the terminally ill. (Chap. 5)
10. Constant monitoring of people's health conditions, using wearable sensors and smart homes, can improve well-being tremendously, at the cost of extensive patient education, deployment of much new and possibly expensive information technology, and potential privacy vulnerabilities. (Chap. 5)
11. Convergent technologies will definitely assist people suffering from mental disabilities, and may do harm to mentally deviant people whose characteristics are simply variant rather than objectively pathological. (Chap. 6)
12. Research and development strategies, both convergent and divergent, can achieve remarkable progress, but only if greatly increased understanding of cognition allows us to implement strategies that are really compatible with the way human minds function. (Chap. 6)
13. With distributed manufacturing methods, capabilities will shift from the hands of the few into the hands of the many, potentially reducing the profits of large corporations and their investors. (Chap. 7)
14. Distributed manufacturing requires shipment of raw materials in small quantities to large numbers of locations, which will increase costs unless methods inspired by nanotechnology can reduce waste, and new systems of transportation can improve efficiency of delivery. (Chap. 7)
15. Online education will lower price and increase accessibility, at the cost of distancing students from their physical subject matter and reducing their opportunities to become members of intellectual communities. (Chap. 8)
16. Three different strategies for siting research and education centers compete: placing them at the most advanced research universities, siting them at geographic locations where several institutions of higher learning can participate, and locating them in regions of the nation where science is weak in order to improve those regions. (Chap. 8)
17. In all fairness, the developing nations have the right to achieve the same levels of economic prosperity as the most technologically developed nations, yet if this happens the natural environment of the world could be destroyed by pollution and resource depletion. (Chap. 9)
18. Increased use of nuclear energy would allow reduction in the use of polluting fossil fuels and thus reduce global warming, at the risk of proliferation of nuclear weapons and accidents. (Chap. 9)
19. It is essential to develop institutional capacities for enhancing the interactions between science and society, but one result may be rendering some older institutions obsolete and generating resistance from people who are committed to older customs. (Chap. 10)
20. The semiconductor industry provides a good model for convergence to other industries, but relying exclusively on strategies developed in its special context may delay development of good alternative models that might arise in very different areas of human activity. (Chap. 10)

The purpose of this list is not to establish an authoritative viewpoint on the future but to illustrate the range of issues that need to be considered. Real convergence will require transcendence of the differences between the perspectives held by human beings with varied backgrounds, interests, and purposes. The proper response to most of these issues is to conduct scientific research and technology development, typically multidimensional, asking many interrelated questions and seeking to integrate answers as they are found.

Conclusion

The experience of the past decade of Converging Technologies meetings shows that there is real value to collecting visionary but professionally grounded predictions and other descriptions of future possibilities. They inspire action, stimulate thought, and anchor debate. Thus it could be valuable to establish a periodic survey of CKTS scientists and engineers, ideally a panel of several hundred with expertise across all convergent fields, probably done each time in two stages. In Stage 1, members of the panel would contribute ideas like those listed above, which would be combined, edited, and selected to maximize diversity by a survey research team that would transform them into a formal questionnaire. In Stage 2, panel members would respond to the questionnaire, as in our 2005 study, rating each idea in terms of the year it might be substantially achieved, and how good that achievement would be for humanity. Conducted over the Internet, this survey process could be quite inexpensive, and the resultant dataset could be analyzed in several ways for multiple publications and purposes.

References

Bainbridge, W.S., Roco, M.C. (eds.): Managing Nano-Bio-Info-Cogno Innovations: Converging Technologies in Society. Springer, Berlin (2006a)

Bainbridge, W.S., Roco, M.C. (eds.): Progress in Convergence: Technologies for Human Wellbeing. New York Academy of Sciences, New York (2006b)

Roco, M.C., Bainbridge, W.S. (eds.): Converging Technologies for Improving Human Performance: Nanotechnology, Biotechnology, Information Technology, and Cognitive Science. Kluwer Academic, Dordrecht (now Springer) (2003)

Roco, M.C., Montemagno, C.D. (eds.): The Coevolution of Human Potential and Converging Technologies. New York Academy of Sciences, New York (2004)

Appendix E
Selected Reports and Books About Convergence

Compiled by William Sims Bainbridge, National Science Foundation

Since the beginning of the National Nanotechnology Initiative (NNI), which contributed to advancing convergence principles, large numbers of experts worldwide have examined technological convergence from many perspectives. The publications described below report the issues raised in their discussions, and their major conclusions. Arranged in chronological order, they begin with early NNI reports that linked that field to biotechnology and information technology, and they include the social and cognitive sciences when ethical and human implications are under consideration. In addition, to provide information plus suggestions of policy issues that require consideration, the listed reports outline research and development agendas within and across many fields of science and technology. Formal mapping of the full NBIC (nano-bio-info-cognitive science) convergence was accomplished by 2003, and the following decade expanded both the scope and the depth of our understanding.

Nanostructure Science and Technology: A Worldwide Study. Siegel, Richard W., Evelyn Hu, and Mihail C. Roco (eds.). 1999. Dordrecht, Netherlands: Kluwer.

This first comprehensive report begins with a decisive sentence: "Nanostructure science and technology is a broad and interdisciplinary area of research and development activity that has been growing explosively worldwide in the past few years" (p. 1). This proclaims not only the vast potential of nanoscale research and development, but through the word "interdisciplinary" defines it from the very beginning as both an instance and a source of convergence, both across traditional fields and around the globe. A starting point for this report was a workshop held in the United States May 8–9, 1997, followed by workshops and site visits in France, Germany, Belgium, the Netherlands, Sweden, Switzerland, the United Kingdom, Japan, and Taiwan. Five chapters focused on somewhat distinct nanotechnology technical areas: (1) synthesis and assembly, (2) dispersions and coatings, (3) high surface area materials, (4) functional nanoscale devices, and (5) bulk behavior of nanostructured materials. A sixth technical chapter foreshadowed nano-bio convergence, identifying a wide range of issues in which nanoscale phenomena in biology (such as protein structure and dynamics) could be included within nanoscale science and technology and achieve synergies with non-biological fields. The final chapter surveys the state of nanoscale research around the world, including data on government funding support and the development of research infrastructure that could be the basis of rapid progress.

Nanotechnology Research Directions. Roco, Mihail C., R. Stanley Williams, and Paul Alivisatos (eds.). 2000. Dordrecht, Netherlands: Kluwer.

Offering a research agenda and specific policy recommendations, as well as a detailed overview of the field, this report resulted from a workshop held January 27–29, 1999. The first four chapters outline research issues specific to nanotechnology, the intellectual and empirical tools for research, and the processes for creating nanostructures. The next five chapters cover application areas, at least two of which related directly to NBIC convergence because they cover nanoelectronics and biomedical applications. The tenth chapter concerns how improved understanding of structures and processes at the nanoscale can enhance our abilities to preserve and improve the natural environment, and the eleventh chapter considers the infrastructure required for success in nanoscale research, development, and education. Participants achieved a high degree of consensus that a "grand coalition" was needed to bring together academic institutions, the private sector, government laboratories, government funding agencies, and professional science and engineering societies in support of the multidisciplinary vision of a national nanotechnology initiative.

Societal Implications of Nanoscience and Nanotechnology. Roco, Mihail C., and William Sims Bainbridge (eds.). 2001. Dordrecht, Netherlands: Kluwer.

This extensive examination of the societal implications of nanotechnology was based on the principle that powerful new technologies must serve human well-being and follow good ethical principles. Based on a large workshop held in Arlington, Virginia, September 28–29, 2000, it consists of two parts. In the first part, five chapters survey the field of nanotechnology and conclude with major recommendations that arose from the groups' consensus, including four that were seen as crucial: (1) A high priority should be given to social and economic research studies in nanotechnology, which would imply a degree of convergence between nanotechnology and the social sciences. (2) An effective mechanism should be developed to inform, educate, and involve the public concerning potential impacts of innovation in this new domain. (3) The knowledge base and institutional infrastructure should be developed to evaluate, on a continuing short-term and long-term basis, the scientific, technological, and societal impacts. (4) A new generation of scientists and workers should be educated in nanoscience and nanotechnology, including giving them the multidisciplinary perspective that nanoscale work naturally requires and that will serve them well in the coming years of more general technological convergence. The second part of the report consists of 37 statements communicating the individual perspectives of participants, written from their own technical and value orientation, many of which link to biomedical or information technologies or to education and ethics.

Converging Technologies for Improving Human Performance. Roco, Mihail C., and William Sims Bainbridge (eds.). 2003. Dordrecht, Netherlands: Kluwer.

This pivotal book-length report resulted from a workshop on Converging Technologies to Improve Human Performance, held at the National Science Foundation, December 2–4, 2001, at which the large number of expert participants deliberated in task forces and plenary sessions. It focused on four major NBIC provinces of science and technology, each of which is advancing rapidly: (1) nanoscience and nanotechnology), (2) biotechnology and biomedicine, including genetic engineering, (3) information technology, including advanced computing and communications, (4) cognitive science, including cognitive neuroscience. It asserted that convergence of diverse technologies is based on the material unity of nature at the nanoscale and on technology integration from that scale and that key transforming tools can achieve advances at the interfaces between previously separate fields of science and technology, accelerate progress within each field through intellectual contributions from the others, and promote the overall unification of science and technology. It demonstrated that developments in systems approaches, mathematics, and computation associated with NBIC convergence will allow us for the first time to understand the natural world and human cognition in terms of complex, hierarchical systems. With this basis, improvement of human performance becomes possible, providing realistic hope of dealing successfully with the challenges of the modern world. The main body of the report is divided into six sections, each reflecting a distinctive theme and beginning with a task-force summary followed by a number of more individual perspectives. The themes are: (1) motivation for advancing convergence and outlooks on major trends, (2) expanding human cognition and communication, (3) improving human health and physical capabilities, (4) enhancing group and societal outcomes, (5), national security, (6) unifying science and education.

The Coevolution of Human Potential and Converging Technologies. Roco, Mihail C., and Carlo D. Montemagno (eds.). 2004. New York: New York Academy of Sciences.

This was the second report emerging from the original NBIC convergence project, stemming from a conference held in Los Angeles, February 5–7, 2003. Its first part consists of three overview chapters, anchoring the conference in the earlier work and looking toward the future: "Science and Technology Integration for Increased Human Potential and Societal Outcomes," "Vision for Converging Technologies and Future Society," and "Collaborating on Convergent Technologies: Education and Practice." The second and larger part consists of 13 research contributions, many of which reported innovative pilot research on new technologies that could be created via convergence of scientific fields. Among the more striking but plausible information technology examples were basing electronic computation not on solid state devices but rather on excitable vesicles composed of membrane-enclosed liquids, and designing solid state computer vision devices in a biomimetic manner along the lines of the human retina. Other contributions examined the ethnical, legal, and managerial dimensions of technological convergence, the evolution of semantic systems, and human performance enhancement.

Converging Technologies – Shaping the Future of European Societies. Nordmann, Alfred (ed.). 2004. Brussels, Belgium: European Commission.

This report considered Converging Technologies to be the first major research initiative of the current century, having four potentially revolutionary qualities: (1) Embeddedness: The new technologies will be the fundamental infrastructure for society, so profound as to be unnoticed much of the time. (2) Unlimited Reach: No aspect of human life will escape the influence of converging technologies, and solutions based on them will be proposed for all human problems; in some cases, however, these may be false solutions. (3) Engineering the Mind and the Body: Inspired by the NBIC motto "Improving Human Performance," and the myriad potential health applications, for better or worse there is the real possibility of transforming human nature. (4) Specificity: In such areas as individually targeted medications based on the genetic code of the individual patient, these new technologies will facilitate specialized design and treatment that reverses the trend toward standardization that became the norm in two centuries ago with the Industrial Revolution. The report observes of the four qualities: "Each of these presents an opportunity to solve societal problems, to benefit individuals, and to generate wealth. Each of these also poses threats to culture and tradition, to human integrity and autonomy, perhaps to political and economic stability" (p. 6). While fundamentally enthusiastic about the vast potential of converging technologies, this report sets out a "European Approach" intended to maximize positive social benefits and guard against harm caused by unsophisticated or irresponsible applications.

Converging Technologies and the Natural, Social and Cultural World. Bibel, Wolfgang (ed.). 2004. Brussels, Belgium: European Commission.

While recognizing the innovative contributions of the original NBIC report, this report to the European Commission suggests a slightly different emphasis: "While Europe does not abhor addressing 'mental, physical and overall human performance' of the individual, it lays an equal emphasis on the social and economic dimensions of sustainable development. Examples for the latter are issues of access to information and knowledge and the resulting impacts on societies' and economies' ability to innovate, of equality and justice with respect to access opportunities, and of individual capabilities to learn or engage socially and politically" (p. 7). The report focuses especially on how issues of social cohesion and public decision-making connect to NBIC through new information technologies and potential advances in cognitive science, identified by the new term "ambient intelligence." Among the more visionary topics discussed is the future potential of artificial intelligence. Eight formal recommendations urge innovation that is consistent with traditional culture, undergoes comprehensive empirical evaluation, and is understood in terms of a "whole society" model of benefit.

Nanotechnology: Societal Implications. Roco, Mihail C., and William Sims Bainbridge (eds.). 2006. Berlin: Springer.

This two-volume report resulted from a workshop held at the National Science Foundation, December 3–5, 2003. The first volume chiefly reports the results of deliberations in ten breakout sessions, one of which had converging technologies as its explicit theme. Two other sessions examined the complex interplay of public policy, governance, ethics, law, international activities, risk, and uncertainty. But the societal implications of NBIC convergence was also a frequent topic in each of the other seven sessions that covered productivity and equity in society, future economic scenarios, the quality of life, national security, interaction with the public, and education. The second volume contains 48 contributions by individual participants, arranged in seven groups, one of which is converging technologies. Again, all of the other sections focus on nanotechnology but have great relevance for complete NBIC convergence: economic impacts, social scenarios, ethics and law, governance, public perception, and education. The goal of all these contributions was to examine trends, identify opportunities, and seek the best ways to maximize benefit for humanity.

Managing Nano-Bio-Info-Cogno Innovations: Converging Technologies in Society.
Bainbridge, William Sims, and Mihail C. Roco (eds.). 2006. Berlin: Springer.

Based on an NBIC conference held February 25–27, 2004, in New York City, this is a diverse collection of 19 essays. The first five provide an introduction to the NBIC concept, outline the emerging policy implications, sketch a roadmap for convergence in the near-term future, consider the implications for an economy based on innovation, and suggest how to measure the progress of convergence. Other chapters examine one or another area that can contribute to and benefit from convergence: education, cyberinfrastructure, developing countries, medicine, law, social science, services science, and technopolitics. Several concern ethics, including neuroethics or issues of human enhancement, and offer conceptual systems that might be applied across multiple traditional areas to achieve cognitive convergence.

Progress in Convergence: Technologies for Human Wellbeing. Bainbridge.
William Sims, and Mihail C. Roco (eds.). 2006. New York: New York Academy of Sciences.

This is a diverse collection of essays by participants in a workshop held in Hawaii, February 24–25, 2005, organized in five main sections: (1) perspectives on convergence, (2) nano-bio-info technology, (3) informatics for convergence, (4), cognitive enhancement, and (5) social and ethical implications. The four perspectives in the first section present the chief orientations that typified convergence at that point in time. The first focused on governance issues, which include how power is exercised in managing resources, how potential conflicts are resolved, and how stakeholders are able to participate in decision processes. The second perspective sought general concepts that could be applied across fields of science, such as laws comparable to conservation of energy in physical processes, potential isomorphisms in material or theoretical configurations across domains, and how random variations may coalesce into similar patterns. The third perspective analyzed the origins and meaning of the consilience goal proposed by Edward O. Wilson, initially applied in sociobiology but transferrable across science and engineering. The fourth perspective suggested that convergence is a new paradigm for higher education, requiring and facilitating major reforms to prepare students for the opportunities and challenges ahead.

Technology Assessment on Converging Technologies. European Technology Assessment Group. 2006. The Hague, Netherlands: Rathenau Instituut.

This is the report from a workshop titled "Converging Technologies in the 21st Century: Heaven, Hell or Down to Earth?" that was held at the European Parliament in Brussels, June 27, 2006. After an opening statement by the workshop chair, presentations were made by authors of a literature study and a vision assessment. While recognizing the importance of the 2001 NBIC workshop, the literature study argues that convergence was already happening naturally across the NBIC fields, and observed: "For decades we have been used to powerful terms such as information revolution and biotechnological revolution. However at the start of this new millennium two revolutionary key technologies came into view: nanotechnology and the cognitive sciences. Nanotechnology is seen as the technology of the twenty-first century and radical breakthroughs are also expected from the cognitive sciences. There is also a growing realisation that these four key technologies are enabling each other in the journey of human progress" (p. 13). The literature study introduced many of the competing visions, and the shorter vision assessment added several more perspectives. After these orienting presentations came a debate among invited experts and an open session including questions from the audience. The debate focused on three main questions: (1) How can we describe the nature and velocity of the emergence of converging technologies, and the forces that are driving it? (2) What impact will converging technologies have on human nature and moral progress, in a context where nations and groups seem to disagree in significant ways? (3) What role should the European Parliament play, for example in organizing public debate, setting a research agenda, or working through the various pan-European organizations?

Converging Applications Enabling the Information Society. van Lieshout, Marc, Axel Zweck. et al. 2006. Delft, Netherlands: Nederlandse Organisatie voor Toegepast Natuurwetenschappelijk Onderzoek.

This report is based on two 2005 workshops, "Building a Vision for a European Research Strategy on Converging Technologies" held July 13 in Amsterdam, The Netherlands, and "Validation of Converging Technologies for Enabling the Information Society," held October 26–27 in Seville, Spain. The first half of this report considers how each of four other fields is converging with information and communications technologies (ICT) in innovative ways that will accelerate the emergence of the information society: (1) cognitive science, (2) biotechnology, (3) nanotechnology, and (4) materials sciences. Two additional chapters survey the current status of these convergences across major advanced societies, systematically comparing scientific activities and impacts. The concluding chapter focuses on European efforts in the global context, observing: "Based on the information contained in their national R&D programmes, it can be ascertained that virtually all EU25 countries have recognized the importance played by ICT, nanotechnology, biotechnology, and the new sciences. They are deemed as leading areas of importance within the R&D sector and acknowledged as strategic domains for future development, likely to influence a vast array of aspects of both the economic and social life" (p. 127).

Appendix E   535

Nanoconvergence. Bainbridge, William Sims. 2007. Upper Saddle River, New Jersey: Prentice-Hall.

This monograph provides a sociological perspective on NBIC convergence, initiated by the emergence of nanotechnology, based on material unity at the nanoscale and technology integration from that scale. As often happened in the history of science and technology during a relatively short period, a very wide range of visions coalesced, both practical and impractical, competed, with the result that rapid and demonstrable progress was energized by human enthusiasm and guided by human intelligence. After a chapter on nanotechnology that set the stage for convergence, chapters show how the three other fields were ready to join the unification process. Concluding chapters examine the social processes through which scientists and engineers cooperate, and suggest a set of eight concepts that can be combined in different ways to frame theories applicable across domains: configuration, information, variation, conservation, interaction, indecision, cognition, and evolution.

Konvergierende Technologien und Wissenschaften: Der Stand der Debatte und politischen Aktivitäten zu »Converging Technologies« (Converging Technologies and Sciences: The State of the Debate and Political Activities about "Converging Technologies"). Coenen, Christopher. 2008. Berlin: Bureau for Technology Assessment at the German Parliament.

This background report for the German parliament begins with observations that technological convergence became the focus of unclear but implicitly political debates, such as that between utopians who saw in convergence a way of achieving radical human enhancement, and those dystopians who used it as the opportunity to issue strident warnings about the harmful consequences of ungoverned technological experimentation. Such extreme viewpoints distract from the very real public policy issues concerning technological convergence that have now been debated in many nations, somewhat differently in Europe from in the United States; this report seeks to survey the landscape of legitimate discussion. The report recognizes the key position of nanotechnology in the broader process of convergence, uses the German equivalent of the word political where an English-speaker might use policy or governmental, and acknowledges that from the very beginning serious attention was given to the meaning of convergence for *people*: "The U.S. NBIC convergence initiative was developed as part of the political activities on the ethical, social, and legal implications of nanotechnology, a field in which the question of dealing with futuristic visions played a central role from the outset" (p. 12). The report also stresses the importance of information and communications technologies in achieving convergence, views on convergence as the logical starting point of a much broader political and societal discussion of the future prospects for science and technology, and expresses the European goal to develop valid methods for forward-looking technology assessment or vision assessment. The main text of the report is organized in six chapters: (1) an overview and introduction, (2) the genesis and content of the alternative convergence ideas, (3) a deeper vision analysis of the convergence debates, (4) consideration of how convergence processes, research results, and academic debates are structuring the field, (5) comparative international analysis of political activities on the convergence issue, and (6) a summary outlining policy options and research needs in the German and European context.

Security Applications for Converging Technologies: Impact on the Constitutional State and the Legal Order. Teeuw, Wouter B., and Anton H. Vedder (eds.). 2008. The Hague, Netherlands: Wetenschappelijk Onderzoek- en Documentatiecentrum.

Writing in the context of the Dutch legal system but published in English, this book-length report explores the implications of NBIC convergence in the fields of security, law, crime prevention, and police. After an introduction, four chapters introduce the four NIBC areas, then another chapter explains the principles of convergence. Chapter 7 is pivotal to this report, examining in depth three "cases," general application areas illustrating the potentially revolutionary impact of convergence. Case 1 concerns monitoring the location and condition of humans and objects, potentially with the option to intervene if something goes wrong, using sensors attached to or implanted inside them. Examples include monitoring the location and behavior of prisoners and triggering a "knee-lock" device if they attempt to run away, monitoring the situation of vulnerable persons and defending them if they become endangered, and securing a valuable possession such as deactivating the engine of a stolen car. Case 2 covers forensic research, typically using new nanotechnology-enabled devices to detect and instantly identify evidence at a crime scene that exists only in trace quantities, using "lab-on-a-chip" devices. Case 3 involves profiling, which may use NBIC technologies to screen masses of people for those with characteristics that may indicate they are dangerous, and identification methods, which may combine biometrics, genetics, and computer vision. The first goal of the study is to identify which technologies might be feasible, then to sketch scenarios for the implementation of those that might prove useful, then to begin to discuss their social, ethical, and legal implications as a prelude to making policy decisions about them.

Knowledge Politics and Converging Technologies. Luce, Jacquelyne, and Liana Giorgi (eds.). 2009. Special issue of The European Journal of Social Science Research, 22(1): March.

This special issue of a journal grew out of a workshop held in Brussels, Belgium, May 6–7, 2008. The editors explained, "The presentations at the May workshop largely focused on practices of anticipatory governance and the analysis of historical and emerging forms of deliberative engagement and policy-making. The articles presented here are concerned with the ways in which discussions about knowledge and technological possibilities unfold and the manners in which social and political debates are circumscribed by pre-existing modes of tackling controversial issues." Steve Fuller offers a social epistemological perspective, suggesting NBIC is based on a fluid conception of human nature in which enhancing evolution may be considered. Adam Briggle argues that the U.S. President's Council on Bioethics was a valuable innovation in knowledge politics that can be applied to convergence. Other articles discuss emerging ethical issues like intellectual property rights, risk assessment, and the role of citizens' panels in deliberations about science and technology policy.

The Fourth Paradigm: Data-Intensive Scientific Discovery. Hey, Tony, Stewart Tansley, and Kristin Tolle (eds.). 2009. Redmond, Washington: Microsoft Research.

The title of this collection of essays about convergence based on information technology came from a visionary talk given in 2007 by the late Jim Gray, for whom this is in part a memorial volume. Gray argued that science has already gone through three stages of evolution, each representing a different paradigm: (1) A thousand years ago, science emerged as an empirical endeavor, describing natural phenomena. (2) In the past few hundreds of years, science became more theoretical, employing models and generalizations to interpret the deeper meaning of empirical findings. (3) In recent decades, science was again transformed as computers were used to simulate complex systems. Now, Gray proposed, a new paradigm is emerging: (4) Data exploration by means of information technology can unify empirical description, theoretical explanation, and computer simulation. This fourth paradigm is a very positive way to view the current trend toward "Big Data," in which scientifically relevant datasets of much greater size and complexity are being managed, often remotely in the "cloud," by information technology.

Nanotechnology Research Directions for Societal Needs in 2020. Roco, Mihail C., Chad A. Mirkin, and Mark C. Hersam (eds.). 2011. Berlin: Springer.

This major report, based on four forums held March through July, 2010, charts the course of nanotechnology research over the following decade (2010–2020) in ways that support NBIC convergence and serve human well-being. The first chapter celebrates the accomplishments of the first decade of the National Nanotechnology Initiative, as inspiration for the second decade. The next three chapters cover the fundamental tools for progress, based on progress in theory, empirical measurement, and manufacturing of nanoscale components. The three following chapters focus on areas of societal impact relevant for all NBIC fields, concerning environmental safety and sustainability. Five chapters survey opportunities for applications in bio-medicine, nanoelectronics, nanophotonics, nanostructured materials, and high-performance materials. Two summary chapters unify and conclude the report, one on development of the needed human and physical infrastructure, and one on innovative and responsible governance. As the book's preface explains, progress should be achieved along four dimensions of human concern: knowledge, material improvement for example in medicine and economics, global cooperation, and moral advance achieving enhanced quality of life and social equity.

The Third Revolution: The Convergence of the Life Sciences, Physical Sciences, and Engineering. Sharp, Phillip A. et al. 2011 Washington, D.C.: Massachusetts Institute of Technology.

This report considers convergence from the perspective of the health sciences research community. The conceptual framework begins by identifying two previous scientific revolutions, first the use in the mid-twentieth century of molecular and cellular biology to understand cells and diseases, second the more recent attempt to understand health and disease in the context of the organism's entire genome, which of course requires advanced information technology. The third revolution, which this report proclaims is just beginning, "involves combining molecular and cellular biology with genomics, engineering, and knowledge of the physical sciences" (p. 8). Among specific examples given are computational biology to model the immune response, imaging technology to diagnose eye diseases, and using nanotechnology to achieve targeted chemotherapy delivery, attacking tumors with powerful chemicals without damaging healthy cells in the patient's body. The report places such specific benefits in a larger societal context by noting how innovation promotes economic development, the challenge to our healthcare system from the changing demographics of an aging population, and more generally the high returns from Federal research funding. Among the specific recommendations are establishing a "convergence ecosystem" (p. 28) connecting multiple disciplines, revising the peer-review process to make it more interdisciplinary, and educating researchers to work in cross-disciplinary fields.

Leadership in Science and Technology. Bainbridge, William Sims (ed.). 2012. Thousand Oaks, California: Sage.

This hundred-chapter, two-volume reference work employs well-established theories and well-documented case studies to bring together insights from many fields of science and engineering to identify principles of innovative leadership and project management. In addition to containing some chapters that explicitly discuss NBIC convergence, the entire work promotes the unification of all fields of science and engineering, with a special emphasis on social science as a means for accomplishment of this goal. The first volume consists of review essays in four general areas: (1) discipline-based social scientific approaches to leadership in science and technology, (2) key concepts relevant to leadership in all areas of R&D, (3) environmental contexts in which leadership and innovation occur, and (4) tactics and tools that leaders may use to achieve progress. The second volume offers a diversity of multidisciplinary case studies that illustrate issues and opportunities in four broad realms: (1) discovery and scientific debate, (2) collaboratories through which teams can combine forms of expertise and other resources, (3) technology development projects that require convergence, and (4) education approached from multiple perspectives. Across both volumes, the hope is that concepts and techniques developed in one area can be transferred to other areas, and combining people, ideas, and resources across areas will achieve the most rapid advances for knowledge discovery and human well-being.

"Report on the Convergence of Biotechnology and Nanotechnology." Enzing, Christien. 2012. Paris: Organisation for Economic Co-operation and Development.

This brief memorandum summarizes issues for consideration by the Working Party on Nanotechnology of the OECD, incidentally providing an up-to-date catalog of questions of wider relevance for broad scientific and technological convergence. After providing the intellectual and historical background, highlighting information technology as well as nanotechnology, this memorandum considers research about how converges actually occurs, lists a number of national and international efforts to promote convergence between nanotechnology and biotechnology, and then considers the potential ethical, social, and economic consequences. The author debates whether convergence is really a radical idea, concluding that it can indeed be revolutionary but also may lead to subsequent divergence: "convergence is an integral part of scientific practice: it refers to an intermediate stage of the scientific and technological process that tends to diversify again" (p. 30).

Additional summaries are available in a previous volume, ***Progress in Convergence: Technologies for Human Wellbeing***. Bainbridge, William Sims, and Mihail C. Roco (eds.). 2006. New York: New York Academy of Sciences.

Appendix F
The Wilson Center Video Interviews: "Leading Scientists Discuss Converging Technologies"

Contact: *David Rejeski and Aaron Lovell, Wilson Center*

As part of this study on convergence, the National Science Foundation and the Woodrow Wilson International Center for Scholars (Washington, DC) are collaborating to conduct and publish a series of video interviews with renowned scientists and scholars who work at the intersections of the key technologies of the early twenty-first century that are discussed in this report. The scientists help to define convergence by posing practical problems and providing concrete examples of convergence approaches that promise practicable solutions and accelerated innovation, based on their own experiences.

The interviews listed below are being made available on the website of the Science & Technology Innovation Program at the Wilson Center. The length of the interviews ranges from 5 ½ min to about 12 min.

The Wilson Center Convergence Interviews website is http://www.wilson-center.org/convergence.

List of Wilson Center Interviews on Convergence[1]

- *George Whitesides*, Professor of Chemistry at Harvard University
- *Bruce Tonn*, Professor of Political Science at the University of Tennessee-Knoxville
- *Clement Bezold*, Founder and Chairman of the Institute for Alternative Futures

---

[1] The order of the list is the order that the interviews were recorded; positions and affiliations are as of the date of the report.

- *Mark Lundstrom*, Professor of Electrical and Computer Engineering at Purdue University
- *Piotr Grodzinski*, Director of the National Cancer Institute Office of Cancer Nanotechnology Research
- *Aude Oliva*, Principal Research Scientist, Computer Science and Artificial Intelligence Lab, Massachusetts Institute of Technology
- *Eli Yablonovitch*, Director, Center for Energy-Efficient Electronics Science at the University of California, Berkeley
- *Sangtae Kim*, Founder and Director, ProWD Sciences, Inc.
- *R. Stanley Williams*, HP Senior Fellow and Vice President, Hewlett-Packard Laboratories
- *Robert G. Urban*, Head of Johnson & Johnson Boston Innovation Center
- *James Murday*, Associate Director, Physical Sciences, University of Southern California Washington, DC, Office of Research Advancement
- *Jian Cao*, Professor of Mechanical, Civil, and Environmental Engineering, Northwestern University
- *Lee Cronin*, Professor of Chemistry, Nanoscience, and Chemical Complexity, University of Glasgow

Appendix G
Glossary of Selected Terms and Concepts[1]

3D printing The production of an object or component in three dimensions by a desktop device analogous to a computerized printer because it takes information input and produces physical output; often equated with additive manufacturing and stereolithography, but potentially including other processes to shape, remove, or transform materials.

Additive manufacturing Building parts by adding material on an existing structure, such as layer-by-layer deposition from digital computer-aided design files in stereolithography.

Antagonistic pleiotropy An hypothesis that arose in genetics but now is applied to other fields, suggesting that typically any major development will have multiple competing effects, some positive and others negative.

Anticipatory governance A convergent method of policy and decision-making based in foresight of and preparation for plausible future scenarios, integration of social science and humanities research with physical science and engineering, and engagement of publics in deliberations.

Assistive technology Robots, computer interfaces, or other engineering solutions designed to compensate for a user's disabilities, but with the potential to help any user in accomplishing unusual tasks.

Big data The use of very large troves of information in science and engineering, unprecedented in their demands on computers and the human mind, requiring development of new means to access, aggregate, analyze, and interpret their significance.

Biotechnology Technologies that are oriented toward biological structures, whether interacting directly with living organisms or merely applying principles derived from them.

Boundary organizations Neutral parties that facilitate and supply technical information to public policy decision-making processes involving two or more

---

[1] These are informal definitions provided by the editors of selected terms and concepts used in this document.

technical areas that have not yet fully converged but have significant implications for each other.

Citizen engineering Technological design and development that incorporates contributions from non-specialists, for example, open-source software projects created by communities of volunteers.

Citizen science The serious involvement of non-specialists in scientific research, often in the role of long-term volunteers, and increasingly acting as partners in the research rather than merely providing free labor.

Cloud The storage and processing of data at locations remote from the user and usually in systems unknown to the user, establishing a new and as-yet unassessed relationship between users and the organizations operating the cloud.

Cognitive computing A new paradigm in which computational systems and algorithms spontaneously recognize context and intent without having to be programmed by experts in arcane software languages, but rather learning successfully in an uncertain and changing environment.

Cognitive science Rigorous research on animal and human mental functions, based on the convergence of previously separate sciences, notably cognitive psychology and neuroscience, but also including artificial intelligence and some branches of anthropology, philosophy, and education.

Cognitive Society, The A society characterized by an ubiquity of convergent cognitive technologies that are leveraged to enhance human decision-making, well-being, and public health.

Cognome The set of principles controlling mental functions in the individual and in the community to which the individual belongs; named by analogy with the genome that defines a genetic structure.

Collaboratory A facility for research or development where individuals, groups, and organizations can cooperate while preserving much of their autonomy, often requiring a special investment outside the scope of any one of the participants.

Common core Standards defining what students in one or more disciplines are expected to learn, ideally designed to be relevant to the real world, reflecting the knowledge and skills that students will need for success in their careers.

Complexity The property of a system consisting of many parts that are not uniform in their structure or interactions, such that the total information content is high, long-term predictability is low, and management is difficult.

Computer regimentation The conjectured imposition of harsh control over humans by machines, as in the traditional factory assembly line or current offices if most work is done through rigidly designed software.

Consilience An intellectual harmony between superficially different methods of analysis that lead to the same result, and therefore to the successful diffusion of an idea from one field to another where it proves to apply without major revision, thus becoming a conceptual link between the fields.

Convergence The creative union of sciences, technologies, and peoples, focused on mutual benefit; this is a process requiring increasing linkages across traditionally separate disciplines, areas of relevance, and across multiple levels of abstraction and organization.

Convergence consultant A hypothetical professional specialty in which experts on multidisciplinary communication and the management of technological convergence would advise corporations and individuals on how to take best advantage of the new opportunities.

Convergence culture A unifying set of values, terms, methods, and theories that enable a team or network to communicate effectively applying convergence principles, and achieve success in projects that span multiple disciplines and domains of relevance.

Convergence–divergence process A sequence of advances achieving significant progress, in four stages: (1) creative assembling of contributions from multiple fields, (2) system integration for known uses, (3) innovation, and (4) spin-off outcomes that lead to emerging new things and uses.

Co-robot A semi-autonomous, intelligent machine designed to cooperate closely with human beings in accomplishing the goals of the humans.

Creative destruction An hypothesis in economics stating that technological innovation has a net positive impact on the job market and the economy more generally, despite the fact that innovation renders old technologies obsolete and thus causes some unemployment and investment losses.

Cultural cognition Thinking entirely within the conventional limits of a culture or subculture, causing people to interpret new evidence in a biased way that reinforces their predispositions.

Cultural lag The delay following the societal impact of a new technology, as the institutions of the society take time to adjust, typically requiring cultural innovations and causing temporary social problems.

Cultural science An emerging science of the shared concepts and practices of large social groups, analogous to cognitive science in both structure and origins, based on convergence across sociology, political science, cultural anthropology, linguistics, and related fields.

Culture and personality A subfield of anthropology and social psychology that examines the complex relationships between the norms, beliefs, and values of a society, and the cognitive or behavioral propensities of its typical citizens.

Cyberinfrastructure The computational and networking hardware and software available for research and development; the definition is sometimes expanded to include the human-computer interface and the information itself.

Cyber-physical system A tightly integrated network of devices that combine information processing with physical processes, such as sensor nets that monitor conditions in a city, or a partnership of robots and environmentally embedded components to accomplish factory production or facility maintenance in a semi-autonomous manner.

Data mining Use of machine learning, mass information retrieval, automatic clustering, natural language processing, and other advanced techniques to derive meaningful information from vast existing sets of raw data.

Demographic transition theory A classic social science theory that currently is very much in doubt, holding that the industrial revolution unleashed an only temporary explosion in the human population, because culture would naturally adjust over time until the birth rate and death rate were again in balance.

Design ethnography Using methods like those developed by cultural anthropologists to empirically develop technology design criteria and evaluate prototypes with individual human subjects or in functioning groups of early adopters.

Digital divide Lack of effective access to information technology by economically disadvantaged people or by those lacking the skills to use it effectively.

Digital government Initially, the use of information technologies to improve communications within government agencies and from government to the citizenry, but now evolving into systems for participatory decision-making about public policies and services.

Distributed manufacturing Production at many local facilities, often in very small and customized batches, enabled by process flexibility, modularity, in-process metrology, predictive sciences and technologies, and human–machine interaction.

Earth-scale platform The environment for human activities oriented toward convergence, including global natural systems, communication systems, and the global economy.

Emerging technologies Newly defined and newly feasible areas of engineering application, often requiring new scientific and economic paradigms, and investment in new industries.

Foundational tools Relatively small-scale instruments, techniques, and assemblers that start from a basic element—such as atom, bit, DNA, or synapse—and generate an integrated system.

Geoengineering Projects that have the goal of transforming some significant portion of the natural environment, such as air quality, the oceans, the biological organisms that live in the ecology, or climate change.

Global risk assessment Analysis of the potential net costs of an event, such as a natural accident or an episode of technology innovation, for humanity as a whole, rather than just for the immediate decision-makers.

Grand challenge A broad, difficult, yet well-defined goal for scientific research and technological development that can serve as the focus for several years of work by a significant number of teams advancing separate but compatible projects.

High-throughput sequencing The bio-info-nano convergence enabling inexpensive gene sequencing of individuals and efficient assessment of the diversity of cells in a biological or medical sample.

Holistic interdependence The principle that the elements of a system influence each other, more powerfully the more tightly coupled the system becomes, with the implication that globalization requires increasingly holistic scientific conceptions of nature and human action.

Hubsite A physical or virtual nexus for communication and coordination, for example, a center used by a number of local universities to support their research and education in a convergence field.

Human factors A multidisciplinary offshoot of psychology that identifies aspects of human psychology and its context that must be taken into account in the design and development of new technology.

Human potential The possibilities for improvement in the ability of human beings to have the intelligence, wisdom, and health to achieve new beneficial goals.

Human-scale platform The set of systems that enable convergence characterized by the interactions between individuals, between humans and machines, and between humans and the environment.

Inclusive development Economic and social progress that benefits all people while minimizing the impact of human activities on Earth's climate and ecosystems.

Inclusive governance Management of the conduct and applications of research and development that ensures effective involvement from all relevant stakeholders, agencies, and experts.

Industrial ecology A science for assessing and minimizing the impact of industry on the environment, using such methods as material flow analysis, life-cycle analysis, and design for environment.

Information infrastructure The system of well-organized data resources available for research and development; the term emphasizes the information content of the system, but performance also depends upon the hardware and the human–computer interface.

Information technology A very general term that highlights electronic computing and communications technologies but also includes more traditional means for managing meaningful information.

Innovation spiral A circular but progressive process over time in which information and innovation in one area stimulates development and innovation in another area, which feeds back to stimulate more innovation in the original area, and so forth.

Institutional infrastructure The social institutions required for science, engineering, and other collective efforts, such as professional associations, industry partnerships, and government agencies.

Intellectual property rights Legal regimes, notably patents and copyrights, that give innovators a temporary monopoly on exploitation of their innovations, which have become increasingly problematic as scientific and technological changes call their basic principles into question.

Intelligent fiber One of several terms referring to sensors and other electronic devices embodied in fabrics rather than boxes, including interactive textile devices and other forms of next-generation wearable technology.

Manufacturing process DNA The design of manufacturing processes in a manner like the genetic code, allowing the rapid switching in and out of production steps and system components, and allowing precise control over product parameters with very little effort.

Medicalization Defining a challenge in terms of a disease model requiring a standard cure, which often can be inappropriate, as in cases when a person's native characteristics do not harmonize with the dominant culture, which might be solved by building a subculture designed for people having that characteristic.

Metaparadigm A conceptual structure developed to combine the conceptual structures possessed by two or more separate fields, thus a paradigm about the combination of existing paradigms, which may require revision of the component paradigms.

Mindfulness Receptiveness, expanded perspectives, and context-relevant interpretations that can lead to more long-range, discriminating, and holistic innovation.

MOOC Massively open online education course, reaching a large number and diversity of students over the Internet, central to a new movement in higher education, and especially suitable for some forms of technical training that do not require specialized laboratories or co-presence of students and teachers. (Pronounced "mook," like "moon.")

Multifunctional manufacturing Versatile and usually small production facilities intended for brief production runs, even customizing of each item, capable of switching easily from one kind of product to another, ideally both efficient and inexpensive.

Multiscaling Scientific research and engineering development across a wide range of dimensional sizes, enabled by convergence tools and theories.

Nanoelectronics Electronic components with at least one dimension less than roughly 100 nm, posing challenges of reliability and production methods and issues of how to avoid or exploit quantum effects.

Nano-geobiochemistry The field of study that examines how naturally occurring nanoscale materials behave in the Earth's environment, chiefly in terms of their biochemical implications.

Nanomaterial Natural, incidental, or manufactured materials with any external dimension in the nanoscale (about 1–100 nm) or having internal structure (aggregate or agglomerate of particles or grains) or surface structure in the nanoscale.

Nanoscience The study of phenomena, processes, and objects structured on the nanoscale, roughly with dimensions from 1 to 100 nm, which may be manufactured or natural, and within the provinces of fields as diverse such as genetics, geology, chemistry, and electrical engineering.

Nanotechnology Technology at the nanoscale, which is about 1–100 nm, including respective processes, components, materials, devices, and systems at the nanoscale.

Nanotherapeutic medicines Drugs delivered exactly at the targeted cells where they are needed in the body, for example, to attack cancer cells, by means of packaging the drugs within nanoparticles, with consequent reduction of side-effects.

Online community A network or subculture of people sharing similar interests or connected in a social network, who communicate primarily via the Internet or comparable technological means.

Open access Essentially free access to scientific journals and databases, without the need for a costly subscription or organization membership, to facilitate both multidisciplinarity among professional scientists and citizen scientists.

Open science A research approach emphasizing collaborations across fields, creative incorporation of contributions from non-specialists, and such mechanisms as virtual organization collaboratories and online citizen science.

Open systems dependability A methodology for continuous operation of huge, complex, and ever-changing systems that recognizes that failures cannot be entirely prevented but must be minimized in their effects, in a context that gives a high priority to accountability.

Appendix G   549

Open-source methodology Inspired by open-source software development projects in which the programming code is widely available for modification and improvement, this approach seeks to achieve greater innovation and broader benefit by conducting many kinds of science and engineering in a public manner.

Orchestration A form of dynamic leadership applied both before and during a complex endeavor to coordinate the productive interactions of people and ideas to achieve the most creative results.

Paradigm In science and related endeavors, a more-or-less coherent conceptual framework that defines a field in terms of concepts, theories, and methods.

Participatory design The involvement of representative users in the design and development of new technologies, often expressed in terminology like user-centered design, value-sensitive design, evaluation studies, and focus groups.

Pasteur's quadrant Use-inspired basic research, in distinction from pure research that lacks immediate application and applied research that fails to explore fundamental principles, named after the work of Louis Pasteur.

Personal genomics Genetic sequencing of ordinary individuals, rendered inexpensive by convergence of information and biological technologies, contributing to health by alerting the individual to risks of specific ailments and side-effects of medications.

Person-centered education Individualized education, designed for the particular needs and learning style of the student, which could be facilitated through convergence of cognitive science and information technology with the field of science or engineering being learned, and which takes on new urgency in the context of the proliferation of online courses that are not inherently personalized.

Planetary boundary One of several limits, such as a specific concentration of carbon dioxide in the atmosphere, that must not be violated, defining a safe operating space for humanity with respect to the Earth system.

Platforms for convergence The general categories of enablers for scientific, technological, and societal convergence, including toolsets as well as complex systems on several scales.

Polycentric governance Managing resources simultaneously at multiple scales, or in multiple overlapping but distinct realms.

Post-Fordism The doctrine that manufacturing industries should move beyond assembly-line production of uniform products in large facilities, named in contrast to the system exemplified a century ago in production of Ford Motor Company vehicles.

Post-industrial society A concept from the 1970s that saw a major economic phase change occurring, as manufacturing became a smaller part of the economy, and as services grew to become the dominant sector, believed to increase the significance of science and information technology; the implications of this shift remain uncertain today.

Post-normal science A problematic form of research under conditions when facts are uncertain, values are in dispute, stakes are high, and decisions are urgent.

Precision agriculture An approach for rigorous site-specific management of crop production, maximizing efficiency and productivity, while minimizing environmental damage, employing geospatial data techniques and other convergence technologies.

Precompetitive research Science and technology development in which competing companies share the funding and intellectual property of early-stage work, for example, through industrial consortia and collaboratories, after which they can separately develop patentable innovations.

Proactive convergence Unification across fields of science or technology that is conducted intentionally, including some form of decision analysis in the convergence approach, thus future-oriented and potentially open-ended.

Reactive convergence A limited form of unification across fields of science or technology, often triggered by coincidental factors, based on ad hoc collaborations of partners or individual fields for predetermined and limited goals.

Regenerative medicine Restoring the cells and organs of the human body to support normal functioning, whether by stimulating the body's own repair mechanisms or by culturing person-specific replacement tissues initially outside the body.

Responsible governance Management of science, technology, and the implementation of their results in a manner that addresses environmental, health, safety, ethical, legal, social, and equity concerns.

Services science Research intended to discover fundamental principles and management guidelines for activities carried out in service industries, especially in the movement to redefine information processing as a service provided by an enterprise, rather than in terms of hardware owned by the user.

Smart homes Integration of information technology and other NBIC developments in dwellings to achieve health monitoring, assistance with daily living activities, and prevention of illness and injury.

Social cognition Processing of information by a social group or network rather than by a single individual, often involving a division of labor in which different people understand different parts of the puzzle.

Social intelligence The ability of a social group or network to solve challenging problems, which may be more or less than the ability of an individual member, and could be enhanced by convergence of the problem-relevant field with cognitive science.

Societal convergence The unification of humanity, facilitated by convergence of all fields of science and engineering, in peaceful and prosperous transcendence of traditional geographic, political, and cultural barriers, while maintaining healthy diversity.

Societal-scale platform The set of convergence enablers characterized by the activities and systems that link individuals and groups on several larger scales; it consists of collective activities, organizations, and procedures, including the various forms of governance.

Stakeholder A person or group having an interest in the outcome of a policy or implementation action, usually in the context of public decision-making.

Sustainable development Progress that meets the needs of people today without compromising the ability of future generations to meet their own needs with respect to social, economic, and environmental conditions.

Appendix G 551

Synergy The phenomenon in which a system of component parts has a different effect from the sum of those components, at the extreme a form of convergence creating a tightly integrated system from which none of the components can be removed without destroying the system.

Systemic convergence Unification that goes beyond a few fields of science and technology to include major aspects of culture and society, driven by higher purposes and often guided by convergence organizations.

Technological unemployment The loss of jobs when technological changes put people out of work, which may or may not in practice be balanced by the creation of new jobs required by the new technology.

Time banking A form of community-based volunteering in which participants provide and receive services, following a bookkeeping system based on hours of work rather than money, managed through an information and communication system.

Transdisciplinary When disparate knowledge converges to produce new knowledge without necessarily uniting the contributing fields.

Transformative governance Creative management having a result-oriented, project-oriented focus and advancing multidisciplinary innovation, such that a sector of the economy or society may change markedly.

Universal access The principle that the benefits of science and technology should be as widely available as possible, sometimes used in connection with disabled or disadvantaged populations, but broadly applicable.

Valley of death The gap between laboratory research and real-world applications that is difficult to bridge and often results in a failure to benefit from new discoveries, and innovations because late-stage application steps like scalability and commercialization cannot be completed.

Virtual organization A form of cooperation in some way outside traditional forms, usually employing electronic means of communication and thus not requiring a physical meeting place, but often also involving multiple groups and individuals who do not belong to a unified authority structure.

Visionary governance Future-oriented management of innovation that looks beyond immediate results, including long-term planning and anticipatory, adaptive policies.

Vision-inspired basic research Rigorous scientific research seeking to solve fundamental problems, motivated more by fresh ideas than by existing intellectual traditions or immediate practical applications.

Wellness A conceptualization of human health that emphasizes prevention of disease more than cure, and proactive enhancement of general physical and cognitive well-being more than merely dealing with problems reactively when they arise

Appendix H
List of Acronyms

AI	Artificial intelligence
AICT	Advanced Institutes of Convergence Technology of Seoul National University
AM	Additive manufacturing
ARPA-E	Advanced Research Projects Agency, Energy (DOE)
ARPA-ED	Advanced Research Projects Agency, U.S. Department of Education
ASU	Arizona State University
ATE	Advanced Technological Education (NSF program for community colleges)
ATR	Advanced Telecommunications Research Institute International (Japan)
BCI	Brain–computer interaction
BMI	Brain–machine interface
BRAIN	Brain Research through Advancing Innovative Neurotechnologies initiative proposed by administration for start in FY2014
BSE	Bovine spongiform encephalopathy (mad cow disease)
CAD	Computer-assisted design
CAISE	Center for Advancement of Informal Science Education
CAS	Chinese Academy of Sciences
CC	Community college
CCNE	Centers of Cancer Nanotechnology Excellence (NIH program)
CCS	Carbon capture and storage
CCSB	Center for Cancer Systems Biology (at MIT)
CEA	Atomic Energy and Alternative Energies Commission of France
CKT	Converging knowledge and technology
CKTS	Convergence of knowledge and technology with/to benefit/society
CMOS	Complimentary metal-oxide semiconductor
CNPP	Cancer Nanotechnology Platform Partnerships (NIH program)
CNRS	National Scientific Research Center of France
CNS	Center for Nanotechnology in Society
CNSE	Center for Nanoscale Science and Engineering (University at Albany)
CPU	Central processing unit

CREATIV	Creative Research Awards for Transformative Interdisciplinary Ventures program (NSF)
DARPA	Defense Advanced Research Projects Agency (U.S.)
DBP	Disinfection byproduct
DFE	Design for environment
DHS	U.S. Department of Homeland Security
DIMP	Desktop integrated manufacturing platform
DM	Desktop manufacturing
DNA	Deoxyribonucleic acid
DOC	U.S. Department of Commerce
DOD	U.S. Department of Defense
DOE	U.S. Department of Energy
DOI	U.S. Department of the Interior
DRAM	Dynamic random-access memory
DRL	Research on Learning in Formal and Informal Settings (NSF)
EC	European Commission
ECAST	Expert and Citizen Assessment of Science and Technology (network)
ECCS	Electrical, Communications and Cyber Systems, NSF Div, Engineering Dir
EEG	Electroencephalography
EHR	Electronic health records
EHS	Environmental, health, and safety (issues)
ELSI	Ethical, legal, and social implications (of a technology area)
EPA	U.S. Environmental Protection Agency
ERC	Engineering Research Center (NSF program)
ETRI	Electronics and Telecommunications Research Institute (Korea)
EU	European Union
EUV	Extreme ultraviolet
FCRP	Focused Center Research Program (SRC/US government program)
FDA	U.S. Food and Drug Administration
FES	Functional electrical stimulation
FET	Field-effect transistor
FET	Frontier emerging technologies
fJ	Femtojoules
fMRI	Functional magnetic resonance imaging
GCRP	Global Change Research Program
GDP	Gross Domestic Product
GEOSS	Global Earth Observation System of Systems
GHG	Greenhouse gas(ses)
GM	Genetically modified
GPU	Graphics processing unit
GSCST	Graduate School of Convergence Science and Technology (Korea)
HBCU	Historically Black Colleges and Universities
HRCT	High-resolution computed tomography
HRI	Human–robot interaction

Appendix H

HSC	Health Science Center
ICBP	Integrative Cancer Biology Program (at MIT)
ICT	Information and communication technologies
ICTAS	Institute for Critical Technology and Applied Science (Virginia Tech)
IEA	International Energy Agency
IGERT	Integrative Graduate Education and Research Traineeship Program (NSF)
IMEC	Formerly the Interuniversity Microelectronics Centre
INDEX	Institute for Nanoelectronics Discovery and Exploration (part of NRI)
INSPIRE	Support Promoting Interdisciplinary Research and Education program (NSF)
IP	Intellectual property
IPCC	Intergovernmental Panel on Climate Change
ISE	Informal science education
IT	Information technology
ITRI	Industrial Technology Research Institute (Taiwan)
JST	Japan Science and Technology Agency
K–12	Kindergarten through high school education, covering students roughly 5–18 years of age
K–14	Kindergarten through community college or technical school education
K–20	Kindergarten through graduate school education
KAIST	Korea Advanced Institute of Science and Technology
KANC	Korea Advanced Nano Center
KEITI	Korea Environmental Industry and Technology Institute
KERI	Korea Electrotechnology Research Institute
KI	David H. Koch Institute for Integrative Cancer Research at MIT
KIAS	Korea Institute for Advanced Study
KIGAM	Korea Institute of Geology, Mining & Materials
KIMM	Korea Institute of Machinery and Materials
KIST	Korea Institute of Science and Technology
KISTEP	Korea Institute of Science and Technology Evaluation & Planning
KIT	Karlsruhe Institute of Technology (Germany)
LCA	Life-cycle analysis
MANA	International Center for Materials Nanoarchitectonics (Japan)
MCDA	Multicriteria decision analysis
MEG	Magnetoencephalography
MEMS	Microelectromechanical
MEXT	Ministry of Education, Culture, Sports, Science and Technology (Japan)
MFA	Material flow analysis
MIT	Massachusetts Institute of Technology
MMO	Massively multiplayer online (role-playing game)
MOOC	Massively open online courses
MOSFET	Metal-oxide semiconductor field-effect transistor
MOST	Ministry of National Science and Technology (China)
MRI	Magnetic resonance imaging

MRS	Magnetic resonance spectroscopy
MRSEC	Materials Research Science and Engineering Center (NSF Program)
MURI	Multidisciplinary University Research Initiative (DOD Program)
NACK	Nanotechnology Applications and Career Knowledge education network (Pennsylvania State University)
NASA	U.S. National Aeronautics and Space Administration
NBIC	Nano-, Bio-, Info-, Cognitive (science and technology fields also, significant fields of convergence)
NBIC2	(See NBIC) "Beyond" (more than) convergence of nano-, bio-, info-, and cognitive technologies
NCI	U.S. National Cancer Institute (NIH)
NCLT	National Center for Learning and Teaching (Northwestern Center)
NCN	Network for Computational Nanotechnology
NEGF	Non-equilibrium Green's function
NEMS	Nanoelectromechanical
NGO	Nongovernmental organization
NIH	U.S. National Institutes of Health
NIMS	National Institute for Materials Science (Japan)
NISE Net	Nanoscale Informal Science and Engineering Network (U.S.)
NIST	U.S. National Institute for Standards and Technology
NITRD	National Information Technology Research and Development program
NNI	National Nanotechnology Initiative
NNIN	National Nanotechnology Infrastructure Network
NOAA	National Oceanic and Atmospheric Administration (U.S.)
NRC	National Research Council of the National Academies (U.S.)
NRI	Nanoelectronics Research Initiative (collaborative SRC research program with NSF and NIST)
NRI	National Robotics Initiative
NSE	Nanoscale science and engineering
NSEC	Nanoscale Science and Engineering Center (NSF Program)
NSEE	Nanoscale Science and Engineering Education
NSET	Nanoscale Science, Engineering, and Technology (S. Korean program)
NSF	U.S. National Science Foundation
NSTA	National Science Teachers Association
NSTC	National Science and Technology Council (of the Executive Office of the President)
NUE	Nanotechnology Undergraduate Education (NSF Program)
ODM	Original design manufacturer
OECD	Organisation for Economic Co-operation and Development
OMOP	Observational Medical Outcomes Partnership
ONAMI	Oregon Nanoscience and Microtechnologies Institute
OSTP	Office of Science and Technology Policy of the Executive Office of the President (U.S.)
P4	Predictive, participative, preventive, and personal medicine
PARC	Palo Alto Research Center (Xerox)

PB	Planetary boundaries
PC	Personal computer
PCAST	President's Council of Advisors on Science and Technology (U.S.)
PCM	Phase change materials
PEN	Program of Excellence in Nanotechnology (NIH Program)
PES	Public Engagement with Science
PET	Positron emission tomography
PI	Principal investigator
PNNL	Pacific Northwest National Laboratory
PNS	Post normal science
POU	Point of use
PSM	Phase shift mask
PSOC	Physical Sciences Oncology Center (NIH, National Cancer Institute, Program)
PV	Photovoltaic
R&D	Research and development
RD&D	Research, development, and demonstration
RISA	Regional Integrated Sciences and Assessment (NOAA program)
RNA	Ribonucleic Acid
S&E	Science and engineering
S&T	Science and technology
S.NET	Society for the Study of Nanoscience and Emerging Technologies
SBIR	Small Business Innovation Research program (multiple U.S. agencies)
SCI-C	Cognitive Society Initiative (proposed)
SciDAC	Scientific Discovery through Advanced Computing Institutes (DOE)
SEES	Science, Engineering, and Education for Sustainability (NSF)
SHG	Self-help group
SI^2	Software Infrastructure for Sustained Innovation (NSF)
SIA	Semiconductor Industry Association
siRNA	Small interfering RNA
SKKU	Sungkyunkwan University, Korea
SME	Small- and/or medium-scale enterprise
SRAM	Static random-access memory
SRC	Semiconductor Research Corporation
STC	Science and Technology Center (NSF program)
STDP	Spike-timing-dependent plasticity
STEM	Science, Technology, Engineering, and Mathematics (education)
STTR	Small Business Technology Transfer program (multiple U.S. agencies)
SUNY	State University of New York
SwA	Software assurance
SWAMP	SwA Market Place initiative (DHS)
SWAN	Southwest Academy of Nanoelectronics (part of NRI)
TC	Technical College
tCMS	Transcranial magnetic stimulation
TCNC	Tumor Cell Networks Center (at MIT)

TCR	T-cell receptor
UIC	University of Illinois, Chicago
UIUC	University of Illinois at Urbana-Champaign
UN	United Nations
UNC	University of North Carolina
USDA	U.S. Department of Agriculture
USGCRP	U.S. Global Change Research Program
USGS	U.S. Geological Survey
VLSI	Very large-scale integration
VOC	Volatile organic compound
VoI	Value of information
WPB	Working Party on Biotechnology of the OECD
WPN	Working Party on Nanomaterials of the OECD
WTEC	World Technology Evaluation Center.

Printed by Publishers' Graphics LLC